Environmental Chemistry: A Global Perspective

Environmental Chemistry:
A Global Perspective

Editor: Aisha McCullough

NYRESEARCH
P R E S S

New York

Published by NY Research Press
118-35 Queens Blvd., Suite 400,
Forest Hills, NY 11375, USA
www.nyresearchpress.com

Environmental Chemistry: A Global Perspective
Edited by Aisha McCullough

International Standard Book Number: 978-1-63238-839-1 (Hardback)

Cataloging-in-Publication Data

Environmental chemistry : a global perspective / edited by Aisha McCullough.
     p. cm.
Includes bibliographical references and index.
ISBN 978-1-63238-839-1
1. Environmental chemistry. 2. Green chemistry. 3. Chemistry. 4. Ecology. I. McCullough, Aisha.
TD193 .E58 2022
577.14--dc23

Contents

Preface

The scientific study of the chemical and biochemical phenomena that occur in natural places falls under the discipline of environmental chemistry. It deals with the study of the sources, transport, reactions, effects and fates of chemical species in the air, soil and water environment. It is also concerned with the effects of human and biological activity on these. This interdisciplinary science involves aquatic, atmospheric and soil chemistry. Such activities may have an impact at a local or a global scale. Environmental chemistry plays a crucial role in the identification and detection of pollutants. It also helps in characterizing the nature and source of such pollutants. The book studies, analyzes and upholds the pillars of environmental chemistry and its utmost significance in modern times. It includes some of the vital pieces of work being conducted across the world, on various topics related to this field. This book is appropriate for students seeking detailed information in this area as well as for experts.

Various studies have approached the subject by analyzing it with a single perspective, but the present book provides diverse methodologies and techniques to address this field. This book contains theories and applications needed for understanding the subject from different perspectives. The aim is to keep the readers informed about the progresses in the field; therefore, the contributions were carefully examined to compile novel researches by specialists from across the globe.

Indeed, the job of the editor is the most crucial and challenging in compiling all chapters into a single book. In the end, I would extend my sincere thanks to the chapter authors for their profound work. I am also thankful for the support provided by my family and colleagues during the compilation of this book.

Editor

Synthesis, Characterization, and Catalytic Activity of Pd(II) Salen-Functionalized Mesoporous Silica

Rotcharin Sawisai, Ratchaneekorn Wanchanthuek, Widchaya Radchatawedchakoon, and Uthai Sakee

Creative Chemistry and Innovation Research Unit, Center of Excellence for Innovation in Chemistry (PERCH-CIC), Department of Chemistry, Faculty of Science, Mahasarakham University, Mahasarakham 44150, Thailand

Correspondence should be addressed to Uthai Sakee; uthai.s@msu.ac.th

Academic Editor: Bartolo Gabriele

Salen ligand synthesized from 2-hydroxybenzaldehyde and 2-hydroxy-1-naphthaldehyde was used as a palladium chelating ligand for the immobilization of the catalytic site. Mesoporous silica supported palladium catalysts were prepared by immobilizing Pd(OAc)$_2$ onto a mesoporous silica gel through the coordination of the imine-functionalized mesoporous silica gel. The prepared catalysts were characterized by X-ray diffraction (XRD), scanning electron microscopy (SEM), energy dispersive X-ray (EDX), inductivity couple plasma (ICP), nitrogen adsorption-desorption, and Fourier transform infrared (FT-IR) spectroscopy. The solid catalysts showed higher activity for the hydroamination of C-(tetra-O-acetyl-β-D-galactopyranosyl)allene with aromatic amines compared with the corresponding homogenous catalyst. The heterogeneous catalytic system can be easily recovered by simple filtration and reused for up to five cycles with no significant loss of catalytic activity.

1. Introduction

The synthesis of nitrogen containing compounds is of great importance in basic research and industrial processes due to these important structures playing vital roles in drug research and development [1, 2]. Among the several methods to form C–N bond from N–H bond, hydroamination of unsaturated C=C bonds has the benefit of being cost effective and feasible, and it is widely used by many researchers worldwide [3–7]. Allenes are a unique class of organic compounds with two cumulated double bonds [8]. There has been much interest in using them as starting material for the synthesis of ally-lamines. Only a small number of transition-metal catalyzed intermolecular hydroamination reactions of allenes have been described [9–14]. Previously, we reported the addition of amines to allene using Pd(OAc)$_2$, from which a mixture of the desired allylic amine and a minor amount of dienic amine was obtained [15]. Following this, the gold catalytic hydroamination of allene was investigated to give the desired allylic amine with a moderate to good yield [16]. However, this reported method suffers from disadvantages, such as long reaction time, expensive catalyst, and unstable operating state. The practical application of the catalyst in liquid phase reactions is hindered by both high cost and difficulties in catalyst separation and recycling. The common catalysts for these reactions are generally homogeneous systems, which cause difficulties in recycling the catalyst and aggregation of metal nanoparticles [17]. Thus, a heterogeneous system for the metal catalyst is a reasonable step to avoid these problems, and it is feasible to overcome such limitations by employing Pd on a solid support [18]. When considering this, silica fulfils many of the required aspects as an appropriate solid support for the deposition of palladium catalysts [19, 20]. There have been several methods established towards the preparation of supported Pd-catalysts [21]. Silica, due to its inexpensive, comfortable, accessibility, high thermal resistant, high pore volume, and narrow pore size distribution characteristics, has been the focus of significant consideration compared to other operative supports like polymeric ionic liquid, alumina, MCM-41, SBA-15, and so on [22, 23]. Among the various N-based ligands, Schiff bases are well known as ligands for the complexation of metal ions and as catalysis because of their easy preparation and being simple and cost effective starting materials with high resistance under a variety of conditions. Therefore, many researchers have developed several approaches to immobilize palladium Schiff-base complexes on solid supports [24]. To date, only a few palladium(II) complexes attached to functionalized mesoporous silica supports have been synthesized and successfully applied in organic reactions. However, no catalytic study has focused on the hydroamination of allene using palladium supported on mesoporous materials as catalysts. In our ongoing efforts to develop greener organic reactions, we now report a simple strategy for the preparation and characterization of a novel Salen-functionalized mesoporous silica supported Pd Schiff-base complex as a reusable catalyst and its application for the hydroamination of *C*-(tetra-*O*-acetyl-β-D-galactopyranosyl)allene (**1**) with aromatic amines.

2. Experimental

2.1. General. All the starting materials, reagents, solvents, and eluents were commercial and used as purchased without purification. Flash column chromatography was performed using a silica gel 60 (230–400 mesh). Allene (**1**) was prepared according to a previously described method [25, 26].

2.2. Material Characterization. The prepared catalyst was characterized by various techniques, such as SEM, SEM-EDX, XRD, FT-IR, and BET. XRD measurements were performed with a D8 Advance Bruker diffractometer with the Cu-Kα_1 and Cu-Kα_2 radiations over a wide 2θ range of 0°–80°. Scanning electron micrographs (SEM) and energy dispersive X-ray (EDX) analyses were operated on a JEOL JSM-6460LV microscope. The IR spectrum was obtained using a FT-IR spectrophotometer (Bruker, Tensor 27). The surface area and porosity were measured with a Quantachrome Autosorb 1-MP. The palladium content was analyzed by an inductively coupled plasma optical emission spectrometer (ICP-OES, Optima 3000 DV, Perkin Elmer, EUA).

2.3. Catalyst Preparation. The catalyst was easily prepared from commercially available starting materials as shown in Scheme 1.

2.3.1. General Procedure for Preparation of Imine-Functionalized Silica Gel (Salen-Silica): SiO$_2$@ Imine. A commercially available silica gel (230–400 mesh) was dried by heating at 120°C for 12 h prior to use.

Method A. Salicylaldehyde or 3-hydroxy-2-naphthaldehyde (42.86 mol) and then 3-aminopropyltriethoxysilane (APTES) (10 mL, 42.86 mol) were added to toluene (60 mL). A yellowish color showed immediately due to the formation of the imine. The resulting solution was stirred at a refluxing temperature under a nitrogen atmosphere for 7 h. After cooling, the solid silica gel (20 g) was added to the mixture and stirred

at a refluxing temperature under a nitrogen atmosphere for 7 h, and then the solids were filtered and washed thoroughly with toluene until the washings were colorless. The solid product was dried in air at 120°C overnight before being used in the next step, and the resulting materials were denoted as SiO$_2$@ imineSA and SiO$_2$@ imineNA for salicylaldehyde and 2-hydroxy-1-naphthaldehyde, respectively.

Method B. Silica gel (20 g) was added to 60 mL of toluene. To this slurry, 42.86 mole of APTES was added and the resultant slurry was heated at 110°C under a nitrogen atmosphere for 7 h, and then salicylaldehyde or 2-hydroxy-1-naphthaldehyde (42.86 mol) was added to the slurry and stirred at a refluxing temperature under a nitrogen atmosphere for 7 h. Following this, the solids were filtered and washed thoroughly with toluene until the washings were colorless. The solid product was dried in air at 120°C overnight and denoted as SiO$_2$@ imineSB and SiO$_2$@ imineNB for salicylaldehyde and 2-hydroxy-1-naphthaldehyde, respectively.

2.3.2. General Procedure for Immobilization of Pd(OAc)$_2$ onto Salen-Silica

Method I. To a round-bottomed flask, palladium acetate (415.68 mg, 1.73 mmol) and acetone (100 mL) were added. The solution was stirred at room temperature for 30 min under a nitrogen atmosphere and then 5 g of SiO$_2$@ imineSA was added. The mixture was stirred at room temperature for 24 h. The deposited solids were filtered from the solvent and washed repeatedly through the Soxhlet extraction with ethanol and acetone until the washings were colorless, and they were dried overnight at 120°C and denoted as SiO$_2$@ imineSA-Pd-I.

Method II. To a round-bottomed flask, palladium acetate (415.68 mg, 1.73 mmol) and acetone (100 mL) were added. The solution was stirred at room temperature for 30 min under a nitrogen atmosphere and then 5 g of Salen-silica was added. The mixture was refluxed while being stirred for 4 h. The deposited solids were separated from the solvent by filtration and washed repeatedly for Soxhlet extraction with ethanol and acetone, until the washings were colorless, and they were dried overnight at 120°C and denoted as SiO$_2$@ imineSA-Pd-II, SiO$_2$@ imineSB-Pd-II, SiO$_2$@ imineNA-Pd-II, and SiO$_2$@ imineNB-Pd-II.

*2.4. Procedure for Hydroamination of Allene (**1**).* All reactions were run under an air atmosphere. A reaction tube was charged with allene (**1**) (0.27 mmol), amine (0.81 mmol), Pd catalyst (90 mg, 5 mol%), trifluoroacetic acid (20 mol%), and THF (0.5 mL). The reaction mixture was allowed to proceed while being stirred at room temperature. The progress of the reaction was monitored by TLC. After completion of the reaction, the catalyst was filtered off and the solvent was evaporated under reduced pressure, after which the residue was purified by flash column chromatography on silica gel to yield the product. The solid compounds were characterized by ^{1}H and ^{13}C NMR spectroscopy.

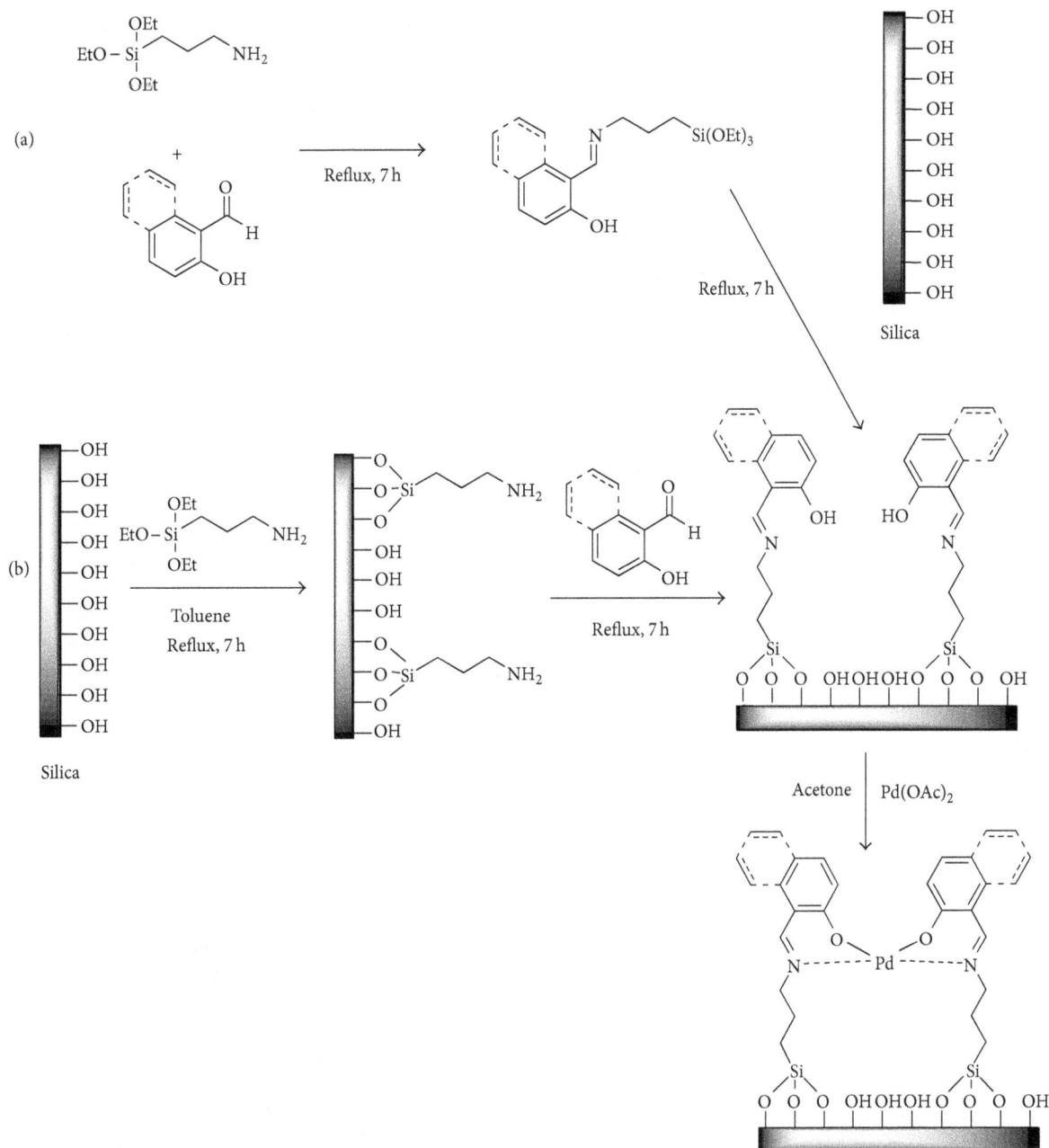

SCHEME 1: Synthetic strategy for the preparation of the silica gel supported Pd catalyst. (a) Homogeneous route for mesoporous silica functionalization (method A) and (b) heterogeneous route for mesoporous silica functionalization (method B).

3a: ^{1}H-NMR (400 MHz, CDCl$_3$): δ 7.17 (2H, apt, J 8.4 and 7.2), 6.71 (1H, apt, J 8.4 and 7.2), 6.60 (2H, d, J 7.6), 5.98 (1H, dt, J 15.6 and 5.2), 5.82 (1H, q, J 15.6 and 5.2), 5.37 (1H, m), 5.28 (1H, dd, J 10.4 and 6.0), 5.10 (1H, dd, J 10.4 and 3.6), 4.81 (1H, apt, J 6.0 and 5.2), 4.00–4.16 (3H, m), 3.75–3.95 (3H, m), 1.97–2.15 (12H, 4 × s); ^{13}C-NMR (100 MHz, CDCl$_3$): δ 170.52, 170.19, 170.06, 169.87, 147.64, 134.47, 129.27, 123.49, 117.87, 113.14, 72.53, 68.31, 68.29, 68.01, 61.83, 45.62, 20.73, 20.69, 20.66.

3b: ^{1}H-NMR (400 MHz, CDCl$_3$): δ 6.79 (2H, d, J 8.8), 6.60 (2H, d, J 8.8), 5.97 (1H, dt, J 15.6 and 5.2), 5.85 (1H, q,

J 15.6 and 5.2), 5.37 (1H, m), 5.31 (1H, dd, J 10.4 and 6.0), 5.13 (1H, dd, J 10.4 and 3.2), 4.81 (1H, apt, J 6.0 and 5.2), 4.00–4.15 (3H, m), 3.75–3.87 (3H, m), 3.74 (3H, s), 1.99–2.15 (12H, 4 × s); ^{13}C-NMR (100 MHz, CDCl$_3$): δ 170.53, 170.23, 170.13, 170.08, 152.44, 141.82, 134.79, 123.41, 114.92, 114.57, 72.50, 68.40, 68.33, 68.04, 68.01, 61.87, 45.94, 20.72, 20.70, 20.66.

3c: ^{1}H-NMR (400 MHz, CDCl$_3$): δ 7.09 (1H, apt, J 8.4 and 8.0), 6.15–6.30 (3H, m), 5.99 (1H, dt, J 15.6 and 5.2), 5.87 (1H, q, J 15.6 and 5.6), 5.38–5.40 (1H, m), 5.29 (1H, dd, J 10.4 and 6.0), 5.11 (1H, dd, J 10.4 and 3.2), 4.80 (1H, apt, J 6.0 and 5.2),

4.00–4.15 (4H, m), 3.82 (2H, d 5.2), 3.75 (3H, s), 1.96–2.16 (12H, 4 × s); ^{13}C-NMR (100 MHz, CDCl$_3$): δ 170.53, 170.20, 170.04, 169.89, 160.84, 149.10, 134.52, 130.04, 123.49, 106.18, 102.82, 99.28, 72.52, 68.29, 68.07, 68.00, 61.86, 55.01, 45.61, 20.73, 20.68, 20.63.

3d: ^{1}H-NMR (400 MHz, CDCl$_3$): δ 6.87 (1H, dt, J 7.6 and 1.6), 6.78 (1H, dd, J 8.0 and 1.2), 6.69 (1H, dt, J 7.6 and 1.2), 6.58 (1H, dd, J 8.0 and 1.2), 6.02 (1H, dt, J 15.6 and 5.2), 5.87 (1H, q, J 15.6 and 5.6), 5.37 (1H, d, J 3.2), 5.29 (1H, dd, J 10.4 and 6.0), 5.12 (1H, dd, J 10.4 and 3.2), 4.82 (1H, apt, J 6.0 and 5.2), 4.00–4.16 (3H, m), 3.75–3.86 (3H, m), 3.85 (3H, s), 1.98–2.13 (12H, 4 × s); ^{13}C-NMR (100 MHz, CDCl$_3$): δ 170.47, 170.20, 170.04, 169.88, 146.95, 137.59, 134.84, 123.30, 121.23, 117.01, 109.57, 72.62, 68.29, 68.09, 68.05, 61.88, 55.41, 45.47, 20.75, 20.72, 20.68, 20.65.

3e: ^{1}H-NMR (400 MHz, CDCl$_3$): δ 6.99 (2H, d, J 8.4), 6.55 (2H, d, J 8.4), 6.00 (1H, dt, J 15.6 and 5.2), 5.86 (1H, q, J 15.6 and 5.6), 5.37 (1H, d, J 3.2), 5.31 (1H, dd, J 10.4 and 6.0), 5.11 (1H, dd, J 10.4 and 3.6), 4.79 (1H, apt, J 6.0 and 5.6), 4.00–4.16 (4H, m), 3.82 (2H, d J 5.2), 2.22 (3H, s), 1.99–2.13 (12H, 4 × s), ^{13}C-NMR (100 MHz, CDCl$_3$): δ 170.50, 170.19, 170.04, 169.86, 145.37, 134.73, 129.76, 127.09, 123.34, 113.35, 72.59, 68.24, 68.05, 68.00, 61.83, 46.03, 20.70, 20.67, 20.34.

3f: ^{1}H-NMR (400 MHz, CDCl$_3$): δ 7.06 (1H, t, J 7.6), 6.55 (1H, d, J 7.6), 6.38–6.48 (2H, d, m), 6.01 (1H, dt, J 15.6 and 5.2), 5.87 (1H, q, J 15.6 and 5.2), 5.39 (1H, dd, J 3.6 and 1.6), 5.31 (1H, dd, J 10.4 and 6.0), 5.12 (1H, dd, J 10.4 and 3.6), 4.81 (1H, apt, J 6.0 and 5.2), 4.00–4.15 (4H, m), 3.84 (2H, d, J 5.2), 2.27 (3H, s), 1.97–2.14 (12H, 4 × s); ^{13}C-NMR (100 MHz, CDCl$_3$): δ 170.48, 170.16, 170.01, 169.83, 147.67, 138.98, 134.61, 129.13, 123.31, 118.75, 113.88, 110.19, 72.57, 68.26, 68.02, 67.96, 61.80, 45.65, 21.54, 20.70, 20.65.

3g: ^{1}H-NMR (400 MHz, CDCl$_3$): δ 8.09 (2H, d, J 8.8), 6.58 (2H, d, J 8.8), 6.01 (1H, dt, J 15.6 and 5.2), 5.88 (1H, q, J 16.0 and 5.2), 5.38–5.40 (1H, m), 5.30 (1H, dd, J 10.4 and 5.6), 5.11 (1H, dd, J 10.4 and 2.8), 4.82 (1H, apt, J 6.0 and 5.2), 3.75–4.18 (6H, m), 1.93–2.14 (12H, 4 × s); ^{13}C NMR (100 MHz, CDCl$_3$): δ 170.64, 170.17, 170.08, 169.77, 152.98, 138.32, 132.15, 126.36, 124.85, 111.39, 72.24, 68.45, 68.23, 67.95, 67.82, 61.73, 44.89, 20.73, 20.69, 20.64.

3h: ^{1}H-NMR (400 MHz, CDCl$_3$): δ 7.55 (1H, d, J 8.0), 7.40 (1H, s), 7.32 (1H, apt, J 8.8 and 8.0), 6.90 (1H, d, J 8.0), 5.99 (1H, dt, J 15.6 and 5.2), 5.92 (1H, q, J 15.6 and 5.2), 5.37–5.40 (1H, m), 5.31 (1H, dd, J 10.4 and 5.6), 5.10 (1H, dd, J 10.4 and 2.8), 4.82 (1H, apt, J 6.0 and 5.2), 4.00–4.19 (4H, m), 3.93 (2H, d, J 5.2), 1.96–2.17 (12H, 4 × s); ^{13}C-NMR (100 MHz, CDCl$_3$): δ 170.61, 170.18, 170.07, 169.85, 149.44, 148.44, 132.92, 129.81, 124.40, 119.04, 112.38, 106.52, 72.39, 68.35, 68.25, 67.92, 61.73, 45.31, 20.72, 20.69, 20.66.

3i: ^{1}H-NMR (400 MHz, CDCl$_3$): δ 7.84 (2H, d, J 8.8), 6.59 (2H, d J 8.8), 5.97 (1H, dt, J 15.6 and 5.2), 5.85 (1H, q, J 15.6 and 5.2), 5.35–5.40 (1H, m), 5.29 (1H, dd, J 10.4 and 5.6), 5.10 (1H, dd, J 10.4 and 3.2), 4.81 (1H, apt, J 6.0 and 5.2), 4.0–4.17 (4H, m), 3.93 (2H, d, J 4.8), 2.50 (3H, s), 1.96–2.16 (12H, 4 × s); ^{13}C-NMR (100 MHz, CDCl$_3$): δ 196.43, 170.53, 170.15, 170.03, 169.78, 151.74, 133.05, 130.78, 130.73, 127.01, 124.12, 111.73, 111.66, 72.32, 68.35, 67.92, 67.85, 61.74, 44.83, 25.98, 20.70, 20.65, 20.60.

2.5. Recycling Studies. Recycling studies were carried out using the procedure for the hydroamination of allene (1) (0.27 mmol) with aniline (0.81 mmol), Pd catalyst (SiO$_2$@ imineNB-Pd-II, 90 mg, 5 mol%), trifluoroacetic acid (20 mol%), and THF (0.5 mL). After the reaction, the catalyst was separated from the reaction mixture by filtration through a sintered glass funnel. The separated catalyst was washed with dichloromethane (3 × 1 mL), dried in a vacuum, and reused again to check its recycling efficiency.

3. Results and Discussion

3.1. Synthesis and Characterization of Catalyst. The preparation of the palladium immobilized on the Salen-silica is shown in Scheme 1. Salen was covalently bonded to the commercially available silica gel (230–400 mesh) giving the Salen-silica. It was readily prepared by two different approaches. The first route for the functionalized mesoporous silica used the homogenous system (method A) for the preparation of the silylating agent by the reaction of 3-aminopropyltriethoxysilane (APTES) with salicylaldehyde or 3-hydroxy-2-naphthaldehyde. The product was condensed with the silanol group of the solid silica gel to give the imine-functionalized silica gel (Scheme 1(a)). The solid powders from salicylaldehyde and 3-hydroxy-2-naphthaldehyde were designated as SiO$_2$@ imineSA and SiO$_2$@ imineNA, respectively. In the second route, the heterogeneous system (method B) involved the reaction of the silica gel with the APTES to give aminosilica followed by a condensation with salicylaldehyde or 3-hydroxy-2-naphthaldehyde to provide the Salen-silica (Scheme 1(b)). The solid powders from the reaction of salicylaldehyde and 3-hydroxy-2-naphthaldehyde were designated as SiO$_2$@ imineSB and SiO$_2$@ imineNB, respectively. Finally, the precursor SiO$_2$@ imineSA was applied for the synthesis of the Pd complex by reacting with Pd(OAc)$_2$ in acetone at room temperature for 24 h (method I) to give the product, designated as SiO$_2$@ imineSA-Pd-I. The Pd complexes of SiO$_2$@ imineSA, SiO$_2$@ imineSB, SiO$_2$@ imineNA, and SiO$_2$@ imineNB were also obtained by reacting with Pd(OAc)$_2$ in acetone under reflux conditions for 4 h (method II) to provide the resulting products, and they were designated as SiO$_2$@ imineSA-Pd-II, SiO$_2$@ imineSB-Pd-II, SiO$_2$@ imineNA-Pd-II, and SiO$_2$@ imineNB-Pd-II, respectively. The characterization of the SiO$_2$@ imine and Salen-silica supported Pd complex were done on the basis of their properties, such as via FT-IR, XRD, SEM, SEM-EDX, ICP-OES, and N$_2$ adsorption-desorption spectra data.

The silanol ($\equiv$Si–OH) groups on the surface of the silica showed a significant qualification through silylation. Figure 1 shows the FT-IR spectra of (I) SiO$_2$, (II) SiO$_2$@ imineSA, (III) SiO$_2$@ imineSB, (IV) SiO$_2$@ imineSA-Pd-II, (V) SiO$_2$@ imineSB-Pd-II, and (VI) SiO$_2$@ imineSA-Pd-II. According to Figure 1, the characteristic silica bands combined with a silica backbone can be clearly detected in all spectra. The characteristic Si–O–Si bands at 1057 and 799 cm^{-1}, present in all samples, are assigned to the silica network. The spectrum of the free silica displays a typical broad band at 3361 cm^{-1} due to the vibration of the H bond of the silanol

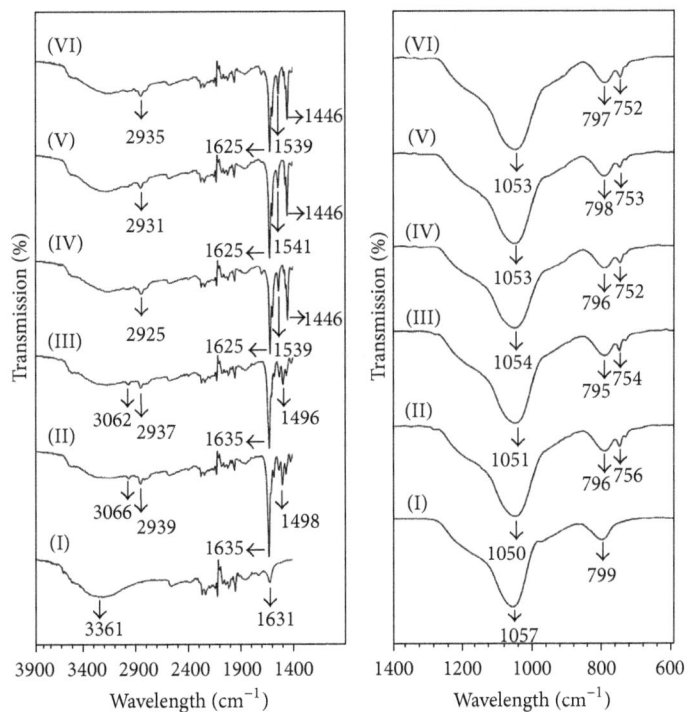

(a)

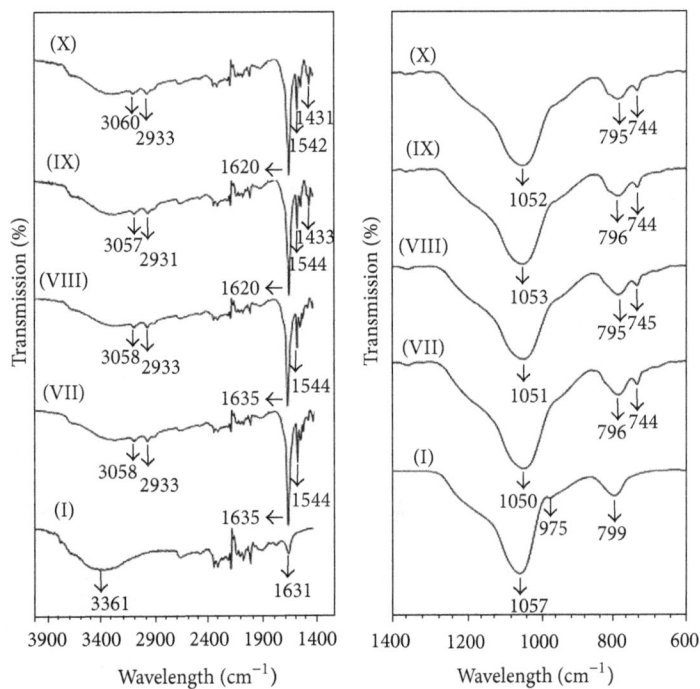

(b)

FIGURE 1: (a) FT-IR of (I) SiO$_2$, (II) SiO$_2$@ imineSA, (III) SiO$_2$@ imineSB, (IV) SiO$_2$@ imineSA-Pd-II, (V) SiO$_2$@ imineSB-Pd-II, and (VI) SiO$_2$@ imineSA-Pd-II. (b) FT-IR of (I) SiO$_2$, (VII) SiO$_2$@ imineNA, (VIII) SiO$_2$@ imineNB, (IX) SiO$_2$@ imineNA-Pd-II, and (X) SiO$_2$@ imineNB-Pd-II.

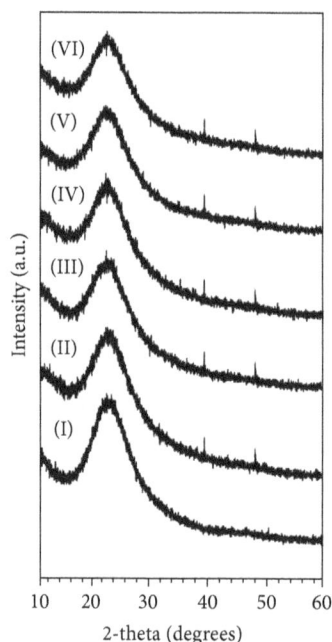

FIGURE 2: XRD patterns of (I) SiO$_2$, (II) SiO$_2$@ imineSA, (III) SiO$_2$@ imineSB, (IV) SiO$_2$@ imineSA-Pd-II, (V) SiO$_2$@ imineSB-Pd-II, and (VI) SiO$_2$@ imineSA-Pd-II.

group and water of the siloxane backbone. The spectra of all the samples (Figures 1(a) and 1(b)) of the modified silica were identified by the presence of two weak bands at 3058–3066 and 2931–2939 cm^{-1} due to the vibration of the aromatic and aliphatic C-H groups, respectively. Meanwhile, the bending vibration of the Si–OH on the free silica at 975 cm^{-1} significantly decreased after grafting with APTES. The Schiff bases display strong bands at 1635 and 1498 cm^{-1} due to the azomethine (C=N) and phenolic (C–O) stretching modes, respectively. The band at 1635 cm^{-1}, upon the reaction of the Salen complex with Pd, shifted to a lower frequency (1625 cm^{-1} for SiO$_2$@ imineSA-Pd and SiO$_2$@ imineSB-Pd; 1625 cm^{-1} for SiO$_2$@ imineNA-Pd and SiO$_2$@ imineNB-Pd), which indicated the formation of a Pd complex. The powder X-ray diffraction patterns of the parent silica gel (SiO$_2$) and palladium immobilized Salen-silica are shown in Figure 2.

The patterns have a broad peak of $2\theta = 22.4°$, which indicated that the material had low crystallinity or the amorphous nature of the silica. A few decreases in the intensity were identified for the Pd(II)-Salen-silica that confirmed the immobilization of the Salen-silica. This decrease in intensity could be due to the covering of the pores in the silica surface during the metalation.

In good agreement with the XRD, the SEM images of the free silica and supported Pd(II)-Schiff-base@SiO$_2$ samples (Figure 3) showed the ordered channel structure of the mesoporous materials, which is stored during the complex grafting. The silica gel did not change significantly in size or gathered state under the supported processes in method II, but the presence of palladium caused a significant decrease

in the silica particle size from the preparation with method I. Although the palladium particles and their dispersions are not clearly visible from the SEM spectra, considering the results of the SEM-EDX (Figure 4) and the ICP-OES, it was possible to define the palladium attachment on the surface of the silica supports. The presence of palladium in the Pd(II)-Schiff-base@SiO$_2$ samples could be clearly seen from Figure 4. The amount of adhered palladium was evaluated by ICP-OES, as shown in Table 1.

The results from the N$_2$ adsorption-desorption isotherms of the silica-based samples and the respective textural parameters containing the specific surface area (SBET), the pore diameter, and the total pore volume (V_{total}) are summarized in Table 1. The isotherms of all samples showed typical type IV patterns corresponding to the IUPAC definitions of porosity (Figure 5). The specific surface area provided by the BET method for the free SiO$_2$ was 411.16 m^{2}g^{-1}, which upon functionalization by methods A and B with salicylaldehyde and 3-hydroxy-2-naphthaldehyde gave SiO$_2$@ imineSA, SiO$_2$@ imineSB, SiO$_2$@ imineNA, and SiO$_2$@ imineNB, which were reduced to 340.58, 366.92, 273.49, and 268.89 m^{2}g^{-1}, respectively. In addition, the nitrogen adsorption-desorption studies indicated that a significant decrease in pore size by the imine-functionalization of the silica channels was monitored. There was not much difference in the surface areas of the various catalysts (entries 6–10, Table 1), with values between ca. 302 and 358 m^{2}g^{-1}. Different surface areas and pore volumes can support different activities of the porous solid catalysts, but there was no support due to different pore volumes and BET surface areas in this study.

FIGURE 3: SEM image of (a) SiO_2, (b) SiO_2@ imineSA, (c) SiO_2@ imineSA-Pd-II, (d) SiO_2@ imineSB, (e) SiO_2@ imineSB-Pd-II, and (f) SiO_2@ imineSA-Pd-I, (g) SiO_2@ imineNA, (h) SiO_2@ imineNA-Pd-II, (i) SiO_2@ imineNB, and (j) SiO_2@ imineNB-Pd-II.

3.2. Catalytic Hydroamination. After structural characterization of the prepared Pd(II) immobilized on Salen-functionalized silica, SiO_2@ imineSA-Pd-II, SiO_2@ imineSA-Pd-I, SiO_2@ imineSB-Pd-II, SiO_2@ imineNA-Pd-II, and SiO_2@ imineNB-Pd-II, its catalytic activity was investigated via the hydroamination of allene (1). The starting material was prepared as previously described. The reaction between the allene (1) and the aniline (2a) was selected as the model

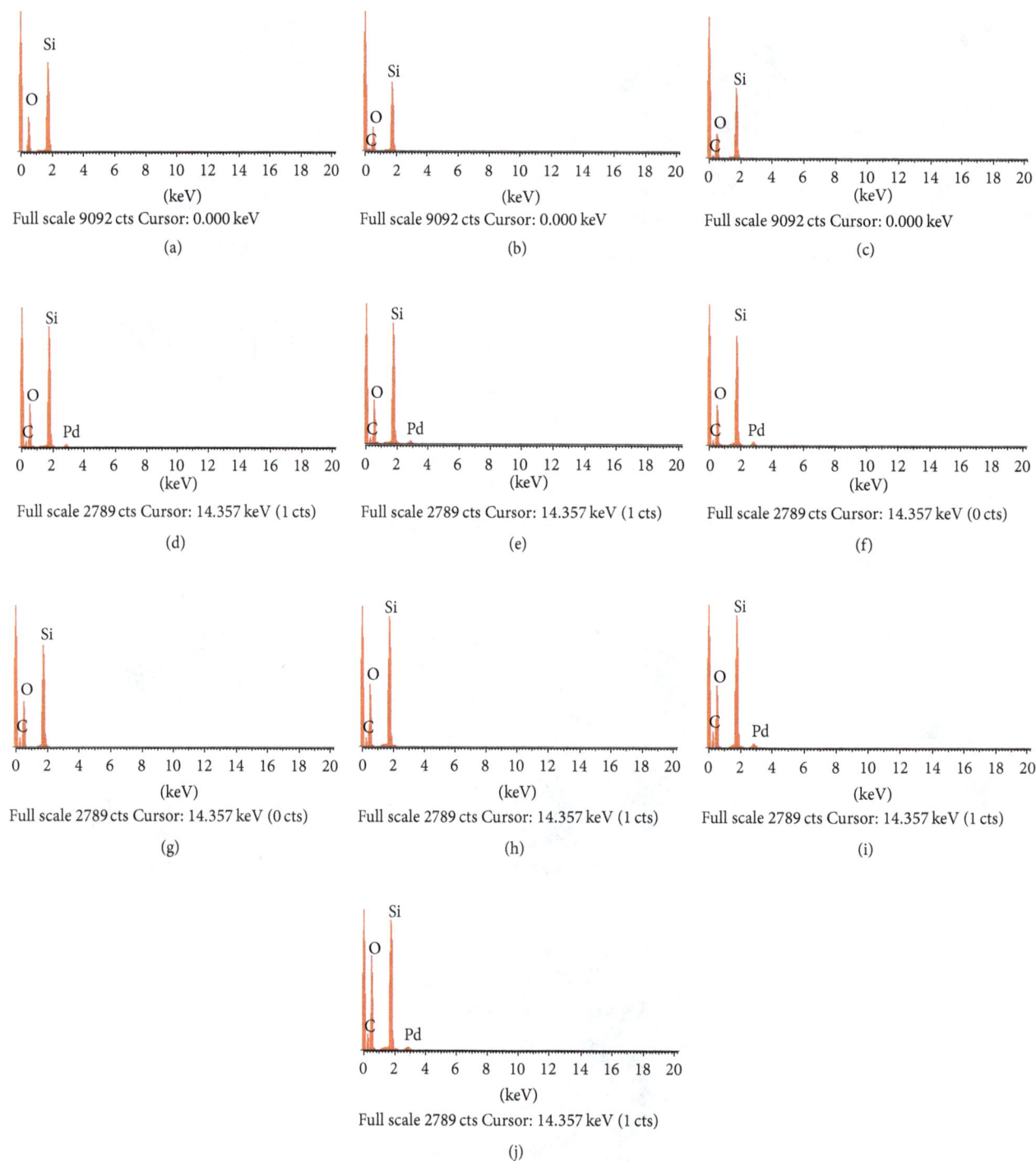

FIGURE 4: EDS spectrum of (a) SiO_2, (b) SiO_2@ imineSA, (c) SiO_2@ imineSB, (d) SiO_2@ imineSA-Pd-II, (e) SiO_2@ imineSB-Pd-II, (f) SiO_2@ imineSA-Pd-I, (g) SiO_2@ imineNA, (h) SiO_2@ imineNB, (i) SiO_2@ imineNA-Pd-II, and (j) SiO_2@ imineNB-Pd-II.

reaction using THF as the solvent and TFA as the additive with 5 mol% palladium catalyst, as shown in Table 2.

The reaction was successfully operated at room temperature under air without any special precautions. To verify the effect of the imine-functionalization on the catalysis, SiO_2@ imineSA-Pd-II, SiO_2@ imineSA-Pd-I, SiO_2@ imineSB-Pd-II,

SiO_2@ imineNA-Pd-II, SiO_2@ imineNB-Pd-II, and homogeneous [Pd(OAc$_2$)] were compared as catalysts (with the same palladium content) under identical conditions. It is necessary to note that SiO_2@ imineSA-Pd-I exhibited the same yield as indicated by the homogeneous [Pd(OAc$_2$)] (entries 1 and 3, Table 2). Then, when comparing with

TABLE 1: Surface area, total pore volume, pore size distribution, and amount of Pd loaded on the surface of the silica-based materials.

Entry	Materials	S_{BET}[a] $(m^2 g^{-1})$	Total pore volume[b] $(cm^3 g^{-1})$	Mean pore diameter[c] (nm)	Pd content[d] $\pm$ SD[e] $(mg\ g^{-1})$
1	SiO_2	411.16	0.860	4.18	NA[f]
2	SiO_2@ imineSA	340.58	0.503	2.95	NA[f]
3	SiO_2@ imineSB	366.92	0.589	3.21	NA[f]
4	SiO_2@ imineNA	273.49	0.430	3.14	NA[f]
5	SiO_2@ imineNB	268.89	0.416	3.10	NA[f]
6	SiO_2@ imineSA-Pd-II	358.63	0.505	2.82	30.74 $\pm$ 0.42
7	SiO_2@ imineSA-Pd-I	302.64	0.429	2.83	33.6 $\pm$ 1.52
8	SiO_2@ imineSB-Pd-II	365.38	0.539	2.95	29.58 $\pm$ 1.63
9	SiO_2@ imineNA-Pd-II	322.64	0.416	2.58	25.85 $\pm$ 0.72
10	SiO_2@ imineNB-Pd-II	305.41	0.407	2.65	25.12 $\pm$ 0.77

[a]BET method used in N_2 sorption. [b]Single-point pore volume at $P/P_o = 0.975$. [c]Adsorption average pore diameter by BET method. [d]Determined by ICP-OES analysis. [e]Average of triplicates $\pm$ SD. [f]Not applicable.

TABLE 2: Optimization of reaction conditions of C-(tetra-O-acetyl-β-D-galactopyranosyl)allene (**1**) with aniline (**2a**).[a]

Entry	Catalyst	Yield[b] (%) 3a
1	$Pd(OAc)_2$	30
2	SiO_2@ imineSA-Pd-II	17
3	SiO_2@ imineSA-Pd-I	34
4	SiO_2@ imineSB-Pd-II	16
5	SiO_2@ imineNA-Pd-II	50
6	SiO_2@ imineNB-Pd-II	60

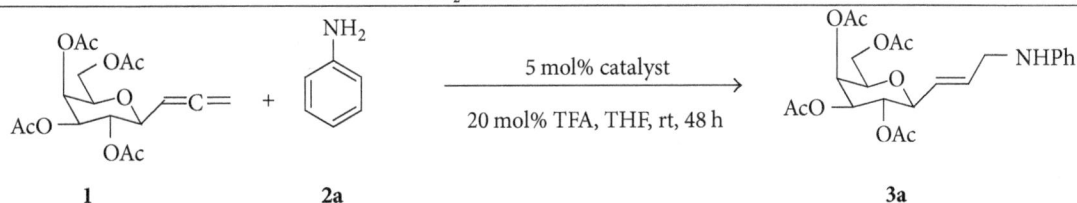

[a]All reactions were carried out with conditions: C-(tetra-O-acetyl-β-D-galactopyranosyl)allene (**1**) (0.27 mmol) and aniline (0.81 mmol) in 0.5 mL of solvent for 48 h. [b]Isolated yields.

the imine-functionalization of silica from salicylaldehyde, the activity was significantly decreased with imineSA-Pd-II and SiO_2@ imineSB-Pd-II (entries 2 and 4, Table 2). However, in the presence of the imine-functionalization of silica from 3-hydroxy-2-naphthaldehyde, SiO_2@ imineNA-Pd-II, and SiO_2@ imineNB-Pd-II, the reaction proceeded with an improved yield of the desired product (**3a**) (entries 5 and 6, Table 2). In addition, the formation of diallylated acetate and diallylated amine was not observed as minor products. The recycling ability of SiO_2@ imineNB-Pd-II for the hydroamination of **1** with aniline was evaluated. After the recovery of the catalyst, it was reused directly without further purification. The catalyst could be successfully recycled for five consecutive cycles for the hydroamination of **1** with aniline and no significant loss of catalytic activity was

observed after the fifth cycle. Additionally, palladium leaching in SiO_2@ imineNB-Pd-II was determined. The palladium content of the clear filtrates obtained upon filtration after the reaction was < 0.2 ppm, from ICP-OES analyses.

To test the extent and restrictions of our catalyst, SiO_2@ imineNB-Pd-II, we also studied the hydroamination of **1** with substituted anilines (**2b–i**) bearing an electron-poor or an electron-rich group on the benzene ring provided to the corresponding allylamines (**3b–i**) (Table 3).

According to previous reports, the results also showed that the onset potential for the heterogeneous catalyst was more positive than under the homogenous condition. This could be due to the reduced formation of diallylated amine and diallylated acetate as byproducts.

TABLE 3: Hydroamination of C-(tetra-O-acetyl-β-D-galactopyranosyl)allene (**1**) with aromatic amines[a].

Entry	Amine	Product (%)[b]
1	MeO—⟨⟩—NH$_2$ **2b**	**3b** (26)
2	MeO⟨⟩NH$_2$ **2c**	**3c** (32)
3	MeO⟨⟩NH$_2$ **2d**	**3d** (41)
4	H$_3$C—⟨⟩—NH$_2$ **2e**	**3e** (76)
5	H$_3$C⟨⟩NH$_2$ **2f**	**3f** (78)
6	O$_2$N—⟨⟩—NH$_2$ **2g**	**3g** (18)
7	O$_2$N⟨⟩NH$_2$ **2h**	**3h** (18)
8	O=C⟨⟩NH$_2$ **2i**	**3i** (23)

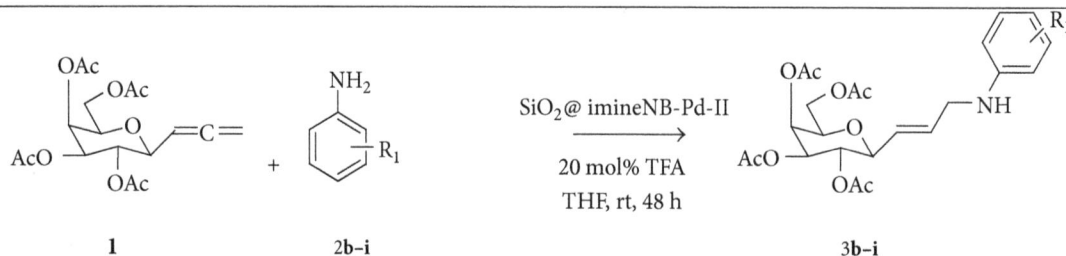

| **1** | **2b–i** | SiO$_2$@ imineNB-Pd-II, 20 mol% TFA, THF, rt, 48 h | **3b–i** |

[a]Reaction condition: C-(tetra-O-acetyl-β-D-galactopyranosyl)allene (**1**) (0.27 mmole) and aniline (0.81 mmole) in 0.5 mL of solvent for 48 h; SiO$_2$@ imineNB-Pd-II (5 mol%); TFA (20 mol%). [b]Isolated yields.

4. Conclusion

A novel mesoporous silica supported palladium(II)-based heterogeneous catalyst has been successfully synthesized by immobilizing Pd(OAc)$_2$ onto a Salen-functionalized silica gel via coordination. The supported catalyst exhibited good activity for the hydroamination of allene (**1**) with a variety of aromatic amines. It could be easily recovered by simple separation even after the catalyst was reused for a fifth cycle with no significant loss of catalytic activity.

Conflicts of Interest

The authors declare that they have no conflicts of interest.

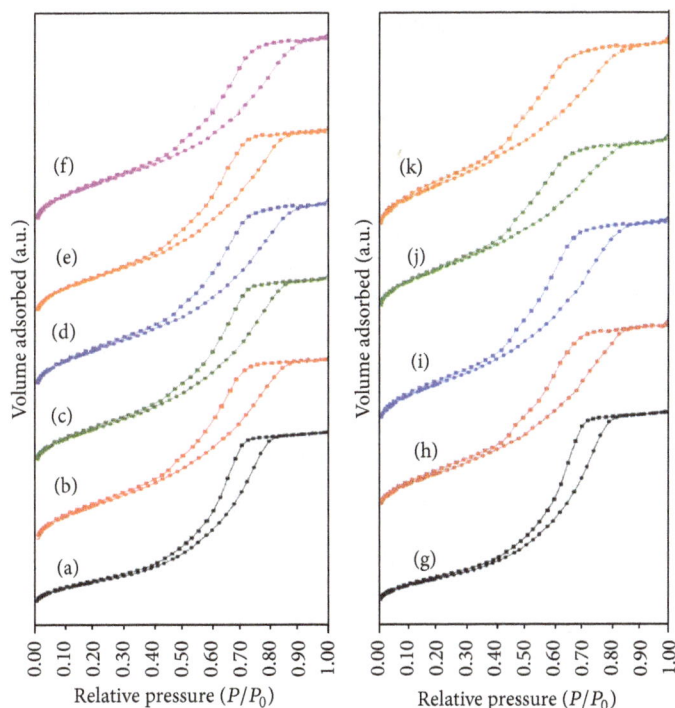

FIGURE 5: N_2 adsorption-desorption isotherms of the silica-based materials: (a) SiO_2, (b) SiO_2@ imineSA, (c) SiO_2@ imineSB, (d) SiO_2@ imineSA-Pd-II, (e) SiO_2@ imineSB-Pd-II, (f) SiO_2@ imineSA-Pd-I, (g) SiO_2, (h) SiO_2@ imineNA, (i) SiO_2@ imineNB, (j) SiO_2@ imineNA-Pd-II, and (k) SiO_2@ imineNB-Pd-II.

Acknowledgments

This work was supported by Mahasarakham University, Human Resource Development in Science Project (Science Achievement Scholarship of Thailand (SAST)), the Center of Excellence for Innovation in Chemistry (PERCH-CIC).

References

[1] L. Costantino and D. Barlocco, "Privileged structures as leads in medicinal chemistry," *Current Medicinal Chemistry*, vol. 13, no. 1, pp. 65–85, 2006.

[2] C. D. Duarte, E. J. Barreiro, and C. A. M. Fraga, "Privileged structures: A useful concept for the rational design of new lead drug candidates," *Mini-Reviews in Medicinal Chemistry*, vol. 7, no. 11, pp. 1108–1119, 2007.

[3] T. E. Müller and M. Beller, "Metal-initiated amination of alkenes and alkynes," *Chemical Reviews*, vol. 98, no. 2, pp. 675–703, 1998.

[4] P. W. Roesky and T. E. Müller, "Enantioselective catalytic hydroamination of alkenes," *Angewandte Chemie - International Edition*, vol. 42, no. 24, pp. 2708–2710, 2003.

[5] F. Alonso, I. P. Beletskaya, and M. Yus, "Transition-metal-catalyzed addition of heteroatom-hydrogen bonds to alkynes," *Chemical Reviews*, vol. 104, no. 6, pp. 3079–3159, 2004.

[6] M. Kawatsura and J. F. Hartwig, "Palladium-catalyzed intermolecular hydroamination of vinylarenes using arylamines," *Journal of the American Chemical Society*, vol. 122, no. 39, pp. 9546-9547, 2000.

[7] L. Huang, M. Arndt, K. Gooßen, L. J. Gooßen, and H. Heydt, "Late transition metal-catalyzed hydroamination and hydroamidation," *Chemical Reviews*, vol. 115, no. 7, pp. 2596–2697, 2015.

[8] K. M. Brummond and J. E. DeForrest, "Synthesizing allenes today," *Synthesis*, no. 6, pp. 795–818, 2007.

[9] P. J. Walsh, A. M. Baranger, and R. G. Bergman, "Stoichiometric and catalytic hydroamination of alkynes and allene by zirconium bisamides Cp2Zr(NHR)2," *Journal of the American Chemical Society*, vol. 114, no. 5, pp. 1708–1719, 1992.

[10] J. S. Johnson and R. G. Bergman, "Imidotitanium complexes as hydroamination catalysts: Substantially enhanced reactivity from an unexpected cyclopentadienide/amide ligand exchange," *Journal of the American Chemical Society*, vol. 123, no. 12, pp. 2923-2924, 2001.

[11] R. O. Ayinla and L. L. Schafer, "Bis(amidate) titanium precatalyst for the intermolecular hydroamination of allenes," *Inorganica Chimica Acta*, vol. 359, no. 9, pp. 3097–3102, 2006.

[12] L. Besson, J. Goré, and B. Cazes, "Palladium-catalyzed addition of malonate type compounds to allenes via a hydropalladation process," *Tetrahedron Letters*, vol. 36, no. 22, pp. 3853–3856, 1995.

[13] M. Al-Masum, M. Meguro, and Y. Yamamoto, "The two component palladium catalyst system for intermolecular hydroamination of allenes," *Tetrahedron Letters*, vol. 38, no. 34, pp. 6071–6074, 1997.

[14] J. Zhang, C.-G. Yang, and C. He, "Gold(I)-catalyzed intra- and intermolecular hydroamination of unactivated olefins," *Journal of the American Chemical Society*, vol. 128, no. 6, pp. 1798-1799, 2006.

[15] C. Khamwong and U. Sakee, "Palladium-catalyzed hydroamination of C-(tetra-O-acetyl-β-d-galactopyranosyl)allene," *Carbohydrate Research*, vol. 346, no. 2, pp. 334–339, 2011.

[16] C. Khamwong, S. Kruanetr, and U. Sakee, "Gold(III)-catalyzed intermolecular hydroamination of c-(tetra-O-acetyl-β-D-galactopyranosyl)allene," *Letters in Organic Chemistry*, vol. 9, no. 9, pp. 650–654, 2012.

[17] N. Selander and K. J. Szabó, "Catalysis by palladium pincer complexes," *Chemical Reviews*, vol. 111, no. 3, pp. 2048–2076, 2011.

[18] L. Yin and J. Liebscher, "Carbon-carbon coupling reactions catalyzed by heterogeneous palladium catalysts," *Chemical Reviews*, vol. 107, no. 1, pp. 133–173, 2007.

[19] A. Corma, "From microporous to mesoporous molecular sieve materials and their use in catalysis," *Chemical Reviews*, vol. 97, no. 6, pp. 2373–2419, 1997.

[20] N. Linares, A. E. Sepúlveda, J. R. Berenguer, E. Lalinde, and J. Garcia-Martinez, "Mesoporous organosilicas with Pd(II) complexes in their framework," *Microporous and Mesoporous Materials*, vol. 158, pp. 300–308, 2012.

[21] M. Čapka, M. Czakoová, E. Hillerová, E. Paetzold, and G. Oehme, "[2-(3-Trimethoxysilylthio)ethyl]diphenylphosphine—a new agent for transition metal immobilization," *Journal of Molecular Catalysis A: Chemical*, vol. 104, no. 2, pp. 123–125, 1995.

[22] A. Ghorbani-Choghamarani, F. Nikpour, F. Ghorbani, and F. Havasi, "Anchoring of Pd(ii) complex in functionalized MCM-41 as an efficient and recoverable novel nano catalyst in C-C, C-O and C-N coupling reactions using Ph_3SnCl," *RSC Advances*, vol. 5, no. 42, pp. 33212–33220, 2015.

[23] M. Opanasenko, P. Štěpnička, and J. Čejka, "Heterogeneous Pd catalysts supported on silica matrices," *RSC Advances*, vol. 4, no. 110, pp. 65137–65162, 2014.

[24] Y. He and C. Cai, "Polymer-supported macrocyclic Schiff base palladium complex: an efficient and reusable catalyst for Suzuki cross-coupling reaction under ambient condition," *Catalysis Communications*, vol. 12, no. 7, pp. 678–683, 2011.

[25] Y.-H. Zhu and P. Vogel, "Synthesis of a C-disaccharide analog of the Thomsen-Friedenreich (T) epitope," *Synlett*, no. 1, pp. 79–81, 2001.

[26] L. Kröger, D. Henkensmeier, A. Schäfer, and J. Thiem, "Novel O-glycosyl amino acid mimetics as building blocks for O-glycopeptides act as inhibitors of galactosidases," *Bioorganic and Medicinal Chemistry Letters*, vol. 14, no. 1, pp. 73–75, 2004.

2

Local Molecular Reactivity of the Colored Dansylglycine in Water and Dioxane Studied through Conceptual DFT

Juan Frau[1] **and Daniel Glossman-Mitnik**[1,2]

[1]*Departament de Química, Universitat de les Illes Balears, Palma de Mallorca 07122, Spain*
[2]*Laboratorio Virtual NANOCOSMOS, Departamento de Medio Ambiente y Energía, Centro de Investigación en Materiales Avanzados, Miguel de Cervantes 120 Complejo Industrial Chihuahua, 31136 Chihuahua, CHIH, Mexico*

Correspondence should be addressed to Daniel Glossman-Mitnik; daniel.glossman@cimav.edu.mx

Academic Editor: Philippe Dugourd

This study evaluated a fixed long-range corrected range-separated hybrid (RSH) density functional associated with the Def2TZVP basis set alongside the SMD solvation model for the computation of the structure, molecular properties, and chemical reactivity of the M8 intermediate melanoidin pigment in the presence of water and dioxane. The preference of the active sites pertinent to radical, nucleophilic, and electrophilic attacks is made through linking them with the electrophilic and nucleophilic Parr functions, Fukui function indices, and condensed dual descriptor which are chemical reactivity descriptors that arise from conceptual density functional theory. The study confirmed the results from previous works showing that the MN12SX density functional is the most appropriate in predicting the chemical reactivity of this molecule in both solvents.

1. Introduction

Melanoidins are the final product of the Maillard reaction that starts with formation of a Schiff base between a reducing sugar and the amino group of a peptide or protein. Some of these molecules can be isolated, and it has been observed that they possess interesting properties that can be useful not only as dyes for coloring foods and for dye-sensitized solar cells (DSSC) but also for the design and development of drugs for the pharmaceutical industry.

One of these isolated molecules is called dansylglycine (or 2-[[5-(dimethylamino)naphthalen-1-yl]sulfonylamino] acetic acid) which has interesting properties as a colored molecule and as a fluorescent dye in water and dioxane, and we believe that it could be of interest to study their molecular reactivity by using the ideas of conceptual DFT [1–5], in the same way of our previous works [6–10].

Thus, in this computational study, we will assess a powerful density functional in calculating the molecular properties and structure of the dansylglycine pigment in the presence of water and dioxane as solvents. Following the same ideas of previous works, we will consider a fixed RSH functional instead of the optimally tuned RSH density functionals that have attained great success [11–18].

2. Theoretical Background

If we consider the KID (for Koopmans in DFT) procedure presented in our previous works [6–10] together with a finite difference approximation, then the global reactivity descriptors can be calculated starting from the orbital energies of the HOMO and LUMO: electronegativity χ [19, 20], global hardness η [19, 20], electrophilicty ω [21], electro-donating ω^- [22], electroaccepting power ω^+ [22], and net electrophilicity $\Delta\omega^\pm$ [23].

Applying the same ideas related to the KID procedure, the local reactivity descriptors can be expressed as follows: nucleophilic Fukui function: $f^+(\mathbf{r}) = \rho_{N+1}(\mathbf{r}) - \rho_N(\mathbf{r})$ [19], electrophilic Fukui function: $f^-(\mathbf{r}) = \rho_N(\mathbf{r}) - \rho_{N-1}(\mathbf{r})$ [19], dual descriptor: $\Delta f(\mathbf{r}) = (\partial f(\mathbf{r})/\partial N)_{v(\mathbf{r})}$ [24–30], nucleophilic Parr function: $P^-(\mathbf{r}) = \rho_s^{rc}(\mathbf{r})$ [31, 32], and electrophilic Parr function: $P^+(\mathbf{r}) = \rho_s^{ra}(\mathbf{r})$ [31, 32], where $\rho_{N+1}(\mathbf{r})$,

TABLE 1: Global reactivity descriptors for the dansylglycine molecule calculated with the MN12SX/Def2TZVP model chemistry in water and in dioxane.

	Electronegativity (α)	Chemical hardness (η)	Electrophilicity (ω)	Electrodonating power (ω^-)	Electroaccepting power (ω^+)	Net electrophilicity ($\Delta\omega^{\pm}$)
Water	4.1950	3.6287	2.4248	4.2184	2.7183	6.9366
Dioxane	4.1427	3.6924	2.3240	4.0868	2.5974	6.6842

TABLE 2: Maximum wavelength absorption (λ_{max}) of the dansylglycine molecule in water and dioxane calculated from the HOMO-LUMO gap and from TDDFT results in comparison with experimental value.

	λ_{max}(HL)	Δ(HL)	λ_{max}(TDDFT)	Δ(TDDFT)
Water	342	2	362	22
Dioxane	336	3	354	15

$\rho_N(\mathbf{r})$, and $\rho_{N-1}(\mathbf{r})$ are the electronic densities at point $\mathbf{r}$ for the system with $N+1$, N, and $N-1$ electrons, respectively, and $\rho_s^{rc}(\mathbf{r})$ and $\rho_s^{ra}(\mathbf{r})$ are related to the atomic spin density (ASD) at the $\mathbf{r}$ atom of the radical cation or anion of a given molecule, respectively [33].

3. Settings and Computational Methods

Following the lines of our previous work [6–10], the computational studies were performed with the Gaussian 09 [34] series of programs with density functional methods as implemented in the computational package. The basis set used in this work was Def2SVP for geometry optimization and frequencies, while Def2TZVP was considered for the calculation of the electronic properties [35, 36]. All the calculations were performed in the presence of water and dioxane as solvents by doing integral equation formalism-polarized continuum model (IEF-PCM) computations according to the solvation model density (SMD) solvation model [37].

4. Results and Discussion

In the same way, as in our recent works on melanoidins [6–10], it has been found that the model chemistry formed by the connection between the MN12SX density functional and the Def2TZVP basis set is the best for justifying the fulfilling of the KID procedure for the dansylglycine molecule, both in the presence of water and dioxane. This, instead of presenting the comparison of the values of the orbital energies with the ionization potential I and the electron affinity A for different density functionals, we are showing the results for the global descriptors calculated with the MN12SX density functional in Table 1 comparing the values that arise from both solvents.

It is interesting to observe that the SMD solvation model is powerful enough to allow discriminating between the values for the global descriptors in both solvents.

In the past, various TDDFT studies of molecules of different sizes have used optimally tuned RSH density functionals

with great success [11–18]. Table 2 provides a comparison between the results involved in the ground-state approximation derived from the HOMO-LUMO gap together with TDDFT results and the experimental value of 340 nm in water and 339 nm in dioxane. It is clearly observed from the results presented in tables that the MN12SX density functional in connection with the Def2TZVP basis set is capable of reproducing accurately the excitation energies of the dansylglycine molecule starting from the HOMO-LUMO gap in both solvents.

Having verified that the MN12SX/Def2TZVP model chemistry is a good choice for the calculation of the global reactivity descriptors and the prediction of the maximum absorption wavelength from the ground state calculation of the HOMO and LUMO, we now present the molecular structures of dansylglycine in water and in dioxane in Figure 1. Meanwhile, the calculated bond lengths and bond angles for both cases are shown in Tables S1–S4 of the Supplementary Information (SI).

The calculations of the condensed Fukui functions and dual descriptor are done by using the Chemcraft molecular analysis program to extract the Mulliken population analysis (MPA), natural population analysis (NPA), and Hirshfeld population analysis (HPA) atomic charges [38] beginning with single-point energy calculations involving the MN12SX density functional that uses the Def2TZVP basis set in line with the SMD solvation model and water and dioxane utilized as solvents.

Considering the potential application the dansylglycine molecule as an antioxidant, it is of interest to get insights into the active sites for radical attack. A graphical representation of the radical Fukui function f^0 calculated with the MN12SX/Def2TZVP model chemistry in water and dioxane is presented in Figure 2.

It can be concluded that the most favorable site for the radical attack will be N7 in both cases, being the chemical reactivity in water a bit larger than in dioxane.

A graphical representation of the dual descriptor Δf_k of the dansylglycine molecule calculated with the MN12SX/Def2TZVP model chemistry in water and dioxane is presented in Figure 3.

It can be easily seen from the results for the dual descriptor $\Delta f(\mathbf{r})$ in Figure 3 that C13 will be the preferred site for a nucleophilic attack and that this atom will act as an electrophilic species in a chemical reaction. In turn, it can be appreciated that N6 will be more prone to electrophilic attacks and that these atomic sites will act as nucleophilic species in chemical reactions that involve the dansylglycine molecule in either solvent. As for the case of radical attacks, in these cases, the chemical reactivity in water will be larger

(a)

(b)

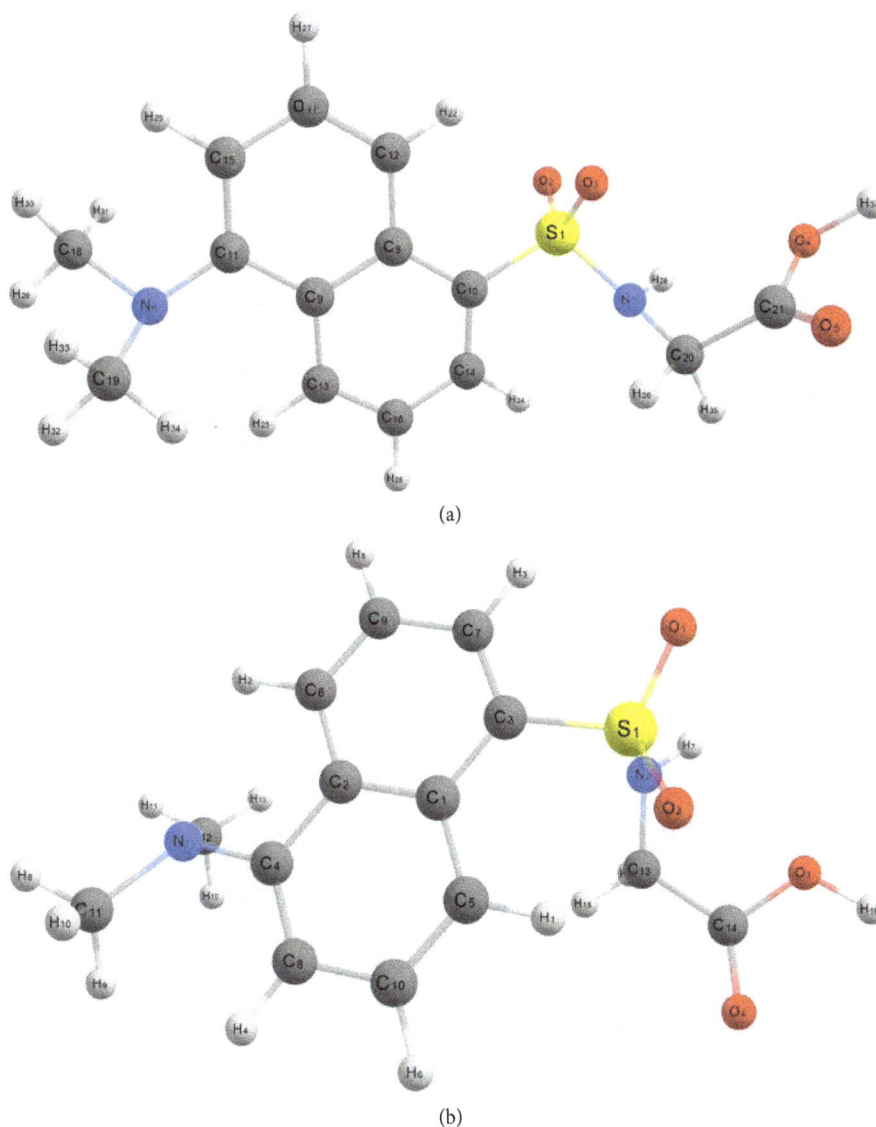

FIGURE 1: A schematic representation of the optimized structure of the dansylglycine molecule in water (a) and in dioxane (b) calculated with the MN12SX density functional showing the numbering of the atoms.

when considering dioxane as a solvent. The results for the condensed descriptor Δf_k show the maximum positive value over C13 and the minimum negative value over N6 in both solvents using either population analysis.

Finally, the condensed electrophilic and nucleophilic Parr functions P_k^+ and P_k^- over the atoms of the dansylglycine molecule in water and in dioxane have been calculated by extracting the Mulliken and Hirshfeld (or CM5) atomic charges using the Chemcraft molecular analysis program [38] starting from single-point energy calculations of the ionic species with the MN12SX density functional using the Def2TZVP basis set in the presence of the solvents according to the SMD solvation model. The maximum value of P_k^+ is located over C13 (0.3616 for MPA and 0.2143 for HPA), and the maximum value of P_k^- is located over N6 (0.4804 for MPA and 0.3719 for HPA) for the calculation in the presence of water, while in dioxane the results are similar with $P_k^+ = 0.519$ for MPA and 0.2081

for HPA and $P_k^- = 0.3814$ for MPA and 0.2962 for HPA. Thus, there is a nice agreement between the results describing the local reactivity of the dansylglycine molecule provided by Fukui functions, dual descriptor Δf (or its condensed counterpart), and the Parr functions.

5. Conclusions

A fixed RSH density functional, namely, MN12SX, was examined to determine whether it fulfills the empirical KID procedure. The assessment was conducted by comparing the values from HOMO and LUMO calculations to those generated by the ΔSCF technique for the dansylglycine molecule in water and in dioxane. This is a compound which is of academic as well as industrial interest. The study has confirmed that the range-separated hybrid meta-NGA density functional MN12SX is the most suited in meeting

(a)

(b)

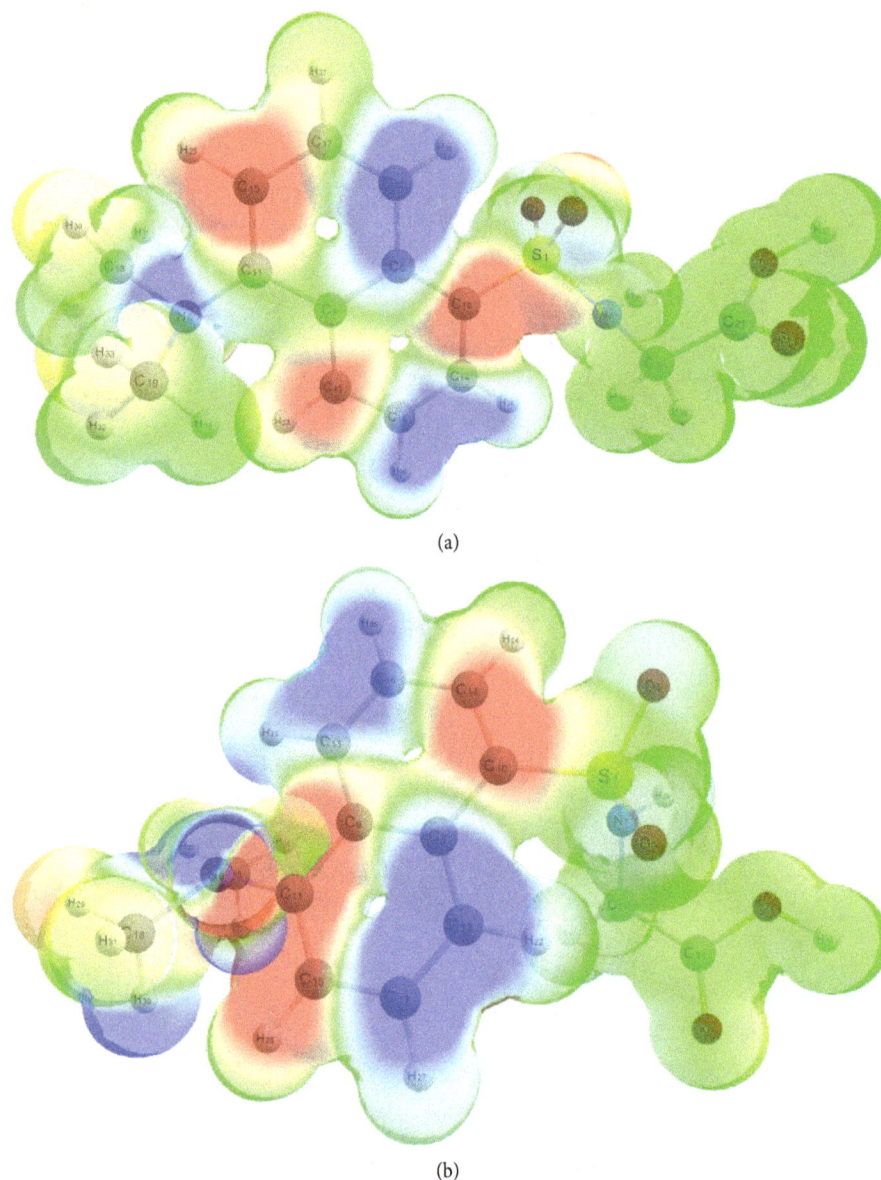

FIGURE 2: A graphical representation of the radical Fukui function f^0 of the dansylglycine molecule calculated with the MN12SX/Def2TZVP model chemistry in water (a) and dioxane (b).

this goal as it has been observed in previous works. Thus, this density functional can be a suitable alternative to density functionals where their behavior is optimally tuned using a gap-fitting procedure. It also exhibits the desirable prospect of benefiting future studies aimed at understanding the chemical reactivity of the colored products with larger molecular weights that are obtained when reducing sugars react with proteins and peptides.

From the results of this study, it can be concluded that the model chemistry formed by the connection of the MN12SX density functional and the Def2TZVP basis set is powerful enough to describe adequately the chemical reactivity of the dansylglycine molecule in both solvents through the calculation of the global and local DFT-based reactivity descriptors, including Fukui functions, Parr functions, and the dual descriptor calculations.

Furthermore, it is also possible to predict the maximum absorption wavelength for the dansylglycine in water an in dioxane with considerable accuracy. The prediction involves the MN12SX density functional beginning with the HOMO-LUMO gap instead of TDDFT calculations. Such a finding is particularly crucial considering the likelihood of it being used to inform the alternative determination method on the color that larger systems have such as prosthetic chromophore groups. This becomes necessary in circumstance when it is not possible to carry out TDDFT calculations as in the case of molecules of larger size like peptides and proteins.

(a)

(b)

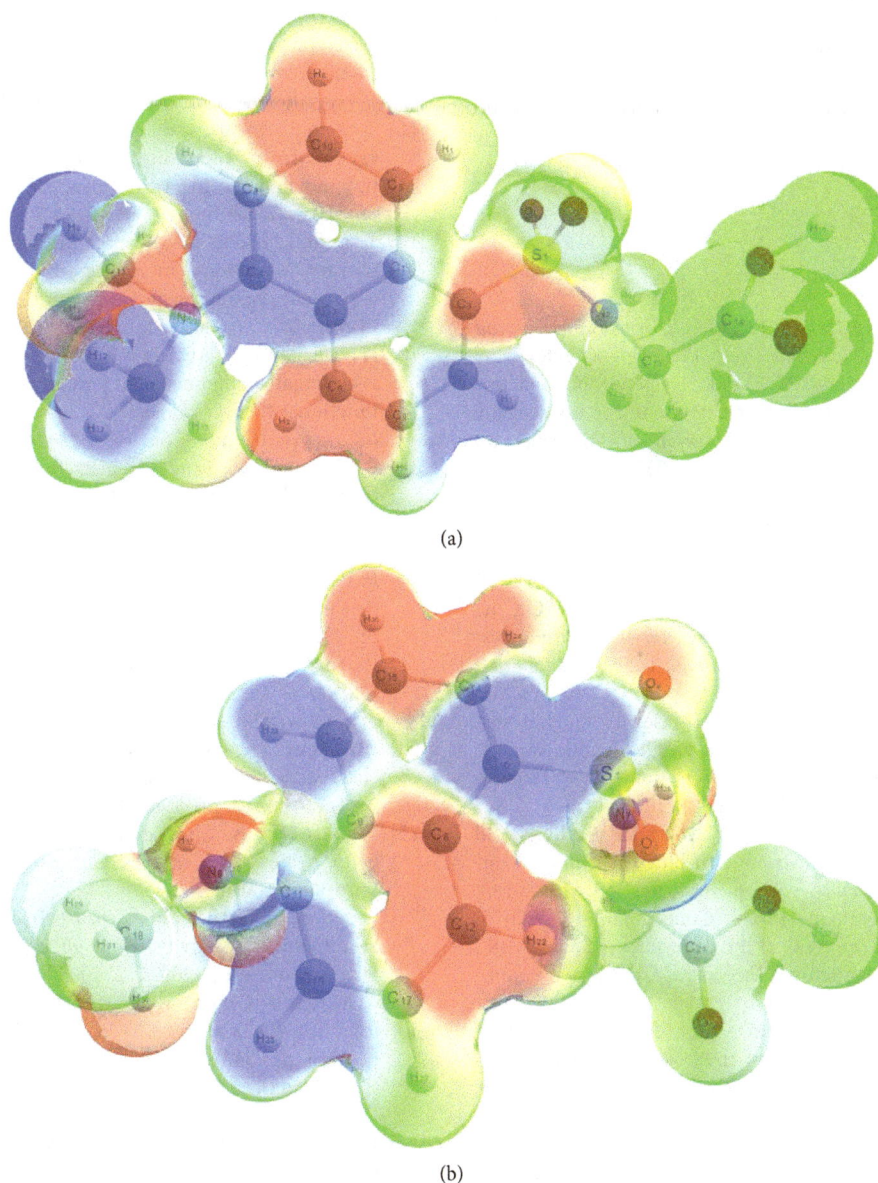

FIGURE 3: A graphical representation of the dual descriptor $\Delta f(\mathbf{r})$ of the dansylglycine molecule calculated with the MN12SX/Def2TZVP model chemistry in water (a) and dioxane (b).

Conflicts of Interest

The received funding did not lead to any conflicts of interest regarding the publication of this manuscript, and there are no other conflicts of interest regarding this submitted paper.

Acknowledgments

This work has been partially supported by CIMAV, SC, and Consejo Nacional de Ciencia y Tecnología (CONACYT, Mexico) through Grant no. 219566/2014 for Basic Science Research and Grant no. 265217/2016 for a Foreign Sabbatical Leave. Daniel Glossman-Mitnik conducted this work while a Sabbatical Fellow at the University of the Balearic Islands from which support is gratefully acknowledged. This work was cofunded by the Ministerio de Economía y Competitividad (MINECO) and the European Fund for Regional Development (FEDER) (CTQ2014-55835-R).

Supplementary Materials

Table S1: calculated bond lengths (in Å) of the dansylglycine molecule with the MN12SX density functional using water as a solvent simulated with the SMD solvation model. Table S2: calculated bond angles (in °) of the dansylglycine molecule with the MN12SX density functional using water as a solvent simulated with the SMD solvation model. Table S3:

calculated bond lengths (in Å) of the dansylglycine molecule with the MN12SX density functional using dioxane as a solvent simulated with the SMD solvation model. Table S4: calculated bond angles (in °) of the dansylglycine molecule with the MN12SX density functional using dioxane as a solvent simulated with the SMD solvation model. (*Supplementary Materials*)

References

[1] R. Parr and W. Yang, *Density-Functional Theory of Atoms and Molecules*, Oxford University Press, Oxford, UK, 1989.

[2] T. Mineva, E. Sicilia, and N. Russo, "Density-functional approach to hardness evaluation and its use in the study of the maximum hardness principle," *Journal of the American Chemical Society*, vol. 120, no. 35, pp. 9053–9058, 1998.

[3] T. Mineva, N. Russo, and E. Sicilia, "Solvation effects on reaction profiles by the polarizable continuum model coupled with the gaussian density functional method," *Journal of Computational Chemistry*, vol. 19, no. 3, pp. 290–299, 1998.

[4] G. De Luca, E. Sicilia, N. Russo, and T. Mineva, "On the hardness evaluation in solvent for neutral and charged systems," *Journal of the American Chemical Society*, vol. 124, no. 7, pp. 1494–1499, 2002.

[5] E. Sicilia, N. Russo, and T. Mineva, "Correlation between energy, polarizability, and hardness profiles in the isomerization reaction of HNO and ClNO," *Journal of Physical Chemistry A*, vol. 105, no. 2, pp. 442–450, 2001.

[6] J. Frau and D. Glossman-Mitnik, "Molecular reactivity and absorption properties of Melanoidin blue-G1 through conceptual DFT," *Molecules*, vol. 23, no. 3, pp. 559–615, 2018.

[7] J. Frau and D. Glossman-Mitnik, "Conceptual DFT study of the local chemical reactivity of the dilysyldipyrrolones A and B intermediate melanoidins," *Theoretical Chemistry Accounts*, vol. 137, no. 5, p. 1210, 2018.

[8] J. Frau and D. Glossman-Mitnik, "Conceptual DFT study of the local chemical reactivity of the colored BISARG melanoidin and its protonated derivative," *Frontiers in Chemistry*, vol. 6, no. 136, pp. 1–9, 2018.

[9] J. Frau and D. Glossman-Mitnik, "Molecular reactivity of some Maillard reaction products studied through conceptual DFT," *Contemporary Chemistry*, vol. 1, no. 1, pp. 1–14, 2018.

[10] J. Frau and D. Glossman-Mitnik, "Computational study of the chemical reactivity of the Blue-M1 intermediate melanoidin," *Computational and Theoretical Chemistry*, vol. 1134, pp. 22–29, 2018.

[11] A. Karolewski, T. Stein, R. Baer, and S. Kümmel, "Communication: tailoring the optical gap in light-harvesting molecules," *Journal of Chemical Physics*, vol. 134, no. 15, pp. 151101–151105, 2011.

[12] A. Karolewski, L. Kronik, and S. Kümmel, "Using optimally tuned range separated hybrid functionals in ground-state calculations: consequences and caveats," *Journal of Chemical Physics*, vol. 138, no. 20, article 204115, 2013.

[13] J. V. Koppen, M. Hapka, M. M. Szczeniak, and G. Chalasinski, "Optical absorption spectra of gold clusters Au(n) (n = 4, 6, 8,12, 20) from long-range corrected functionals with optimal tuning," *Journal of Chemical Physics*, vol. 137, no. 11, article 114302, 2012.

[14] L. Kronik, T. Stein, S. Refaely-Abramson, and R. Baer, "Excitation gaps of finite-sized systems from optimally tuned range-separated hybrid functionals," *Journal of Chemical Theory and Computation*, vol. 8, no. 5, pp. 1515–1531, 2012.

[15] S. Refaely-Abramson, R. Baer, and L. Kronik, "Fundamental and excitation gaps in molecules of relevance for organic photovoltaics from an optimally tuned range-separated hybrid functional," *Physical Review B*, vol. 84, no. 7, pp. 075144–075148, 2011.

[16] T. Stein, L. Kronik, and R. Baer, "Prediction of charge-transfer excitations in coumarin-based dyes using a range-separated functional tuned from first principles," *Journal of Chemical Physics*, vol. 131, no. 24, article 244119, 2009.

[17] T. Stein, L. Kronik, and R. Baer, "Reliable prediction of charge transfer excitations in molecular complexes using time-dependent density functional theory," *Journal of the American Chemical Society*, vol. 131, no. 8, pp. 2818–2820, 2009.

[18] H. Sun and J. Autschbach, "Electronic energy gaps for π-conjugated oligomers and polymers calculated with density functional theory," *Journal of Chemical Theory and Computation*, vol. 10, no. 3, pp. 1035–1047, 2014.

[19] R. Parr and W. Yang, "Density functional approach to the frontier-electron theory of chemical reactivity," *Journal of the American Chemical Society*, vol. 106, no. 14, pp. 4049-4050, 1984.

[20] P. Geerlings, F. De Proft, and W. Langenaeker, "Conceptual density functional theory," *Chemical Reviews*, vol. 103, no. 5, pp. 1793–1873, 2003.

[21] R. Parr, L. Szentpaly, and S. Liu, "Electrophilicity index," *Journal of the American Chemical Society*, vol. 121, no. 9, pp. 1922–1924, 1999.

[22] J. Gázquez, A. Cedillo, and A. Vela, "Electrodonating and electroaccepting powers," *Journal of Physical Chemistry A*, vol. 111, no. 10, pp. 1966–1970, 2007.

[23] P. Chattaraj, A. Chakraborty, and S. Giri, "Net electrophilicity," *Journal of Physical Chemistry A*, vol. 113, no. 37, pp. 10068–10074, 2009.

[24] C. Morell, A. Grand, and A. Toro-Labbé, "New dual descriptor for chemical reactivity," *Journal of Physical Chemistry A*, vol. 109, no. 1, pp. 205–212, 2005.

[25] C. Morell, A. Grand, and A. Toro-Labbé, "Theoretical support for using the descriptor," *Chemical Physics Letters*, vol. 425, no. 4–6, pp. 342–346, 2006.

[26] C. Cárdenas, N. Rabi, P. Ayers, C. Morell, P. Jaramillo, and P. Fuentealba, "Chemical reactivity descriptors for ambiphilic reagents: dual descriptor, local hypersoftness, and electrostatic potential," *Journal of Physical Chemistry A*, vol. 113, no. 30, pp. 8660–8667, 2009.

[27] A. Toro-Labbé, *Theoretical Aspects of Chemical Reactivity*, Elsevier Science, Amsterdam, Netherlands, 2007.

[28] P. Ayers, C. Morell, F. De Proft, and P. Geerlings, "Understanding the Woodward–Hoffmann rules by using changes in electron density," *Chemistry–A European Journal*, vol. 13, no. 29, pp. 8240–8247, 2007.

[29] C. Morell, P. Ayers, A. Grand, S. Gutiérrez-Oliva, and A. Toro-Labbé, "Rationalization of the Diels–Alder reactions through the use of the dual reactivity descriptor," *Physical Chemistry Chemical Physics*, vol. 10, no. 48, pp. 7239–7246, 2008.

[30] C. Morell, A. Hocquet, A. Grand, and B. Jamart-Grégoire, "A conceptual DFT Study of hydrazino peptides: assessment of the nucleophilicity of the nitrogen atoms by means of the dual descriptor $\Delta f(\mathbf{r})$," *Journal of Molecular Structure*, vol. 849, no. 1–3, pp. 46–51, 2008.

[31] L. R. Domingo, P. Pérez, and J. Sáez, "Understanding the local reactivity in polar organic reactions through electrophilic and nucleophilic Parr functions," *RSC Advances*, vol. 3, no. 5, pp. 1486–1494, 2013.

[32] E. Chamorro, P. Pérez, and L. R. Domingo, "On the nature of Parr functions to predict the most reactive sites along organic

polar reactions," *Chemical Physics Letters*, vol. 582, pp. 141–143, 2013.

[33] L. R. Domingo, M. Ríos-Gutiérrez, and P. Pérez, "Applications of the conceptual density functional theory indices to organic chemistry reactivity," *Molecules*, vol. 21, no. 6, p. 748, 2016.

[34] M. J. Frisch, G. W. Trucks, H. B. Schlegel et al., *Gaussian 09 Revision D.01*, Gaussian Inc., Wallingford, CT, USA, 2018.

[35] F. Weigend and R. Ahlrichs, "Balanced basis sets of split valence, triple zeta valence and quadruple zeta valence quality for H to Rn: design and assessment of accuracy," *Physical Chemistry Chemical Physics*, vol. 7, no. 18, pp. 3297–3305, 2005.

[36] F. Weigend, "Accurate Coulomb-fitting basis sets for H to R," *Physical Chemistry Chemical Physics*, vol. 8, no. 9, pp. 1057–1065, 2006.

[37] A. Marenich, C. Cramer, and D. Truhlar, "Universal solvation model based on solute electron density and a continuum model of the solvent defined by the bulk dielectric constant and atomic surface tensions," *Journal of Physical Chemistry B*, vol. 113, no. 8, pp. 6378–6396, 2009.

[38] G. A. Zhurko, *Chemcraft: Graphical Program for Visualization of Quantum Chemistry Computations*, Russian Federation, Ivanovo, Russia, https://chemcraftprog.com.

Ion Diffusion Behavior between Fracturing Water and Shale and Its Potential Influence on Production

Yinghao Shen,[1,2] Zhaopeng Zhu,[3] Peng Shi,[4] Hongkui Ge,[1] and Zhihui Yang[1]

[1] *China University of Petroleum, Beijing, China*
[2] *Texas Tech University, Lubbock, TX, USA*
[3] *PetroChina Jilin Oilfield Company, Songyuan, China*
[4] *Research Institute of Shaanxi Yanchang Petroleum "Group" Co., Ltd., Xian, Shaanxi, China*

Correspondence should be addressed to Yinghao Shen; yinghao.shen@ttu.edu

Academic Editor: Jae Ryang Hahn

Water imbibition, conductivity measurements, and ion identification were performed to investigate ion diffusion behavior between slick water and shale for large-scale hydraulic fracturing. The results indicated that there was strong ion exchange between water and shale. The ion concentration in water increases with fracture complexity and is dependent on the salinity of fracturing fluids. This implies that fracturing effects could be forecast from flow-back fluid ion concentrations after large-scale slick water fracturing. Higher levels of ion diffusion imply the presence of larger fracturing areas and higher level of fracture density for a similar reservoir. The mechanism of ion diffusion and the corresponding effects on IOR (increased oil recovery) based on a field example are discussed.

1. Introduction

The increasing demand for energy has prompted a need to find economical ways of developing unconventional resources, such as shale gas, globally. Shale oil and gas reservoirs are of relatively low permeability and porosity, and most hydrocarbons are stored in their tight matrixes [1, 2]. Large-scale hydraulic fracturing serves as an effective means of exploiting shale formation. In view of the performance of shale hydraulic fracturing methods, a large amount of fracturing fluid (generally more than 30%) is retained in shale formations after flow back [3]. The interaction between water and shale plays an important role in large-scale hydraulic fracturing [4–6]. Water has an enormous influence on the mechanical behaviors [7], effective flow channels [8], flow-back methods [9], and gas production patterns [10] of unconventional reservoirs. Fracturing fluids imbibition process and way in which slick water enters formations have received considerable attention [11, 12]. High capillary forces are recognized as a major force because there are abundance nanopores and always ultralow initial water saturation in shale [13, 14]. Series experiments have been conducted to

investigate related physical processes [15–18]. Fractal theory has been used in interaction studies and gas proving that ultracomplex microstructures render interactions between fracturing fluids and shale matrixes much more complex than those of conventional reservoirs [19, 20].

The chemical action between slick water and shale has recently attracted attentions. Osmotic pressure is recognized as an important driving force for water imbibition and shale oil production [21]. The abundant content of clay minerals serves as foundation for clay-chemical effects. Recently, a model considering osmotic effects has shown that chemical actions play a key role in interactions between slick water and shale [22]. Typically, salinity control is one important way to enhance oil recovery based on the oilfield chemistry [23, 24]. However, the impacts of fracturing fluid salinity on IOR (increasing oil recovery) have not received enough attention. Ion exchange between slick water and shale must be investigated to understand the chemical processes involved.

In this study, we conducted water imbibition, conductivity measurement, and ion identification experiments to examine patterns of the ion exchange between slick water and shale. Slick water is here defined as the main fracturing

TABLE 1: Mineral concentrations (wt%) of the shale samples as determined by X-ray Diffraction.

Sample	Shale
Calcite	6
Quartz	44
Dolomite	3
Pyrite	3
Feldspar	2
Illite	23
Chlorite	12
Illite/smectite	7

fluid for shale oil and gas development [25, 26], and the main component is water, the percentage of which always exceeds 98%. The aim of this study was to determine ion diffusion performance between shale and slick water in Longmaxi Shale in China. The effects of fractures in rock, slick water with different salinity, and different clay content are considered. Related mechanisms and potential applications are discussed based on actual field studies conducted in a shale gas field of southern China.

2. Experiment

2.1. Sample Description and Preparation. Shale samples in this study are from the Longmaxi Shale Formation of Lower Silurian in the Sichuan Basin of China. The samples were cut from one large outcrop, the size of which is approximately 80 cm × 80 cm × 80 cm. The shale sample is composed of a relatively high proportion of quartz. Clay mineral type and content were tested by X-ray diffraction (XRD) using an Empyrean diffractometer provided by a third test company. Corresponding results are shown in Table 1. Backscattered electron images were obtained for the thin slices of the shale samples using NanoLab 650 (FEI, USA), a highly accurate Scanning Electron Microscope (SEM) stationed at China University of Petroleum, Beijing. According to XRD results, quartz and clay minerals are the most abundant minerals in the shale samples as confirmed by Figure 1. The main clay minerals found are illite and chlorite. A small amount of smectite is found in interlayers, denoting the presence of few swelling clay minerals.

The porosity of the shale samples ranges from 6% to 8%. Porosity was measured using a helium porosimeter developed by Core lab. The permeability of the shale samples ranges from 4×10^{-4} mD to 7×10^{-4} mD as measured from an ultralow permeability measurement instrument (YRD-CP200 type). The permeability was measured as pulse-decay permeability with a confining pressure 8 MPa and pore pressure 5 MPa. The rocks exhibit a bimodal pore-size distribution of both micropores and nanopores that vary in size from 30 μm to 60 μm and from 1.7 nm to 20 nm, respectively [27].

Figure 1 shows shale backscattered electron images. The main components are quartz and clay. The image shows well-developed pores in organic matter, a major characteristic of

Longmaxi Shale. Direct measurements of pore sizes within the organic matter show pore sizes of the nanoscale.

2.2. Experimental Procedures. The experiment equipment mainly includes a conductivity meter, burette, and balance. Water electrical conductivity was measured based on a multifunctional conductivity measurement produced by METTLER-TOLEDO (Type: SevenExcellence S700). The electrode used is a normal solution conductivity electrode with a precision of 0.1 mS/cm–2000 mS/cm (±0.5%) and a suitable temperature of −30°C to 130°C. The chemical agent used included silver nitrate solution, K_2CrO_4 indicator, hydrochloric acid, phenolphthalein, methyl orange, EDTA standard solution, NaOH standard solution, calcon-carboxylic acid, and ammonium hydroxide.

The environmental temperature was set to 25°C. The tests were conducted under atmosphere pressure (0.1 MPa). The tests progressed as follows:

(1) Measure the sample weight and place the sample into different liquids.

(2) Measure the conductivity of the liquid with time.

(3) Use the titrimetric method to measure the ion content.

(4) Measure the final sample weight after water imbibition.

To investigate the interaction between water and shale, we used three different schemes to investigate different factors influencing ion diffusion such as contact area, liquid salinity, and fracture density. The samples described blow were collected from one outcrop of the Longmaxi formation, the properties of which are discussed in Section 2.1.

Case 1. Samples of the same size (1 cm × 2 cm × 3 cm) are placed in distilled water, in low salinity water and in high salinity water. We conduct a test following the measure steps listed above.

Case 2. We obtain samples from one large sample. The curved samples are of two sizes: 1 cm × 1 cm × 1 cm and 1 cm × 2 cm × 3 cm. Rock volumes are the same between 6 samples of the former size and 1 sample of the latter size. However, the six small samples include three fractures (Figure 2). In a similar fashion, 12 small samples, 18 small samples, and the corresponding same-sized large sample are, respectively, tested on as well.

Case 3. We obtain three samples from one large sample. The first includes two pronounced fractures. The second has one smaller fracture and its sample volume is the same as that of the former. The third has one fracture as well and differs from the second in that its volume is slightly smaller than the second (Figure 3).

3. Results

The conductivity change found for similar shale samples in the different solutions is shown in Figure 4. The conductivity

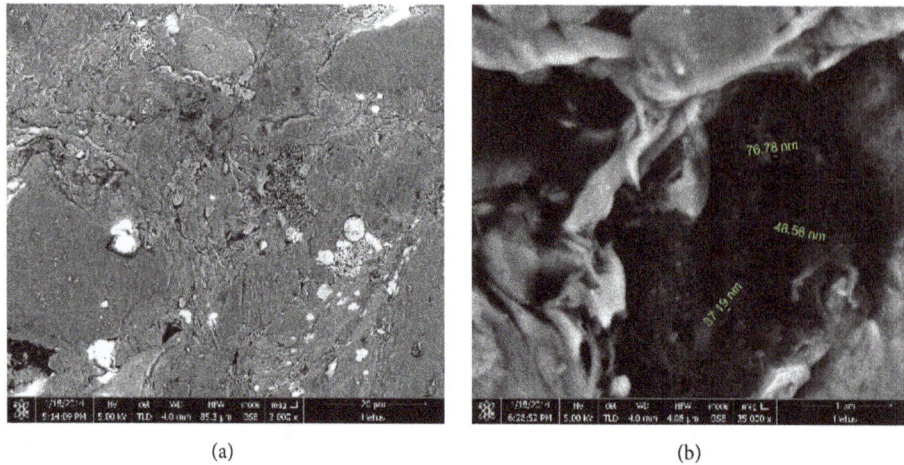

(a)   (b)

FIGURE 1: Backscattered electron images of shale samples: (a) SEM image of a lower resolution; (b) SEM image of a higher resolution.

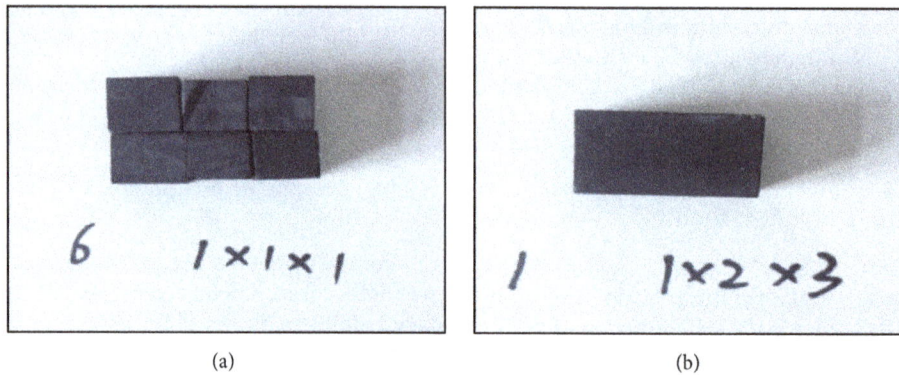

(a)   (b)

FIGURE 2: Shale samples of Case 2. (a) Six small samples ($1\,cm \times 1\,cm \times 1\,cm$); (b) one large sample ($1\,cm \times 2\,cm \times 3\,cm$) of the same volume with six small samples in (a).

increased with time as we placed the samples into the liquids. The conductivity of distilled water and of low salinity slick water increased significantly. However, the conductivity of high salinity slick water increased only slightly. Conductivity changes for distilled water, low salinity water, and high salinity water are $284\,\mu S/cm$, $194\,\mu S/cm$, and $125\,\mu S/cm$ after 8 hours, respectively.

The final conductivity changes and imbibed water volume observed are shown in Figure 5. The imbibed water volumes for distilled water, low salinity water, and high salinity water are $0.35\,cm^3$, $0.29\,cm^3$, and $0.22\,cm^3$ after 8 hours, respectively. We can see that ions have a considerable effect on water imbibition patterns, as the shale samples are separated by one larger sample. Clearly, the imbibed water volume is proportional to the conductivity change, proving the existence of chemical action between water and shale. Stronger osmotic effects between the sample and water were achieved under higher levels of salinity difference. Osmotic effects increasing the water intake capacities of shale and ion concentrations serve as evidence of this process.

Conductivity changes for the different samples are shown in Figure 6. First, the conductivity of the same sample(s) increased significantly early on and rates declined with time.

Second, the conductivity curve of the three samples shows that the larger sample exhibits a higher level of conductivity for one specific time. Third, the conductivity of the samples simulating fracturing is much greater than that of the whole sample, though the whole volume is the same, indicating that the fracturing considerably influences the interaction between fracturing fluids and shale. For example, the conductivity of 6 samples ($1\,cm \times 1\,cm \times 1\,cm$) is greater than that of one sample ($1\,cm \times 2\,cm \times 3\,cm$). Additionally, conductivity levels increased with sample values.

Figure 7 shows ion content levels for the different conditions. The main ions shown here are Cl^- and total cations. Cl^- is set as an evaluation object because it is easy to test in the field. We find that ion content levels increase with the number of samples involved and such changes corresponded to changes in conductivity. Cl^- content levels are recorded as $201.45\,mg/L$, $403.2\,mg/L$, and $723.15\,mg/L$ for 6 samples, 12 samples, and 18 samples, respectively. Total cation contents levels are recorded as $133.45\,mg/L$, $250.45\,mg/L$, and $514.03\,mg/L$, respectively.

Figure 8 shows the change in conductivity for three samples cut from one large sample. The trend of conductivity change observed is the same as that observed for the other

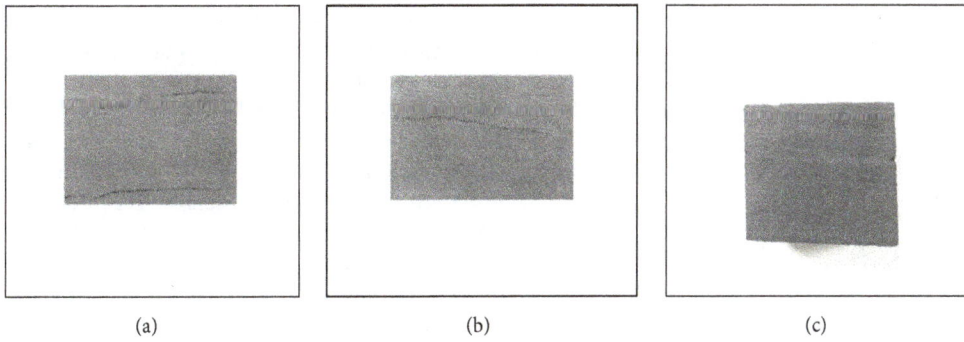

(a)   (b)   (c)

FIGURE 3: Shale samples of Case 3. (a) Sample with significant fractures. (b) Sample with limited fractures: the volume is the same as that of the sample shown in (a). (c) Sample with limited fractures: the volume is smaller than that of the sample shown in (a) and (b).

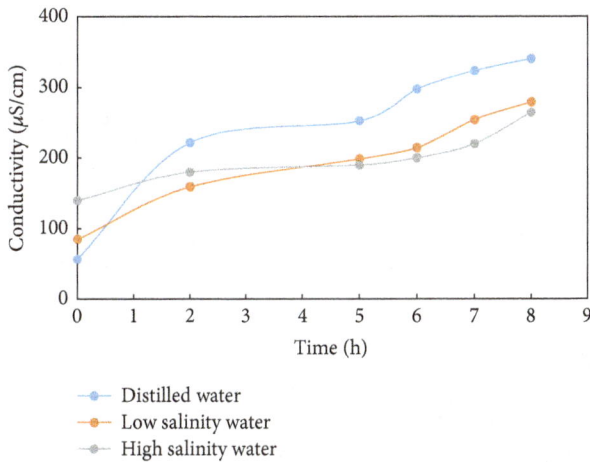

FIGURE 4: Conductivity changes for different solutions for Case 1.

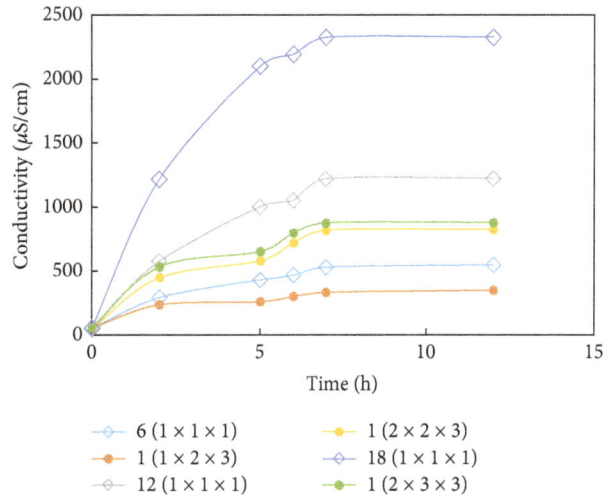

FIGURE 6: Conductivity changes for different samples for Case 2.

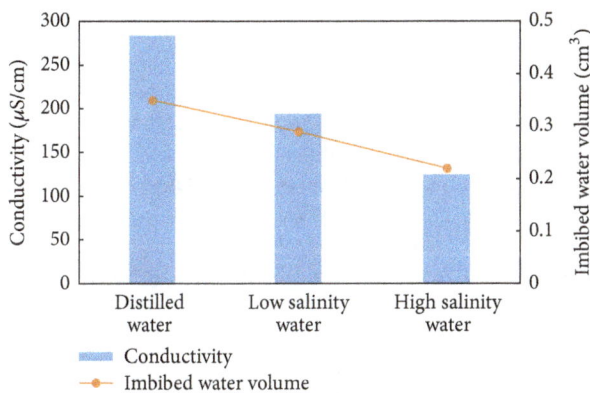

FIGURE 5: Final conductivity changes and imbibed water volumes for Case 1.

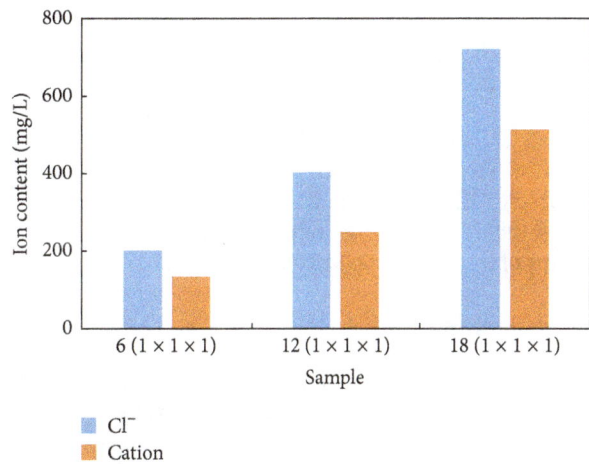

FIGURE 7: Ion content of different samples for Case 2.

samples. The one with more fractures is more conductive than the other two, showing that natural fractures play a significant role in water imbibition. Here, sample 2 and sample 3 present minor differences. The conductivity levels of the three samples are recorded as 1,020 μS/cm, 666 μS/cm,

and 612 μS/cm after 11 hours for example 1, example 2, and example 3, respectively.

Corresponding ion concentrations shown in Figure 9 exhibit the same conductivity trends. Cl$^-$ concentrations are

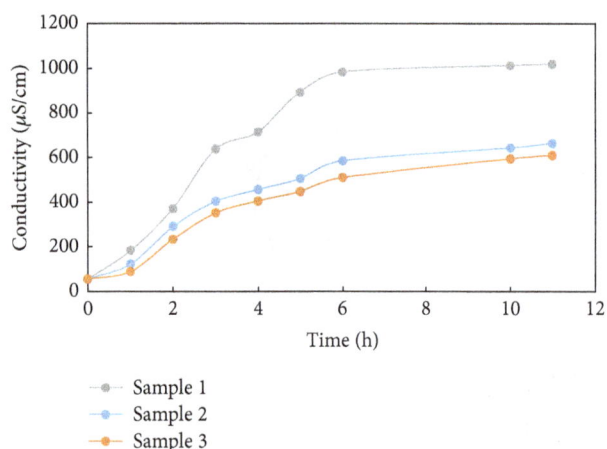

FIGURE 8: Conductivity changes for the different samples for Case 3.

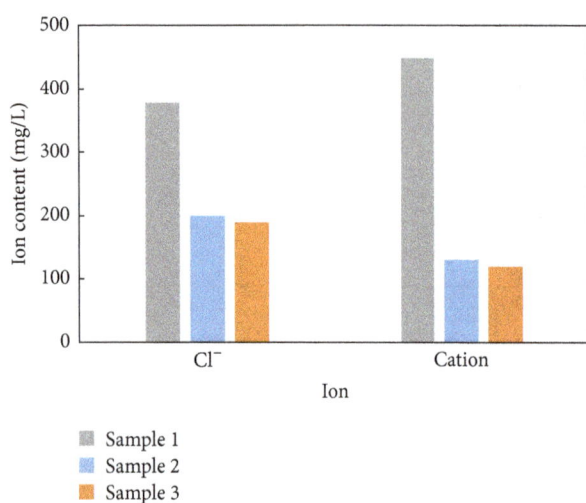

FIGURE 9: Ion content for the different samples for Case 3.

recorded as 378.2 mg/L, 199.73 mg/L, and 188.91 mg/L for sample 1, sample 2, and sample 3, respectively. Ion concentrations of sample 1 are much higher than those of sample 2 and sample 3. It is evident that the difference between sample 2 and sample 3 is minor as well, as these samples are similar in terms of fracture complexity and volume.

4. Discussion

Conductivity is defined as the physical quantity representing the ability to conduct electricity from an electrolyte. In most cases, the conductivity value reflects TDS (total dissolved solids) in a solution. Conductivity is traditionally used to measure ion concentrations in industrial solutions. Thus, ion exchange is the main mechanism that operates in the interaction between the slickwater and formation rock. Changes in conductivity in fluids serve as an indication of ion concentration change.

In large-scale hydraulic fracturing formations, ions from formations are derived from two main sources: ions captured by clay minerals and ions in formation water. When fluids come into contact with a formation after hydraulic fracturing, ion exchange occurs between fluids and the formation. The driving force of ion exchange is mainly concentration diffusion resulting from concentration differences between formation fluids and fracturing fluids and from ion exchange controlled by clay minerals. The second driving force is dependent on ion exchange properties of clay minerals. Clay minerals adsorb some cations and anions and these ions remain exchangeable. Generally, exchangeable cation includes Ca^{2+}, Mg^{2+}, H^+, K^+, $(NH_4)^+$, and Na^+ and exchangeable anions include $(SO_4)^{2-}$, Cl^-, and $(NO_3)^-$.

According to our experiments and to ion diffusion theory, ion diffusion between slick water and shale formations is characterized as follows.

(1) Salinity differences between formations and slick water serve as the main driving force for ion diffusion.

The samples we used in the experiments were drawn from the Longmaxi formation in southern China. The formation is composed of typical marine deposit shale. The surrounding deposit environment affords the formation of high salinity relative to the continental shale. The salinity of fracturing fluids is typically lower than that of a formation to decrease liquid friction and to thus improve the penetrability and volume fracturing. Water in the formation is highly saline and clay minerals adsorb more ion. This salinity difference serves as the driving force to ion diffusion. Figure 3 shows that as the difference in salinity levels increases, ion diffusion processes become more pronounced. Furthermore, ion diffusion is always accompanied by water adsorption, whereby higher salinity differences lead to higher amounts of water volume imbibed for Case 1.

(2) The higher cation exchange capacity (CEC) of shale serves as a key driving force for fracturing fluids imbibition and ion exchange.

Shale formations typically include more clay minerals than regular sandstone reservoirs. Clay content levels of Longmaxi Shale in the oriented shale gas field amount to approximately 40%. The CEC value increases with increases in clay content and especially as smectite and ion exchange potential levels increase. Water is imbibed into the interlayers of clay minerals, accelerating ion diffusion.

(3) Contact areas and fracture complexity are critical factors that influence the ion diffusion in a specific environment.

Salinity and clay content and time patterns are stable in a specific reservoir. When slick water is pumped into a reservoir, the interaction between slick water and formations is mainly controlled by the contact area, which is supported by Figures 5 and 6. Large-scale slick water fracturing generates higher levels of volume fractures. Higher levels of fracture density adds the contact areas between water and rock (Figures 5 and 7). In turn, ion diffusion is more pronounced at contact areas and fracture complexity levels in turn increase. During and after fracturing, ion diffusion occurs through interactions between fracturing fluids and formation rock and this effect is reflected at the ion concentrations in flowback fluids.

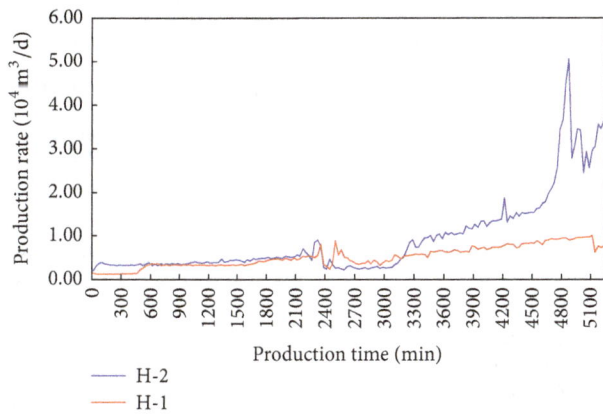

FIGURE 10: Postfracturing production rate for H1 and H2.

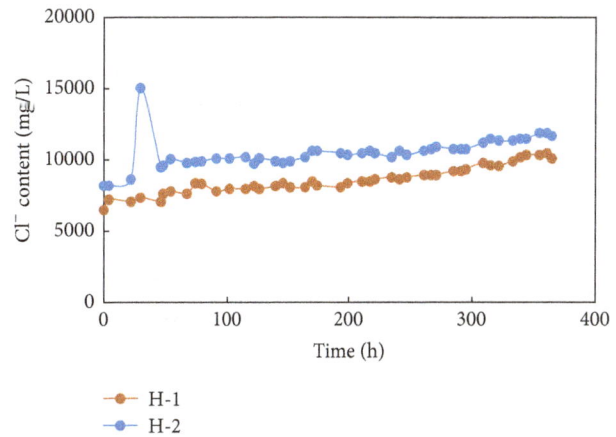

FIGURE 11: Postfracturing Cl⁻ content levels in H1 and H2.

According to our previous study, ion concentrations in flow-back fluids are central to evaluating fracturing complexities after hydraulic fracturing. We use two shale gas wells in China as a case to explain the relationship between ion concentration and productivity after hydraulic fracturing. Platform-H is a development platform in southern China. The development formation is Longmaxi Shale. The depth of high-quality shale of the platform is around 3530 m and the thickness of the formation is between 50.6 m and 53.7 m. Reservoir pressure gradient is about 1.96, which indicates a high level of reservoir pressure. The total organic content (TOC) of the formation is 4.90%~6.28% according to well log interpretations. Gas content levels in the shale, as tested by geochemistry logging, range from 5.59~7.43 m^3/t. Porosity is approximately 7.34%. There are two horizontal wells in this platform, and we refer to them as H-1 and H-2. The horizontal section of H-1 is 1,450 m, where the ratio of high-quality shale is 98.28%. Staged slick water fracturing is performed in the H-1 well, its stage number is 22, and its cluster number is 64. Volumes of fracturing fluid and proppants used amount to 41,340 m^3 and 1,660 m^3, respectively. The horizontal section of the H-2 well is 1,415 m, where the ratio of high-quality shale is 100%. Staged slick water fracturing is performed in the H-2 well. Its stage number is 22 and its cluster number is 65. Volumes of fracturing fluid and proppants used amount to 38,510 m^3 and 1,739 m^3, respectively. The above data show that the same drilling and fracturing techniques are applied to these two wells in the same platform. Fracturing methods and scales used are the same for the two wells. Figures 10 and 11 show the daily production levels and Cl⁻ content of flow-back fluids.

Figure 10 shows that the production rates of the two wells become more similar early on at approximately 0.5 × 10^4 m^3/d. The production rate of H-2 is much larger than that of H-1 after 50 hours. The production rate of H-2 is approximately three times that of H-1, showing that the two wells have significantly different hydraulic fracturing effect. Figure 11 shows the Cl⁻ content of flow-back fluid after fracturing retrieved through field ion concentration detection, which is used after fracturing and which is easy to apply. The results show that Cl⁻ content levels for the wells vary.

The Cl⁻ content of H-2 is approximately 20% greater than that of H-1. As the salinity of the formation and salinity of corresponding fracturing fluids are the same for the two wells, the Cl⁻ content level found corresponds to the contact area between fracturing fluids and the formation, showing that the fracture system induced by hydraulic fracturing in H-2 may be more complex than that induced in H-1. This field sample and our experiments show that the ion concentrations serve as an important auxiliary tool for evaluating the hydraulic fracturing effects.

5. Conclusion

Conductivity is a reflection of ion concentration change, which can be used to investigate interactions between fluids and rock. Ion concentrations in flow-back fluids are critical for evaluating fracturing complexity levels after hydraulic fracturing. Ion exchange between water and shale is controlled by water salinity, contact areas, and fracture density. Our field study shows that ion concentrations serve as an auxiliary tool method for estimating fracture complexity after hydraulic fracturing for shale gas wells in marine shale formations. The high salinity of marine shale brings about significant levels of ion exchange between formations and fracturing fluid. Ion concentrations can be used to determine the contact area between fracturing fluids and formations. Fracture systems are more complex when ion concentrations of flow-back rates are higher under similar conditions.

Conflicts of Interest

The authors declare no conflicts of interest.

Authors' Contributions

Each author made a contribution to the present paper. Yinghao Shen and Hongkui Ge developed the research questions. Zhaopeng Zhu collected the field data. Zhihui Yang conducted the experiments.

Acknowledgments

This work was supported by the National Natural Science Foundation of China (Grants 51604287 and 51490652) and Science Foundation of China University of Petroleum-Beijing at Karamay (no. RCYJ2016B-01-001).

References

[1] D. M. Jarvie, R. J. Hill, T. E. Ruble, and R. M. Pollastro, "Unconventional shale-gas systems: the Mississippian Barnett Shale of north-central Texas as one model for thermogenic shale-gas assessment," *AAPG Bulletin*, vol. 91, no. 4, pp. 475–499, 2007.

[2] Z. Pan, Y. Ma, L. D. Connell, D. I. Down, and M. Camilleri, "Measuring anisotropic permeability using a cubic shale sample in a triaxial cell," *Journal of Natural Gas Science and Engineering*, vol. 26, pp. 336–344, 2015.

[3] H. Dehghanpour, H. A. Zubair, A. Chhabra, and A. Ullah, "Liquid intake of organic shales," *Energy and Fuels*, vol. 26, no. 9, pp. 5750–5758, 2012.

[4] W. Yuan, X. Li, Z. Pan, L. D. Connell, S. Li, and J. He, "Experimental investigation of interactions between water and a lower silurian chinese shale," *Energy and Fuels*, vol. 28, no. 8, pp. 4925–4933, 2014.

[5] J. Li, X. Li, K. Wu et al., "Water sorption and distribution characteristics in clay and shale: effect of surface force," *Energy and Fuels*, vol. 30, no. 11, pp. 8863–8874, 2016.

[6] H. H. Liu, P. G. Ranjith, D. T. Georgi, and B. T. Lai, "Some key technical issues in modelling of gas transport process in shales: a review," *Geomechanics and Geophysics for Geo-Energy and Geo-Resources*, vol. 2, no. 4, pp. 231–243, 2016.

[7] D. Zhang, R. Pathegama Gamage, M. Perera, C. Zhang, and W. Wanniarachchi, "Influence of water saturation on the mechanical behaviour of low-permeability reservoir rocks," *Energies*, vol. 10, no. 2, p. 236, 2017.

[8] Y. Shen, H. Ge, C. Li et al., "Water imbibition of shale and its potential influence on shale gas recovery—a comparative study of marine and continental shale formations," *Journal of Natural Gas Science and Engineering*, vol. 35, pp. 1121–1128, 2016.

[9] Y. Shen, H. Ge, S. Su, D. Liu, Z. Yang, and J. Liu, "Imbibition characteristic of shale gas formation and water-block removal capability," *Scientia Sinica Physica, Mechanica & Astronomica*, 2017.

[10] Y. Shen, H. Ge, M. Meng, Z. Jiang, and X. Yang, "Effect of water imbibition on shale permeability and its influence on gas production," *Energy Fuels*, vol. 31, no. 5, pp. 4973–4980, 2017.

[11] A. Hayatdavoudi, M. A. Boamah, A. Tavnaei, K. G. Sawant, and F. Boukadi, "Post Frac Gas Production through Shale Capillary Activation," in *Proceedings of the SPE Production and Operations Symposium*, Oklahoma City, Oklahoma, USA.

[12] A. Settari, R. Sullivan, and R. Bachman, "The Modeling of the Effect of Water Blockage and Geomechanics in Waterfracs," in *Proceedings of the SPE Annual Technical Conference and Exhibition*, San Antonio, Tex, USA.

[13] M. Sharma and S. Agrawal, "Impact of Liquid Loading in Hydraulic Fractures on Well Productivity," in *Proceedings of the SPE Hydraulic Fracturing Technology Conference*, The Woodlands, Tex, USA.

[14] J. Cai, X. Hu, B. Xiao, Y. Zhou, and W. Wei, "Recent developments on fractal-based approaches to nanofluids and nanoparticle aggregation," *International Journal of Heat and Mass Transfer*, vol. 105, pp. 623–637, 2017.

[15] X. Wang, S. Xiao, Z. Zhang, and J. He, *Effect of Nanoparticles on Spontaneous Imbibition of Water into Ultraconfined Reservoir Capillary by Molecular Dynamics Simulation. Energies*, 10, 506, 2017.

[16] B. Roychaudhuri, T. T. Tsotsis, and K. Jessen, "An experimental investigation of spontaneous imbibition in gas shales," *Journal of Petroleum Science and Engineering*, vol. 111, pp. 87–97, 2013.

[17] A. Javaheri, H. Dehghanpour, and J. M. Wood, "Tight rock wettability and its relationship to other petrophysical properties: A Montney case study," *Journal of Earth Science*, vol. 28, no. 2, pp. 381–390, 2017.

[18] Q. Meng, H. Liu, and J. Wang, "A critical review on fundamental mechanisms of spontaneous imbibition and the impact of boundary condition, fluid viscosity and wettability," *Advances in Geo-energy Research*, vol. 1, pp. 1–17, 2017.

[19] J. Cai, B. Yu, M. Zou, and L. Luo, "Fractal characterization of spontaneous co-current imbibition in porous media," *Energy & Fuels*, vol. 24, no. 3, pp. 1860–1867, 2010.

[20] J. C. Cai and B. M. Yu, "A discussion of the effect of tortuosity on the capillary imbibition in porous media," *Transport in Porous Media*, vol. 89, no. 2, pp. 251–263, 2011.

[21] P. Fakcharoenphol, B. Kurtoglu, H. Kazemi, S. Charoenwongsa, and Y.-S. Wu, "The effect of osmotic pressure on improve oil recovery from fractured shale formations," in *Proceedings of the SPE USA Unconventional Resources Conference 2014*, pp. 456–467, Woodlands, Tex, USA, April 2014.

[22] J. Wang and S. S. Rahman, "An Investigation of Fluid Leak-off Due to Osmotic and Capillary Effects and Its Impact on Micro-Fracture Generation during Hydraulic Fracturing Stimulation of Gas Shale," in *Proceedings of the EUROPEC 2015*, Madrid, Spain.

[23] E. Sadatshojaei, M. Jamialahmadi, F. Esmaeilzadeh, and M. H. Ghazanfari, "Effects of low-salinity water coupled with silica nanoparticles on wettability alteration of dolomite at reservoir temperature," *Petroleum Science and Technology*, vol. 34, no. 15, pp. 1345–1351, 2016.

[24] J. Yang, Z. Dong, Z. Yang, and M. Lin, "Wettability alteration by salinity and calcium bridge in a crude oil/brine/rock system," *Petroleum Science and Technology*, vol. 33, no. 19, pp. 1660–1666, 2015.

[25] P. Handren and T. Palisch, "Successful hybrid slickwater-fracture design evolution: an east texas cotton valley taylor case history," *SPE Production and Operations*, vol. 24, no. 3, pp. 415–424, 2009.

[26] Z. Zhang and X. Li, *Numerical Study on the Formation of Shear Fracture Network*, vol. 9, 299, Energies, 4 edition, 2016.

[27] T. Cao, Z. Song, S. Wang, X. Cao, Y. Li, and J. Xia, "Characterizing the pore structure in the Silurian and Permian shales of the Sichuan Basin, China," *Marine and Petroleum Geology*, vol. 61, pp. 140–150, 2015.

4

Application of Starch-Stabilized Silver Nanoparticles as a Colorimetric Sensor for Mercury(II) in 0.005 mol/L Nitric Acid

Penka Vasileva,[1] Teodora Alexandrova,[1] and Irina Karadjova[2]

[1]Department of General and Inorganic Chemistry, Faculty of Chemistry and Pharmacy, Laboratory of Nanoparticle Science and Technology, University of Sofia "St. Kliment Ohridski", 1 J. Bourchier Blvd., 1164 Sofia, Bulgaria
[2]Department of Analytical Chemistry, Faculty of Chemistry and Pharmacy, University of Sofia "St. Kliment Ohridski", 1 J. Bourchier Blvd., 1164 Sofia, Bulgaria

Correspondence should be addressed to Penka Vasileva; pvasileva@chem.uni-sofia.bg

Academic Editor: Roberto Comparelli

A sensitive and selective Hg^{2+} optical sensor has been developed based on the redox interaction of Hg^{2+} with starch-coated silver nanoparticles (AgNPs) in the presence of $0.005 \, mol \, L^{-1}$ HNO_3. The relative intensity of the localized surface plasmon absorption band of AgNPs at 406 nm is linearly dependent on the concentration of Hg^{2+} with positive slope for the concentration range 0–$12.5 \, \mu g \, L^{-1}$ and negative slope for the concentration range 25–$500 \, \mu g \, L^{-1}$. Experiments performed demonstrated that metal ions (Na^+, K^+, Mg^{2+}, Ca^{2+}, Pb^{2+}, Cu^{2+}, Zn^{2+}, Cd^{2+}, Fe^{3+}, Co^{2+}, and Ni^{2+}) do not interfere under the same conditions, due to the absence of oxidative activity of these ions, which guarantees the high selectivity of the proposed optical sensor towards Hg^{2+}. The limits of detection and quantification were found to be $0.9 \, \mu g \, L^{-1}$ and $2.7 \, \mu g \, L^{-1}$, respectively, and relative standard deviations varied in the range 9–12% for Hg content from 0.9 to $12.5 \, \mu g \, L^{-1}$ and 5–9% for Hg levels from 25 to $500 \, \mu g \, L^{-1}$. The method was validated by analysis of CRM Estuarine Water BCR505. A possible mechanism of interaction between AgNPs and Hg^{2+} for both concentration ranges was proposed on the basis of UV-Vis, TEM, and SAED analyses.

1. Introduction

Monitoring of toxic metals in aquatic ecosystems is an important analytical task as far as these contaminants adversely affect the environment and have serious medical effects on human health. One of the most harmful pollutants among them is Hg, which is still released in the environment and widely distributed in air, water, and soil [1]. At very low concentrations, Hg affects human's health, causing a variety of diseases to the heart, kidneys, brain, and nervous and endocrine systems [2] Naturally occurring levels of mercury in groundwater and surface water are less than $0.5 \, \mu g \, L^{-1}$, although local mineral deposits may produce higher levels in groundwaters. Essential quality standard for Hg maximum permissible limit of $1 \, \mu g \, L^{-1}$ has been adopted at EU level and requires regular monitoring of Hg content in drinking waters. It is well known that Hg exists in natural waters as different species: Hg^0, methyl-Hg, and inorganic Hg(II); however, the dominant toxic species in drinking waters is Hg(II). Various

instrumental methods and techniques have been developed for Hg determination at low environmentally relevant concentrations like atomic absorption/emission spectrometry (AAS/AES) [3], atomic fluorescence spectrometry (AFS) [4, 5], and high-performance liquid chromatography (HPLC) [6, 7]. In spite of being very sensitive and precise for Hg determination, these methods often require a time-consuming sample preparation step as well as expensive instrumentation. Various colorimetric assays (based on the use of sensitive chromophores or fluorophores [8–11], polymers [12, 13], oligonucleotides [14, 15], DNA [16, 17], and metal nanoparticles [18–20]) have been developed and reported in the literature as convenient and simple alternative methods for the detection of target analytes without the requirement of sophisticated apparatus.

Metal nanoparticles have unique properties and applications in numerous fields, which are attributed to the collective dipole oscillation known as Surface Plasmon Resonance (SPR) [21]. This phenomenon makes them very desirable for colorimetric sensing of Hg^{2+} ions because the interaction between the nanoparticles and the analyte changes the intensity and/or position of the absorption band in the visible spectrum, which often might be observed with the naked eye [22]. The limitations observed for these systems are mainly connected with poor selectivity, high detection limit for Hg(II), complicated synthesis of the probe materials, or complicated analytical procedures.

In this study, we present a new colorimetric assay for Hg^{2+} ions in $0.005\,mol\,L^{-1}$ HNO_3 using starch-stabilized silver nanoparticles (AgNPs). A change in the absorbance strength is expected as a result of the redox interaction between AgNPs and either Hg^{2+} ions or NO_3^{-} ions. The Hg concentration determines which of these two redox reactions dominates as the two oxidants compete with each other for Ag oxidation. This way, detection of very low environmentally relevant Hg contents is possible. Several sensing systems have been already reported based on the interaction between AgNPs and Hg(II) ions [23–32]; however, detailed study of Hg behavior in the presence of another competitive oxidant is rarely performed and discussed. A dual functional sensor for determination of Hg and H_2O_2 has been developed based on a similar approach: addition of H_2O_2 to a mixture of AgNPs and Hg(II) ions [33]. The method presented in this study, however, differs not only as a mechanism of the process, but also as a behavior of Hg^{2+} ions at very low concentrations (below $25\,\mu g\,L^{-1}$) towards AgNPs in the presence of NO_3^{-} ions as a second oxidant. A simple and fast analytical procedure for determination of Hg in drinking waters is developed and verified by the analysis of a certified reference material.

2. Materials and Methods

2.1. Apparatus. UV-Vis absorption spectra were recorded on an Evolution 300 spectrometer (Thermo Scientific, USA) within the 200–800 nm range using quartz cuvettes with 1 cm optical path length. High-purity water was used as a reference sample for background absorption. The morphology and particle sizes were examined using a high-resolution transmission electron microscope (TEM, JEOL JEM-2100 operating at an accelerating voltage of 200 kV). A volume of $5\,\mu L$ AgNPs suspension was placed on a carbon-covered copper grid for TEM and air-dried. The histogram of AgNPs size distribution and the mean diameter of nanoparticles were determined by counting at least 200 nanoparticles from the different TEM images using ImageJ software. Some structural details of the nanoparticles were analyzed using the high-resolution TEM image and SAED pattern. The zeta (ζ) potential of nanoparticles was measured with a ZetaSizer Nano ZS (Malvern) instrument.

2.2. Chemicals. All chemicals used were of analytical-reagent grade and all aqueous solutions were prepared in high-purity water (Millipore Corp., Milford, MA, USA). Silver nitrate ($AgNO_3$, 99.8%), soluble starch, sodium hydroxide (NaOH, 99%), nitric acid (HNO_3, 65%), salts of the different cations studied (NaCl, KCl, $MgCl_2$, $CaCl_2$, $Pb(NO_3)_2$, $ZnCl_2$, $CuCl_2$,

$NiCl_2$, $CdCl_2$, $CoCl_2$, and $FeCl_3$) (from Merck, Germany), and pharmaceutical grade D-(+) glucose (from Alfa Aesar, Germany) were used. Stock Hg standard solution, Trace CEPT™, $998\,\mu g\,mL^{-1}$ in $2\,mol\,L^{-1}$ HNO_3 (Sigma-Aldrich, USA), was used to prepare a working standard solution of $1000\,\mu g\,L^{-1}$ Hg^{2+} in $0.01\,mol\,L^{-1}$ HNO_3. Standard solutions for Hg within the concentration range of 0–$1000\,\mu g\,L^{-1}$ were prepared weekly by serial dilution of this solution in $0.01\,mol\,L^{-1}$ HNO_3. All diluted Hg solutions were stored in dark glass flasks and kept refrigerated at $4°C$.

2.3. Synthesis and Characterization of Silver Nanoparticles. The synthesis of AgNPs follows a green synthetic procedure as described in our previous study [34]. The silver nanoparticles were obtained through a reduction reaction of silver nitrate with D-glucose as a reducing agent in the presence of starch as a stabilizer and suitable sodium hydroxide amount as a reaction catalyst. Briefly, 24 mL of 0.001 M $AgNO_3$ and 48 mL of 0.2% solution of starch were mixed and left for at least 15 minutes to form a complex under an ultrasonic treatment (ultrasonic bath, power 100 W, frequency 38 MHz). After that, $720\,\mu L$ of 0.1 M D-glucose was added and sonicated for 5 minutes. The reaction was started by the addition of 3.6 mL of 0.1 M NaOH and continued for one hour at a constant temperature ($30°C$) in an ultrasonic bath to ensure the homogeneous formation of the silver nanoparticles.

The as-prepared AgNPs were purified and concentrated three times by ultracentrifugation (90 min, 14,000 rpm). The dispersion obtained was denoted as a stock solution of AgNPs and used in the experiments for colorimetric determination of Hg^{2+}. The AgNPs stock solution was kept in a dark glass flask at room temperature and was homogenized in an ultrasonic bath for 30 min prior to each experiment.

2.4. Colorimetric Detection of Hg^{2+} Ions. The colorimetric detection of Hg^{2+} ions via starch-stabilized silver nanoparticles was conducted as follows: an aliquot of $200\,\mu L$ AgNPs stock solution and $300\,\mu L$ high-purity water were consecutively added to a small quartz cuvette, followed by addition of $500\,\mu L$ Hg^{2+} solution with varying concentrations. The resulting mixture was equilibrated by stirring on Vortex for an optimum incubation time and then the UV-Vis spectrum in the wavelength range of 200–800 nm was recorded. In order to investigate the sensitivity of the colorimetric assay towards other ions, starch-stabilized AgNPs were allowed to interact under the same conditions with $50\,\mu mol\,L^{-1}$ solutions of alkali (Na^+, K^+), alkaline earth (Mg^{2+}, Ca^{2+}), Pb^{2+}, and transition-metal ions (Cu^{2+}, Zn^{2+}, Cd^{2+}, Fe^{3+}, Co^{2+}, and Ni^{2+}) (separately for each ion). The resulting solutions were monitored by optical absorption spectroscopy.

2.5. Determination of Hg in Tap/Underground Water. Tap/underground water sample (20 mL) was filtered through a $0.45\,\mu m$ filter and acidified with HNO_3 until reaching pH in the range 2–2.3. Sample aliquot of $500\,\mu L$ was transferred to a quartz cuvette, and $200\,\mu L$ stock solution of AgNPs was added and the mixture was stirred by the Vortex. After the incubation time of 5 min, the UV-Vis absorbance was

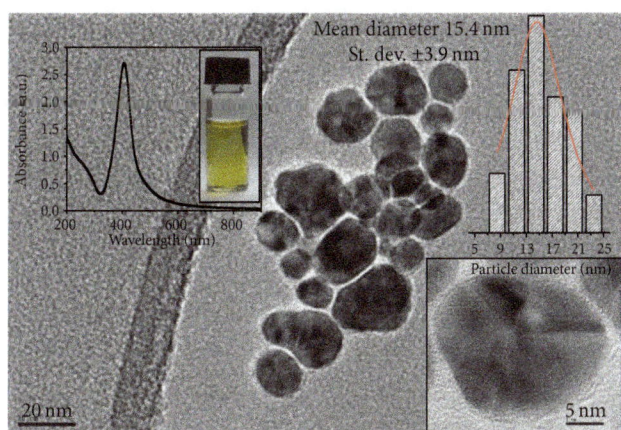

FIGURE 1: TEM image of starch-stabilized silver nanoparticles; insets: UV-Vis absorption spectrum and digital photographs (left), high-resolution TEM image of single nanoparticles, and histogram of nanoparticle size distribution (right).

measured at 407 nm. Parallel sample aliquot of 250 μL is diluted twice with 0.005 mol L^{-1} HNO$_3$ and passed through the procedure described above. The response of this sample (increase or decrease, related to the original one, Figure 4) is used to distinguish the low from the high linear concentration range of Hg and to choose an appropriate calibration curve.

3. Results and Discussion

3.1. Characterization of AgNPs. The UV-Vis absorption spectrum of starch-stabilized AgNPs, recorded at 25°C, is shown in Figure 1 (inset). A single and sharp SPR band appears at 407 nm, which indicates the formation of nanometer-sized particles. This is further confirmed by the TEM observation and size distribution histogram, shown in Figure 1.

The spherical-like AgNPs exhibit a relatively narrow size distribution with a mean diameter of 15.4 ± 3.9 nm. In addition to the nanospheres, some typical polyhedral nanoparticles (multiple twined nanocrystals) can be easily observed. The crystalline nature of AgNPs is clearly observed on the HRTEM image in Figure 1 (inset) and proved by the lattice characterization (e.g., the spacing between the individual lattice fringes of 0.235 nm, which corresponds to (111) plane lattice spacing of pure silver). The colloidal stability of starch-coated AgNPs is confirmed by the ζ-potential value of −25.3 ± 1.3 mV measured in 0.001 mol L^{-1} KCl at pH 6.8.

3.2. The Optimization of Colorimetric Sensing of Hg^{2+}. Several parameters were investigated systematically in order to establish optimal conditions for the direct colorimetric detection of Hg^{2+}. As a first step, the pH value was adjusted taking into account the analysis of real samples and HNO$_3$ which is typically used for water sample preservation. The experiments performed showed that 0.005 mol L^{-1} HNO$_3$ ensured the highest sensitivity and could be accepted as an optimal sample medium. In order to evaluate the optimum contact time, the kinetic of interaction between AgNPs and

Hg^{2+} in the presence of 0.005 mol L^{-1} nitric acid was followed within one hour by measurements of UV-Vis absorbance. Typical evolution of UV-Vis absorbance spectrum with time, due to the interaction of AgNPs with 400 μg L^{-1} Hg^{2+} and respective color change of the AgNPs dispersion, is shown in Figure 2.

The changes that occurred in the LSPR absorption band of AgNPs are reflected on the color of the samples, which can be seen even with the naked eye. It is seen that the sensor's response is significant during the first five minutes of the reaction process and a negligible change in the absorption intensity is observed over time. This fact allows convenient analytical detection of Hg^{2+} within only five minutes.

As a next step, the sensitivity and applicability of starch-coated AgNPs for quantitative determination of Hg^{2+} ions under the defined optimal conditions were studied. The colorimetric response and LSPR band behavior were monitored as a function of Hg^{2+} concentrations, ranging from 0 to 500 μg L^{-1} in the presence of 0.005 mol L^{-1} HNO$_3$ (Figure 3).

As seen from the UV-Vis absorbance spectra (5-minute incubation time), the addition of 0.005 mol L^{-1} HNO$_3$ results in a considerable decrease of the intensity of AgNPs characteristic plasmon band at 407 nm accompanied by a slight blue shift (Figure 3(a)). In addition, a shoulder band appears at the wavelength range of 450–600 nm. The increase of Hg^{2+} concentration from 0 to 12.5 μg L^{-1} in 0.005 mol L^{-1} HNO$_3$ leads to a gradual increase of the intensity of the characteristic plasmon band of AgNPs at 407 nm and its value gradually approximates to the absorption intensity of the blank nanoparticle solution (without both NO$_3^-$ and Hg^{2+}). In addition, the intensity of the shoulder band decreases along with increasing intensity of the main plasmon absorbance band. The spectra show a clear isosbestic point at 445 nm upon addition of Hg^{2+} in 0.005 mol L^{-1} HNO$_3$, demonstrating that the aggregation of AgNPs is directly related to the concentration of Hg^{2+}. Contrariwise, a gradual decrease of the intensity of the characteristic plasmon band of the AgNPs at 407 nm is observed for the Hg concentration range from 25 to 500 μg L^{-1}. The spectra presented in Figure 3(b) also show that the decrease of intensity of the absorbance maximum at 407 nm is accompanied with a slight blue shift, which is strengthened for the higher concentrations of Hg^{2+}. This phenomenon is already reported and described as a change of the refractive index of the particles and the formation of a mercury layer around AgNPs, yielding an amalgam-like structure [25, 35, 36]. It might be suggested that, for the first Hg concentration range (0–12.5 μg L^{-1}), a redox reaction proceeds between zero-valent silver (Ag^0) and either Hg^{2+} or NO$_3^-$ ions. The values of standard electrode potentials of the components in the system confirm this suggestion: E_0 (Ag$^+$/Ag^0) = 0.799 V; E_0 (Hg^{2+}/Hg^0) = 0.854 V; E_0 (NO$_3^-$/NH$_4^+$) = 0.864 V. Because the standard electrode potential of NO$_3^-$/NH$_4^+$ is commeasurable with that of Hg^{2+}/Hg^0, two competitive oxidizing agents are involved in the studied sensing system. The most probable explanation for the decrease of LSPR band intensity in the presence of 0.005 mol L^{-1} HNO$_3$ and further increase upon addition of

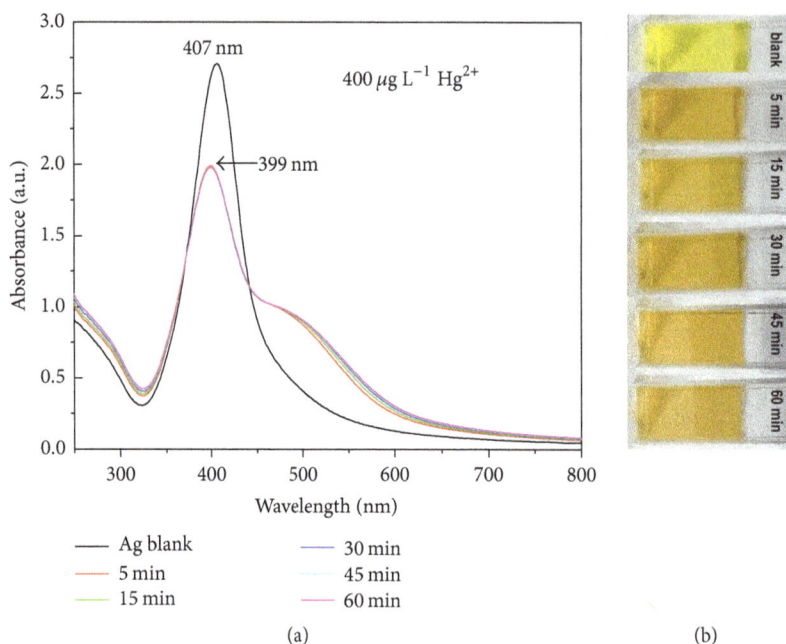

FIGURE 2: (a) Evolution of UV-Vis absorbance spectrum of AgNPs and (b) color change of the AgNPs dispersion upon the addition of $400\,\mu g\,L^{-1}\,Hg^{2+}$ in the presence of $0.005\,mol\,L^{-1}\,HNO_3$.

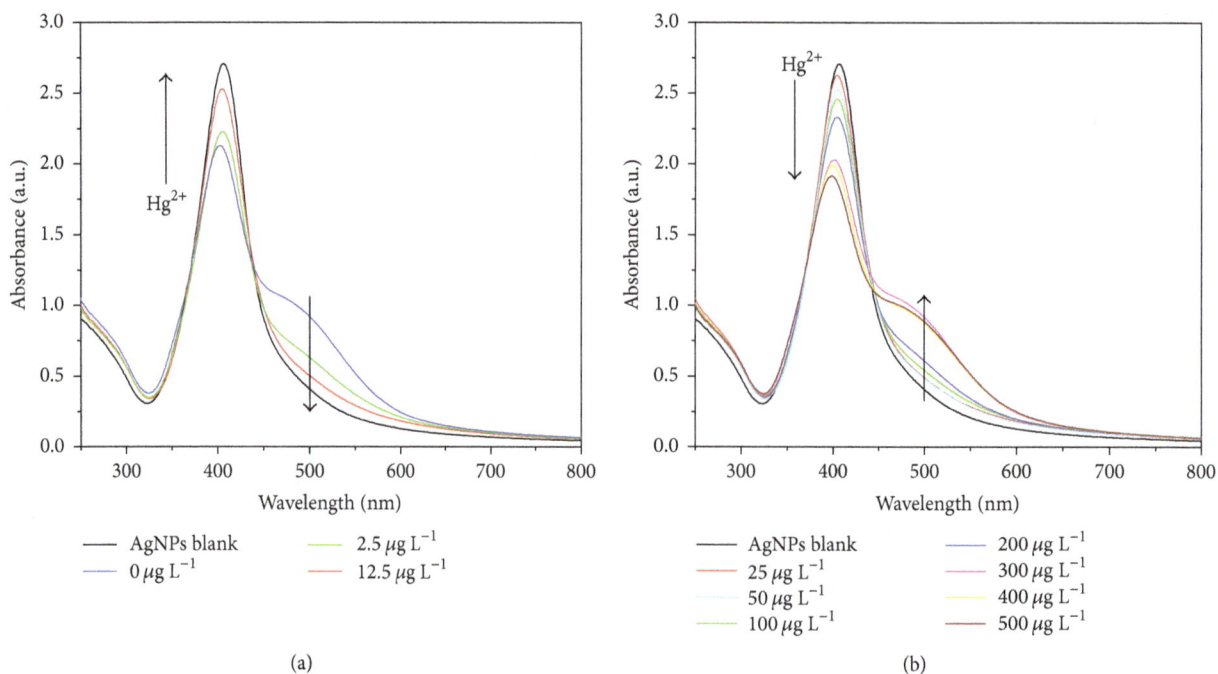

FIGURE 3: UV-Vis absorption responses of starch-stabilized AgNPs recorded 5 min after the addition of Hg^{2+} with various concentrations: (a) $0-12.5\,\mu g\,L^{-1}\,Hg^{2+}$ and (b) $25-500\,\mu g\,L^{-1}\,Hg^{2+}$ in the presence of $0.005\,mol\,L^{-1}\,HNO_3$.

Hg^{2+} (Figure 3(a)) is that, at low Hg^{2+} concentrations, the oxidative effect of NO_3^- ions towards surface silver atoms is dominant. In the presence of higher concentrations of Hg^{2+} (Figure 3(b)), the surface of nanoparticles is protected by the layer of Ag-Hg-amalgam due to the sorption and reduction of positively charged Hg^{2+} on the surface of negatively charged AgNPs followed by amalgamation. In this way, the surface of AgNPs is inaccessible for oxidation by NO_3^-. Evidently, within the range of $25-500\,\mu g\,L^{-1}\,Hg^{2+}$, the main redox interaction is between the AgNPs and Hg^{2+}. Such

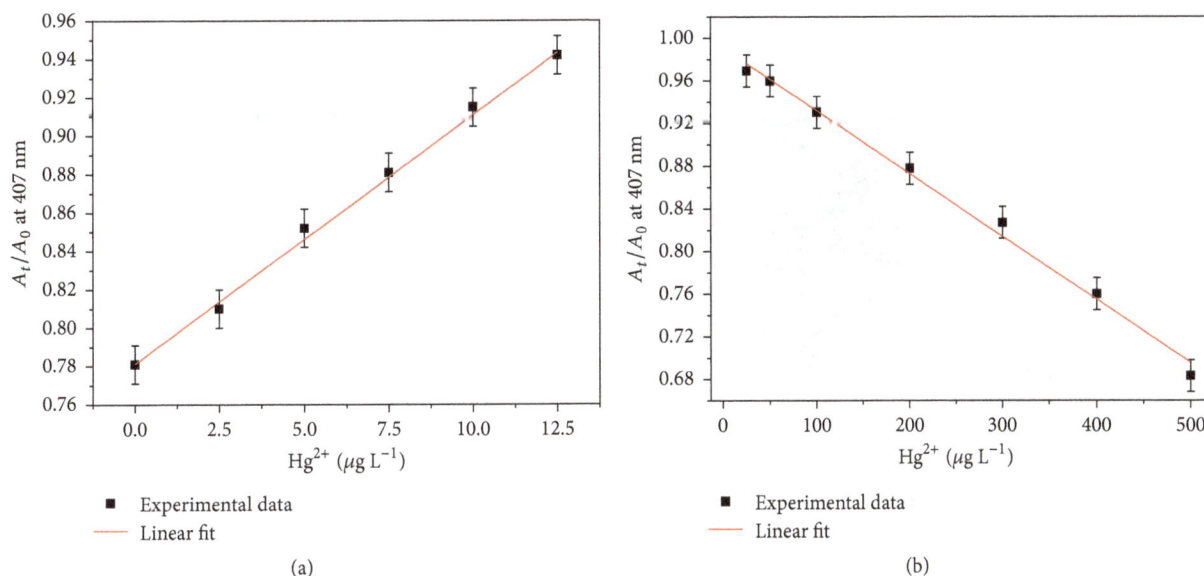

FIGURE 4: Plot of A_t/A_0 as a function of the Hg^{2+} concentration over the ranges of (a) 0–12.5 $\mu g\,L^{-1}$ and (b) 25–500 $\mu g\,L^{-1}$ in the presence of 0.005 mol L^{-1} HNO_3.

behavioral dissimilarities of the analyte (Hg^{2+}) for different concentration ranges have not been observed and reported in the previous studies on the AgNPs-based optical sensing system for Hg^{2+} colorimetric detection. We have to point out, however, that none of these reports mention the acidity of the reaction media, which most probably determines the oxidizing power of reagents in the system.

For quantitative determination of Hg^{2+}, the change of the intensity of LSPR band maximum of silver nanoparticles at 407 nm upon the addition of analyte with various concentrations was estimated as a ratio A_t/A_0, where A_0 corresponds to the intensity of the absorbance maximum of blank AgNPs solution (without both NO_3^- and Hg^{2+} ions) and A_t corresponds to the intensity of the absorbance maximum of silver nanoparticles 5 min after the addition of Hg^{2+} standard solutions (Figure 4).

As indicated in Figures 4(a) and 4(b), linear correlations exist between the relative value of the absorbance maximum intensity and the concentration of Hg^{2+} over the concentration ranges 0–12.5 $\mu g\,L^{-1}$ ($A = 0.7814 + 1.30 \times 10^{-2}c$ with $R^2 = 0.995$) and 25–500 $\mu g\,L^{-1}$ ($A = 0.991 - 5.90 \times 10^{-4}c$ with $R^2 = 0.993$), respectively. As a conclusion, the optical sensor studied using starch-stabilized AgNPs ensures a linear response over the concentration range from 0.9 to 12.5 $\mu g\,L^{-1}$ which covers all environmentally relevant concentrations of Hg and might be used for fast screening of Hg in the aquatic environment. The second concentration range from 25 to 500 $\mu g\,L^{-1}$ Hg^{2+} can be successfully applied for the determination of Hg in highly contaminated and rarely found industrial wastewaters.

3.3. Selectivity of Hg^{2+} Optical Sensing by Starch-Coated AgNPs.
From an analytical point of view, it is very important

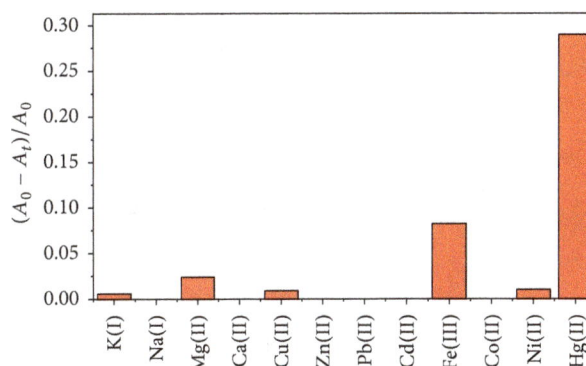

FIGURE 5: Colorimetric response of starch-stabilized AgNPs recorded 5 min after the addition of 5×10^{-5} mol L^{-1} metal ions.

to define the selectivity of the proposed system for colorimetric Hg^{2+} determination. This has been evaluated through the response of the assay to various environmentally relevant metal ions including Na^+, K^+, Mg^{2+}, Ca^{2+}, Pb^{2+}, Cu^{2+}, Zn^{2+}, Cd^{2+}, Fe^{3+}, Co^{2+}, and Ni^{2+} under the same conditions as in the case of Hg^{2+}. The optical response of AgNPs to the tested ions (concentration level of 50 $\mu mol\,L^{-1}$) after 5 min of their addition (separately for each ion) is illustrated in Figure 5. For comparison, the optical response of AgNPs to the Hg^{2+} ions at a concentration level of 2.5 $\mu mol\,L^{-1}$ is also presented.

It is easy to observe that all other metal ions produce a much weaker signal (almost at baseline level) except Fe^{3+} which shows modest interference. The reason is that only Hg^{2+} can be reduced by surface atoms of AgNPs to form stable Ag-Hg amalgam. The addition of Fe^{3+} resulted in a tiny intensity decrease and red shift of the absorption band. This

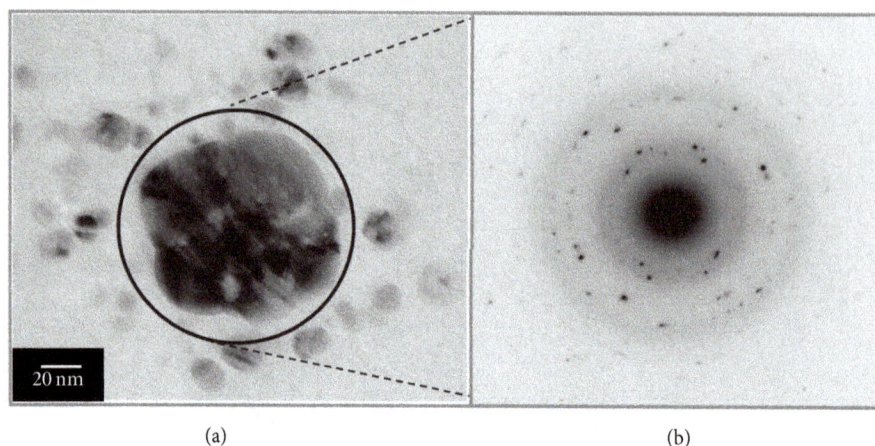

FIGURE 6: (a) TEM image (scale bar is 20 nm) and (b) corresponding SAED pattern of starch-coated silver nanoparticles after treatment by Hg^{2+} solution at a concentration of 500 $\mu g\,L^{-1}$ in the presence of 0.005 mol L^{-1} HNO_3.

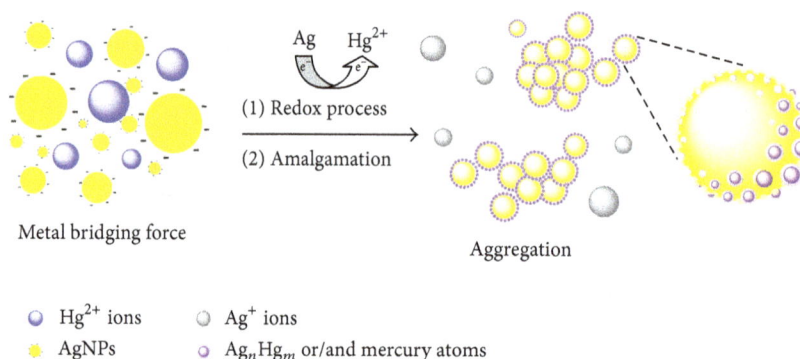

FIGURE 7: The proposed mechanism of the interaction between starch-coated AgNPs and Hg^{2+} solution.

effect could be interpreted in terms of Fe(III) complexation with oxidized species of carbohydrates (starch and glucose) which are sorbed on the surface of silver nanoparticles [37].

3.4. *Mechanism of Interaction between AgNPs and Hg^{2+}.* To elucidate the mechanism of sensing activity of the starch-coated AgNPs towards Hg^{2+}, the nanoparticles were examined before and after Hg^{2+} exposure using TEM with SAED observations. Figure 6 shows TEM micrograph with the corresponding SAED pattern obtained from the agglomerate formed during interaction of AgNPs with Hg^{2+} solution at a concentration of 500 $\mu g\,L^{-1}$.

As can be seen from Figure 6(a), the nanoparticles are of varying sizes and there is a large distribution after Hg^{2+} exposure. The TEM image shows a larger particle, which is surrounded by smaller particles. It seems that larger particles are undergoing Ostwald ripening. A similar observation is already reported for gold nanoparticles utilized for mercury removal from drinking water [38] and for colorimetric detection of Hg^{2+} using the AgNPs embedded in cyclodextrin-silicate composite [39].

The data from the analysis of SAED pattern (Figure 6(b)) are summarized with interpretation accuracy of 1% in Table 1. The analysis shows the existence of Ag_2Hg_3 amalgam (PDF

65-3156) and Ag (PDF 89-3722) as main phases in the aggregated mass formed during the interaction of starch-coated AgNPs with Hg^{2+}. Some impurities of metallic Hg (PDF 01-1017) are also detected.

On the basis of TEM/SAED results, a multistep interaction of Hg^{2+} with the silver nanoparticles could be inferred. The interaction involves (i) the electrostatic attraction between negatively charged silver nanoparticles and positively charged Hg^{2+} species, decreasing the distance between nanoparticles; (ii) adsorption of Hg^{2+} on the surface of AgNPs and their reduction to Hg^0 by the surface Ag atoms (simultaneously obtained Ag^+ diffuse into the solution); (iii) amalgamation of the freshly generated mercury atoms with the surface Ag atoms [25, 32, 39]; (iv) the interaction of Hg^{2+} with AgNPs which decreases surface charges of nanoparticles, leading to their destabilization and aggregation. The latter one is confirmed by the shape evolution of AgNPs observed in Figure 6(a). The suggested mechanism of optical sensing of Hg^{2+} by starch-coated silver nanoparticles is illustrated in Figure 7.

3.5. *Analytical Application.* In order to test the applicability of the sensor developed for Hg^{2+} and total Hg determination, samples of tap water (Sofia) and mineral water (Gorna Bania,

TABLE 1: SAED data of AgNPs aggregates formed after exposure of silver nanoparticles to Hg^{2+} at concentration of 500 $\mu g\,L^{-1}$ in the presence of 0.005 mol L^{-1} HNO_3. $(hkl)_f$: double electron diffraction effects; SAED interpretation: accuracy 1%.

d (Å)	Relative intensity	Ag PDF 89-3722 $a = 4.0855(1)$ Å SG Fm$\overline{3}$m	Ag PDF 87-0598 $a = 2.8862$ Å, $c = 10.000$ Å P6$_3$/mmc	Ag$_2$Hg$_3$ PDF 65-3156 $a = 10.0506$ Å SG I23
2.438	s	—	101	$(410)_f$
2.136	s	—	—	332
1.506	s	—	—	622
1.287	s	$(310)_f$	—	237
1.057	w	—	205	930
0.979	w	$(410)_f$	—	059

s: strong; w: weak.

TABLE 2: Comparison of different methods using silver nanoparticles as a colorimetric sensing probe for Hg^{2+} determination.

Sensing probe	Linear concentration range	Detection limit	Ref.
Starch-stabilized AgNPs	50 nmol L^{-1}–5000 nmol L^{-1}	25 nmol L^{-1}	[23]
Unmodified AgNPs stabilized with extract of soap-root plant	10–100 μmol L^{-1}	2.2 μmol L^{-1}	[26]
Gum kondagogu-stabilized AgNPs	50–500 nmol L^{-1}	50 nmol L^{-1} (LOQ)	[31]
Citrate-capped AgNPs	0.02 nmol L^{-1}–0.9 μmol L^{-1}	—	[33]
1-Dodecanethiol-capped Ag nanoprisms upon the presence of iodides	10–4000 nmol L^{-1}	3.3 nmol L^{-1}	[40]
Poly(vinylpyrrolidone)-stabilized AgNPs	1 nmol L^{-1}–30 μmol L^{-1}	1 nmol L^{-1}	[41]
Carrageenan-functionalized Ag/AgCl NPs	1–100 μmol L^{-1}	1 μmol L^{-1}	[42]
Starch-coated AgNPs in the presence of 0.005 mol L^{-1} HNO_3	4.5–2500 nmol L^{-1}	4.5 nmol L^{-1}	This work

Kniagevo) were spiked at levels close to the permissible limit (drinking water) of 1 $\mu g\,L^{-1}$. Total Hg content in these samples was defined preliminarily by ICP-MS and results for all samples were below 0.05 $\mu g\,L^{-1}$ Hg. Recoveries achieved using the described procedure are in the range 93–97%, thus confirming the possibility of fast Hg^{2+} screening in drinking waters using the proposed sensor based on starch-coated AgNPs. The limits of detection (LOD) and limits of quantification (LOQ) were evaluated on the basis of repeated analysis of blank (AgNPs). The calculations were based on 3σ and 10σ criteria using the linear regression equations and slopes of calibration graphs for Hg^{2+} (Figure 4). The defined values for LOD (0.9 $\mu g\,L^{-1}$) and LOQ (2.7 $\mu g\,L^{-1}$) show that the proposed sensor is not suitable for surface water monitoring but might be successfully used for fast *on-site* control of the quality of sources for drinking water. Within-batch precision strongly depends on the analyte concentration in the measuring solution: 9–12% for Hg^{2+} in the range 0.9–12.5 $\mu g\,L^{-1}$ and 5–9% for Hg^{2+} in the range over 25–500 $\mu g\,L^{-1}$. Table 2 further summarizes the linear ranges and detection limits of various Hg^{2+} detection methods based

on silver nanoparticles as a colorimetric sensing probe. It is evident that the proposed method ensures higher or equal sensitivity with those of earlier reported colorimetric AgNPs-based sensors [23, 26, 31, 33, 40–42].

For partial validation of the procedure, CRM Estuarine Water BCR505 was analyzed after solid phase extraction (10-fold Hg enrichment) [43]. Three sample aliquots of 800 μL were analyzed according to the proposed analytical procedure. The result of 0.73 $\pm$ 0.08 nmol L^{-1} Hg was in reasonable agreement with the (additional material information) value of 0.69 nmol kg^{-1} Hg (138 $\mu g\,L^{-1}$).

4. Conclusions

A simple, fast, and low cost analytical procedure is developed for easy and sensitive quantification of Hg^{2+} in the presence of 0.005 mol L^{-1} HNO_3 by using starch-coated AgNPs as a LSPR-based optical sensor. The Hg^{2+} sensing is based on the optical response (change in the absorbance strength of LSPR band) of silver nanoparticles depending on the Hg^{2+} concentration. Possible mechanism of interaction between

AgNPs and Hg^{2+} was proposed. An accurate and reliable determination of Hg is achieved in two concentration ranges: $0.9–12.5\ \mu g\,L^{-1}$ and $25–500\ \mu g\,L^{-1}$. The limits of detection and quantification achieved were $0.9\ \mu g\,L^{-1}$ and $2.7\ \mu g\,L^{-1}$, respectively, and relative standard deviations varied in the range 9–12% for Hg content from 0.9 to $12.5\ \mu g\,L^{-1}$ and 5–9% for Hg levels from 25 to $500\ \mu g\,L^{-1}$. The LSPR-based optical sensor for Hg(II) might be used for simple and fast *on-site* screening of sources for abstraction of drinking water and for Hg determination in wastewaters.

Conflicts of Interest

The authors declare that there are no conflicts of interest regarding the publication of this paper.

Acknowledgments

The authors acknowledge the support by the Horizon 2020 program of the European Commission (project Materials Networking).

References

[1] D. W. Boening, "Ecological effects, transport, and fate of mercury: a general review," *Chemosphere*, vol. 40, no. 12, pp. 1335–1351, 2000.

[2] P. Holmes, K. A. F. James, and L. S. Levy, "Is low-level environmental mercury exposure of concern to human health?" *Science of the Total Environment*, vol. 408, no. 2, pp. 171–182, 2009.

[3] H. Erxleben and J. Ruzicka, "Atomic absorption spectroscopy for mercury, automated by sequential injection and miniaturized in lab-on-valve system," *Analytical Chemistry*, vol. 77, no. 16, pp. 5124–5128, 2005.

[4] L.-P. Yu and X.-P. Yan, "Flow injection on-line sorption preconcentration coupled with cold vapor atomic fluorescence spectrometry and on-line oxidative elution for the determination of trace mercury in water samples," *Atomic Spectroscopy*, vol. 25, no. 3, pp. 145–153, 2004.

[5] M. J. Bloxham, S. J. Hill, and P. J. Worsfold, "Determination of mercury in filtered sea-water by flow injection with on-line oxidation and atomic fluorescence spectrometric detection," *Journal of Analytical Atomic Spectrometry*, vol. 11, no. 7, pp. 511–514, 1996.

[6] M. Lombardo, I. Vassura, D. Fabbri, and C. Trombini, "A strikingly fast route to methylmercury acetylides as a new opportunity for monomethylmercury detection," *Journal of Organometallic Chemistry*, vol. 690, no. 3, pp. 588–593, 2005.

[7] L. Liu, Y.-W. Lam, and W.-Y. Wong, "Complexation of 4,4′-di(*tert*-butyl)-5-ethynyl-2,2′-bithiazole with mercury(II) ion: synthesis, structures and analytical applications," *Journal of Organometallic Chemistry*, vol. 691, no. 6, pp. 1092–1100, 2006.

[8] A. Caballero, R. Martínez, V. Lloveras et al., "Highly selective chromogenic and redox or fluorescent sensors of Hg^{2+} in aqueous environment based on 1,4-disubstituted azines," *Journal of the American Chemical Society*, vol. 127, no. 45, pp. 15666–15667, 2005.

[9] H. Zheng, Z.-H. Qian, L. Xu, F.-F. Yuan, L.-D. Lan, and J.-G. Xu, "Switching the recognition preference of rhodamine B spirolactam by replacing one atom: design of rhodamine B thiohydrazide for recognition of Hg(II) in aqueous solution," *Organic Letters*, vol. 8, no. 5, pp. 859–861, 2006.

[10] Y. Zhao and Z. Zhong, "Tuning the sensitivity of a foldamer-based mercury sensor by its folding energy," *Journal of the American Chemical Society*, vol. 128, no. 31, pp. 9988–9989, 2006.

[11] H. Wang, Y. Wang, J. Jin, and R. Yang, "Gold nanoparticle-based colorimetric and "turn-on" fluorescent probe for mercury(II) ions in aqueous solution," *Analytical Chemistry*, vol. 80, no. 23, pp. 9021–9028, 2008.

[12] X. Liu, Y. Tang, L. Wang et al., "Optical detection of mercury(II) in aqueous solutions by using conjugated polymers and label-free oligonucleotides," *Advanced Materials*, vol. 19, no. 11, pp. 1471–1474, 2007.

[13] I.-B. Kim and U. H. F. Bunz, "Modulating the sensory response of a conjugated polymer by proteins: an agglutination assay for mercury ions in water," *Journal of the American Chemical Society*, vol. 128, no. 9, pp. 2818–2819, 2006.

[14] S.-J. Liu, H.-G. Nie, J.-H. Jiang, G.-L. Shen, and R.-Q. Yu, "Electrochemical sensor for mercury(II) based on conformational switch mediated by interstrand cooperative coordination," *Analytical Chemistry*, vol. 81, no. 14, pp. 5724–5730, 2009.

[15] Z. Zhu, Y. Su, J. Li et al., "Highly sensitive electrochemical sensor for mercury(II) ions by using a mercury-specific oligonucleotide probe and gold nanoparticle-based amplification," *Analytical Chemistry*, vol. 81, no. 18, pp. 7660–7666, 2009.

[16] D. Zhang, M. Deng, L. Xu, Y. Zhou, J. Yuwen, and X. Zhou, "The sensitive and selective optical detection of mercury(II) ions by using a phosphorothioate DNAzyme strategy," *Chemistry—A European Journal*, vol. 15, no. 33, pp. 8117–8120, 2009.

[17] M. Hollenstein, C. Hipolito, C. Lam, D. Dietrich, and D. M. Perrin, "A highly selective DNAzyme sensor for mercuric ions," *Angewandte Chemie—International Edition*, vol. 47, no. 23, pp. 4346–4350, 2008.

[18] M. Rex, F. E. Hernandez, and A. D. Campiglia, "Pushing the limits of mercury sensors with gold nanorods," *Analytical Chemistry*, vol. 78, no. 2, pp. 445–451, 2006.

[19] Y. Wang, F. Yang, and X. Yang, "Colorimetric detection of mercury(II) ion using unmodified silver nanoparticles and mercury-specific oligonucleotides," *ACS Applied Materials and Interfaces*, vol. 2, no. 2, pp. 339–342, 2010.

[20] Y.-R. Kim, R. K. Mahajan, J. S. Kim, and H. Kim, "Highly sensitive gold nanoparticle-based colorimetric sensing of mercury(II) through simple ligand exchange reaction in aqueous media," *ACS Applied Materials and Interfaces*, vol. 2, no. 1, pp. 292–295, 2010.

[21] D. V. Talapin, J.-S. Lee, M. V. Kovalenko, and E. V. Shevchenko, "Prospects of colloidal nanocrystals for electronic and optoelectronic applications," *Chemical Reviews*, vol. 110, no. 1, pp. 389–458, 2010.

[22] Y.-L. Hung, T.-M. Hsiung, Y.-Y. Chen, Y.-F. Huang, and C.-C. Huang, "Colorimetric detection of heavy metal ions using label-free gold nanoparticles and alkanethiols," *The Journal of Physical Chemistry C*, vol. 114, no. 39, pp. 16329–16334, 2010.

[23] Y. Fan, Z. Liu, L. Wang, and J. Zhan, "Synthesis of starch-stabilized Ag nanoparticles and Hg^{2+} recognition in aqueous media," *Nanoscale Research Letters*, vol. 4, no. 10, pp. 1230–1235, 2009.

[24] G. V. Ramesh and T. P. Radhakrishnan, "A universal sensor for mercury (Hg, Hg^{I}, Hg^{II}) based on silver nanoparticle-embedded polymer thin film," *ACS Applied Materials & Interfaces*, vol. 3, pp. 988–994, 2011.

[25] E. Sumesh, M. S. Bootharaju, Anshup, and T. Pradeep, "A practical silver nanoparticle-based adsorbent for the removal of Hg^{2+} from water," *Journal of Hazardous Materials*, vol. 189, no. 1-2, pp. 450–457, 2011.

[26] K. Farhadi, M. Forough, R. Molaei, S. Hajizadeh, and A. Rafipour, "Highly selective Hg^{2+} colorimetric sensor using green synthesized and unmodified silver nanoparticles," *Sensors and Actuators B: Chemical*, vol. 161, no. 1, pp. 880–885, 2012.

[27] G. Maduraiveeran and R. Ramaraj, "Enhanced sensing of mercuric ions based on dinucleotide-functionalized silver nanoparticles," *Analytical Methods*, vol. 8, no. 44, pp. 7966–7971, 2016.

[28] A. Jeevika and D. R. Shankaran, "Functionalized silver nanoparticles probe for visual colorimetric sensing of mercury," *Materials Research Bulletin*, vol. 83, pp. 48–55, 2016.

[29] Y. Ma, Y. Pang, F. Liu, H. Xu, and X. Shen, "Microwave-assisted ultrafast synthesis of silver nanoparticles for detection of Hg^{2+}," *Spectrochimica Acta Part A: Molecular and Biomolecular Spectroscopy*, vol. 153, pp. 206–211, 2016.

[30] Z. Guo, G. Chen, G. Zeng et al., "Ultrasensitive detection and co-stability of mercury(II) ions based on amalgam formation with Tween 20-stabilized silver nanoparticles," *RSC Advances*, vol. 4, no. 103, pp. 59275–59283, 2014.

[31] L. Rastogi, R. B. Sashidhar, D. Karunasagar, and J. Arunachalam, "Gum kondagogu reduced/stabilized silver nanoparticles as direct colorimetric sensor for the sensitive detection of Hg^{2+} in aqueous system," *Talanta*, vol. 118, pp. 111–117, 2014.

[32] S. S. Ravi, L. R. Christena, N. Saisubramanian, and S. P. Anthony, "Green synthesized silver nanoparticles for selective colorimetric sensing of Hg^{2+} in aqueous solution at wide pH range," *Analyst*, vol. 138, no. 15, pp. 4370–4377, 2013.

[33] G.-L. Wang, X.-Y. Zhu, H.-J. Jiao, Y.-M. Dong, and Z.-J. Li, "Ultrasensitive and dual functional colorimetric sensors for mercury (II) ions and hydrogen peroxide based on catalytic reduction property of silver nanoparticles," *Biosensors and Bioelectronics*, vol. 31, no. 1, pp. 337–342, 2012.

[34] P. Vasileva, B. Donkova, I. Karadjova, and C. Dushkin, "Synthesis of starch-stabilized silver nanoparticles and their application as a surface plasmon resonance-based sensor of hydrogen peroxide," *Colloids and Surfaces A: Physicochemical and Engineering Aspects*, vol. 382, no. 1–3, pp. 203–210, 2011.

[35] P. Mulvaney, "Surface plasmon spectroscopy of nanosized metal particles," *Langmuir*, vol. 12, no. 3, pp. 788–800, 1996.

[36] T. Morris, H. Copeland, E. McLinden, S. Wilson, and G. Szulczewski, "The effects of mercury adsorption on the optical response of size-selected gold and silver nanoparticles," *Langmuir*, vol. 18, no. 20, pp. 7261–7264, 2002.

[37] S. Komulainen, J. Pursiainen, P. Perämäki, and M. Lajunen, "Complexation of Fe(III) with water-soluble oxidized starch," *Starch*, vol. 65, no. 3-4, pp. 338–345, 2013.

[38] K. P. Lisha, Anshup, and T. Pradeep, "Towards a practical solution for removing inorganic mercury from drinking water using gold nanoparticles," *Gold Bulletin*, vol. 42, no. 2, pp. 144–152, 2009.

[39] S. Manivannan and R. Ramaraj, "Silver nanoparticles embedded in cyclodextrin-silicate composite and their applications in Hg(II) ion and nitrobenzene sensing," *Analyst*, vol. 138, no. 6, pp. 1733–1739, 2013.

[40] L. Chen, X. Fu, W. Lu, and L. Chen, "Highly sensitive and selective colorimetric sensing of Hg^{2+} based on the morphology transition of silver nanoprisms," *ACS Applied Materials and Interfaces*, vol. 5, no. 2, pp. 284–290, 2013.

[41] L. Li, L. Gui, and W. Li, "A colorimetric silver nanoparticle-based assay for Hg(II) using lysine as a particle-linking reagent," *Microchimica Acta*, vol. 182, no. 11-12, pp. 1977–1981, 2015.

[42] K. B. Narayanan and S. S. Han, "Highly selective and quantitative colorimetric detection of mercury(II) ions by carrageenan-functionalized Ag/AgCl nanoparticles," *Carbohydrate Polymers*, vol. 160, pp. 90–96, 2017.

[43] E. K. Mladenova, I. G. Dakova, D. L. Tsalev, and I. B. Karadjova, "Mercury determination and speciation analysis in surface waters," *Central European Journal of Chemistry*, vol. 10, no. 4, pp. 1175–1182, 2012.

5

Research on Property of Multicomponent Thickening Water Fracturing Fluid and Application in Low Permeability Oil Reservoirs

Chengli Zhang,[1] Peng Wang 🆔,[1] and Guoliang Song 🆔[2]

[1]College of Petroleum Engineering, Northeast Petroleum University, Daqing, Heilongjiang 163318, China
[2]College of Mathematics and Statistics, Northeast Petroleum University, Daqing, Heilongjiang 163318, China

Correspondence should be addressed to Peng Wang; 2537298882@qq.com

Academic Editor: Gulaim A. Seisenbaeva

The clean fracturing fluid, thickening water, is a new technology product, which promotes the advantages of clean fracturing fluid to the greatest extent and makes up for the deficiency of clean fracturing fluid. And it is a supplement to the low permeability reservoir in fracturing research. In this paper, the study on property evaluation for the new multicomponent and recoverable thickening fracturing fluid system (2.2% octadecyl methyl dihydroxyethyl ammonium bromide (OHDAB) +1.4% dodecyl sulfonate sodium +1.8% potassium chloride and 1.6% organic acids) and guar gum fracturing fluid system (hydroxypropyl guar gum (HGG)) was done in these experiments. The proppant concentration (sand/liquid ratio) at static suspended sand is up to 30% when the apparent viscosity of thickening water is 60 mPa·s, which is equivalent to the sand-carrying capacity of guar gum at 120 mPa·s. When the dynamic sand ratio is 40%, the fracturing fluid is not layered, and the gel breaking property is excellent. Continuous shear at room temperature for 60 min showed almost no change in viscosity. The thickening fracturing fluid system has good temperature resistance performance in medium and low temperature formations. The fracture conductivity of thickening water is between 50.6 μm^2·cm and 150.4 μm^2·cm, and the fracture conductivity damage rate of thickening water is between 8.9% and 17.9%. The fracture conductivity conservation rate of thickening water is more than 80% closing up of fractures, which are superior to the guar gum fracturing fluid system. The new wells have been fractured by thickening water in A block of YC low permeability oil field. It shows that the new type thickening water fracturing system is suitable for A block and can be used in actual production. The actual production of A block shows that the damage of thickening fracturing fluid is low, and the long retention in reservoir will not cause great damage to reservoir.

1. Introduction

Fracturing is one of the main measures for low-permeability reservoir reconstruction. The performance of fracturing fluid is an important factor affecting the increasing production of oil field after construction [1, 2]. The conventional fracturing fluid has the disadvantages of residues, low flow-back rate, and unrecoverable and causes great harm to reservoir [3].

No damage or negative fracturing fluid represents the direction of fluid development in the fracturing industry. At present, clean fracturing fluid is a representative system of low-damage fracturing fluid [4]. Schlumberger's engineers with fluid experts with Eni-Agip have launched a clean fracturing fluid or surfactant fracturing fluid. Since the product has been put into the market, it has been popularized rapidly. The three largest countries and regions at present are Canada, the Gulf of Mexico, and the east of the United States. So far, the fracturing fluid system has been constructed for about 6000 wells and has achieved good economic benefits. According to the data [5], the first surface-active agent used by Schlumberger company is a cationic compound containing long-chain alkyl, and the combined salt is an organic sodium salt and an inorganic potassium salt. After that, Schlumberger has strengthened the further research and development of the clean fracturing fluid, and has produced a more mature series of the surfactant fracturing fluid. The surface active agent not only has a small molecular weight, but also has a large molecular weight [6, 7]. It is not only cationic, but also other types, and the salt also involves a variety of organic and inorganic salts. BJ Services also developed a clean fracturing fluid system for AquaClear and ElastraFrac. The ElastraFrac gel is a heterogeneous aggregate (three-dimensional network structure) composed of environmentally friendly anionic surfactants and various salts, with a good temperature resistance and up to 120°C.

In 2009, Xinquan et al. published a report on the clean fracturing fluid of viscoelastic surfactant, and pointed out that the viscoelastic surface active agent formed a colloidal vermicular structure in the presence of brine, thus changing the viscoelasticity of the solution [8, 9]. In 2012, Bo et al. reported the performance of the temperature resistant and clean fracturing fluid, and proposed that this fracturing fluid is a kind of viscoelastic surfactant fracturing fluid with good temperature resistance, which can resist 150°C for a short time [10]. In 2013, Xiaojuan et al. reported the performance and application of the FRC-1 clean fracturing fluid system, and reported the clean water base fracturing fluid containing viscoelastic surfactant 2.5%, special stabilizer 0.1%, and chloride salt water 4%. According to its field application in Changqing oil field, it is proposed that the overall performance of FRC-1 clean fracturing fluid can meet the needs of fracturing construction below 60°C [11]. The field is easy to operate, has little damage to the reservoir, easy to return and to break, and does not leave residue. The average daily production increment is 1.34 times. In 2015, Manxue and Yifei reported on the development and application of low injury and clean fracturing fluid VES-1, the thickening agent of the fracturing fluid could be dispersed evenly in 1-2 minutes, and it could form a gel with a temperature resistance of 80°C [12]. After shearing for 60 min at 80°C, the viscosity of the fracturing fluid is still greater than that of 90 mPa·s. In 2017, Yongjun et al. reported the technical performance of leVES-70 clean fracturing fluid. The main agent of VES-70 fracturing fluid was C16 or C18 alkyl three methyl quaternary ammonium salt, which was mixed with organic acids, isopropanol, and other auxiliaries. Its composition is simple, so it is easy to mix. It has excellent viscoelasticity and temperature resistance, shear resistance, gel breaking, and reflow performance [13].

But, application of many fields proves that all the current clean fracturing fluids have residues without exception. These residues are bound to cause serious clogging of the strata and filling layers to greatly reduce the permeability, and the cumulative damage can reach more than 90%, which greatly reduces the effect of fracturing and cannot achieve continuous mixing. This situation is especially prominent for low pressure and low permeability reservoirs, which often leads to fracturing failure.

In order to solve the above problems, a new type of multicomponent thickening water clean fracturing fluid system has been studied. The main component of the thickening fracturing fluid system (2.2% octadecyl methyl dihydroxyethyl ammonium bromide (OHDAB) +1.4% dodecyl sulfonate sodium +1.8% potassium chloride and 1.6% organic acids) can be quickly cross-linked to carry sand, when it encounters crude oil and formation water, it can break gel without adding a gel breaker. After breaking glue, no residue is found, and the viscosity of gel breaker is <5 mPa·s. In addition, the system can be continuously mixed to shorten the fracturing cycle and can carry acid (HCL and HF) of concentration 1%–9% and organic acid 1%–3%, to achieve combination of acidizing and sand fracturing. The flow-back liquid can be recovered directly, and the same thickener can be recycled and reused. It saves water consumption and is suitable for horizontal wells to improve efficiency, to meet the needs of saving energy and environmental protection and large-scale factory operation.

2. Gelling and Gel Breaking Mechanism

The clean fracturing fluid, new multicomponent and recoverable thickening water, is mainly composed of a variety of special surfactants. The application of the composite liquid mutually contains molecular force between material and attachment, and the complex becomes transparent liquid substances, namely thickener. When the thickener is exposed to aqueous solution, the surfactant molecules are released and diffused rapidly under the influence of polar substances and aggregated to form wormlike micelles. The equilibrium system and micelle charge of the special organic ions make the micelles grow. A salt of the variable length wormlike micelles entangled, thus forming a uniform mesh space-like floc-like in the system, so as to form a gel [14–16].

In the gel system, there are mainly physical interactions between the surfactant and the salt, which are different from guar gum in the polymer fracturing system connected with each other in chemistry. Therefore, when the surfactant fracturing fluid meets the appropriate amount of formation water and oil gas, the intermolecular interaction distance between the surfactant and salt will increase. Furthermore, the entanglement state of wormlike micelles is destroyed, even the wormlike micelles disintegrated into simple micelles, and these can make the gel system automatically break [17–22]. The mechanism of gelling and breaking of thickening water is shown in Figure 1.

3. Performance Evaluation of the New Type Thickening Fracturing Fluid

3.1. Experimental Equipment and Reagents. HAD-CQ2A double deflector long-term conductivity tester (Beijing Heng Aude instrument limited company), SNB-2 digital viscometer (Shanghai Jingke day U.S. Trade Co. Ltd.), mixer, 250 mm cylinder, and electric thermostatic water bath (DGSY Beijing) were used.

The new type thickening water fracturing fluid is mainly composed of viscoelastic surfactant solution. The main agent of this experiment is the ionic surfactant with concentration of 2.2% octadecyl methyl dihydroxyethyl ammonium bromide (OHDAB). The role of the additives in thickening water fracturing fluid mainly depends on the cationic group adsorbed on the surfactant, which reduces the repulsive force between

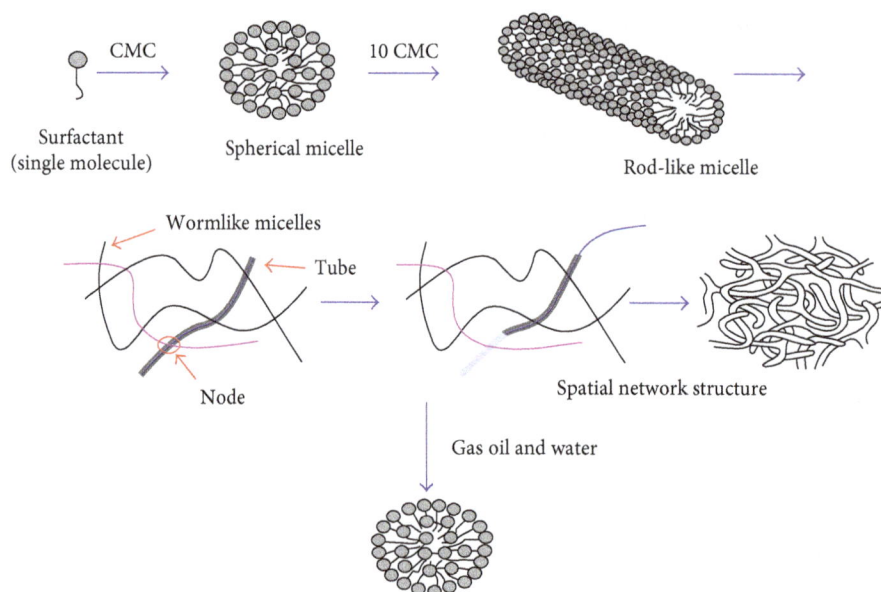

FIGURE 1: Schematic diagram of the mechanism of thickening water.

TABLE 1: Comparison of static sand-carrying capacity of two kinds of fracturing fluids.

| Sand/liquid ratio (%) | Settling velocity (mm·s^{-1}) | | | | | |
| | 60 (mPa·s) | | 90 (mPa·s) | | 120 (mPa·s) | |
	Thickening water	Guar gum	Thickening water	Guar gum	Thickening water	Guar gum
5	0.02973	0.04136	0.02461	0.03162	0.02016	0.02914
10	0.07823	0.09161	0.07012	0.08136	0.06812	0.07216
20	0.11986	0.14161	0.11014	0.13264	0.10011	0.12234
30	0.14832	0.17214	0.14036	0.16213	0.13613	0.15614
40	0.19613	0.21463	0.18214	0.20436	0.17614	0.19632

the cationic groups and increases the micelle growth. In this experiment, 1.4% dodecyl sulfonate sodium, 1.8% potassium chloride, and 1.6% organic acids are selected as the auxiliaries. The guar gum system used in this experiment is the most commonly used hydroxypropyl guar gum (HGG) in China.

3.2. The Sand-Carrying Capacity

3.2.1. Determination of Static Sand-Carrying Capacity. The new type thickening water gel and guar gum fracturing fluid gel are packed in No. 1 and No. 2 cylinder, and then the proppant with different sand ratio (volume ratio) is added. The static sand-carrying capacity is measured by observing the proppant in two cylinders at the time of settlement, and the results are shown in Table 1 and Figure 2.

It can be seen from Table 1 and Figure 2 that with the increase of sand ratio, the settling velocity of sand in thickening water and guar gum increases. And with the increase of viscosity of fracturing fluid, the sand falling velocity decreases gradually. Compared with thickening water and guar gum under the same viscosity, the sand dropping rate shows that the sand-carrying capacity of condensed water is better than that of guar gum. The sand-carrying capacity of guar gum at 120 mPa·s is equivalent to the carrying capacity of 60 mPa·s of thickened water.

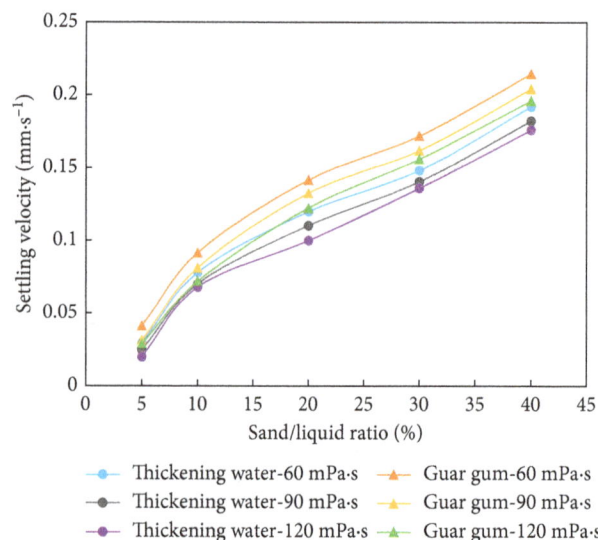

FIGURE 2: Comparison of static sand-carrying capacity of two kinds of fracturing fluids.

3.2.2. Determination of Dynamic Sand-Carrying Capacity. The dynamic sand-carrying capacity of thickening water (120 mPa·s) and guar gum (120 mPa·s) at different sand

TABLE 2: Comparison of dynamic sand-carrying capacity of two kinds of fracturing fluids.

Sand/liquid ratio (%)	Thickening water	Guar gum	Clean water
5	No layered	No layered	Layered
10	No layered	No layered	Layered
20	No layered	No layered	Layered
30	No layered	A little precipitation	Layered
40	No layered	Layered	Layered

TABLE 3: Comparison of gel breaking property of two kinds of fracturing fluids.

Time (min)	Apparent viscosity (mPa·s)					
	50 (°C)		60 (°C)		70 (°C)	
	Thickening water	Guar gum	Thickening water	Guar gum	Thickening water	Guar gum
10	100.3	115.6	80.3	98.2	50.2	70.2
30	80.6	96.5	60.1	75.2	26.2	50.6
60	40.2	62.4	30.6	45.8	12.3	30.7
120	10.2	20.1	10.1	19.6	2.6	16.8
240	4.3	10.6	1.9	8.7	1.3	5.6

ratios was determined at the stirring rate of 100 r/min and 70°C of temperature. The results are shown in Table 2.

It can be seen from Table 2 that the thickening water fracturing fluid can effectively carry sand at the sand ratio of 40% at 70°C, which indicates that the dynamic fracturing performance of the clean fracturing fluid is good. The guar gum precipitates at a sand ratio of 30%, and stratification occurs when the sand ratio is 40%, indicating that the effective sand-carrying ratio must be controlled below 40%. Therefore, when the temperature is 70°C, the dynamic sand-carrying capacity of the thickening water is better than guar gum, and the maximum sand ratio can reach 40%.

After thickening fracturing fluid is subjected to shear action, the network structure formed between micelles is destroyed, resulting in reduced number of micelles. And, the shear does not destroy the structure of the small molecule, but only reduces the degree of polymerization of the micelles; when the shear action stops, the micelles in the solution are rewound through the association to restore the reticular structure, and the viscosity is also restored. However, under the external shear effect, the network structure of gum fracturing fluid is destroyed, and the polymer molecular chain also shortens and degrades, and the viscosity of the system decreases. When the shear effect disappears, the molecule cannot recover the original viscosity through cross-linking. Therefore, compared with the polymer system, the carrying capacity of the thickened water fracturing fluid system is better.

3.3. *The Gel Breaking Property.* The experiment is divided into two groups. The volume ratio of simulated oil to thickening water (120 mPa·s) and guar gum fracturing fluid (120 mPa·s) is 1 to 20. The viscosity in different temperature and shear time was measured by the rotating viscometer at the speed of 7.5 r/min, and the results are shown in Table 3 and Figure 3.

The viscosity of water at normal temperature is 1.3 mPa·s. It can be seen from Table 3 and Figure 3 that the

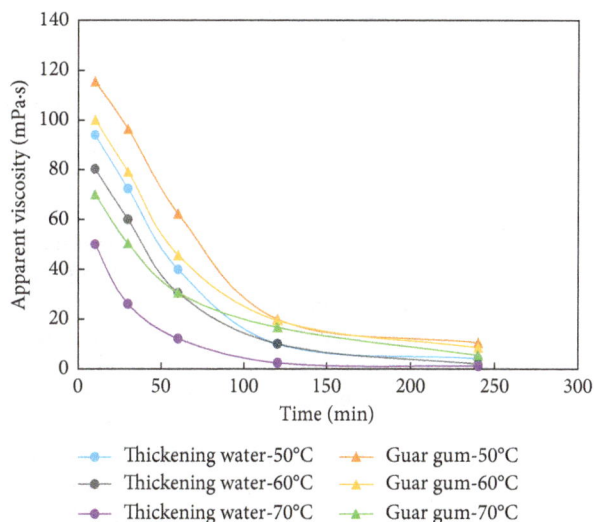

FIGURE 3: Comparison of gel breaking property of two kinds of fracturing fluids.

viscosity of guar gum fracturing fluid under the three temperature conditions is higher than thickening water fracturing fluid, indicating that thickening water is easier to return than guar gum after fracturing. At 70°C, the viscosity of the thickening water fracturing fluid is very close to the viscosity of the water after breaking the gel. In the experiment, it was found that after the system was broken, the insoluble matter could not be observed, while the guar gum system appeared with different degree of insoluble matter. All the above experimental phenomena show that clean fracturing fluid has excellent gel breaking performance and completely flow-back after gel breaking, which has little damage to formation.

The gel breaking property of fracturing fluid directly affects the damage degree to formation. Gum fracturing fluid needs to add a certain amount of the gel breaker, so that fracturing fluid can break glue and it is difficult to control the

TABLE 4: Comparison of three properties of two kinds of fracturing fluids.

Closing pressure (MPa)	Thickening water fracturing fluid system			Guar gum fracturing fluid system		
	Fracture conductivity ($\mu m^2 \cdot cm$)	Diversion retention rate (%)	Diversion damage rate (%)	Fracture conductivity ($\mu m^2 \cdot cm$)	Diversion retention rate (%)	Diversion damage rate (%)
5	150.4	82.3	17.6	120.1	61.2	40.2
10	130.6	83.1	15.3	100.6	52.3	30.7
20	100.3	84.5	12.1	80.6	63.2	27.6
30	80.1	84.7	10.1	50.1	64.7	16.1
40	50.6	86.2	8.9	26.3	65.1	10.8

Note. Fracture conductivity: after the fracture is closed, the proppant filling zone can pass through the reservoir fluid. Diversion retention rate: determination of the ratio between the conductivity of fractured support and the diversion capacity of formation water measured by hydraulic fracturing after fracturing. Diversion damage rate: the decrease of the diversion capacity of the support fracture induced by the fracturing fluid relative to the formation water.

degree of gel breaking. Unlike the gum fracturing fluid, it is not necessary to add the gel breaking agent in the thickened water clean fracturing fluid, but after the hydrocarbon compounds are encountered, the compound can be transformed into a spherical micelle by solubilizing to the micelles formed by the surfactant, and the mesh structure disintegrates to make it lose the viscoelasticity. Or, under the action of dilution of formation water, the content of surfactant is lower than the critical micelle concentration CMC, which makes the solution lose viscoelasticity and automatically breaks glue.

3.4. The Fracture Conductivity, Diversion Damage Rate, and Diversion Retention Rate.

The main experimental instrument is the conductivity tester. The pressure difference and flow rate at each end of each fluid are measured quantitatively when the fracture is supported by simulation. And, when the liquid viscosity and fracture model size are replaced by Darcy's formula, the flow conductivity tester can automatically calculate the diversion capacity value under different closed pressure conditions [23, 24]. The viscosity of the two fracturing fluids is 120 mPa·s, and the fracture body model is selected as a multistage fracture system. The statistical results are shown in Table 4 and Figures 4 and 5.

It can be seen from Table 4 and Figures 4 and 5 that with the increase of the closing pressure, the flow conductivity increases gradually, the diversion retention rate increases gradually, and the diversion damage rate decreases gradually. The flow conductivity of the thickening fracturing fluid system is between 50.6 and 150.4 $\mu m^2 \cdot cm$, which is obviously higher than that of the guanidine gum fracturing fluid system. The diversion damage rate of the thickening water fracturing fluid system is between 8.9% and 17.6%, which is obviously lower than that of the guanidine gum fracturing fluid system. It is also known that the conductivity retention rate of the thickening hydraulic fracturing fluid system is above 80% after fracture closure, and the retention rate is high.

Firstly, when the two fracturing fluids are used as prefluid, respectively, the viscosity loss of the thickening water is less than the gum fracturing fluid because of the shear recovery of the thickening water, which is more likely to achieve the fracturing effect. Secondly, when the two fracturing fluids are used as sand-carrying fluid, respectively, the

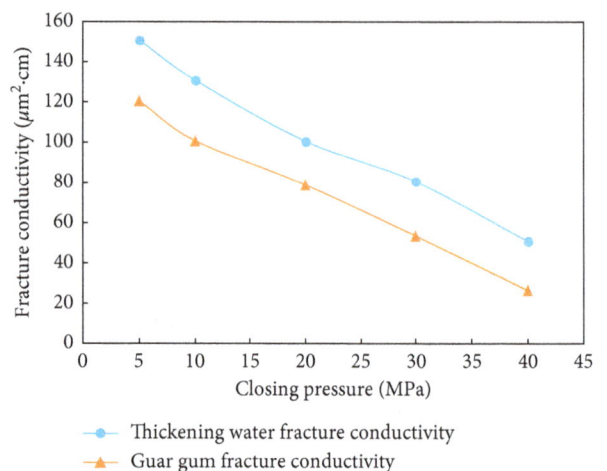

FIGURE 4: Comparison of fracture conductivity of two kinds of fracturing fluids.

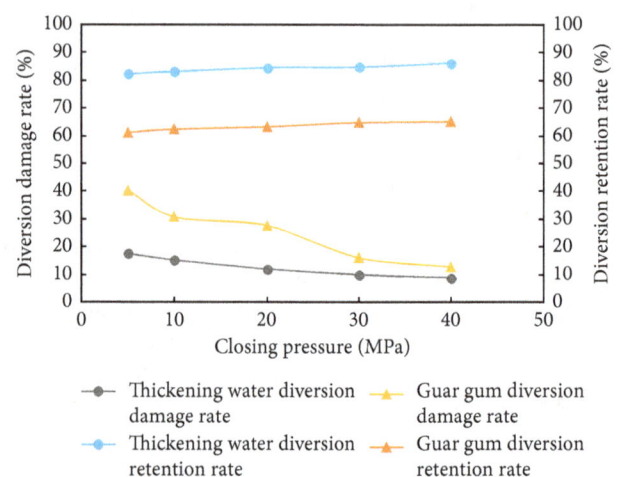

FIGURE 5: Comparison of diversion retention rate and diversion damage rate of two kinds of fracturing fluids.

carrying capacity of the thickening fracturing fluid is better than gum fracturing fluid. When the viscosity of the thickening water fracturing fluid is restored through the wellhead hole, it is beneficial to the transportation of the proppant, and the proppant can enter the depth of the crack.

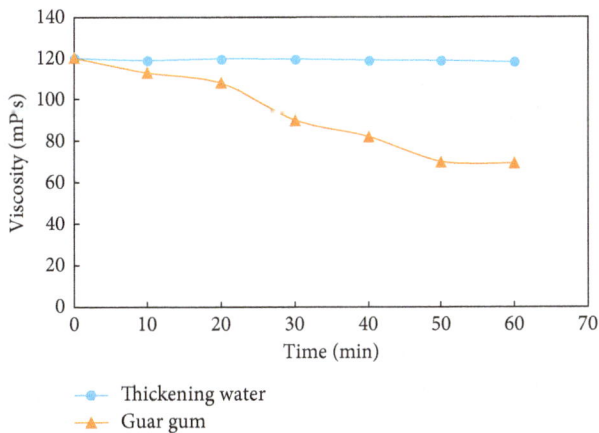

FIGURE 6: Comparison of viscosity changes with time: two kinds of fracturing fluids.

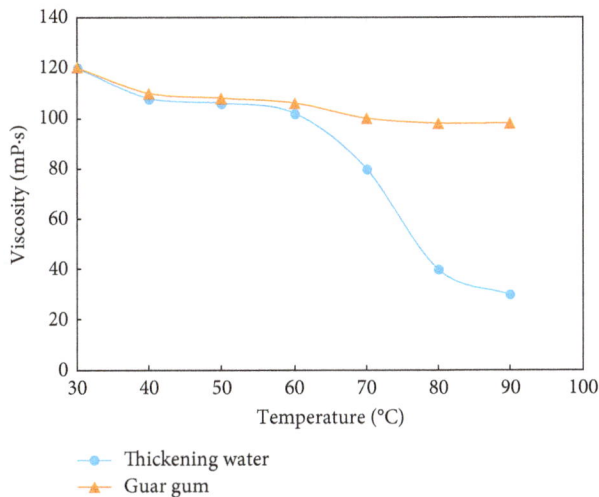

FIGURE 7: Comparison of viscosity changes with temperature: two kinds of fracturing fluids.

After the fracture is closed, the capacity of the support filling belt to the reservoir fluid is higher than the gum fracturing fluid. Finally, the thickening water fracturing fluid has good gel breaking performance in the formation, and the formation damage rate after controlled pressure reentry is small.

3.5. The Antishearing and Heat Resistance. The viscosity changes of two kinds of fracturing fluids under continuous shear for 60 min at room temperature (25 r/min) were measured by a rotating viscometer at 750 rpm. The viscosity of two kinds of fracturing fluids at 30~90°C under rotational speed 75 r/min with a rotational viscometer is shown in Figures 6 and 7.

From Figure 6, we can know that the thickening water fracturing fluid viscosity does not change with the shear time and that viscoelastic clean fracturing fluid keeps the shear stable; this characteristic is decided by the thickening mechanism of it.

The viscoelastic micelle system in the condensed water fracturing fluid has a strong self-healing property, and at the same shear rate, the system reaches a dynamic equilibrium of micelle destruction and repair. Therefore, the system can maintain the degree that will not be significantly decreased in long time under shear. In the guar gum fracturing fluid system, the polymer chains are sheared and the viscosity of the polymer decreases with the decrease of molecular weight.

From Figure 7, it can be seen that the viscosity drop of the thickened hydraulic fracturing fluid system increases more than 70 degrees, which indicates that the fracturing fluid system has good temperature resistance performance in medium and low temperature formations.

When the temperature rises, the solubility of the surfactant molecules in the water increases, thus, increasing the aggregation number of the micelles of the surfactant and speeding up the irregular movement of the micelles. This movement of micelles helps the intertwining between the rod-like micelles, and this movement maintains the viscosity of the gelatin. As the temperature continues to rise and when the temperature exceeds the critical micelle temperature, the micelle molecules dissociate and destroy the structure of some small micellar molecules. The viscosity of the thickened water system is not in a dynamic equilibrium state, so the viscosity of the gel decreases.

4. Field Application

4.1. Block Overview. The area of A block in YC low permeability oil field is 2.03 km^2, and the underground pore volume is 196.3×104 m^3. The geological reserve of the target layer is 132.1×10^4 t, the average single well shooting sandstone thickness is 14.32 m, the effective thickness is 10.23 m, and the average effective permeability is 12.3×10^{-3} μm^2; A block adopts diamond inverted nine-point well pattern, and there are 83 wells, including 21 injection wells and 62 production wells, and the distance between the injection well and the production well is 300 meters. A block is a typical low porosity and low permeability block, but it has potential for development. In 2016, the new multicomponent thickening water fracturing fluid was used in 3 horizontal wells in A block, and the fracturing effect was improved obviously compared with the previous fracturing production and guar gum system.

4.2. Technical Index of Field Application. The new multicomponent thickened water clean fracturing fluid is a viscoelastic surfactant-based fracturing fluid system. The main component, liquid thickener, can be quickly cross-linked and carries sand when mixed with water. Without adding new equipment, continuous mixing operation can be realized, and the glue can be broken when oil and formation water are encountered. The fracturing fluid can carry different types and concentrations of acid solution. It has the technical features of no residue and low damage, and can be put into operation without swabbing and draining.

(1) The new type thickening water can carry acid (HCL and HF) concentration of 1%–9% and organic acid

TABLE 5: Comparison of construction process.

Process	Thickening water	Guar gum	Remark
Before fracturing construction	Preparation and thickener	Preparing and cleaning tank, fracturing fluid material to field distribution	Guar gum fracturing fluid is easy to deteriorate by long time of placement
Fracturing construction	Continuous mixing, how much design is required, and how much it can be matched	Construction according to the design requirements	The amount of thickening water can be determined according to the actual amount of field construction, and there is no problem of liquid waste
After fracturing construction	Sand washing, well washing, production	Injection, sand washing, well washing, swabbing, and production	
Overall evaluation	Thickening water clean fracturing fluid system is more practical than guar gum fracturing fluid system		

TABLE 6: Comparison of operation cycle.

Time	Guar gum	Thickening water	Remarks
Preparation tank	1-2 days	1-2 days	Truckage
Tank cleaning	3-4 h	0 h	
Liquid distribution	10 h	0 h	There is no need for advance solution for thickening water
Fracture	1–1.5 h	1–1.5 h	The continuous mixing of the thickened water mixture is used in the field, and the amount can be determined according to the actual construction amount of the site
Liquid discharge	2–5 days	1 day	The thickening water can not be discharged and put into production directly
Total	4.5–8 days	1.5–3.5 days	Save 3-4 days

1%–3% to achieve acidification combined with sand fracturing.

(2) The temperature resistance and shear resistance: temperature of new type thickening water is not higher than 75°C and 40°C fracturing fluid continuous shear viscosity of >100 mPa·s, no change of viscosity.

(3) The new type of gel breaking thickening water without adding breaking agent, gel breaking in crude oil and formation water, no residue after gel breaking, and gel breaking liquid viscosity of <5 mPa·s.

(4) The sand-carrying capacity: 40 centigrade and 35% sand ratio, the static state can be suspended more than 60 min.

(5) The core damage rate: the total is less than 20%.

(6) The antiswelling rate of clay is >89%.

4.3. Comparison of the Effect of Guar Gum Fracturing Fluid. Guar gum fracturing fluid ensures the success of high-temperature deep fracturing. However, when guar gum fracturing fluid enters the pipeline at high speed and enters the pipeline at high speed during construction, it will cause severe shearing degradation and produce permanent viscosity loss. Moreover, the compound chain of guar gum fracturing fluid is connected with the cross-linked chain between the chains, resulting in the body shape. The result is

that the gel breaking is incomplete, so that the residue left behind after breaking the gel will remain in the crack, which will seriously reduce the permeability of proppant filling layer and damage the production layer, resulting in poor fracturing effect. In addition, the cross-linking fracturing fluid is not completely broken, causing sand fracturing well to scour, prolonging operation time, and increasing operating cost. At present, the guar gum fracturing fluid system has high content of water insoluble substance (the national standard requires less than 12% of water insoluble substance) and repeating damage to the formation. The construction process and operation cycle of the two kinds of fracturing fluid are shown in Tables 5 and 6.

4.4. Analysis of Fracturing Effect. The development effect of the new multicomponent thickened water fracturing hydraulic fracturing 3 wells in A area is shown in Table 7, as shown in Figures 8–10.

4.5. Comparison with Conventional Fracturing Effect. At present, the average daily production of 3 wells is 10.3 m³, with an average daily oil production of 9.27 t. And, 3 horizontal wells were selected as contrast wells in the same block, and the average daily production of 3 wells was 8.2 m³, with an average daily oil production of 7.28 t. By comparison, the daily production of the test wells increased by 2.1 m³ compared to the selected wells in the block, and the daily production increased by 1.89 t. It indicates that the

TABLE 7: The effect of 3 wells fractured by thickening water.

Block	Well	Physical parameters				Oil test		Put into operation		
		Shale content (%)	Porosity (%)	Permeability (md)	Oil saturation (%)	Daily oil (t)	Daily water (t)	Daily liquid (m³)	Daily oil (t)	Water cut (%)
A	A1	10.3	11.5	9.2	62.2	55.3	1.2	11.9	10.8	9.2
	A2	10.4	10.0	8.7	63.8	42.6	1.0	8.6	7.8	9.4
	A3	15.1	9.3	6.9	65.0	52.1	0.8	9.5	8.6	9.3

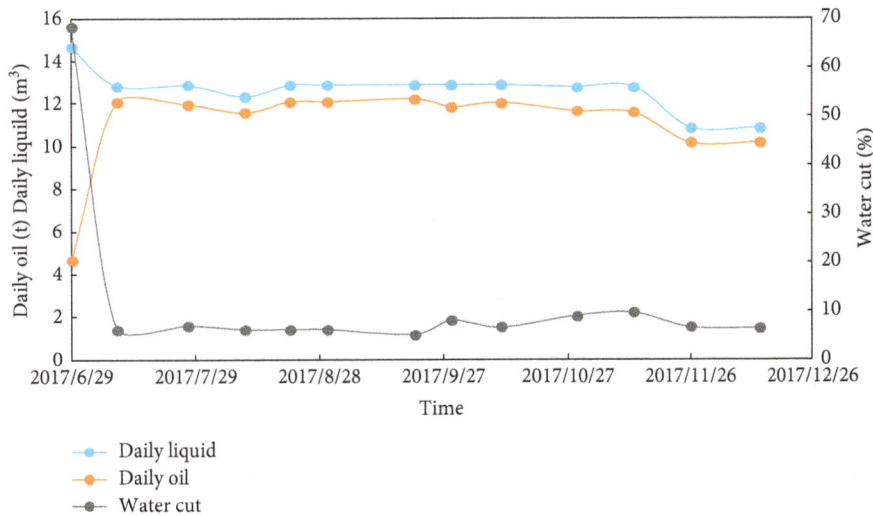

FIGURE 8: Production curve of A1 well.

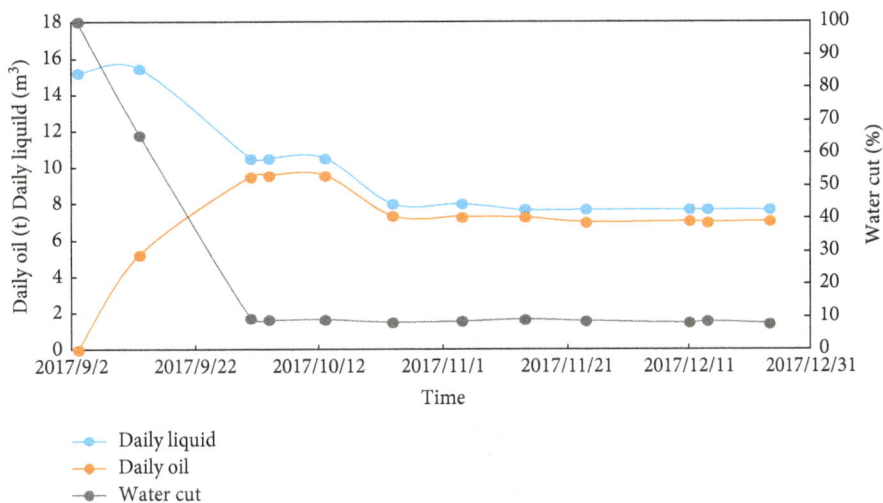

FIGURE 9: Production curve of A2 well.

acidic clean fracturing fluid of thickening water has a good adaptability to A block, which will pave the way for further promotion Table 8.

As a new type of fracturing fluid, clean fracturing fluid have the characteristics of cleaning, low damage, thoroughly gel breaking, little residues, and high flow-back rate. So, it will have good application prospect in low permeability reservoirs. In particular, the reprocessing liquid can be directly recovered, and the same thickener can be added to recycle, which reduces the reservoir damage and saves the water consumption. It is very suitable for improving the efficiency and saving energy for the horizontal well, as well as protecting environment and meeting the demand of large-scale factory operation.

5. Conclusions

(1) When the shear action stops in new type thickening water, the micelles in the solution are rewound through the association to restore the reticular

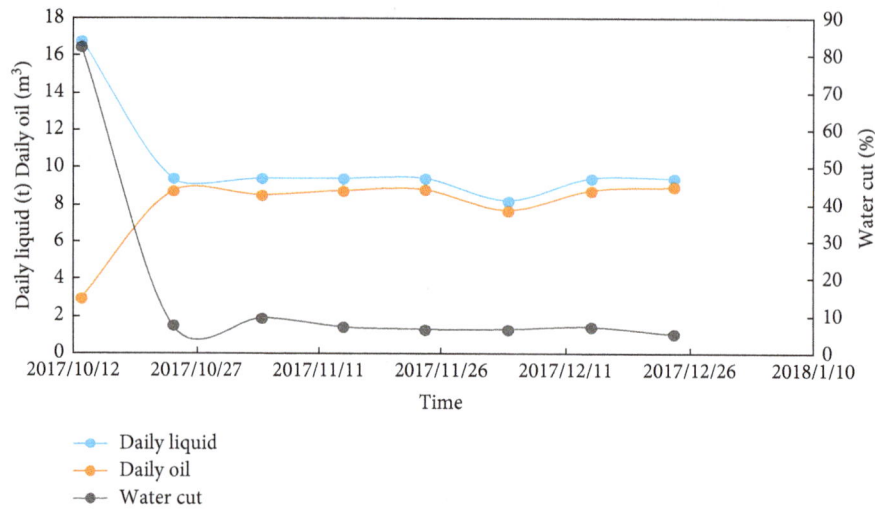

FIGURE 10: Production curve of A3 well.

TABLE 8: Comparison of effect between thickening water fracturing wells and conventional fracturing wells.

Well types	Well	Physical parameters				Oil test		Put into operation		
		Shale content (%)	Porosity (%)	Permeability (md)	Oil saturation (%)	Daily oil (t)	Daily water (t)	Daily liquid (m³)	Daily oil (t)	Water cut (%)
Thickening water fracturing well	3	13.2	8.3	10.3	68.3	50.2	1.0	10.3	9.27	10.0
Guar gum fracturing well	3	14.2	9.4	11.5	60.1	45.2	1.2	8.2	7.38	10.0

structure, and the viscosity is also restored. The sand/liquid ratio at static suspended sand is up to 30% when the apparent viscosity of the new system is 60 mPa·s, which is equivalent to the sand-carrying capacity of guar gum at 120 mPa·s.

(2) When the viscosity of the thickening water is restored through the wellhead hole, it is beneficial to the transportation of the proppant, and the fracture conductivity is between 50.6 μm^2·cm and 150.4 μm^2·cm, and the fracture conductivity damage rate is between 8.9% and 17.9%. The fracture conductivity conservation rate is more than 80% closing up of fractures, which are superior to guar gum.

(3) The new system does not need to be added to the adhesive; after the hydrocarbon compounds are encountered, the compound can be transformed into a spherical micelle by solubilizing to the micelles formed by the surfactant and the reticular structure disintegrates to make it lose the viscoelasticity. Under the action of dilution of formation water, the content of surfactant is lower than the critical micelle concentration CMC, which makes the solution lose viscoelasticity and automatically break the gel.

(4) The viscoelastic micelle system in the multicomponent thickening water has a strong self repair. Under the same shear rate, the system achieves a dynamic balance of the micellar damage and repair. At room temperature, the shear rate is 60 min, and the viscosity is almost unchanged. As the temperature continues to rise and when it exceeds the critical micelle temperature, the micelle molecules dissociate and destroy the structure of some small micellar molecules. The viscosity system is not in a dynamic equilibrium state, so the viscosity of the gel decreases. It shows that the new system has good temperature resistance in middle and low temperature formation.

(5) The actual production of A block shows that the damage of thickening fracturing fluid is low, and the long retention in reservoir will not cause great damage to the reservoir. Therefore, it cannot be pumped directly into production and continuous mixing, shortening the overall testing oil cycle. The flow-back liquid can be recovered directly, and the same thickener can be recycled and reused.

Conflicts of Interest

The authors declare that they have no conflicts of interest.

Acknowledgments

This work is financially supported by the National Natural Science Foundation of China under Grant no. 51504069. The foundation provides the author with many financial help, such as the cost of experimental materials, the layout of the articles, and so on.

References

[1] B. Zeng, L. Cheng, C. Li et al., "Evaluation of development effect of fractured horizontal wells in extra low permeability reservoirs," *Journal of Petroleum*, vol. 31, no. 5, pp. 792–796, 2010.

[2] S. Hao, W. Li, and C. Guo, "Difficulties and breakthrough of horizontal well drilling technology in ultra low permeability shallow reservoir," *Journal of Petroleum Exploration in China*, vol. 22, no. 5, pp. 16–19, 2017.

[3] Y. An, L. V. Yi, L. Lu, and S. Hu, "Study on inflow performance of fractured horizontal well in ultra-low permeability reservoir," *Journal of Special Oil and Gas Reservoir*, vol. 19, no. 3, pp. 90–92, 2012.

[4] G. A. Al-Muntasheri, F. Liang, and K. L. Hull, "Nanoparticle-enhanced hydraulic-fracturing fluids: a review," *SPE Production and Operations*, vol. 32, no. 2, pp. 186–195, 2017.

[5] M. Samuel, R. J. Card, E. B. Nelson et al., *Polymer-Free Liquid for Hydraulic Fracturing. SPE38622*, Society of Petroleum Engineers, Houston, TX, USA, 1997.

[6] M. S. Dahanayake, J. Yang, J. H. Y. Niu et al., "Viscoelastic surfactant fluids and related methods of use," US Patent US6482866B1, Schlumberger Technology Corporation, Sugar Land, TX, USA, 2002.

[7] Q. Qu, E. B. Nelson, D. M. Willberg et al., "Compositions containing aqueous viscosifying surfactants and methods for applying such compositions in subterranean formations," US Patent US6435277B1, Schlumberger Technology Corporation, Sugar Land, TX, USA, 2002.

[8] L. Xinquan, Y. Mingxin, Z. Jinyu et al., "Viscoelastic surfactant (VES) fracturing fluid," *Oil Field Chemistry*, vol. 18, no. 3, pp. 273–277, 2009.

[9] C. Fu, W. Anpei, L. Fengxia et al., "Research progress of clean fracturing fluid abroad," *Journal of Southwest Petroleum Institute*, vol. 24, no. 5, pp. 65–67, 2009.

[10] J. Bo, Z. Deng, L. Dongping et al., "Properties of temperature resistant VES fracturing fluid S C F," *Oil Field Chemistry*, vol. 20, no. 4, pp. 332–334, 2012.

[11] R. Xiaojuan, L. Shuren, L. Zhihang et al., "The performance and application of FRC-1 clean fracturing fluid system," *Fine Petrochemical Progress*, vol. 1, p. 53, 2013.

[12] W. Manxue and L. Yifei, "Development and application of low injury and clean fracturing fluid V E S-1," *Petroleum and Natural Gas Chemical*, vol. 33, no. 3, pp. 188–192, 2015.

[13] L. Yongjun, F. Bo, and F. D. Ye, "Research on the performance of VES-70 viscoelastic clean fracturing fluid," *Oil Field Chemistry*, vol. 21, no. 2, pp. 120–123, 2015.

[14] Z. Yan, C. Dai, M. Zhao et al., "Progress in research and application of clean fracturing fluid," *Journal of Oilfield Chemistry*, vol. 32, no. 1, pp. 142–145, 2015.

[15] C. Williams, P. Mclfresh, M. Khodaverdian et al., "Non-ionic fracture fluids can recover 90% permeability after proppant run," *Offshore*, vol. 61, no. 10, pp. 76–80, 2001.

[16] O. Contreras, M. Alsaba, G. Hareland, M. Husein, and R. Nygaard, "Effect on fracture pressure by adding iron-based and calcium-based nanoparticles to a nonaqueous drilling fluid for permeable formations," *Journal of Energy Resources Technology*, vol. 138, no. 3, p. 032906, 2016.

[17] H. Hofmann, T. Babadagli, and G. Zimmermann, "Numerical simulation of complex fracture network development by hydraulic fracturing in naturally fractured ultratight formations," *Journal of Energy Resources Technology*, vol. 136, no. 4, p. 042905, 2014.

[18] X. Li, J. Zhang, P. He et al., "Laboratory experimental study on clean fracturing fluid for thickening water in North oilfield," *Journal of China Petroleum and Chemical Engineering Standard Quality*, vol. 9, p. 268, 2012.

[19] O. A. Bustos, K. R. Heiken, M. E. Stewar et al., "Application of a viscoelastic surfactant-based CO_2 compatible fracturing fluid in the frontier formation, big horn basin, Wyoming," *SPE Production and Operations*, vol. 966, 2007.

[20] K. N. Hughes, N. R. Santos, R. E. A. Urbina et al., "New viscoelastic surfactant fracturing fluids now compatible with CO_2 drastically improve gas production in rockies," in *Proceedings of SPE International Symposium and Exhibition on Formation Damage Control*, Lafayette, LA, USA, February 2008.

[21] M. Samuel, R. Card, E. Nelson et al., "Polymer-free fluid for hydraulic fracturing," *SPE Production and Operations*, vol. 622, 1997.

[22] J. Wang, H. Liu, T. Liu et al., "Water reuse fracturing fluid," *Journal of Drilling and Drilling Technology in Northeast Oil and Gas Field*, vol. 39, no. 3, pp. 339–342, 2017.

[23] Y. Su and G. Lin, "Tight sandstone reservoir hydraulic sand propped fracture conductivity," *Journal of Daqing Petroleum Geology and Development*, vol. 9, pp. 2–6, 2017.

[24] A. Qajar, Z. Xue, A. J. Worthen et al., "Modeling fracture propagation and cleanup for dry nanoparticle-stabilized-foam fracturing fluids," *Journal of Petroleum Science and Engineering*, vol. 146, pp. 210–221, 2016.

6

The Impact of Ce-Zr Addition on Nickel Dispersion and Catalytic Behavior for CO$_2$ Methanation of Ni/AC Catalyst at Low Temperature

Minh Cam Le, Khu Le Van, Thu Ha T. Nguyen, and Ngoc Ha Nguyen

Theoretical and Physical Chemistry Division, Faculty of Chemistry, Hanoi National University of Education, Hanoi 1000, Vietnam

Correspondence should be addressed to Minh Cam Le; camlm@hnue.edu.vn

Academic Editor: Anton Kokalj

The CO$_2$ methanation was studied over 7 wt.% nickel supported on Ce$_{0.2}$Zr$_{0.8}$O$_2$/AC to evaluate the correlation of the structural properties with catalytic performance. The catalysts were investigated in more detail by means of X-ray diffraction (XRD), transmission electron microscopy (TEM), and scanning electron microscopy (SEM). A sample of 7 wt.% nickel loading supported on activated carbon (AC) was also prepared for comparison. The results demonstrated that the ceria-zirconia solid solution phase could disperse and stabilize the nickel species more effectively and resulted in stronger interaction with nickel than the parent activated carbon phase. Therefore, 7% Ni/Ce$_{0.2}$Zr$_{0.8}$O$_2$/AC catalyst exhibited higher activity for CO$_2$ reduction than 7% Ni//AC. It can attain 85% CO$_2$ conversion at 350°C and have a CH$_4$ selectivity of 100% at a pressure as low as 1 atm. The high activity of prepared catalysts is attributed to the good interaction between Ni and Ce$_{0.2}$Zr$_{0.8}$O$_2$ and the high CO$_2$ adsorption capacity of the activated carbon as well.

1. Introduction

Increasing emissions of carbon dioxide arising from the widespread production of energy from fossil fuels is a critical matter regarding greenhouse gases effect and, thus, global warming [1, 2]. Technologies including possible reduction or conversion of CO$_2$ give valuable advantages for protecting the environment by recycling CO$_2$ effectively based on the catalytic methanation [3–5]. Conversion of carbon oxides into methane

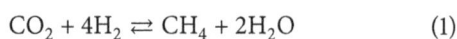

$$CO_2 + 4H_2 \rightleftarrows CH_4 + 2H_2O \qquad (1)$$

is a exothermic reaction with $\Delta H° = -165$ kJ/mol. The exothermic character of the methanation process causes problems with respect to an exact control of the reaction temperature, which can result in a further increased conversion of CO$_2$ [6]. Therefore, the development of catalysts for methanation of carbon dioxide is the key factor. Recently, results of Beuls et al. [7] and Jacquemin et al. [8] give evidences that at low temperature (<200°C) and atmospheric pressure the reaction takes place with very high selectivity.

Various metal-based catalysts have been studied for the CO$_2$ methanation reaction such as Fe [9], Ru [10], Co [11], Rh [12, 13], Pd [14, 15], Pt [16], and Ni [16, 17] supported on several oxides (SiO$_2$ [18], TiO$_2$ [19], Al$_2$O$_3$ [20, 21], ZrO$_2$ [22, 23], CeO$_2$ [24], and Ce-Zr mixed oxides [25, 26]) or porous materials (HZSM-5 [27], HUSY [28, 29]). Although the noble metals (Ru, Rh, and Pd) exhibit better activity, they are too expensive for a large-scale industrial application; therefore nonnoble metal-based catalysts are often preferred. Among group VIII metals, the nickel-based catalysts have covered the larger part of published works [30–35] due to their high catalytic activity, high selectivity for methane, and relatively low price. The main problems of Ni-based catalysts are the deactivation due to carbon deposition and poor stability at high temperature [29, 34]. Therefore, great efforts have been made to develop an effective promoted Ni-based catalyst which exhibits both high activity and high thermal stability in CO$_2$ methanation.

Firstly, adding catalyst promoters, Trovarelli et al. [36, 37], who compared the catalytic activity of several Rh-based catalysts using different types of supports, CeO$_2$, SiO$_2$, Ta$_2$O$_2$, and Nb_2O_5, found that the catalytic activity and thermal stability of the catalyst could be improved by using CeO_2 or ZrO_2 as the support. Rynkowski et al. [38] reported that Ni (or Ru) supported on Al_2O_3 (or SiO_2) which is promoted with CeO_2 possessed an improved activity for CO_2 hydrogenation into methane. The CO_2 methanation reaction using Ni supported on Ce-Zr mixed oxides catalysts was for the first time investigated by Ocampo et al. [39–41]. They found that these catalysts exhibited excellent levels of activity, selectivity, and stability for CO_2 methanation. Liu et al. [34] found that CeO_2 promoted the dispersion of metal Ni on the support and prevented the nickel species from sintering leading to the high activity and good stability. In addition, the presence of oxygen vacancies on the support, such as CeO_2, will create the additional driving force for the CO_2 conversion to CO in reducing atmosphere. Results from [42] seem to indicate that $Ce_xZr_{1-x}O_2$ ($0.5 < x < 0.8$) solid solution has a superior performance in terms of overall reduction and total oxygen storage.

Secondly, choosing a porous support, Wei and Jinlong [43] had written an overview about methanation of carbon dioxide. The article focuses on recent developments in catalytic materials, novel reactors, and reaction mechanism for methanation of CO_2. The authors demonstrated that the different interactions that can be established between the metal and the support shall influence the catalytic properties of the active metal sites. Jwa et al. [33] who studied the hydrogenation of carbon oxides (CO and CO_2) into methane over Ni/β-zeolite catalysts have the same result. In order to increase catalytic activity of the methanation, it is necessary to enhance CO_2 supply at the surface of the catalyst. Some researchers have studied nickel supported on porous alumina [44] or MCM41 [45, 46] catalysts and their results showed that the porous structure of the supports improved the dispersion of the nickel species on their surfaces and prevented the nickel species from sintering. Recently, activated carbon has been investigated by various research groups because of its large surface area, surface functionalization, and low energy requirements for regeneration. Their results indicated that activated carbon (AC) is a promising adsorbent for CO_2, at ambient conditions [47–49]. Vargas et al. [47] studied carbon dioxide adsorption at 273 K on three series of activated carbon monoliths prepared by impregnation of African palm shells. Their results showed that the carbon monoliths obtained can adsorb as much CO_2 as 5.8 mmol CO_2 g^{-1} at 1 bar and 273 K. Wickramaratne et al. [48, 49] indicated that the activated carbon spheres exhibited very high CO_2 uptake of 8.9 and 4.55 mmol/g at 0°C and 25°C under atmospheric pressure, respectively. In the work of Li et al. [50] pine cone shell-based activated carbons were used to adsorb CO_2. The results indicated good CO_2 adsorption of performance of activated carbon with a high adsorption capacity of 7.63 mmol g^{-1} and 2.35 mmol g^{-1} at 0°C under 1 and 0.15 bar pressure, respectively. In our previous work [51], the activation of carbon dioxide (CO_2) by catalytic systems comprising a transition metal (Co, Cu, and Ni) on an activated carbon (AC) support was investigated using a combination of different theoretical calculation methods: Monte Carlo simulation, DFT and DFT-D, molecular dynamics (MD), and a climbing image nudged elastic band (CI-NEB) method. The results obtained indicate that CO_2 is easily adsorbed by Ni/AC. Usually, catalytic reaction properties can be affected by the catalyst composition and structure (e.g., specific surface area, pore size distribution, pore size, and structure). As is generally known, the support with high surface area will make the dispersion of active sites more easily and consequently a higher active surface area is generated. Highly dispersed supported nickel catalysts have been widely used in the hydrogenation of CO_2 to methane. Activated carbon, which is characterized by large specific surface areas ($>1000\,m^2\,g^{-1}$) and developed pore structures, has exhibited good catalytic properties, thus making it of great interest to researchers in the field of catalysis. The nickel supported on activated carbon used for CO_2 catalytic hydrogenation had not been reported to date; we believe that activated carbon is a good support in modifying the surface properties to promote the nickel catalyst activity for hydrogenation of CO_2.

In this article, the Ni/AC and Ni/$Ce_{0.2}Zr_{0.8}O_2$/AC catalysts with 7 wt.% nickel loadings were prepared by the incipient wetness impregnation. In these catalysts, nickel species are considered as active sites supported on Tra Bac activated carbon (AC) or on $Ce_{0.2}Zr_{0.8}O_2$/AC (mixed oxides $Ce_{0.2}Zr_{0.8}O_2$ deposited on AC). The catalysts and the supports were characterized by XRD, SEM, TEM, H2-TPR, and nitrogen adsorption-desorption. The activity and the CH_4 selectivity of the catalyst samples for the CO_2 methanation were also performed by a continuous flowing microreactor apparatus. $Ce_{0.2}Zr_{0.8}O_2$ mixed oxide was chosen because it promoted the dispersion of the nickel species on the supports and prevented the nickel species from sintering, leading to the high activity and the good stability. Activated carbon can act as a storage source of both H_2 and CO_2 and it helps in making the dispersion of nickel on the surface much easier. The highly dispersed nickel species are easily reduced and they are responsible for the high catalytic performance and for reducing the inactive carbon deposition. The goals of this study are to report the effects of CeO_2-ZrO_2 promoter and of the pore structure of activated carbon on the dispersion of nickel species, as well as the catalytic performances for CO_2 methanation. The possible reasons for the effect of $Ce_{0.2}Zr_{0.8}O_2$ promoter on the catalytic activity of the Ni/AC catalyst were given.

2. Experimental

2.1. Catalyst Preparation

2.1.1. Preparation of Ni Catalyst with Activated Carbon as the Support. The Ni/AC was prepared by incipient wetness impregnation method at a nickel loading of 7 wt%. Typically, 1.74 g nickel nitrate hexahydrate, $Ni(NO_3)_2 \cdot 6H_2O$ (99.0%, Merck), was dissolved in 30 mL distilled water. Then 5 g activated carbon support (coconut shell activated carbon was provided by Tra Bac factory, Vietnam) that was previously washed, crushed, and sieved to a size of 0.65–1 mm was added. The samples, subsequently, were dried in an oven at 60°C for 10 hours and continuously at 100°C for another

5 hours. Finally, the samples were calcined in N_2 environment at 400°C for 4 hours and then stored for further characterizations. The catalyst samples were denoted as 7 Ni/AC for weight percentage of 7% Ni.

2.1.2. Preparation of Ni Catalyst with $Ce_{0.2}Zr_{0.8}O_2/AC$ as the Support.

The mixed oxide $Ce_{0.2}Zr_{0.8}O_2$ was prepared using hydrothermal method as in the work of Pham et al. [52]. Typically, 1.6 mmol $Ce(NO_3)_3 \cdot 6H_2O$ (98.5%, Merck, Darmstadt, Germany) and 6.4 mmol $ZrOCl_2 \cdot 8H_2O$ (99.0%, Merck) were dissolved with 16 mmol urea-CH_4N_2O (98%, Merck) in 80 mL H_2O. The solution was then stirred until complete solubility. The obtained solution was poured into an autoclave, which was then maintained at 160°C for 24 h. The obtained light-yellow precipitate was washed with distilled water until constant pH and then dried at 80°C and finally calcined at 500°C for 4 hours.

$Ce_{0.2}Zr_{0.8}O_2$ was deposited on AC by suspension method: 5 g AC grains were immersed in 30 mL aqueous slurry of 20 wt% of $Ce_{0.2}Zr_{0.8}O_2$ powder, 20 vol% molten (70°C) Brij 56 (Sigma Aldrich, Steinheim, Germany), and 2.8 M HNO_3 and then dried and air blown. This coating and drying process was performed five times before calcination at 200°C for 4 h. The amount of $Ce_{0.2}Zr_{0.8}O_2$ on AC was determined by weighting the sample before (m_o) and after (m_t) the loading. The wt% loading was calculated as follows: wt% $= ((m_t - m_o)/m_o) \times 100$.

Nickel (the active phase, with the loading of 7 wt%) was deposited on the $Ce_{0.2}Zr_{0.8}O_2/AC$ samples by wet impregnation. Suitable amount (1.74 g) of $Ni(NO_3)_2 \cdot 6H_2O$ (99.0 wt%, Merck) was dissolved in 30 mL distilled water; then the $Ce_{0.2}Zr_{0.8}O_2/AC$ support was immersed in the prepared solution for 5 min. The wet pellets were dried until becoming completely dry. This procedure was repeated until all the solution ran out. Finally, the impregnated samples were heated at a heating rate of 3°C/min till 200°C and maintained at 200°C for 4 h. The catalyst sample was then symbolized as 7 Ni/CeZrAC.

2.2. Characterization of Catalysts.

X-ray powder diffraction (XRD) patterns of the samples were obtained in a X-ray diffractometer (D8 Advance-Bruker) using Cu Kα radiation with a wavelength of 0.154 nm from 10° to 70° with a step size of 0.03°. The data were compared to reference data from JCPDS or ICDD. The particle size calculations were performed using the Scherrer equation.

Brunauer–Emmett–Teller (BET) specific surface areas, average pore diameter, and pore volume of the samples were determined by N_2 adsorption-desorption isotherm at 77 K using the BET (Brunauer–Emmett–Teller) method in a Micromeritics Tristar 3000 instrument. Before each measurement, the sample was degassed at 523 K for 4 hours.

The scanning electron microscopy (SEM) studies of the catalysts were performed on a scanning electron microscope (Hitachi S-4800) apparatus with an accelerating voltage of 10.0 kV. The samples were placed onto a metallic support and covered with a thin platinum film.

Transmission electron microscopy (TEM) studies were performed on a JEOL JEM-2000FX II instrument operated at 80.0 kV. All samples were suspended in ethanol by ultrasonication. The suspension was deposited on a copper grid with carbon film for TEM measurements.

Temperature-programmed reduction (TPR) measurements were carried out with a Micromeritics AutoChem 2920 instrument in a quartz U tube microreactor. Prior to the reduction the sample (app. 40 mg) was purged with Ar (50 mL min^{-1}) for 1 hour at 423 K to remove physically adsorbed water and then cooled down to room temperature. Afterwards, the sample is reduced in the flow of 10 vol% H_2/Ar (50 mL min^{-1}) at a heating rate of 10 K min^{-1} up to 973 K. The consumption of hydrogen was detected with a thermal conductivity detector (TCD) during the TPR procedure.

2.3. Evaluation of Catalysts.

The gas phase hydrogenation of CO_2 to methane was carried out in a continuous-flow fixed-bed quartz reactor with an internal diameter of 1.5 mm under normal atmospheric pressure. A thermocouple was inserted into the catalyst bed to measure the reaction temperature. Typically for each run about 0.3 grams of catalyst pellets (similar size of 40–60 mesh) was loaded into a quartz reactor and reduced in situ under continuous flow of pure H_2 at the rate of 30 mL min^{-1}. The reduction temperature was programmed to increase from room temperature to 600°C and maintained at 600°C for 4 h. After reduction, the temperature was decreased to 100°C under the same hydrogenation flow and the catalyst was subsequently exposed to the feed gases CO_2/H_2/He with a molar ratio of $CO_2 : H_2 : He = 1 : 4 : 5$ at a gas hourly space velocity (GHSV) of 4000 mL·g$_{cat}^{-1}$·h^{-1} under atmospheric pressure. Catalytic activity was measured at 100, 200, 250, 300, 350, 400, and 450°C. At each temperature, after the stabilization of the catalytic system, three measures of CO, CO_2, and CH_4 were taken and an average value was calculated. The effluent gases were passed through a cold trap to condense water before being analyzed. The water level in the cold trap was low enough to prevent absorption of any gases. The analysis of evolved gases was conducted using an online GC (Trace-GC-RGA, Thermo Scientific) equipped with a thermal conductivity detector (TCD). The HayeSep Q capillary column (8′ × 1/8″ SS) is capable of separating CO_2 and C_1-C_2 paraffin and the Molecular Sieve 5A plot capillary column is capable of separating O_2, N_2, CH_4, and CO. During catalytic testing, carbon balances were calculated and were repeatedly between 97 and 99%.

Activity-selectivity data were obtained at steady-state conditions after 1 h of time on stream, at reaction temperatures. CO_2 conversion values (X_{CO_2}) were calculated by mass-balance method:

$$\text{Conversion of } CO_2 \text{ (\%)} = \frac{\text{moles of } CO_2 \text{ reacted}}{\text{moles of } CO_2 \text{ supplied}} \times 100$$

$$\text{Selectivity for methane (\%)} = \frac{\text{moles of } CH_4 \text{ formed}}{\text{moles of } CO_2 \text{ reacted}} \times 100$$

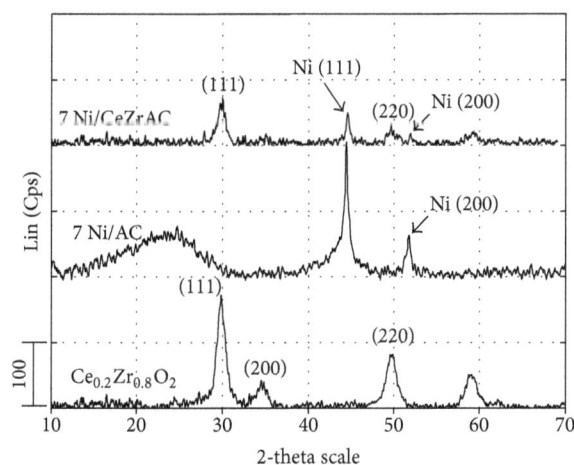

FIGURE 1: XRD patterns of $Ce_{0.2}Zr_{0.8}O_2$ and 7 Ni/AC and 7 Ni/CeZrAC reduced at 600°C for 4 hours.

Selectivity for CO (%)

$$= \frac{\text{moles of CO formed}}{\text{moles of } CO_2 \text{ reacted}} \times 100,$$

(2)

CH_4 formation rate is reported as number of molecules formed per unit time and per catalyst weight ($mmol\, h^{-1}\, g_{cat}^{-1}$).

Each data set was obtained with an accuracy of ±4%, from an average of two independent measurements.

3. Results and Discussion

3.1. XRD Characterization. The identification of the crystalline phases was carried out by XRD. The XRD patterns of 7 Ni/AC after reduction are presented in Figure 1. After reduction at 600°C in hydrogen atmosphere for 4 hours, all the reduced samples showed prominent peaks of metallic Ni at the $2\theta = 44.2°$ and 51° which are indexed to (111) and (200) diffractions planes, respectively. This matched the standard data for a cubic structure Ni (JCPDS 96-151-2527). No other peaks appear. It can be seen that the thermal treatment in H_2 at 600°C is sufficient for producing bulk Ni crystallites from nickel containing species.

The XRD patterns of mixed oxide $Ce_{0.2}Zr_{0.8}O_2$ and 7 Ni/CeZrAC after reduction were also displayed in Figure 1. The formation of $Ce_{0.2}Zr_{0.8}O_2$ on the AC was proved by XRD characterization. Phase transitions occurring in $Ce_xZr_{1-x}O_2$ depend on their composition. For pure ZrO_2 (Z) the diffraction peaks at $2\theta \approx 30.3°$, 34.6°, and 35.4° can be assigned to the tetragonal ZrO_2 structure (JCPDS 79-1769) and for pure CeO_2 (C) the diffraction peaks at $2\theta \approx 28.7°$ and 33.1° can be assigned to the cubic CeO_2 structure (JCPDS 65-5923). The work done by Hori et al. [53] showed that the tetragonal $Ce_xZr_{1-x}O_2$ phase appears with Ce < 50 mol% (C20Z), whereas above 50 mol%, a cubic $Ce_xZr_{1-x}O_2$ (C50Z and C80Z) phase is formed. With 80% Zr in our samples, the ceria peak originally at $2\theta \approx 28.6°$ is now at $2\theta \approx 30.0°$, which overlaps with a zirconia line at $2\theta \approx 30.1°$. In addition, we

detected a doublet at $2\theta \approx 34.3°$ and 34.8° which is close to the tetragonal zirconia peaks at $2\theta \approx 34.6°$ and 35.4°. The peak at $2\theta \approx 34.9°$ is clearly a tetragonal zirconia line, shifted down 0.5° due to doping by small amounts of Ce. The peak with $2\theta \approx 34.3°$ is in a region where a cubic ceria-zirconia line (shifted up from $2\theta \approx 33.1°$, in pure CeO_2) overlaps with a tetragonal zirconia-ceria line (shifted down from $2\theta \approx 34.6°$, in pure ZrO_2). This shift is indicative of change in lattice parameter, and it is evident that CeO_2 and ZrO_2 form a solid solution. The powder possesses the diffraction peaks at $2\theta = 30, 34.9, 49.7,$ and 58.5° related to the reflection planes (111), (200), (220), and (311) of $Ce_{0.2}Zr_{0.8}O_2$, respectively, showing the replacement of Zr atoms for Ce. Our measured lattice parameters are similar to those from reference materials [53]. Duwez and Odell [54] also obtained a tetragonal zirconia phase when high compositions of zirconia were used (around 80%), but for the sample containing 25% Zr they still obtained a cubic phase.

XRD pattern of 7 Ni/CeZrAC shows characteristic peaks of metallic Ni ($2\theta = 44.2°$ and 51°). There are also other diffraction peaks (at the 2θ values of 30, 49.7, and 59.6°) matching the standard data for a tetragonal mixed oxide $Ce_{0.2}Zr_{0.8}O_2$ (ICDD Card number 80-0785). The existence of ZrO_2 and CeO_2 or other species were not observed in XRD pattern of 7 Ni/CeZrAC. The presence of only tetragonal structure in the $Ce_{0.2}Zr_{0.8}O_2$ sample and in 7 Ni/CeZrAC as well indicates that Ce and Zr are highly homogeneously distributed. The approximate average crystallite sizes of mixed oxide $Ce_{0.2}Zr_{0.8}O_2$ in 7 Ni/CeZrAC sample and of pure $Ce_{0.2}Zr_{0.8}O_2$ were calculated by Scherrer's equation that indicates similar values. The approximate average crystallite sizes of Ni in catalyst samples were calculated from the (111) peak at 44,2° in the XRD patterns and Scherrer equation and are presented in Table 1. It can be seen that the Ni species dispersed well on the AC surface due to a high specific area of the support. However, the results show that the Ni particle size in 7 Ni/CeZrAC (17,39 nm) is smaller than that in 7 Ni/AC (21.82 nm) with pure AC as the support. This observation suggests that the dispersion of Ni species increases for the 7 Ni/CeZrAC catalyst due to the character of structural promoter of $Ce_xZr_yO_z$ mixed oxide. Since the XRD patterns exhibit identical 2θ angles, it can be said that the samples are completely reduced to metallic nickel without any detection of NiO_x phases, and the experimental procedure did not alter significantly the main crystalline phases of the samples.

3.2. N_2 Adsorption-Desorption Analysis (Table 2). BET surface areas, pore volume, and pore diameter of AC and of 7 Ni/AC reduced in H_2 at 600°C for 3 hours were listed in Table 2. It could be seen that activated carbon has a microporous structure and a developed specific surface area of 1159 $m^2\, g^{-1}$ with high microporous content (micro surface area is ~1139 m^2/g and microporous volume is ~0.5025 cm^3/g). The addition of nickel species resulted in slight decreases in surface areas and pore volume of the sample. This could be mainly attributed to a partial blockage of micropores by nickel species and the variation in mass

TABLE 1: Lattice parameters from XRD results and crystallite size of Ni (D) from Scherrer equation.

Catalyst samples		a	b	c	D (nm)
7 Ni/AC	Cubic	3.5350	3.5350	3.5350	21.82
7 Ni/CeZrAC:					
(i) Ni particle	Cubic	3.5240	3.5240	3.5240	17.39
(ii) $Ce_{0.2}Zr_{0.8}O_2$	Tetragonal	3.6325	3.6235	5.2288	8.78
Pure $Ce_{0.2}Zr_{0.8}O_2$	Tetragonal	3.6325	3.6325	5.2288	7.43

TABLE 2: Textural properties of AC, 7 Ni/AC, and 7 Ni/CeZrAC catalysts.

Catalyst samples	S_{BET} (m^{2}g^{-1})	S_{mic} (m^{2}g^{-1})	S_{ext} (m^{2}g^{-1})	V_{mic} (cm^{3}g^{-1})	V_{mes} (cm^{3}g^{-1})	$\overline{D}$ (nm)
AC	1159	1139	20	0.5025	0.0323	1.85
7 Ni/AC	923	908	15	0.3943	0.0302	1.84
7 Ni/CeZrAC	705	547	158	0.2404	0.3433	3.29

S_{BET}: BET surface area; S_{mic}, S_{ext}, and V_{mic}: micropore, external surface area, and micropore volume, calculated from t-plot method; V_{mes}: mesopore volume, calculated from BJH (Barrett-Joyner-Halenda) method; $\overline{D}$: Average pore width, calculated from $4V/S_{BET}$.

density of the catalyst. A decrease in the external surface area with the Ni loading was also observed, which could suggest that Ni species may deposit on the external surface of the support. However, for all samples, the active sites of the catalysts are accessible to the reactant molecules.

The textural properties of 7 Ni/CeZrAC which was reduced in H_2 at 600°C for 4 hours are also presented in Table 2. There was a strong decrease in surface area of AC when $Ce_{0.2}Zr_{0.8}O_2$ was introduced. It may be due to the deposition of $Ce_{0.2}Zr_{0.8}O_2$ on the AC surface, which blocked micropores leading to a strong decrease in micro surface area and a simultaneous increase in external surface areas and mesoporous volume. The wide pore diameter can provide favorable conditions for the reactant molecules to diffuse and transfer in the catalyst and it may be one reason for the better performance of the 7 Ni/CeZrAC in comparison to that of the 7 Ni/AC.

3.3. SEM and TEM Images. Morphologies of pure AC (Figure 2(a)) of mixed oxide $Ce_{0.2}Zr_{0.8}O_2$ (Figure 2(b)) and of catalyst sample 7 Ni/AC after reduction at 600°C for 4 hours (Figures 2(c) and 2(d)) were analyzed by SEM at 300 nm and 1.0 μm scales. For 7 Ni/AC it can be seen that the surface of the sample exhibits a high density block structure. SEM image showed well the existence of large cavities over the catalyst texture, likely originated from activated carbon surface (Figure 2(a)).

SEM images of 7 Ni/CeZrAC after reduction at 600°C for 4 hours were shown in Figures 2(e) and 2(f). A homogeneous distribution of spherical particles was obtained when $Ce_{0.2}Zr_{0.8}O_2$ was deposited on the AC surface. At higher magnification (Figure 2(f)), the catalyst showed morphology with spherical particles of about 50–60 nm. Further, the Ni particles could not be seen obviously on the support in SEM images, suggesting a better dispersion of Ni crystallite species that were doped into the ceria-zirconia solid solution.

TEM images of 7 Ni/AC and 7 Ni/CeZrAC after reduction at 600°C in H_2 atmosphere for 4 hours were shown in Figure 3

where black spots are Ni particles. It could be seen that Ni particles are well dispersed over $Ce_{0.2}Zr_{0.8}O_2$ layer which was deposited on the activated carbon support. The introduction of $Ce_{0.2}Zr_{0.8}O_2$ improved the formation of smaller particles. It can see easily that the particle's sizes from TEM results are in good agreement with XRD and BET analysis.

3.4. The Reducibility of the Catalysts. TPR-H_2 was carried out to study the reduction property of the catalysts. Figure 4 shows the TPR profiles of 7 Ni/CeZrAC and 7 Ni/AC. The TPR profiles of AC and $Ce_{0.2}Zr_{0.8}O_2$ are also presented for comparison. TPR profiles for studied samples display two distinct reduction bands in the temperature range of 160–350°C which can be attributed to the reduction of nickel species and another broad reduction band in the temperature range of 350–700°C corresponding to the reduction of the supports. In order to gain more insight into the TPR results, the profiles were deconvoluted into several Gaussian peaks. In the reduction profile of $Ce_{0.2}Zr_{0.8}O_2$ three peaks around 401–614°C are attributed to the reduction peaks of the surface oxygen and the bulk oxygen in $Ce_{0.2}Zr_{0.8}O_2$, respectively [26]. According to the literature [26], the existence of reduction peaks at temperatures below 600°C for the CeO_2 is assigned to the presence of surface and subsurface oxygen atoms, which are the main ones responsible for the improved CeO_2 oxygen storage capacity. It can be seen that these peaks are shifted to lower temperatures due to the presence of nickel indicating an existence of interaction between Ni and Ce. Similarly, two peaks around 638–693°C appearing in the profile of activated carbon are assigned for the reduction peaks of surface oxygen and bulk oxygen in AC and/or functional groups in AC. The addition of Ni species shifted these peaks to lower temperatures.

3.4.1. The Reduction of Ni Species in Ni-AC Catalysts. Regarding the reduction peaks of nickel species in 7 Ni/AC catalyst, it can be seen that three obvious reduction bands are observed: the first band (I) at the lowest-temperature with a maximum

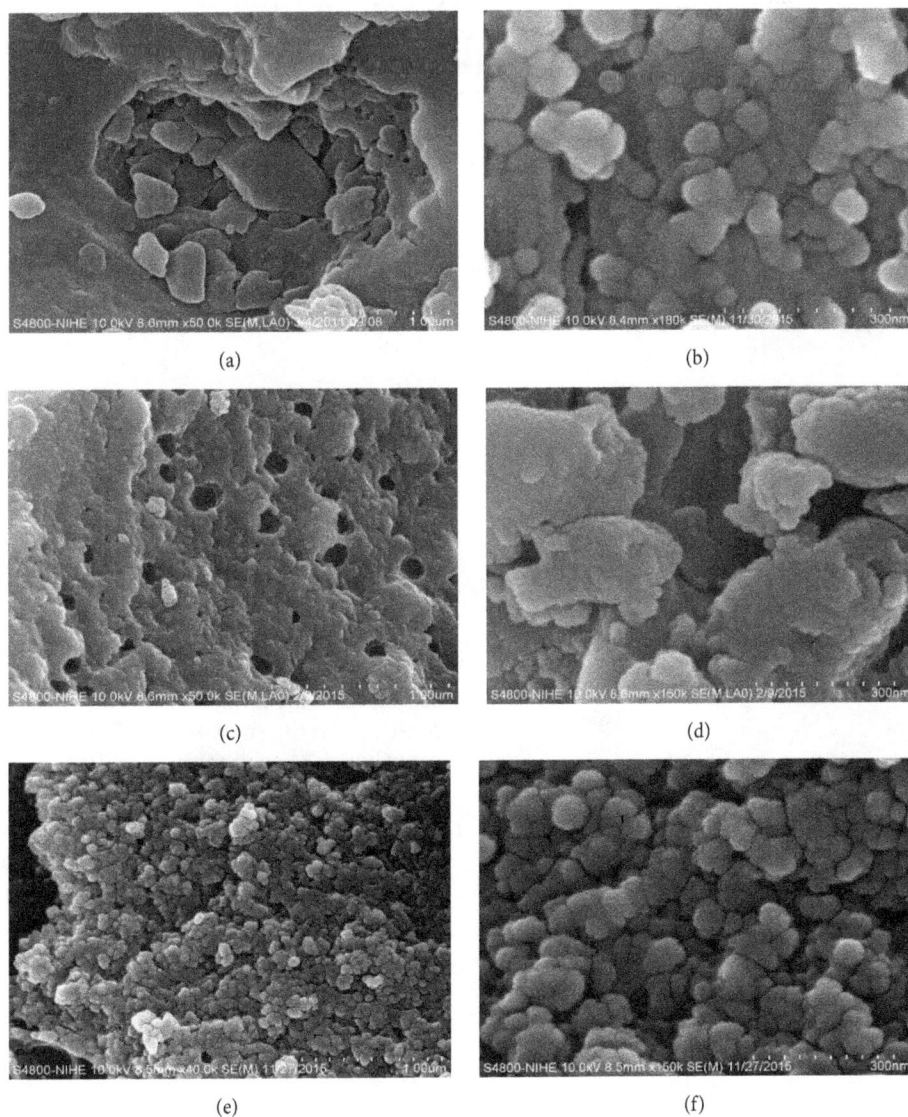

FIGURE 2: SEM images at 1.0 μm and 300 nm scales of AC (a), $Ce_{0.2}Zr_{0.8}O_2$ (b), 7 Ni/AC (c, d), and 7 Ni/CeZrAC (e, f).

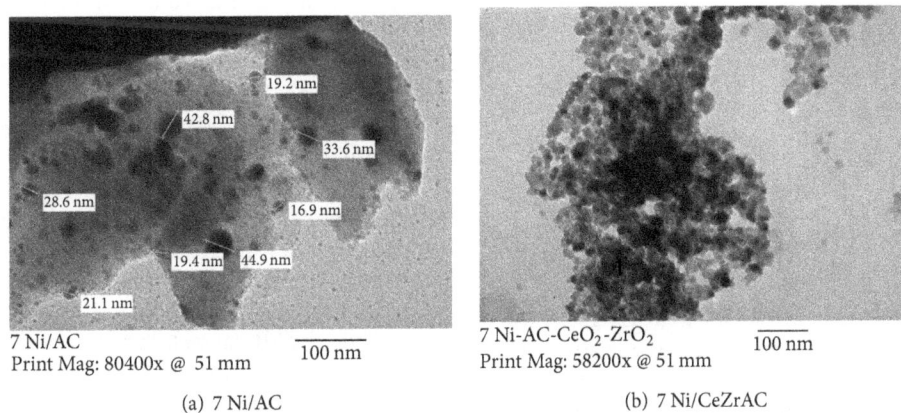

(a) 7 Ni/AC   (b) 7 Ni/CeZrAC

FIGURE 3: TEM images of reduced catalyst samples.

TABLE 3: T_{max} and consumed H_2 for TPR patterns over studied catalysts.

Samples	T_{max} (°C)	H_2 consuming (mmol g^{-1})	Total H_2 consuming for first three peaks (mmol g^{-1})
7 wt% Ni/AC (7 Ni/AC)	231.3	0.33	
	277.6	0.86	2.62
	357.4	1.43	
	489.4	4.97	—
	579.8	1.88	
7 wt% Ni/Ce$_{0.2}$Zr$_{0.8}$O$_2$/AC (7 Ni/CeZrAC)	260.6	0.34	
	290.7	1.66	3.54
	345.1	1.54	
	504.7	2.72	—
	577.2	1.88	
Ce$_{0.2}$Zr$_{0.8}$O$_2$	428.1	0.09	
	548.1	0.24	0.64
	614.2	0.31	
AC	637.7	0.80	1.65
	693.0	0.85	

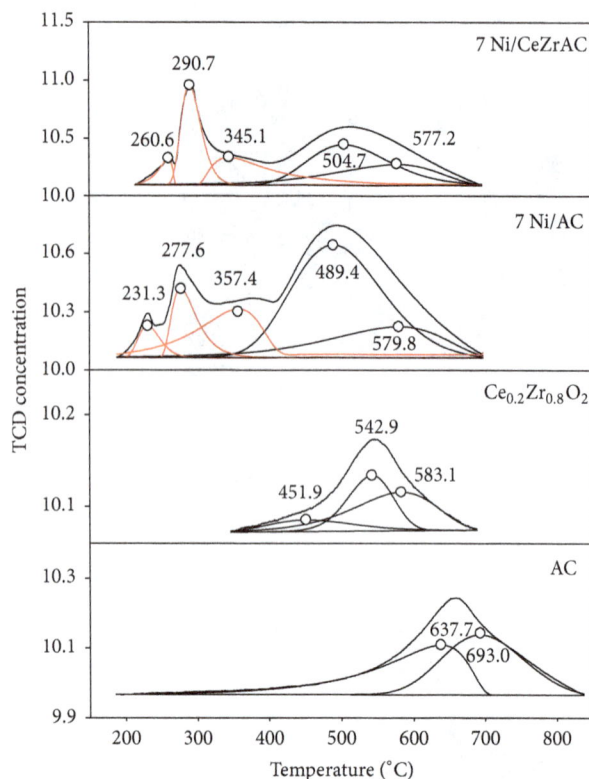

FIGURE 4: TPR-H_2 profiles of 7 Ni/AC and 7 Ni/CeZrAC.

around 228–231°C can be assigned to the well dispersed nickel species in the samples (may be assigned to the relatively free nickel species weakly interacting with support), which are easily reduced [29]. The low temperature band, the second band (II), shows a maximum at about 275–290°C, which may be due to the reduction of dispersed nickel species [55] and the third band (III) with a maximum around 343–357°C may

be related to the reduction of bulk nickel species in intimate contact with the support [56]. The peak positions and their contribution are summarized in Table 3.

3.4.2. The Reduction of Ni Species in Ni/CeZrAC Catalyst. It can be seen that the curve of 7 Ni/CeZrAC is similar to that for 7 Ni/AC but the first three reduction peaks slightly shift toward the higher temperatures than in the 7 Ni/AC, which indicates a higher interaction between nickel species and the support (Ce$_{0.2}$Zr$_{0.8}$O$_2$/AC). Since Ce$_{0.2}$Zr$_{0.8}$O$_2$ and AC are not reducible at the temperature range of 160–350°C (as shown in their TPR-H_2 profiles), these first three peaks are attributed to the reduction of nickel species in the samples. Although the maximum reduction temperatures are slightly higher compared to that in 7 Ni/AC, the concentration of nickel species which can be easily reduced increases for 7 Ni/CeZrAC sample, indicating that the presence of Ce and Zr helps the dispersion of active sites and hence improves the reducibility of the sample. This attribution is in good agreement with that reported by Xu and Wang [57].

H_2 consumption of the supports and catalysts was calculated (Table 3). The obtained results show that the H_2 consumption of 7 Ni/CeZrAC catalyst (3.54 mmol g^{-1}) is higher than sum of 7 Ni/AC (2.62 mmol g^{-1}) and Ce$_{0.2}$Zr$_{0.8}$O$_2$ (0.64 mmol g^{-1}). It has been shown that the metal-support interaction between cerium-zirconium oxides and nickel oxides promotes the reducibility of samples [58]. This intimate metal-support interaction also promotes the dispersion of nickel oxide.

Based on TPR data, all samples were pretreated in H_2 at 600°C for 4 hours before measuring the catalytic activity in CO_2 hydrogenation reactions.

3.5. Catalytic Performance. Prior to the evaluation of the studied catalyst samples, the blank test in the absence of the

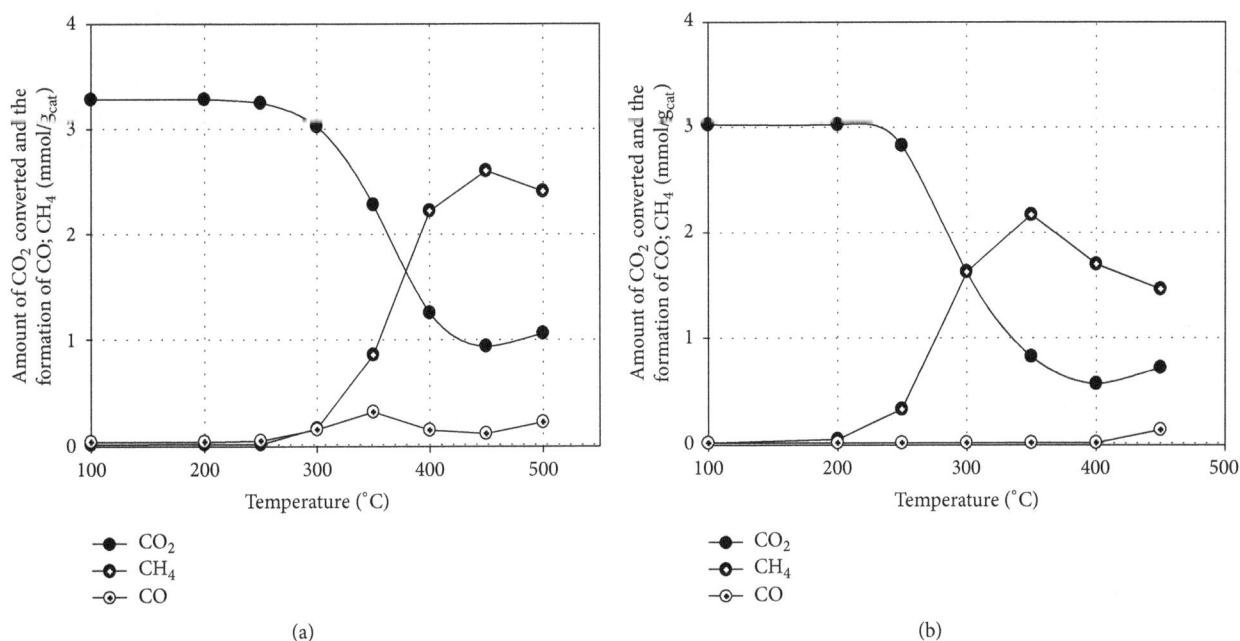

FIGURE 5: CO_2 conversion and CH_4 and CO selectivity versus temperature, at GHSV = 4000 mL/g_{cat}·h, 1 atm for 7 Ni/AC (a) and 7 Ni/CeZrAC (b).

catalyst sample was carried out in the range of 100–500°C and GHSV = $4000 \, mL \, h^{-1} \, g^{-1}$ (STP). The results showed that the blank reactor system was relative inert; only a negligible CO_2 conversion (<1%) could be detected under experimental conditions. Another two tests in the presence of pure AC and $Ce_{0.2}Zr_{0.8}O_2$/AC (without the presence of Ni species), respectively, were also performed under the same experimental conditions. It was found that not any CO_2 conversion and also CO or CH_4 were detected indicating that CH_4 and/or CO would be the products of CO_2 hydrogenation over studied Ni containing catalyst samples.

The catalytic activities of the samples were evaluated by analyzing the CO_2 conversion and CH_4 selectivity. In all experiments carried out only CH_4 and small amount of CO were detected at the outlet of the reactor; the carbon molar balances was about 97%.

Figures 5(a) and 5(b) present the CO_2 conversion and CH_4 and CO formation over 7 Ni/AC and 7 Ni/CeZrAC samples as a function of the reaction temperature.

3.5.1. The CO_2 Conversion.
As seen in Figure 5 for both two samples, the amounts of CO_2 in the gas mixture decreased as temperature increased. The temperature at which the amount of CO_2 started going down was 200°C over 7 Ni/AC and 170°C over 7 Ni/CeZrAC. These phenomena indicated the conversion of CO_2 occurred and the conversion gradually increases with the temperature up to 450°C (over 7 Ni/AC) and 400°C (over 7 Ni/CeZrAC), but as further rise in temperature, the CO_2 conversion starts going down.

3.5.2. The CH_4 and CO Formation.
There were two temperature ranges for product selectivity. At low temperature range

of 200°C to 400°C over 7 Ni/AC and 170°C to 350°C over 7 Ni/CeZrAC, the formation of CH_4 was dominant; no alcohols or other hydrocarbons could be expected to be formed. CO formation was accompanied with CH_4 but very slightly. A further increase in the temperature (up to 500°C) will result in the stable increase in CO formation with the decrease of CO_2 conversion and the decrease of selectivity to methane formation as well. These phenomena are related to the thermodynamic nature of the CO_2 hydrogenation reaction because CO formation can occur mostly by the reversed water gas shift reaction (RWGS) and a small contribution depending on temperature from steam reforming (SR) of methane.

These observations are in good agreement with the work done by Graça et al. [29] and by Janke et al. [10].

3.6. The Effects of the Mixed Oxide $Ce_{0.2}Zr_{0.8}O_2$ Addition.
A comparison in CO_2 conversion over 7 Ni/AC and 7 Ni/CeZrAC was made and shown Figure 6. It can be seen that the addition of solid solution $Ce_{0.2}Zr_{0.8}O_2$ is responsible for improvement of both CO_2 conversion and CH_4 selectivity: the conversion of CO_2 started at a lower temperature (170°C) compared to that of 7 Ni/AC sample (Figure 6), and it reached the maximum conversion value even at only 350°C. In the temperature range of 150–350°C no CO was detected at the outlet indicating a 100% for CH_4 selectivity (Figure 5(b)). This positive effect which shows the improvement of catalyst performance resulted from Ce-Zr well incorporation with Ni and AC surface.

It has been claimed in the literature that (Ce-Zr) species can activate CO_2 molecules and reduce them into CO due to the great mobility of the oxygen atoms. These CO species can be subsequently hydrogenated into methane. In the work

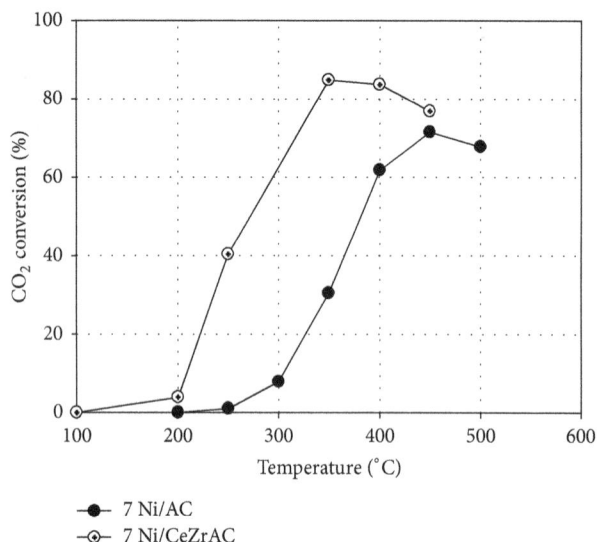

FIGURE 6: The comparison in CO_2 conversion between two Ni/AC samples with and without $Ce_{0.2}Zr_{0.8}O_2$: 7 Ni/AC and 7 Ni/CeZrAC.

done by Sharma et al. [13] Ru-doped ceria, $Ce_{0.95}Ru_{0.05}O_2$, prepared by a combustion method showed higher catalytic activity for CO_2 methanation than 5 wt% Ru/CeO_2 and the conversion of CO_2 and selectivity of CH_4 were 55% and 99%, respectively. By feeding 13% CO_2, 54% H_2, and 33% Ar at 450°C and GHSV = ca. 10,000 h^{-1}, Wang et al. [25] indicated that $Ni/Ce_{0.5}Zr_{0.5}O_2$ catalysts prepared by impregnation method possessed the highest activity for CO_2 hydrogenation. It can attain 73% conversion at 300°C and have a CH_4 selectivity of 100%.

The $Ni/Ce_{0.2}Zr_{0.8}O_2/AC$ catalyst prepared in the present work had an excellent activity for CO_2 methanation at lower temperatures and a high CH_4 selectivity of 100%. TPR-H_2 characterization indicated the intimate interaction between the metal and the support that could promote the reduction of $Ce_{0.2}Zr_{0.8}O_2$, while the strong interaction inhibits the reduction of Ni species. Swalus et al. [21] indicated that nickel supported on AC is able to activate high amount of hydrogen, while Rynkowski et al. [38] reported that $Ce_{0.2}Zr_{0.8}O_2$ could supply the surface oxygen sites for CO_2 adsorption. Also, our previous theoretical study [51] showed that nickel plays an important role in the dissociative adsorption of CO_2. In that work the adsorption of carbon dioxide on AC was studied in two steps: (i) GCMC (grand canonical ensemble Monte Carlo) simulation to determine the most favorable adsorption positions; (ii) these configurations optimized using the DFT and DFT-D2 methods. Results obtained from the GCMC simulation showed that the CO_2 molecule is most favorably physically adsorbed on the surface of AC. The preferred configurations were then optimized to determine the adsorption energy of CO_2 on AC (Eads). The results obtained using the DFT and DFT-D2 methods were −27.3 kJ/mol and −46.9 kJ/mol, respectively, which indicate that CO_2 is easily adsorbed on the AC. When Ni was doped on the AC surface, the CO_2 molecule was chemically adsorbed and the C-O bond is strongly activated after the adsorption of CO_2. The adsorption process of CO_2 did not involve a transition state. Our calculated results showed that CO_2 adsorption and dissociation are the first steps in the mechanism of CO_2. Jacquemin et al. [8] had the same statement that the first step of the mechanism in the methanation reaction could be the chemisorption of CO_2 on the catalyst and followed by the dissociation of CO_2 into CO and O adsorbed on the surface. From obtained results we suggested that using AC as a carrier may lead to a significant increase of the partial pressure of CO_2 on the surface of the catalyst. In other words, the conversion of CO_2 with high efficiency can be carried out at unusually low pressures due to the increased CO_2 partial pressure on the AC surface. Our results give an evidence that at low temperature and at atmospheric pressure, it is possible to obtain methane from hydrogenation of CO_2 when using an adequate catalyst. The work done by Beuls et al. [7] shows the same conclusion, but the catalyst used was $Rh/\gamma-Al_2O_3$.

4. Conclusion

The present work investigated the correlation between structural properties and catalytic performance of 7 Ni/AC and 7 $Ni/Ce_{0.2}Zr_{0.8}O_2/AC$ catalysts for CO_2 methanation reaction. The characterization of the samples by XRD, SEM, TEM, BET, and H_2-TPR techniques indicated that the dispersion of Ni species on the AC or $Ce_{0.2}Zr_{0.8}O_2/AC$ was influenced by the structure of the supports and $Ce_{0.2}Zr_{0.8}O_2/AC$ could stabilize the nickel species more effectively than AC. The characterized results suggested that Ni species interacted with $Ce_{0.2}Zr_{0.8}O_2/AC$ more strongly than that with AC, and compared with the AC support the $Ce_{0.2}Zr_{0.8}O_2/AC$ support had greater ability to facilitate the reduction of Ni species. The "synergistic effect" between the metal active sites (Ni), the promoter ($Ce_{0.2}Zr_{0.8}O_2$), and the support (AC) could promote the activation of adsorbed CO_2; therefore 7 $Ni/Ce_{0.2}Zr_{0.8}O_2/AC$ showed the higher activity toward hydrogenation of CO_2 to methane than 7 Ni/AC. Our results suggest that the use of dissociative chemisorption of CO_2 could probably allow decreasing reaction temperature and the methanation of CO_2 at low temperature could be a solution for the control of increasing emission of CO_2.

Conflicts of Interest

The authors declare that there are no conflicts of interest regarding the publication of this paper.

Acknowledgments

Financial support from Ministry of Training and Education, Vietnam, under Project no. B2013-17-38, is gratefully acknowledged.

References

[1] M. Burkhardt and G. Busch, "Methanation of hydrogen and carbon dioxide," *Applied Energy*, vol. 111, pp. 74–79, 2013.

[2] F. T. Zangeneh, S. Sahebdelfar, and M. T. Ravanchi, "Conversion of carbon dioxide to valuable petrochemicals: an approach

to clean development mechanism," *Journal of Natural Gas Chemistry*, vol. 20, no. 3, pp. 219–231, 2011.

[3] R. Ladera, F. J. Pérez-Alonso, J. M. González-Carballo, M. Ojeda, S. Rojas, and J. L. Fierro, "Catalytic valorization of CO_2 via methanol synthesis with Ga-promoted Cu–ZnO–ZrO$_2$ catalysts," *Applied Catalysis B: Environmental*, vol. 142-143, pp. 241-248, 2013.

[4] G. Bonura, M. Cordaro, C. Cannilla, F. Arena, and F. Frusteri, "The changing nature of the active site of Cu-Zn-Zr catalysts for the CO_2 hydrogenation reaction to methanol," *Applied Catalysis B: Environmental*, vol. 152-153, pp. 152–161, 2014.

[5] P. Sabatier and J. B. Senderens, "Hydrogénation directe des oxydes du carbone eu présence de divers métaux divisés," *Comptes Rendus de l'Académie des Sciences Paris*, vol. 134, p. 689, 1902.

[6] S. Rahmani, M. Rezaei, and F. Meshkani, "Preparation of highly active nickel catalysts supported on mesoporous nanocrystalline γ-Al$_2$O$_3$ for CO_2 methanation," *Journal of Industrial and Engineering Chemistry*, vol. 20, no. 4, pp. 1346–1352, 2014.

[7] A. Beuls, C. Swalus, M. Jacquemin, G. Heyen, A. Karelovic, and P. Ruiz, "Methanation of CO_2: further insight into the mechanism over Rh/γ-Al$_2$O$_3$ catalyst," *Applied Catalysis B: Environmental*, vol. 113-114, pp. 2–10, 2012.

[8] M. Jacquemin, A. Beuls, and P. Ruiz, "Catalytic production of methane from CO_2 and H_2 at low temperature: insight on the reaction mechanism," *Catalysis Today*, vol. 157, no. 1–4, pp. 462–466, 2010.

[9] M. Schoder, U. Armbruster, and A. Martin, "Heterogeneously catalyzed hydrogenation of carbon dioxide to methane at increased reaction pressures," *Chemie-Ingenieur-Technik*, vol. 85, no. 3, pp. 344–352, 2013.

[10] C. Janke, M. S. Duyar, M. Hoskins, and R. Farrauto, "Catalytic and adsorption studies for the hydrogenation of CO2 to methane," *Applied Catalysis B: Environmental*, vol. 152-153, no. 1, pp. 184–191, 2014.

[11] G. Zhou, T. Wu, H. Xie, and X. Zheng, "Effects of structure on the carbon dioxide methanation performance of Co-based catalysts," *International Journal of Hydrogen Energy*, vol. 38, no. 24, pp. 10012–10018, 2013.

[12] A. Karelovic and P. Ruiz, "Mechanistic study of low temperature CO_2 methanation over Rh/TiO$_2$ catalysts," *Journal of Catalysis*, vol. 301, pp. 141–153, 2013.

[13] S. Sharma, Z. Hu, P. Zhang, E. W. McFarland, and H. Metiu, "CO2 methanation on Ru-doped ceria," *Journal of Catalysis*, vol. 278, no. 2, pp. 297–309, 2011.

[14] J.-N. Park and E. W. McFarland, "A highly dispersed Pd-Mg/SiO$_2$ catalyst active for methanation of CO$_2$," *Journal of Catalysis*, vol. 266, no. 1, pp. 92–97, 2009.

[15] H. Y. Kim, H. M. Lee, and J. Park, "Bifunctional mechanism of CO_2 methanation on Pd-MgO/SiO$_2$ catalyst: independent roles of MgO and Pd on CO_2 methanation," *The Journal of Physical Chemistry C*, vol. 114, no. 15, pp. 7128–7131, 2010.

[16] D. C. D. Da Silva, S. Letichevsky, L. E. P. Borges, and L. G. Appel, "The Ni/ZrO$_2$ catalyst and the methanation of CO and CO$_2$," *International Journal of Hydrogen Energy*, vol. 37, no. 11, pp. 8923–8928, 2012.

[17] M. A. A. Aziz, A. A. Jalil, S. Triwahyono, R. R. Mukti, Y. H. Taufiq-Yap, and M. R. Sazegar, "Highly active Ni-promoted mesostructured silica nanoparticles for CO2 methanation," *Applied Catalysis B: Environmental*, vol. 147, pp. 359–368, 2014.

[18] R. Razzaq, H. Zhu, L. Jiang, U. Muhammad, C. Li, and S. Zhang, "Catalytic methanation of CO and CO2 in coke oven gas over Ni-Co/ZrO2-CeO2," *Industrial and Engineering Chemistry Research*, vol. 52, no. 6, pp. 2247–2256, 2013.

[19] C. Deleitenburg and A. Trovarelli, "Metal-support interactions in Rh/CeO2, Rh/TiO2, and Rh/Nb2O5 catalysts as inferred from CO2 methanation activity," *Journal of Catalysis*, vol. 156, no. 1, pp. 171–174, 1995.

[20] K. Zhao, Z. Li, and L. Bian, "CO2 methanation and co-methanation of CO and CO2 over Mn-promoted Ni/Al2O3 catalysts," *Frontiers of Chemical Science and Engineering*, vol. 10, no. 2, pp. 273–280, 2016.

[21] C. Swalus, M. Jacquemin, C. Poleunis, P. Bertrand, and P. Ruiz, "CO2 methanation on Rh/γ-Al2O3 catalyst at low temperature: "in situ" supply of hydrogen by Ni/activated carbon catalyst," *Applied Catalysis B: Environmental*, vol. 125, pp. 41–50, 2012.

[22] M. Cai, J. Wen, W. Chu, X. Cheng, and Z. Li, "Methanation of carbon dioxide on Ni/ZrO$_2$-Al$_2$O$_3$ catalysts: effects of ZrO$_2$ promoter and preparation method of novel ZrO$_2$-Al$_2$O$_3$ carrier," *Journal of Natural Gas Chemistry*, vol. 20, no. 3, pp. 318–324, 2011.

[23] J. Ren, X. Qin, J.-Z. Yang et al., "Methanation of carbon dioxide over Ni-M/ZrO$_2$ (M = Fe, Co, Cu) catalysts: effect of addition of a second metal," *Fuel Processing Technology*, vol. 137, pp. 204–211, 2015.

[24] L. Liu, Z. Yao, B. Liu, and L. Dong, "Correlation of structural characteristics with catalytic performance of CuO/CexZr1–xO2 catalysts for NO reduction by CO," *Journal of Catalysis*, vol. 275, no. 1, pp. 45–60, 2010.

[25] S. Wang, Q. Pan, J. Peng, T. Sun, D. Gao, and S. Wang, "CO2 methanation on Ni/Ce0.5Zr0.5O2 catalysts for the production of synthetic natural gas," *Fuel Processing Technology*, vol. 123, pp. 166–171, 2014.

[26] F. Ocampo, B. Louis, A. Kiennemann, and A. C. Roger, "CO$_2$ methanation over Ni-Ceria-Zirconia catalysts: effect of preparation and operating conditions," *IOP Conference Series: Materials Science and Engineering*, vol. 19, no. 1, Article ID 012007, 2011.

[27] S. Scirè, C. Crisafulli, R. Maggiore, S. Minicò, and S. Galvagno, "Influence of the support on CO2 methanation over Ru catalysts: an FT-IR study," *Catalysis Letters*, vol. 51, no. 3-4, pp. 41–45, 1998.

[28] S. Eckle, H.-G. Anfang, and R. J. Behm, "Reaction intermediates and side products in the methanation of CO and CO$_2$ over supported Ru catalysts in H$_2$-rich reformate gases," *Journal of Physical Chemistry C*, vol. 115, no. 4, pp. 1361–1367, 2011.

[29] I. Graça, L. V. González, M. C. Bacariza et al., "CO$_2$ hydrogenation into CH$_4$ on NiHNaUSY zeolites," *Applied Catalysis B: Environmental*, vol. 147, pp. 101–110, 2014.

[30] S. Tada, T. Shimizu, H. Kameyama, T. Haneda, and R. Kikuchi, "Ni/CeO$_2$ catalysts with high CO$_2$ methanation activity and high CH$_4$ selectivity at low temperatures," *International Journal of Hydrogen Energy*, vol. 37, no. 7, pp. 5527–5531, 2012.

[31] H. Takano, K. Izumiya, N. Kumagai, and K. Hashimoto, "The effect of heat treatment on the performance of the Ni/(Zr-Sm oxide) catalysts for carbon dioxide methanation," *Applied Surface Science*, vol. 257, no. 19, pp. 8171–8176, 2011.

[32] S. Hwang, J. Lee, U. G. Hong et al., "Methanation of carbon dioxide over mesoporous Ni–Fe–Ru–Al2O3 xerogel catalysts: effect of ruthenium content," *Journal of Industrial and Engineering Chemistry*, vol. 19, no. 2, pp. 698–703, 2013.

[33] E. Jwa, S. B. Lee, H. W. Lee, and Y. S. Mok, "Plasma-assisted catalytic methanation of CO and CO_2 over Ni–zeolite catalysts," *Fuel Processing Technology*, vol. 108, pp. 89–93, 2013.

[34] H. Liu, X. Zou, X. Wang, X. Lu, and W. Ding, "Effect of CeO_2 addition on Ni/Al_2O_3 catalysts for methanation of carbon dioxide with hydrogen," *Journal of Natural Gas Chemistry*, vol. 21, no. 6, pp. 703–707, 2012.

[35] J. Liu, W. Bing, X. Xue et al., "Alkaline-assisted Ni nanocatalysts with largely enhanced low-temperature activity toward CO_2 methanation," *Catalysis Science and Technology*, vol. 6, no. 11, pp. 3976–3983, 2016.

[36] A. Trovarelli, C. Deleitenburg, G. Dolcetti, and J. L. Lorca, "CO_2 methanation under transient and steady-state conditions over Rh/CeO_2 and CeO_2-promoted Rh/SiO_2: the role of surface and bulk ceria," *Journal of Catalysis*, vol. 151, no. 1, pp. 111–124, 1995.

[37] A. Trovarelli, C. De Leitenburg, and G. Dolcetti, "CO and CO_2 hydrogenation under transient conditions over Rh–CeO_2: novel positive effects of metal–support interaction on catalytic activity and selectivity," *Journal of the Chemical Society, Chemical Communications*, no. 7, pp. 472–473, 1991.

[38] J. M. Rynkowski, T. Paryjczak, A. Lewicki, M. I. Szynkowska, T. P. Maniecki, and W. K. Jóźwiak, "Characterization of Ru/CeO_2-Al_2O_3 catalysts and their performance in CO_2 methanation," *Reaction Kinetics and Catalysis Letters*, vol. 71, no. 1, pp. 55–64, 2000.

[39] F. Ocampo, B. Louis, L. Kiwi-Minsker, and A.-C. Roger, "Effect of Ce/Zr composition and noble metal promotion on nickel based $Ce_xZr_{1-x}O_2$ catalysts for carbon dioxide methanation," *Applied Catalysis A: General*, vol. 392, no. 1-2, pp. 36–44, 2011.

[40] P. A. U. Aldana, F. Ocampo, K. Kobl et al., "Catalytic CO_2 valorization into CH_4 on Ni-based ceria-zirconia. Reaction mechanism by operando IR spectroscopy," *Catalysis Today*, vol. 215, pp. 201–207, 2013.

[41] F. Ocampo, B. Louis, and A. Roger, "Methanation of carbon dioxide over nickel-based $Ce_{0.72}Zr_{0.28}O2$ mixed oxide catalysts prepared by sol–gel method," *Applied Catalysis A: General*, vol. 369, no. 1-2, pp. 90–96, 2009.

[42] H. C. Yao and Y. F. Y. Yao, "Ceria in automotive exhaust catalysts: I. Oxygen storage," *Journal of Catalysis*, vol. 86, no. 2, pp. 254–265, 1984.

[43] W. Wei and G. Jinlong, "Methanation of carbon dioxide: an overview," *Frontiers of Chemical Science and Engineering*, vol. 5, no. 1, pp. 2–10, 2011.

[44] L. Xu, H. Zhao, H. Song, and L. Chou, "Ordered mesoporous alumina supported nickel based catalysts for carbon dioxide reforming of methane," *International Journal of Hydrogen Energy*, vol. 37, no. 9, pp. 7497–7511, 2012.

[45] J. Zhang, Z. Xin, X. Meng, and M. Tao, "Synthesis, characterization and properties of anti-sintering nickel incorporated MCM-41 methanation catalysts," *Fuel*, vol. 109, pp. 693–701, 2013.

[46] G. Du, S. Lim, Y. Yang, C. Wang, L. Pfefferle, and G. L. Haller, "Methanation of carbon dioxide on Ni-incorporated MCM-41 catalysts: the influence of catalyst pretreatment and study of steady-state reaction," *Journal of Catalysis*, vol. 249, no. 2, pp. 370–379, 2007.

[47] D. P. Vargas, L. Giraldo, and J. C. Moreno-Piraján, "CO_2 adsorption on activated carbon honeycomb-monoliths: a comparison of langmuir and tóth models," *International Journal of Molecular Sciences*, vol. 13, no. 12, pp. 8388–8397, 2012.

[48] A. Samanta, A. Zhao, G. K. H. Shimizu, P. Sarkar, and R. Gupta, "Post-combustion CO_2 capture using solid sorbents: a review," *Industrial & Engineering Chemistry Research*, vol. 51, no. 4, pp. 1438–1463, 2012.

[49] N. P. Wickramaratne and M. Jaroniec, "Activated carbon spheres for CO_2 adsorption," *ACS Applied Materials and Interfaces*, vol. 5, no. 5, pp. 1849–1855, 2013.

[50] K. Li, S. Tian, J. Jiang, J. Wang, X. Chen, and F. Yan, "Pine cone shell-based activated carbon used for CO_2 adsorption," *Journal of Materials Chemistry A*, vol. 4, no. 14, pp. 5223–5234, 2016.

[51] N. N. Ha, N. T. T. Ha, L. Van Khu, and L. M. Cam, "Theoretical study of carbon dioxide activation by metals (Co, Cu, Ni) supported on activated carbon," *Journal of Molecular Modeling*, vol. 21, no. 12, article 322, 2015.

[52] P. T. M. Pham, M. T. Le, T. T. Nguyen, E. Bruneel, and I. Van Driessche, "The influence of deposition methods of support layer on cordierite substrate on the characteristics of a MnO_2–NiO–Co_3O_4/$Ce_{0.2}Zr_{0.8}O_2$/cordierite three way catalyst," *Materials*, vol. 7, no. 9, pp. 6237–6253, 2014.

[53] C. E. Hori, H. Permana, K. Y. S. Ng et al., "Thermal stability of oxygen storage properties in a mixed CeO_2-ZrO_2 system," *Applied Catalysis B: Environmental*, vol. 16, no. 2, pp. 105–117, 1998.

[54] P. Duwez and F. Odell, "Phase relationships in the system zirconia—ceria," *Journal of the American Ceramic Society*, vol. 33, no. 9, pp. 274–283, 1950.

[55] Q. Huang, X. Xue, and R. Zhou, "Influence of interaction between CeO2 and USY on the catalytic performance of CeO2–USY catalysts for deep oxidation of 1,2-dichloroethane," *Journal of Molecular Catalysis A: Chemical*, vol. 331, no. 1-2, pp. 130–136, 2010.

[56] E. Sahle-Demessie, V. G. Devulapelli, and A. A. Hassan, "Hydrogenation of anthracene in supercritical carbon dioxide solvent using Ni supported on Hβ-Zeolite catalyst," *Catalysts*, vol. 2, no. 1, pp. 85–100, 2012.

[57] S. Xu and X. Wang, "Highly active and coking resistant Ni/CeO_2–ZrO_2 catalyst for partial oxidation of methane," *Fuel*, vol. 84, no. 5, pp. 563–567, 2005.

[58] F. Cova, D. G. A. Pintos, A. Juan, and B. Irigoyen, "A first-principles modeling of Ni interactions on CeO_2–ZrO_2 mixed oxide solid solutions," *The Journal of Physical Chemistry C*, vol. 115, pp. 7456–7465, 2011.

7

Biomass Modification Using Cationic Surfactant Cetyltrimethylammonium Bromide (CTAB) to Remove Palm-Based Cooking Oil

Amira Satirawaty Mohamed Pauzan(ID) **and Normala Ahad**

Faculty of Resource Science and Technology, Universiti Malaysia Sarawak, 94300 Kota Samarahan, Sarawak, Malaysia

Correspondence should be addressed to Amira Satirawaty Mohamed Pauzan; mpasatirawaty@unimas.my

Academic Editor: Cesar Mateo

Adsorption based on natural fibre seems to widely used for oily wastewater recovery due to its low cost, simplicity, feasibility, easy handling, and effectiveness. However, oil sorbent based on natural fibre without modification has low adsorption capacity and selectivity. Thus, this paper proposes chemical modification of sago hampas to improve its adsorbent efficiency for the removal of palm-based cooking oil. The chemical modification was performed using a cationic surfactant, cetyltrimethylammonium bromide (CTAB). The chemical and surface properties of both unmodified and modified sago hampas were characterized by Fourier-Transform Infrared (FTIR) and Scanning Electron Microscopy (SEM). Parameters studied for the removal of cooking oil using modified sago hampas were sorption time, adsorbent dosage, and initial pH. The removal capacity was also compared using unmodified sago hampas. The results showed that additional functional groups were introduced on the surface of modified sago hampas. Modified sago hampas also showed a greater porosity than unmodified sago hampas. These properties enhanced the adsorption of palm-based cooking oil onto the surface of modified sago hampas. Modified sago hampas shows better removal of palm-based cooking oil than unmodified sago hampas, where 84.82% and 68.08% removal were achieved by modified and unmodified sago hampas, respectively. The optimum adsorption of palm-based cooking oil was identified at 45 min sorption time, pH 2, and 0.2 g adsorbent dosage.

1. Introduction

Oily wastewater is a big issue in the world since it is a persistent environmental pollutant [1]. Oily wastewater can come from a number of different sources such as waste cooking oil that enter the water from kitchen. River and lake contaminated with oil can have devastating effects on the water environment. Many methods have been introduced to reduce and minimize oily wastewater but the quality of the water seem to be far from satisfactory and contribute to serious physical effect to ecology. Oil is dispersed over the water surface and form oil layer. It could reduce oxygen supply to the water and eventually lead to death of aquatic life forms. There are a few methods that have been developed to reduce and minimize oil contaminant in water including physical, mechanical, biological and photochemical recovery, and filtration but these methods are not very efficient in removing different types of oil and slow in removal of oil, and it is of high cost [2].

Adsorption process have been reported as the most efficient method since it does not require additional chemical and large amount of energy particularly adsorption by activated carbon. However, activated carbon is difficult to regenerate due to its strong interaction with adsorbed molecule [3].

A new method has been introduced which is adsorption using natural fibres for oily wastewater where the method is very effective, simple, and low cost [4]. Many natural fibres have been utilized as oil sorbents such as sugarcane baggase, sawdust, barley straw, rice husk, wool, kapok, and grass [5–8]. These natural fibres are the best oil sorbent since it is low cost, sustainable, and abundant compared to the other sorbent.

Sago hampas is a natural fibre scientifically known as *Metroxylon sagu*. It is the main commodity crop of Sarawak, Malaysia, and potentially can be used to remove oil [9]. Sago comes from genus *Metroxylon* and Palmae family [10]. Sago hampas is highly degradable, has good sorption capacity, chemical-free, and has comparable density with the synthetic sorbent. However, the disadvantages of natural sorbents are lack of hydrophobicity and low buoyancy [11]. This can cause reduction in effectiveness of oil sorption in the aqueous system. It is widely known that unmodified or raw sago hampas with lack of hydrophobicity can be improved by either chemical or physical modifications. The chemical modification of sago hampas by attaching various functional groups is easy to conduct since it contains lignin and cellulose with high amount of hydroxyl functional group that is prone to chemical modification [12].

Although in some recent studies good adsorption performance to remove various metal ions has been reported for chemical modification of biomass [13, 14] or dye removal [15], no studies reported on the chemical modification by cationic surfactant on sago hampas to improve its performance for removal of oily wastewater. This paper aims to investigate the adsorption capacity of cetyltrimethylammonium bromide (CTAB) surfactant-modified sago hampas on palm-based cooking oil at optimum parameters studied which are sorption time, adsorbent dosage, and initial pH. The used of chemically modified sago hampas for oil adsorption is one of the ways that has the potential to reduce oily wastewater in the environment.

2. Materials and Methods

The absorption capacity of palm-based cooking oil by surfactant-modified sago hampas involves two steps. In the first step, the surface properties of sago hampas were modified using CTAB surfactant. Next, optimum removal of palm-based cooking oil using modified sago hampas was studied at three different parameters: sorption time, adsorbent dosage, and initial pH. The detailed steps involved in this study are described next.

2.1. Modification of Sago Hampas. Collected sago hampas from Mukah, Sarawak, was washed repeatedly with distilled water to remove any impurities or dust. It was air-dried to remove excess water and then further dried at about 80°C for 24 h in an air circulating oven. The dried sago hampas was ground to a fine powder and sieved (500 μm) before modification to have reproducible result and uniform modification. The dried sago hampas was pretreated with 1 M KOH solution to improve its binding site. Five grams of sago hampas was immersed into 100 mL 1 M KOH. The mixture was stirred for 24 h at 30°C prior to filtration to achieve good penetration of chemical into the interior of the precursor [16]. Then, the pretreated samples were washed with hot water (80°C), mild acid (0.1 M HCl), and base (0.1 M NaOH) till the effluent water shows neutral pH. Then, the samples were dried at 70°C overnight. After pretreated by KOH, sago hampas was modified using CTAB solution. Two grams of pretreated sago hampas was immersed into 100 mL of 0.1 M CTAB solution and stirred for 24 h [17]. The mixture was filtered and washed repeatedly by distilled water

in order to remove unreacted surfactant. The sample was dried in an oven at 50°C, sieved, and used for characterization and batch adsorption tests.

FTIR (Thermo Scientific/Nicolet iS10) analysis was done on unmodified sago hampas and modified sago hampas to determine the functional groups that are present before and after the modification process. The functional group of pure CTAB and modified sago hampas loaded with oil at optimum condition also has been determined. The unmodified and modified sago hampas was ground with KBr and made into discs. The spectrum was recorded on FTIR in the spectral range 4000–500 cm^{-1} under ambient condition.

The morphology of sago hampas before and after the modification by using cationic surfactant, CTAB, was observed by scanning electron microscopy (SEM) (JSM-6390lA) at 500x and 1000x magnification. The sample was placed on an aluminium stub using double stick. Under a vacuum condition, a thick gold layer was sputtered on the surface by using sputter coater in order to make the sample surface conductive. SEM imaging was used to analyse the morphology of the sample. The difference in the surface structure and porosity was identified between unmodified and modified sago hampas.

2.2. Batch Adsorption Experiments. Batch adsorption experiments were performed, and the mass of palm oil before and after the sorption was measured. In adsorption experiments, 5 mL (4.45 g) of palm oil purchased from local supermarket was used. The initial mass of 250 μm size of adsorbent was measured. Each sample was agitated by a rotary shaker (Heidolph Rotamax 120) at 120 rpm at respective time to reach equilibrium of the solid-solution mixture. The samples were collected and filtered. The sample loaded with oil was air-dried until no further filtered oil was observed. Further drying was conducted at about 60°C for 30 min to evaporate water and *n*-hexane that might be trapped on sago hampas during adsorption process [18]. The final mass of sago hampas was recorded. All experiments were carried out in triplicate for each condition. The adsorption capacity was calculated using the following formula:

$$\text{Adsorption capacity}\,(g/g) = \frac{(S_t - S_o)}{S_o}. \tag{1}$$

The percentage of oil removal will be determined using the following formula:

$$\text{Oil removal}\,(\%) = \frac{(m_{\text{oil}})}{(m_{\text{oil,o}})} \cdot 100, \tag{2}$$

where m_{oil} and $m_{\text{oil,o}}$ are the amount of palm oil extracted from sago hampas (g) and initial amount of palm oil (g), respectively. S_t is the total mass of absorb sample and S_o is the initial weight of sample. This procedure was applied to those of adsorption tests for modified sago hampas and unmodified sago hampas.

2.3. Effect of Sorption Time. Different sorption times which are 2 min, 5 min, 10 min, 15 min, 30 min, 45 min, and 60 min were carried out for sorption experiment. The original pH of palm oil was not adjusted. Adsorbent dosage used was at 0.20 g, and it was spread evenly on top of 5 mL palm oil solution. It was agitated at 120 rpm. The highest adsorption capacity obtained was used to compare the efficiency between unmodified and modified sago hampas.

2.4. Effect of Adsorbent Dosage. Sorption experiment was studied at different dosages of modified sago hampas which are 0.1 g, 0.2 g, 0.3 g, and 0.4 g. The study was carried out for 1 h with 5 mL of palm oil solution. The original pH of palm oil was not adjusted. It was agitated at 120 rpm. The highest adsorption capacity obtained was used to compare the efficiency between unmodified and modified sago hampas.

2.5. Effect of Initial pH. The adsorption of palm oil by the modified sago hampas was carried over pH 0.5, 1, 2, 4, 6, 8, and 10 at room temperature for 1 h. Adsorbent dosage used was at 0.20 g and spread evenly on top of 5 mL palm oil. It was agitated at 120 rpm. A few drops of 0.1 M sodium hydroxide solution (NaOH) or 1 M hydrochloric acid solution (HCl) was added to each flask containing 0.20 g adsorbent and 5 mL palm oil to adjust the pH. A little amount of *n*-hexane was added in palm oil in order to solubilize the HCl in palm oil. The pH was measured by using pH meter. The highest adsorption capacity obtained was used to compare the efficiency between unmodified and modified sago hampas.

2.6. Effect of Modified and Unmodified Sago Hampas. The effect of modified and unmodified sago hampas on the adsorption of palm-based cooking oil was studied to compare the removal efficiency of palm oil by using both unmodified and modified sago hampas. Both experiments were carried out by using the optimum adsorption capacity on modified sago hampas. It was carried out with 0.2 g of adsorbent dosage with 250 μm size that was spread evenly on top of 5 mL palm oil solution. It was agitated at 120 rpm agitation speed for 45 min. It was conducted at room temperature with pH 2 of palm oil.

3. Results and Discussion

3.1. Scanning Electron Microscopy (SEM) Analyses. The morphology of sago hampas before and after modification by using cationic surfactant, CTAB, was observed by scanning electron microscopy (SEM) at 500x and 1000x magnification. At such magnification, the surface structures and porosity can be clearly seen. The surface morphology of unmodified sago hampas was different from modified sago hampas because the modification significantly alters the porosity and physicochemical properties of the materials [19]. After modification by using CTAB, the surface roughness was increased with the formation of pores throughout the structure. The SEM analyses revealed that modified sago hampas contained numerous pores, which can hold oil. The pore distributed is different in sizes. Figure 1 presents the SEM analyses of (a) unmodified sago

hampas and (b) modified sago hampas at 500x and 1000x magnification.

3.2. Fourier-Transform Infrared (FTIR) Analyses. Fourier-transform infrared (FTIR) analysis was used to investigate the effect of the modification on the surface structure of sago hampas. It was done upon unmodified sago hampas and modified sago hampas in order to determine the functional group that is present before and after modification by using CTAB. The functional group that is present in CTAB and modified sago hampas loaded with oil at optimum condition was also being observed.

FTIR spectrum of pure CTAB showed bands at 3020–2800 cm^{-1} attributed to C-H stretching bands of alkylammonium cations. The band at 1472.27 cm^{-1} is an indicative of high organization of the -CH$_2$ chain conformation. The FTIR spectrum of pure CTAB also shows the C-N stretching band at 950–900 cm^{-1}.

Both FTIR spectra of modified and unmodified sago hampas show C-H deformation of the methyl group stretching at 1325–1320 cm^{-1}. The C-O stretching vibration at 1000 cm^{-1} is due to the cellulose backbone [5]. Beyond the fingerprint region, a characteristic broadband in both modified and unmodified occurs in the range of 3600–3100 cm^{-1}, which corresponds to the stretching mode of O-H groups, hydrogen-bonded O-H, and chemisorbed water [16]. FTIR spectrum of modified sago hampas reveals more distribution peaks than that are found in unmodified sago hampas. Modified sago hampas have two sharp bands at 3000–2850 cm^{-1} that refer to the asymmetric and symmetric stretching vibrations of the -CH$_3$ and -CH$_2$ groups in CTAB [20]. The peak at 1700–1500 cm^{-1} shows the C=O of the carboxylic group of celluloses and lignin was seen in unmodified sago hampas but becomes weaker at modified sago hampas due to the alkali pretreatment [21]. These constituents cannot be completely removed in the dilute alkali solution [22]. Moreover, modified sago hampas underwent the -CH$_2$ vibration at 1366.03 cm^{-1} due to the modification by using CTAB [23]. Extra peak that is present at 900.36 cm^{-1} in modified sago hampas shows the presence of C-N stretching vibration support the modification of sago hampas by CTAB [24].

FTIR spectrum for modified sago hampas loaded with oil at optimum condition having peaks at 2920.35 cm^{-1} and 2851.53 cm^{-1} was observed to be stronger than modified sago hampas. It demonstrated the adsorption of palm oil to alkyl chain layer that contributed by CTAB on the surface of modified sago hampas [25]. The weaker peak at 1100–1000 cm^{-1} on modified sago hampas after adsorption of oil is probably because the interaction between the cellulose backbone (C-O) with palm oil at optimum condition. The extra peak between 1800 cm^{-1} and 1700 cm^{-1} on modified sago hampas after adsorption shows the present of C=O which was contributed by ester from palm oil [26].

3.3. Batch Adsorption Experiments. Batch adsorption experiments were carried out with fixed size of modified sago

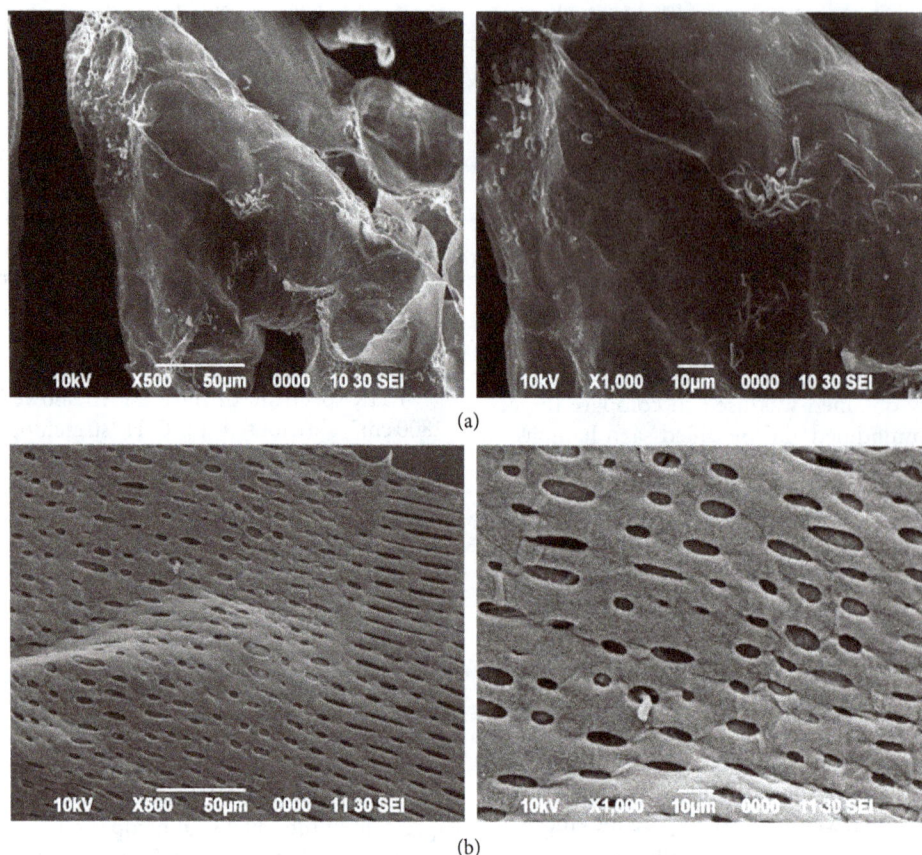

FIGURE 1: (a) 500x and 1000x magnified unmodified sago hampas. (b) 500x and 1000x magnified modified sago hampas.

hampas at $250\,\mu m$. It was agitated at 120 rpm in room temperature at respective time in 5 mL (4.45 g) of palm-based cooking oil. The effect of sorption time, adsorbent dosage, and pH were carried out. The optimum value for each parameter was taken into account to compare the efficiency of unmodified sago hampas and modified sago hampas to remove palm-based cooking oil.

3.4. Effect of Sorption Time.

Figure 2 demonstrates the effect of sorption time on oil removal by modified sago hampas from 2 min to 60 min. Removal of palm oil increases when the sorption time increases. The adsorption capacity increases with the sorption time from the first 2 min. The maximum value for adsorption was reached at 45 min with 13.203 g/g of the adsorption capacity. This effect might be due to adsorption of palm oil on the surface of the sorbent which then starts to break through the inside microscopic voids [27]. There is large amount of available sites that causes a strong hydrophobic interaction between adsorbent and adsorbate. The results also showed the fast and stable nature of the process as only a little difference was observed between the initial and final contact time. When the sorption time is prolonged to 60 min, slow adsorption was observed. This is due to limited available sites of sorbent surfaces for oil entrapment [28]. The use of sorption time over 45 min has no significant effect on the sorption of the palm oils. Therefore, the optimum sorption time was 45 min.

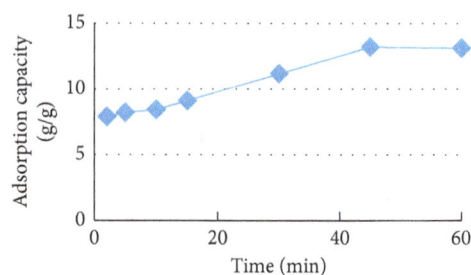

FIGURE 2: Effect of sorption time.

3.5. Effect of Adsorbent Dosage.

The effect of adsorbent dosage on the removal of palm oil was carried out by using different dosages of modified sago hampas. The weight of modified sago hampas used was 0.1 g, 0.2 g, 0.3 g, and 0.4 g. The same size of modified sago hampas at $250\,\mu m$ was used. Figure 3 shows the adsorption capacity and percentage of oil removal of palm-based cooking oil that are plotted on the same axis against adsorbent dosage. It clearly shows that, with raise in adsorbent dosage of modified sago, hampas increases the percentage removal of palm oil and expressed that the existence of large surface is feasible for adsorption with decreasing the maximum amount of adsorbed palm oil per gram of adsorbent [29]. When sorbent dosage increases, the adsorption capacity of sago hampas decreases from 16.307 g/g to 7.647 g/g. This is mainly due to increase of unsaturated oil binding sites [25]. Moreover, the efficiency of

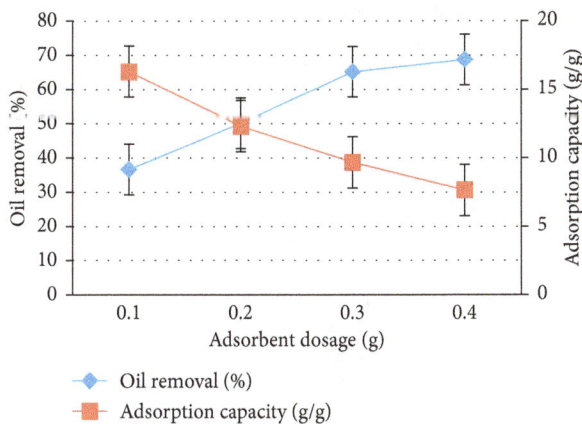

FIGURE 3: Effect of adsorbent dosage.

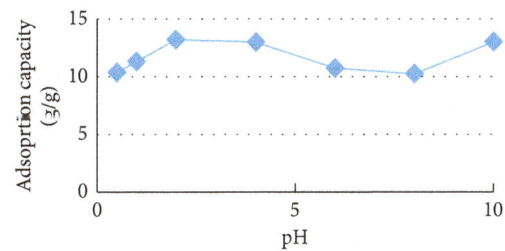

FIGURE 4: Effect of initial pH.

the adsorption of oil also decreased after maximum sorption capacity has been reached. This is because the saturation of sorbent has occurred [30]. Higher adsorbent dosage resulted in high percentage of oil removal of 68.734%. This is due to larger surface area and more adsorption sites available [31]. Considering the removal as well as the adsorption capacity, the intersection part of the plot is chosen as the optimum dosage of adsorbent [32]. Therefore, the adsorbent dosage at 0.2 g was chosen to be compared with unmodified sago hampas.

3.6. Effect of Initial pH. The effect of pH on oil adsorption is shown in Figure 4. The initial pH of the mixture is an important parameter in the adsorption processes since it influences the surface binding sites and surface properties of adsorbent by the concentration of proton or hydroxide ions in solution [33]. Figure 4 clearly presents that the palm oil adsorption process on modified sago hampas is highly dependent on pH.

The greatest adsorption capacities of modified sago hampas were obtained at pH 2. At pH 2, palm oil adsorption increased. Strong acidic condition intensifies oil to form unstable flocs, where modified sago hampas induce physicochemical effect which enhances the adsorption of oil. Adsorption of oil is unlikely to occur through ion exchange. However, spontaneous protonation and deprotonation of surface functional groups still can influence the adsorption. Oil droplets and hydrocarbon often carry negative charge in aqueous solution [34]. Therefore, at pH 2, due to the opposite charge hydrogen ions, the droplet repulsion forces decreased and coalescence occurs apparently serving to rapid demulsification. Moreover, the hydrogen ions could protonate the surface charge of sago hampas and their degree of ionization making it a better residual-oil adsorbent at this condition. However, the adsorption capacity decreased when pH is lower than 2. At pH 1 and 0.5, although hydrogen ions are helpful to induce droplet coalescence and protonization; an excessive charge concentration might delay the deposition and coalescence due to electrostatic repulsive forces between hydrogen ions [35]. Meanwhile, at pH 2, the hydrogen ion concentration is just enough for

electrostatic forces act in favor of faster adsorption. When the pH increases from 2 to 6, the adsorption capacity decreased. This can be explained by the decrease in the dissociation of acid in the solution that reduce the concentration of palm oil available to interact with positively charged adsorbent sites [36].

Under basic condition, the adsorption of palm oil increased. However, this is not corresponding to palm oil removal efficiency since the addition of excessive NaOH in order to increase the pH could result in saponification process [37]. Addition of NaOH thus makes the palm oil react and undergo hydrolysis process that produce glycerol and fatty acid salt called soap, thus making adsorption of palm oil increased. Hence, inspection of Figure 4 indicates that the optimum pH is 2.

3.7. Effect of Modified and Unmodified Sago Hampas. The effect of modified and unmodified sago hampas on the adsorption of palm-based cooking oil was studied to compare the removal efficiency of palm oil by using both unmodified and modified sago hampas. Both experiments were carried out by using the optimum adsorption capacity that has been carried out on modified sago hampas. It was carried out with 0.2 g of adsorbent dosage with 250 μm size and agitated at 120 rpm agitation speed for 45 min. It was conducted at room temperature with pH 2 of palm oil (5 mL).

Modified sago hampas adsorbs large amount of palm oil compared to unmodified sago hampas. The adsorption capacity of modified sago hampas is 18.880 g/g while unmodified sago hampas is 15.147 g/g. The percentage of oil removal for modified sago hampas and unmodified sago hampas is 84.82% and 68.08%, respectively. The percentage of oil removal difference between both modified and unmodified sago hampas is 16.75%, clearly showing that modified sago hampas adsorbs palm-based cooking oil more efficient than unmodified sago hampas. The pretreatment using KOH increases the negatively charged site on the sago hampas, and modification by CTAB reverses the surface potential charge from negative to positive [19]. The modification of sago hampas by CTAB increases the performance to adsorb palm oil. Electrostatic interaction between the negative charge of palm oil at aqueous solution molecules and positive charge introduced by cationic surfactant (CTAB) on the surface of modified sago hampas can be considered as the significant reason for the improved adsorption capacity of the modified sago hampas. The poor adsorption capacity for unmodified sago hampas for removal of palm-based oil might be due to the repulsive electrostatic interactions between the palm oil molecules and the negative sites on the surface of the sago hampas particles [31]. According to [16], the treatment process enhanced the porosity on the surface of sago hampas, thus increasing the number of adsorption site. Rough surface morphology sorbent surfaces are also needed to high adsorption capacity. The morphology and structure have more active sites for oil adsorption in modified sago hampas rather than smooth surfaces in unmodified sago hampas [9].

4. Conclusions

Modified sago hampas by chemical modification using cationic surfactant, cetyltrimethylammonium bromide (CTAB), was found to be an effective sorbent to remove palm-based cooking oil. The characterization of modified sago hampas using FTIR shows the presence of CTAB on the surface of sago hampas. In addition, scanning electron microscopy (SEM) shows that modified sago hampas has higher porosity in the surface morphology compared to unmodified sago hampas. This leads to higher adsorption efficiency of palm-based cooking oil using surfactant modified sago hampas compared to unmodified sago hampas.

Conflicts of Interest

The authors declare that there are no conflicts of interest regarding the publication of this paper.

Acknowledgments

Thanks are extended to the Universiti Malaysia Sarawak for the funding and facilities provided.

References

[1] B. Simonovic, D. Arandjelovic, M. Jovanovic, B. Kovacevic, L. Pezo, and A. Jovanovic, "Removal of mineral oil and wastewater pollutants using hard coal," *Chemical Industry and Chemical Engineering Quarterly*, vol. 15, no. 2, pp. 57–62, 2009.

[2] D. Angelova, I. Uzunov, S. Uzunova, A. Gigova, and L. Minchev, "Kinetics of oil and oil products adsorption by carbonized rice husks," *Chemical Engineering Journal*, vol. 172, no. 1, pp. 306–311, 2011.

[3] H. Kong, S.-C. Cheu, N. S. Othman et al., "Surfactant modification of banana trunk as low-cost adsorbents and their high benzene adsorptive removal performance from aqueous solution," *RSC Advances*, vol. 6, no. 29, pp. 24738–24751, 2016.

[4] S. Ibrahim, H.-M. Ang, and S. Wang, "Removal of emulsified food and mineral oils from wastewater using surfactant modified barley straw," *Bioresource Technology*, vol. 100, no. 23, pp. 5744–5749, 2009.

[5] M. O. Adebajo and R. L. Frost, "Acetylation of raw cotton for oil spill cleanup application: an FTIR and ^{13}C MAS NMR spectroscopic investigation," *Spectrochimica Acta Part A: Molecular and Biomolecular Spectroscopy*, vol. 60, no. 10, pp. 2315–2321, 2004.

[6] T. R. Annunciado, T. H. D. Sydenstricker, and S. C. Amico, "Experimental investigation of various vegetable fibers as sorbent materials for oil spills," *Marine Pollution Bulletin*, vol. 50, no. 11, pp. 1340–1346, 2005.

[7] M. Husseien, A. A. Amer, A. El-Maghraby, and N. A. Taha, "Experimental investigation of thermal modification influence on sorption qualities of barley straw," *Journal of Applied Science Research*, vol. 4, no. 6, pp. 652–657, 2008.

[8] A. Said, A. Ludwick, and H. Aglan, "Usefulness of raw bagasse for oil absorption: a comparison of raw and acylated bagasse and their components," *Bioresource Technology*, vol. 100, no. 7, pp. 2219–2222, 2009.

[9] R. Wahi, L. A. Chuah, T. S. Y. Choong, Z. Ngaini, and M. M. Nourouzi, "Oil removal from aqueous state by natural fibrous sorbent: an overview," *Separation and Purification Technology*, vol. 113, pp. 51–63, 2013.

[10] R. S. Singhal, J. F. Kennedy, S. M. Gopalakrishnan, A. Kaczmarek, C. J. Knill, and P. F. Akmar, "Industrial production, processing, and utilization of sago palm-derived products," *Carbohydrate Polymers*, vol. 72, no. 1, pp. 1–20, 2008.

[11] N. Ali, M. El-Harbawi, A. A. Jabal, and C.-Y. Yin, "Characteristics and oil sorption effectiveness of kapok fibre, sugarcane bagasse and rice husks: oil removal suitability matrix," *Environmental Technology*, vol. 33, no. 4, pp. 481–486, 2012.

[12] S. Y. Quek, D. A. J. Wase, and C. F. Forster, "The use of sago waste for thesorption of lead and copper," *Water Sa*, vol. 24, no. 3, pp. 251–256, 1998.

[13] A. A. Mohammed, S. E. Ebrahim, and A. I. Alwared, "Flotation and sorptive-flotation methods for removal of lead ions from wastewater using SDS as surfactant and barley husk as biosorbent," *Journal of Chemistry*, vol. 2013, p. 6, 2013.

[14] D. Schwantes, A. C. Gonçalves Jr., G. F. Coelho et al., "Chemical modifications of cassava peel as adsorbent material for metals ions from wastewater," *Journal of Chemistry*, vol. 2016, p. 15, 2016.

[15] B. Zhao, W. Xiao, Y. Shang, H. Zhu, H. Zhu, and R. Han, "Adsorption of light green anionic dye using cationic surfactant-modified peanut husk in batch mode," *Arabian Journal of Chemistry*, vol. 10, no. 2, pp. S3595–S3602, 2017.

[16] J. O. Amode, J. H. Santos, Z. Alam, A. H. Mirza, and C. C. Mei, "Adsorption of methylene blue from aqueous solution using untreated and treated (Metroxylon spp.) waste adsorbent: equilibrium and kinetics studies," *International Journal of Industrial Chemistry*, vol. 7, no. 3, pp. 333–345, 2016.

[17] F. Deniz and R. A. Kepekci, "Bioremoval of Malachite green from water sample by forestry waste mixture as potential biosorbent," *Microchemical Journal*, vol. 132, pp. 172–178, 2017.

[18] P. N. Chiaha, H. O. Nwabueze, C. B. Ezekannagha, and C. J. Okenwa, "Kinetic studies on oil sorption using acetylated sugarcane bagasse and groundnut husk," *International Journal of Multidisciplinary Sciences and Engineering*, vol. 8, no. 21, pp. 16–21, 2017.

[19] S. Che Ibrahim, "Treatment of oily and dye wastewater with modified barley straw," Doctoral dissertation, Curtin Uni-

versity, Perth, Australia, 2010.

[20] M. Kozak and L. Domka, "Adsorption of the quaternary ammonium salts on montmorillonite," *Journal of Physics and Chemistry of Solids*, vol. 65, no. 2-3, pp. 441–445, 2004.

[21] M. H. Koh, "Preparation and characterization of carboxymethyl cellulose from sugarcane bagasse," Doctoral dissertation, UTAR, Kampar, Malaysia, 2013.

[22] X. Zhang, F. Wang, and L. Keer, "Influence of surface modification on the microstructure and thermo-mechanical properties of bamboo fibers," *Materials*, vol. 8, no. 10, pp. 6597–6608, 2015.

[23] R. B. Viana, A. B. Da Silva, and A. S. Pimentel, "Infrared spectroscopy of anionic, cationic, and zwitterionic surfactants," *Advances in Physical Chemistry*, vol. 2012, Article ID 903272, 14 pages, 2012.

[24] Ö. Demirbaş and O. Ulus, "Synthesis and characterization of polypropylene and CTAB modified diatomite composites," *International Research Journal of Pure and Applied Chemistry*, vol. 13, no. 4, 2016.

[25] S. Ibrahim, S. Wang, and H. M. Ang, "Removal of emulsified oil from oily wastewater using agricultural waste barley straw," *Biochemical Engineering Journal*, vol. 49, no. 1, pp. 78–83, 2010.

[26] A. G. Peña, F. A. Franseschi, M. C. Estrada et al., "Fourier transform infrared-attenuated total reflectance (FTIR-ATR) spectroscopy and chemometric techniques for the determination of adulteration in petrodiesel/biodiesel blends," *Química Nova*, vol. 37, no. 3, pp. 392–397, 2014.

[27] J. C. Onwuka, E. B. Agbaji, V. O. Ajibola, and F. G. Okibe, "Kinetic studies of surface modification of lignocellulosic Delonix regia pods as sorbent for crude oil spill in water," *Journal of Applied Research and Technology*, vol. 14, no. 6, pp. 415–424, 2016.

[28] H. H. Sokker, N. M. El-Sawy, M. A. Hassan, and B. E. El-Anadouli, "Adsorption of crude oil from aqueous solution by hydrogel of chitosan based polyacrylamide prepared by radiation induced graft polymerization," *Journal of Hazardous Materials*, vol. 190, no. 1-3, pp. 359–365, 2011.

[29] J. Xiong, Y. Qin, E. Islam, M. Yue, and W. Wang, "Phosphate removal from solution using powdered freshwater mussel shells," *Desalination*, vol. 276, no. 1-3, pp. 317–321, 2011.

[30] V. Rajaković-Ognjanović, G. Aleksić, and L. Rajaković, "Governing factors for motor oil removal from water with different sorption materials," *Journal of Hazardous Materials*, vol. 154, no. 1, pp. 558–563, 2008.

[31] R. Ansari, B. Seyghali, A. Mohammad-khah, and M. A. Zanjanchi, "Highly efficient adsorption of anionic dyes from aqueous solutions using sawdust modified by cationic surfactant of cetyltrimethylammonium bromide," *Journal of Surfactants and Detergents*, vol. 15, no. 5, pp. 557–565, 2012.

[32] Q. Liu, L. Zhang, P. Hu, and R. Huang, "Removal of aniline from aqueous solutions by activated carbon coated by chitosan," *Journal of Water Reuse and Desalination*, vol. 5, no. 4, pp. 610–618, 2015.

[33] W. W. Ngah and M. A. K. M. Hanafiah, "Adsorption of copper on rubber (*Hevea brasiliensis*) leaf powder: kinetic, equilibrium and thermodynamic studies," *Biochemical Engineering Journal*, vol. 39, no. 3, pp. 521–530, 2008.

[34] A. M. Djerdjev and J. K. Beattie, "Electroacoustic and ultrasonic attenuation measurements of droplet size and ζ-potential of alkane-in-water emulsions: effects of oil solubility and composition," *Physical Chemistry Chemical Physics*, vol. 10, no. 32, pp. 4843–4852, 2008.

[35] P. Cherukupally, J. H. Hinestroza, R. Farnood, A. M. Bilton, and C. B. Park, "Adsorption mechanisms of emulsified crude oil droplets onto hydrophilic open-cell polymer foams," in *Proceedings of AIP Conference*, p. 170003, Santiago, Chile, September 2017.

[36] G. Gibbs, J. M. Tobin, and E. Guibal, "Sorption of Acid Green 25 on chitosan: influence of experimental parameters on uptake kinetics and sorption isotherms," *Journal of Applied Polymer Science*, vol. 90, no. 4, pp. 1073–1080, 2003.

[37] A. L. Ahmad, S. Bhatia, N. Ibrahim, and S. Sumathi, "Adsorption of residual oil from palm oil mill effluent using rubber powder," *Brazilian Journal of Chemical Engineering*, vol. 22, no. 3, pp. 371–379, 2005.

8

Atmospheric Nitrogen Deposition Associated with the Eutrophication of Taihu Lake

Xi Chen [iD],[1] Yan-hua Wang [iD],[1,2] Chun Ye,[3] Wei Zhou,[4] Zu-cong Cai,[1,2] Hao Yang,[1,2] and Xiao Han[5,6]

[1]School of Geography Science, Nanjing Normal University, Nanjing 210023, China
[2]Jiangsu Center for Collaborative Innovation in Geographical Information Resource Development and Application, Nanjing 210023, China
[3]Chinese Research Academy of Environmental Sciences, Beijing 100012, China
[4]Institute of Soil Science, Chinese Academy of Sciences, Nanjing 210008, China
[5]State Key Laboratory of Atmospheric Boundary Layer Physics and Atmospheric Chemistry, Institute of Atmospheric Physics, Chinese Academy of Sciences, Beijing 100029, China
[6]College of Earth Science, University of Chinese Academy of Sciences, Beijing 100049, China

Correspondence should be addressed to Yan-hua Wang; wangyanhua@njnu.edu.cn

Academic Editor: Adina Negrea

Environmental effects of excessive amounts of atmospheric nitrogen (N) deposition have raised a great deal of attention. In the present study, the characteristics of N deposition and its contribution to water eutrophication were investigated in the Taihu Basin. The results showed that the annual average total deposition (TN), total wet deposition (TN_W), and total dry deposition (TN_D) rates were 6154, 1142, and 5012 kg·km^{-2}, respectively. Moreover, seasonal fluctuations in TN, TN_W, and TN_D deposition were observed, with a higher N deposition rate occurring in spring and summer. Spatially, the distribution of TN and TN_D deposition throughout the Taihu Basin was similar. However, the TN deposition rate declined gradually from the southeast to the northwest, while the TN_W deposition rate increased. A significant positive correlation was also found between the TN deposition contents with rainfall ($R = 0.803, P = 0.01$), rainfall frequency ($R = 0.767, P < 0.01$), and rainfall intensity ($R = 0.659, P < 0.05$). The TN deposition concentration was significantly negatively correlated with rainfall ($R = -0.999, P < 0.01$), rain frequency ($R = -0.805, P < 0.01$), and rainfall intensity ($R = -0.783, P < 0.01$). The riverine input of TN was estimated to be 112,500 t·N·a^{-1}, and the main N pollutants originated from domestic sewage (accounting for 48.88%) and agriculture (accounting for 28.17%). Livestock and aquaculture contributed 90% of the agricultural pollutants. Additionally, TN deposition contributed 14,400 t N·a^{-1} to the lake, which accounted for 12.36% of the annual riverine TN inputs. The TN deposition load already exceeds the eutrophication critical load in theory. Furthermore, the contribution of N deposition to the lake has been increasing in recent years, which may accelerate eutrophication of Taihu Lake.

1. Introduction

Nitrogen (N) deposition mainly originates from the discharge of nitrogen oxides (NO_x), nitrate nitrogen (NO_3^--N), ammonia nitrogen (NH_3), and ammonium nitrogen (NH_4^+-N) from both anthropogenic and natural sources. Ultimately, these compounds return to the surface via wet and dry deposition [1, 2]. Atmospheric N deposition represents an important source of reactive N to the ecosystems [3, 4].

However, excessive N inputs could cause adverse ecological effects, including soil acidification, plant biodiversity reduction, and eutrophication [5–7]. Many literature studies have shown that the concentration of N deposition in water N loads has increased [8–10], and the ecological effects of atmospheric N deposition have received a great deal of attention in recent years [11, 12]. Many methods have been employed to collect N deposition, including ion-exchange resin, micrometeorological integral total N input, and minusing methods [13–16]. Due to the rapid population growth, industrialization, vehicle ownership, and fossil fuel combustion, the NO_x emissions in China have shown a marked increase of 2.8 times from 1980 to 2003 [17, 18]. It is also believed that both the excessive use of chemical N fertilizer and increasing amounts of human, aquaculture, and livestock excrement may have increased NH_3 emissions [19]. The fertilizer production in China in 2010 brought out 37.10 Tg of N. Among them, 75.74% was consumed by domestic agriculture, much more than the total world production and consumption. However, less than half of the N application was taken up by the crops [20]. The majority was discharged into the waterbody or the atmosphere by runoff and volatilization. The average total NH_3 emissions in China is 15 Tg N·a^{-1}, approximately 90% of which is contributed by agricultural activities [21, 22]. Rapid economic development has resulted in a significant increase in reactive N creation worldwide in recent years [19]. It is estimated that the total reactive N produced by anthropogenic activities ranged from 15 Tg in 1860 to 165 Tg in 1995, and global TN deposition is expected to reach 195 Tg in 2050 [17]. In China, the total NO_x emission from anthropogenic activities increased from 8.40 Tg·N·a^{-1} in 1990 to 11.30 Tg·N·a^{-1} in 2000, while the total NH_3 emissions rose from 10.80 Tg·N·a^{-1} to 13.60 Tg·N·a^{-1} [23, 24]. Western Europe, China, and India have had the highest N deposition in the world in recent years [25].

The Taihu watershed has played an important role in the water quantity regulation, industry, agriculture, and tourism. The lake water was under the oligotrophic status in the 1950s [26]. However, it underwent more aggravated eutrophication in the mid-1980s because of the rapid industrial and agricultural development and excessive population growth [27]. Large amounts of nutrients have been discharged into the Taihu Lake via river runoff and N deposition. As a result, the natural environment of Taihu Lake has already deteriorated significantly, and water eutrophication has become a serious problem [10, 27–29]. As a result, many policies have been initiated to improve the water quality of Taihu Lake. Nevertheless, the Taihu Lake water quality has not improved remarkably. Many prior case studies have been conducted to determine the origins and forms of N entering the system [30–33]. From year 2002, a series of investigations for atmospheric N deposition in Taihu Lake have focused on more and more attention and the results were used for preliminary calculation of the contribution of N deposition to the lake [9, 10, 28]. However, studies of the temporal and spatial distribution of N deposition and the contribution to the water eutrophication of Taihu Lake need further focus. Therefore, the present study was conducted to (1) characterize the atmospheric N deposition, (2) make a unified calculation of N migration and transformation in the system, (3) calculate N loads from the inflowing rivers and explore the contribution of N deposition to the water eutrophication, and (4) provide a reference for economic development and environmental governance in the study area.

2. Materials and Methods

2.1. Study Area. The Taihu watershed (29°55'~32°19'N, 118°50'~121°55'E) is located in the lower Yangtze River Delta (Figure 1). The watershed extends across Jiangsu Province (53% of the watershed area), Zhejiang Province (33.40%), Anhui Province (0.1%), and Shanghai (13.50%). Taihu Lake, the third largest freshwater lake in China, is a typical large shallow lake with an area of 2338 km^2 and a mean depth of 1.90 m. More than 200 streams flow radially into the lake. The Taihu watershed is characterized by a typical subtropical monsoon climate, with an annual mean temperature of 16°C and dominant soil types of yellow-brown soil, red soil, and paddy soil. The main crops in the region are wheat and rice. Excessive use of N fertilizer is common, particularly in regions with high population densities, and the average N fertilizer application rate is 570~600 kg·ha^{-1} in the rice-wheat double cropping rotation system [31]. However, only ~35% of fertilizer is absorbed in the season [21, 31]. The rest enters into the environment. For the livestock breeding, free-range chickens and ducks mode is dominant. Approximately 65% of the livestock manure in the region is disposed by the concentrated treatment, while 35% of the undisposed manure is discharged directly into the waterbody [21], resulting in excessive N levels in local aquatic systems [21, 31].

The Taihu watershed has undergone a very high degree of urbanization and became the most important comprehensive industrial base in China [34–37]. As of 2015, the population of the region was 68.27 million, and the GDP per capita in the lake basin was USD 1,325. These increased anthropogenic activities have increased the N deposition rate and aggravated eutrophication [26–39].

2.2. Extraction N Deposition Data and Correlation Analysis. The simulation atmospheric N deposition data (in 2015) were obtained through RAMS-CMAQ (Models 3, USA), and the horizontal resolution was 64 km. The total wet deposition (TN_w), total dry deposition (TN_D), and total nitrogen (TN) rates, which included NO_x (NO, NO_2, NO_3, N_2O, and N_2O_5), NH_3, NO_3^--N, and NH_4^+-N were organized using the MATLAB 8.0 (The MathWorks Company, USA) software. All atmospheric TN_W and TN_D deposition rates were obtained using the Kriging interpolation and mask extraction tool of Spatial Analysis from ArcGIS 10.0 (ESRI, USA). Additionally, SPSS 18.0 (IBM, USA) was used to make a curve estimation between the concentration of TN_W, NH_x-N (including NH_4^+-N and NH_3), and NO_3^--N deposition rate monthly and meteorological conditions. The concentration of wet N deposition had a power-type correlation with rainfall. The Pearson correlation method was used.

2.3. Calculation of N Inputs. Calculation of the N inputs into a river from different pollutant sources is very complicated, and considerable uncertainty in the results exists because of the fluctuating emission coefficients in different regions [29, 30, 40]. Accordingly, determination of the emission coefficient is essential for calculation of the N inputs into the waterbody. In the present study, we calculated the main pollutant source of agricultural (chemical fertilizer, livestock, and aquaculture), domestic

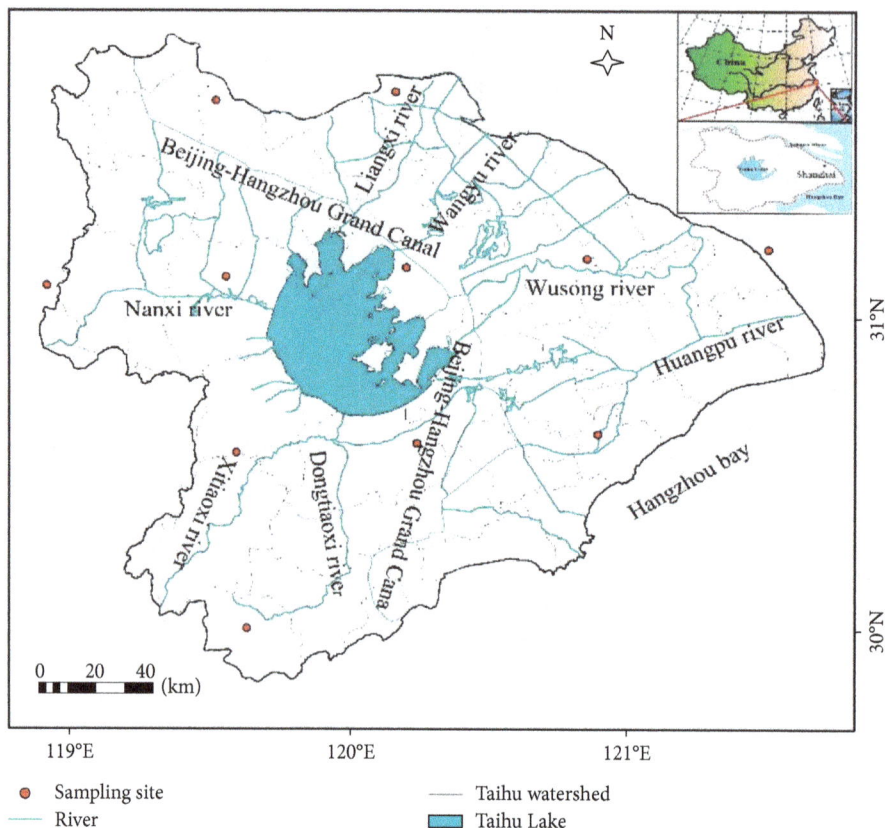

FIGURE 1: Location and administrative divisions of the Taihu watershed.

TABLE 1: Coefficients of various pollutants produced and discharged into rivers.

	Industrial effluents [41]	Urban sewage [42, 43]	Rural-domestic sewage [44]	Runoff fertilizer [38]	Aquaculture [28, 45]
Emission coefficient	—	2.92 kg·N·a^{-1}·capita^{-1}	2.19 kg·N·a^{-1}·capita^{-1}	—	1800 kg·N·(ha·a)$^{-1}$
Coefficient into water	0.65	0.80	0.10	0.05	1.00

TABLE 2: Different N produced and loss coefficients by livestock.

	Pig manure [44]	Pig urine [44]	Cattle manure [42]	Cattle urine [42]	Sheep [29]	Chicken [29, 43]
Emission coefficient (kg·N·a^{-1}·capita^{-1})	2.34	2.17	31.90	29.20	2.28	0.28
Coefficient into water	1.08	50.00	5.60	50.00	10.00	7.80

sewage (urban-domestic sewage and rural-domestic sewage), and industrial effluents (Table 1) by the following equation:

$$N_T = N_A \cdot f_A + N_B \cdot f_B + N_C \cdot f_C + N_E \cdot f_E, \qquad (1)$$

where $N_T\,(\mathrm{t\cdot N\cdot a^{-1}})$ is the total N discharged to surface water, $N_i\,(i = \mathrm{A, B, C, and\ E})$ is the emission coefficient, and $f_i\,(i = \mathrm{A, B, C, and\ E})$ is the coefficient of pollutants into the waterbody [41], where A represents industrial effluent [42], B is urban-domestic sewage [43, 44], C is rural-domestic sewage [44, 45], and E is agricultural pollution [29, 46], and the emission coefficient of livestock was determined as shown in Table 2.

3. Results and Discussion

3.1. Chemical Morphological Characteristics of N Deposition. The deposition rates of TN, TN$_W$, TN$_D$, NO$_3^-$-N, and NH$_4^+$-N were 6514, 1142, 5012, 878, and 1207 kg·km^{-2}·a^{-1}, respectively (Table 3). The TN deposition rate was higher than that observed in previous studies [10, 19, 47]. The main N deposition is dry deposition, which accounted for 81.40%. Additionally, the NH$_4^+$-N/NO$_3^-$-N was 1.4 : 1, indicating that the pollution resources may originate from the agricultural activities, as well as rural and urban sewage [48]. Because of the inadequate sewage treatment system, as well as the high fertilizer application coupled with low absorption, many N nutrients were emitted into the atmosphere [21, 31].

TABLE 3: Monthly N deposition rate around the Taihu watershed (kg·km^{-2}·a^{-1}).

Deposition	Month												Total
	Jan	Feb	Mar	Apr	May	Jun	Jul	Aug	Sep	Oct	Nov	Dec	
TN	321	291	304	747	840	918	612	652	505	310	459	197	6154
NH$_4^+$-N	46	31	55	129	159	248	132	114	78	55	103	59	1207
NO$_3^-$-N	20	39	24	69	82	170	101	194	69	19	72	19	878

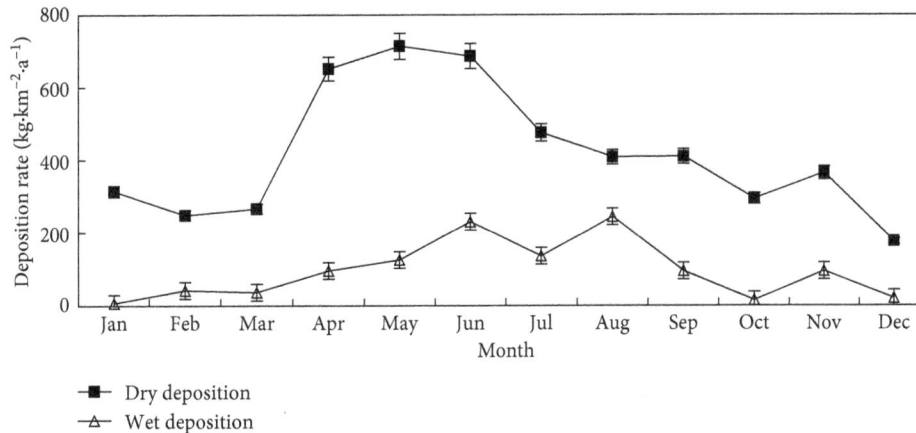

FIGURE 2: Seasonal variations in N deposition flux around the Taihu watershed.

3.2. Spatial and Temporal Distribution Characteristics of N Deposition.

The TN, TN$_D$, and TN$_W$ deposition rates showed seasonal variations in 2015 (Figure 2). The TN$_W$ deposition rate was higher during summer than winter because an El Niño phenomenon occurred in 2015. Cyanobacteria usually grow rapidly in summer. At this time, the high N deposition rate may promote cyanobacterial blooms in Taihu Lake. The wet N deposition rate in July was lower than in June, which was affected by changes in rainfall. However, the continuous heavy rainfall in late June diluted the wet N deposition concentrations in the atmosphere. This phenomenon, coupled with the low rainfall in July (due to the basin being dominated by a subtropical high), caused the wet N deposition rate to decrease rapidly. The high rainfall in August and fertilizer application in the paddy field then led to a remarkable increase in the wet N deposition rate.

The TN$_D$ deposition rate increased rapidly during spring, while it decreased in summer (Figure 2). The NH$_3$ volatilizing from the fertilizer application is emitted into the atmosphere and then deposited back to the ground after 15 days. Application of fertilizer and pesticides may enhance N deposition in spring. Moreover, unstable atmospheric conditions will increase the deposition rate of particulate matter [49]. Some literature studies have shown that the deposition of NO$_x$ and NO$_3^-$-N are both positively correlated with illumination intensity. Consequently, the dry deposition rate may increase due to the rising temperatures in spring. After fertilizer application to oilseed rape and wheat, the amount of dry N deposition increased significantly in November.

The rates of TN$_W$ and TN$_D$ deposition both showed significant spatial distribution in 2015 (Figure 3). Specifically, the TN$_W$ deposition rate decreased gradually from northwest

to southeast, while the TN$_D$ deposition rate increased. The wet deposition rates were highest (1640 kg·km^{-2}·a^{-1}) in Changzhou and Zhenjiang, which are in the northwest of the Taihu Lake Basin. Low levels of deposition (500~770 kg·km^{-2}·a^{-1}·TN$_W$) were observed in Shanghai. The precipitation in the northern Taihu Lake Basin was higher than that in the south in 2015, with the largest precipitation of 1186.7 mm occurring in the Wu-Cheng-Xi-Yu area. The higher rainfall which resulted in more N deposition in this area may be the reason.

The low levels of precipitation in Shanghai resulted in most of the particulate matter returning to the surface via dry deposition. Indeed, the highest dry deposition rate of 10,870 kg·km^{-2}·a^{-1} was observed in Shanghai. Moreover, N deposition formed a high-value band in the cities of Shanghai, Suzhou, Wuxi, and Changzhou because of the higher levels of the industrialization and urbanization. The increased N nutrients accumulated through the discharges from the urban sewage and fossil fuel, especially from vehicle exhaust. Suzhou, which is located on the west of Taihu Lake, has an open terrain and high amount of green area, resulting in lower pollution, and therefore lower N deposition, than other cities.

3.3. Comparisons of N Deposition Values.

Monitoring of the deposition rates of TN$_W$, NH$_4^+$-N, and NO$_3^-$-N revealed values of 1647, 986, and 661 kg·km^{-2}·a^{-1}, respectively, in Nanjing from July 2015 to June 2016 (Figure 4). The simulated deposition rates of TN$_W$, NH$_4^+$-N, and NO$_3^-$-N were 1653, 508, and 1144 kg·km^{-2}·a^{-1}, respectively. Both the monitored and simulated values showed significant seasonal variations in spring and summer. An obvious decrease in the monitored N

(a)

(b)

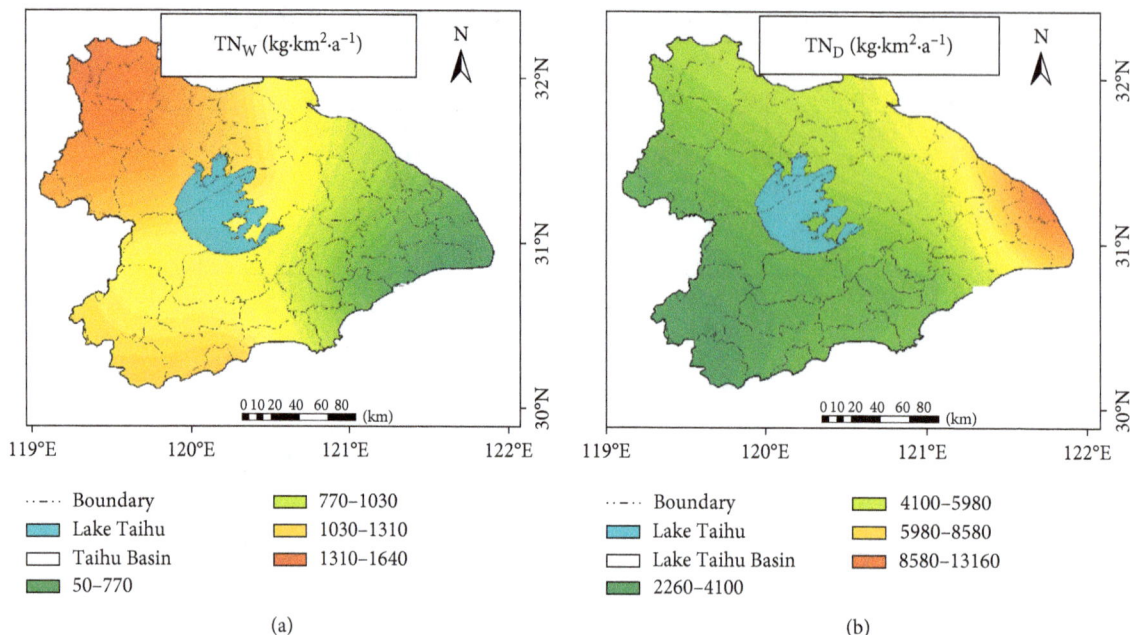

FIGURE 3: Spatial variations in N deposition (kg·km^2·a^{-1}) in the Taihu watershed.

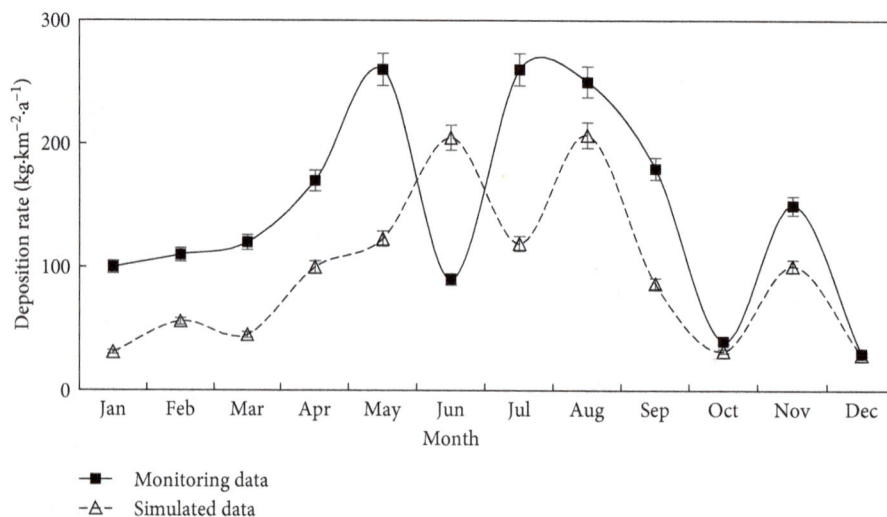

FIGURE 4: Monthly TN deposition (kg·N·km^2·a^{-1}) in Nanjing and the simulated value for the Taihu watershed.

deposition rate in June 2016 was observed. Less precipitation at this time may be the reason. However, the rainy season began in July. In the meantime, the N deposition rate increased remarkably. Overall, there was good consistency between the monitored and simulated values.

Atmospheric total wet inorganic nitrogen (TIN$_w$) deposition rates in the Taihu watershed were compared with those for other areas in China (Figure 5). The TIN$_w$ in the Taihu watershed was higher than in other regions of China from 2001 to 2015. Specifically, the TIN$_w$ deposition rate in the Taihu watershed increased before 2011 and then decreased obviously. The mean annual TIN$_w$ deposition rates in the Taihu watershed, North China Plain, Pearl River Delta, and Western China were 2736, 2352, 2267, and 446 kg·km^{-2} from 2001 to 2015, respectively. However, if the

dry deposition rate was considered, the North China Plain had the highest TN deposition rate because the dry deposition rate is higher in northern than in southern China. Consequently, the mean annual TN deposition rate would reach from 3908 to 4560 kg·km^{-2} in the Taihu watershed. Obviously, the TN deposition load already exceeds the theoretical critical eutrophication load of 491 kg·km^{-2}·a^{-1} [9].

3.4. Influence of Meteorological Conditions on N Deposition. Results of the Pearson correlation analysis showed that meteorological conditions were significantly correlated with TN$_w$ deposition rate (Table 4 and Figure 6). Moreover, a significantly negative correlation was found between the concentration of TN deposition and rainfall ($Y =$

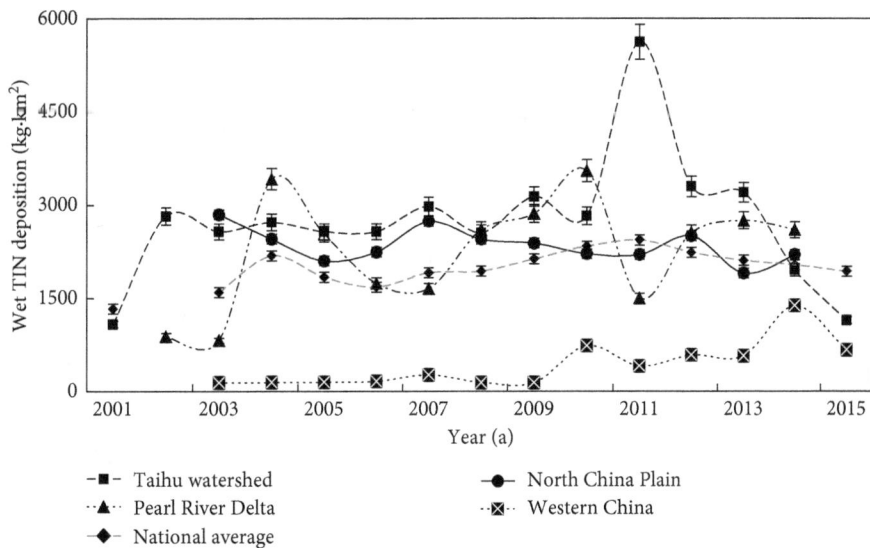

FIGURE 5: Comparisons of atmospheric TIN_w deposition rate listed with other domestic areas.

TABLE 4: Correlation (R) between wet N deposition and meteorological conditions.

	Rainfall	Frequency	Intensity	N deposition	Concentration
Rainfall	1	—	—	—	—
Frequency	0.646*	1	—	—	—
Intensity	0.933**	0.343	1	—	—
N deposition	0.803**	0.767**	0.659*	1	—
Concentration	−0.999**	−0.805**	−0.783**	−0.847**	1

Values are given with their level of significance: ** $p < 0.01$; * $p < 0.05$.

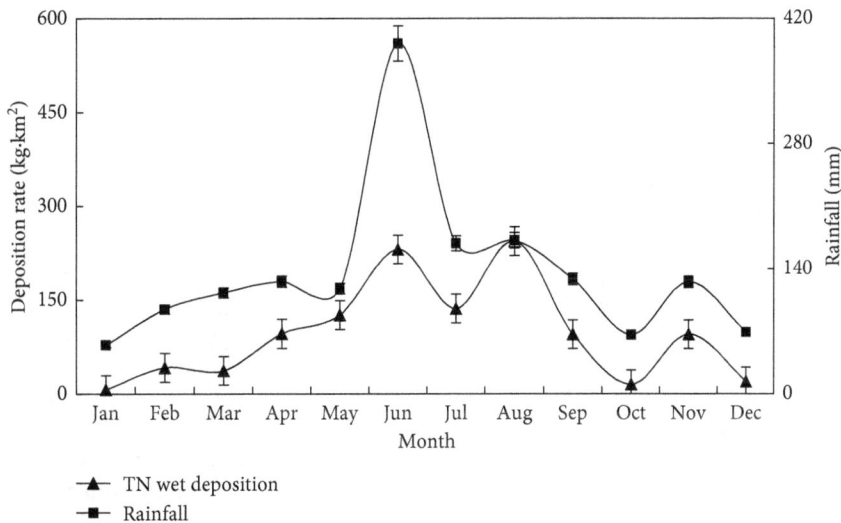

FIGURE 6: Monthly rainfall and TN_w deposition.

$189.268 X^{-0.997}$, $R = -0.999$, $P < 0.01$), rainfall frequency ($R = -0.783$, $P < 0.01$), and rain intensity ($R = -0.783$, $P < 0.01$). However, it was significantly positively correlated with TN deposition. Under the same precipitation scale, the N concentration of rain may gradually decrease. The negative correlation indicates that precipitation can remove N nutrients from the atmosphere and that light rain results in greater removal than heavy rain. The positive correlation between precipitation and N deposition rate explains the

cumulative effect of N nutrients in water bodies after rainfall, as well as the dilution effect of heavy rain.

3.5. *Influence of NH_3 from Agriculture on N Deposition.* Agricultural activities may be the major sources of NH_3 emissions, especially from livestock and fertilizer volatilization. Changshu, which has a good agricultural base in the Taihu watershed, was investigated in this study. A remarkable

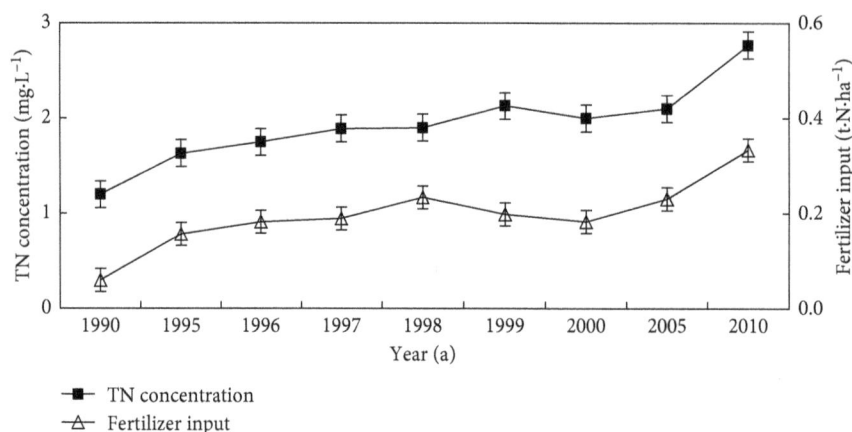

FIGURE 7: Chemical fertilizer input and TN content of water in Changshu from 1990 to 2010.

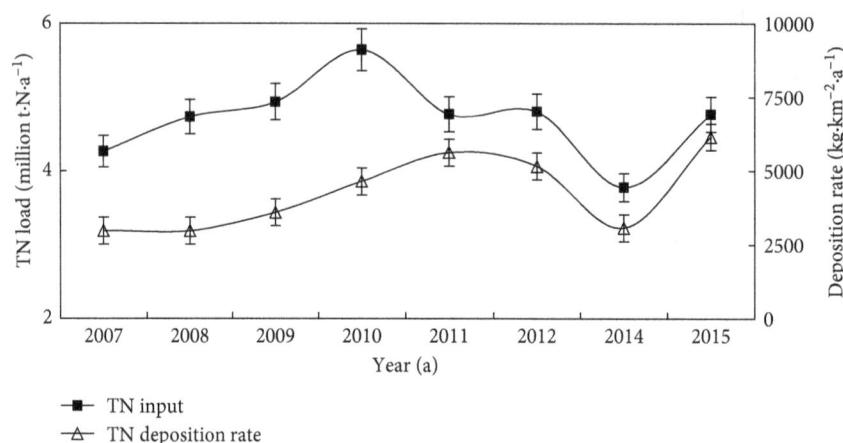

FIGURE 8: TN input into the lake and TN deposition rate in Taihu watershed from 2007 to 2015.

correlation between TN loads of water and N fertilizer applied per hectare was observed (Figure 7). Fertilizer is applied to rice paddies, wheat fields, and oilseed rape fields in March, July, and November. At that time, the wet N deposition rate increased rapidly, demonstrating that ammonia from farmland fertilizer made a significant contribution to N deposition. With the excessive fertilizer application and low absorptivity in this area, the concentration of TN in water bodies has increased in recent years. The mean NH_3 deposition rate was found to be $688 \, kg \cdot ha^{-1}$, accounting for 56.80% of NH_3^+-N from the simulation date. Consequently, the volatilization of NH_3 from fertilizer concentration to N deposition cannot be ignored.

The typical characteristics of livestock and aquaculture are free-ranging in or near the water bodies in this area. About 35% of the manure was discharged into the river directly from undisposed manure [21, 31]. The gross value of production of aquaculture increased from USD 0.87 billion in 2000 to USD 2.13 billion in 2015, while the gross value of livestock production increased from USD 0.5 billion to USD 1 billion. Increasing amounts of N are discharged into aquatic systems and the atmosphere under the imperfect excrement disposal system. If effective measures are not taken, NH_3 volatilized from agricultural systems will continue to increase because of the rapid development of livestock, aquaculture, and excessive fertilizer application.

3.6. Estimated Contribution of N Deposition to N Loads of Water in the Taihu Watershed.

Based on equation (1), the annual riverine input of TN was estimated to be $112,500 \, t \, N \cdot a^{-1}$. The main N pollutants originated from domestic sewage (48.88%) and agriculture (28.17%). Among the agricultural pollutants, livestock and aquaculture contributed 90.00%. However, many studies have shown that the heavy N load in Taihu Lake originated from agricultural activities. In the present study, the TN deposition load into the lake was calculated based on an area of $2338 \, km^2$ and an annual TN deposition load of $14,400 \, t \, N \cdot a^{-1}$. The annual TN deposition accounted for 12.36% of the annual riverine input of TN.

When compared to the case studies conducted from 2007 to 2015 [9, 10, 28, 50–52], the contribution of N deposition to Taihu Lake showed an increasing trend (Figure 8). The eutrophication critical load of atmospheric N deposition, which is the minimal amount of N required to stimulate eutrophication, is lower than $658 \, kg \cdot km^{-2} \cdot a^{-1}$ for Taihu Lake [53]. Additionally, the allowable TN load in the Taihu Lake ecosystem was estimated to be only $491 \, kg \cdot km^{-2} \cdot a^{-1}$

[9], while the TN deposition rate was found to be $6514 \, \text{kg} \cdot \text{km}^{-2} \cdot \text{a}^{-1}$ in the present study. Obviously, the TN deposition load already exceeds the eutrophication critical load in theory. Accordingly, this phenomenon may accelerate the eutrophication process of Taihu Lake. Overall, our results indicate that the contribution of TN deposition to water N load cannot be ignored when the pollution sources are considered.

4. Conclusions

To better understand the spatial-temporal distribution characteristics of N deposition and its estimated contributions to water eutrophication, the N deposition in the Taihu watershed was investigated. The results revealed the following:

(1) Deposition rates of TN, wet deposition, dry deposition, NO_3^--N, and NH_4^+-N were 6514, 1142, 5012, 878, and $1207 \, \text{kg} \cdot \text{km}^{-2} \cdot \text{a}^{-1}$, respectively.

(2) The TN, TN_W, and TN_D deposition had significant temporal and spatial distribution features. Seasonally, both deposition rates were higher in spring and summer. Spatially, the TN_W deposition rate decreased from northwest to southeast while the TN_D deposition rate increased.

(3) Correlation analysis showed that rainfall was significantly correlated with N deposition rate. Rain could clean the atmosphere and that light rain did so more effectively than heavy rain.

(4) The TN deposition contributed to the Taihu lake was $14{,}400 \, \text{t} \, \text{N} \cdot \text{a}^{-1}$, 12.36% of the total annual N input via inflow rivers. The main N pollutants originated from urban domestic sewage and agriculture, especially fertilizer and livestock.

Conflicts of Interest

The authors declare that there are no conflicts of interest regarding the publication of this paper.

Acknowledgments

This work was supported by the "973" Project of the Ministry of Science and Technology of China (Grant no. 2014CB953801) and the National Natural Science Foundation of China (Grant no. 41673107). The authors thank the Changshu Agro-Ecosystem Experimental Station, Chinese Academy of Sciences, for providing statistical data assistance.

References

[1] Y. H. Zhao, L. Zhang, Y. F. Chen et al., "Atmospheric nitrogen deposition to China: a model analysis on nitrogen budget and critical load exceedance," *Atmospheric Environment*, vol. 153, pp. 32–40, 2017.

[2] K. M. Russell, J. N. Galloway, S. A. Macko, J. L. Moody, and J. R. Scudlark, "Sources of nitrogen in wet deposition to the Chesapeake-Bay region," *Atmospheric Environment*, vol. 32, no. 14, pp. 3923–3927, 1998.

[3] J. N. Galloway, F. J. Dentener, D. G. Capone et al., "Nitrogen cycle: past, present and future," *Biogeochemistry*, vol. 70, no. 2, pp. 153–226, 2004.

[4] D. Fowler, M. Coyle, U. Skiba et al., "The global nitrogen cycle in the twenty-first century," *Philosophical Transactions of the Royal Society B: Biological Sciences*, vol. 368, no. 1621, article 20130164, 2013.

[5] A. F. Bouwman, D. P. Van Vuuren, R. G. Derwent, and M. Posch, "A global analysis of acidification and eutrophication of terrestrial ecosystems," *Water, Air, and Soil Pollution*, vol. 141, no. 1–4, pp. 349–382, 2002.

[6] W. D. Bowman, C. C. Cleveland, L. Halada, J. Hresko, and J. S. Baron, "Negative impact of nitrogen deposition on soil buffering capacity," *Nature Geoscience*, vol. 1, no. 11, pp. 767–770, 2008.

[7] C. J. Stevens, N. B. Dise, J. O. Mountford, and D. J. Gowing, "Impact of nitrogen deposition on the species richness of grasslands," *Science*, vol. 303, no. 5665, pp. 1876–1879, 2004.

[8] E. W. Boyer, C. L. Goodale, and N. A. Jaworks, "Anthropogenic nitrogen sources and relationship to riverine nitrogen export in the northeastern USA," *Biogeochemistry*, vol. 57, no. 1, pp. 137–169, 2002.

[9] S. J. Zhai, L. Y. Yang, and W. P. Hu, "Observations of atmospheric nitrogen and phosphorus deposition during the period of algal bloom formation in northern lake Taihu, China," *Environment Management*, vol. 44, no. 3, pp. 542–551, 2009.

[10] H. Yu, L. L. Zhang, and S. W. Yan, "Atmospheric wet deposition characteristics of nitrogen and phosphorus nutrients in Taihu Lake and contributions to the lake," *Research of Environmental Sciences*, vol. 24, no. 11, pp. 1210–1219, 2011.

[11] P. M. Vitousek, J. D. Aber, and R. W. Howarth, "Human alteration of the global nitrogen cycle: sources and consequences," *Ecological Applications*, vol. 7, no. 3, pp. 737–750, 1997.

[12] W. Liu, X. Wang, and Y. Fan, "A review of atmospheric nitrogen deposition and its estimated contributions to nitrogen input of waters," *Environmental Pollution and Control*, vol. 36, no. 5, pp. 88–101, 2014.

[13] C. E. He, X. J. Liu, A. Fangmeier, and F. Zhang, "Quantifying the total airborne nitrogen input into agro-ecosystems in the north China plain," *Agriculture, Ecosystems and Environment*, vol. 121, no. 4, pp. 395–400, 2007.

[14] M. E. Fenn, M. A. Poth, and M. J. Arbaugh, "A through fall collection method using mixed bed ion exchange resin columns," *Scientific World Journal*, vol. 2, pp. 122–130, 2014.

[15] D. H. Guo and Y. Y. Zhang, "The study of atmospheric dry depositions buffer action for precipitation," *Journal of Hubei University: Natural Science*, vol. 1, pp. 96–100, 1987.

[16] D. Fowler, M. Coyle, C. Flechard et al., "Advances in micrometeorological methods for the measurement and interpretation of gas and particle nitrogen fluxes," *Plant and Soil*, vol. 228, no. 1, pp. 117–129, 2011.

[17] D. G. Streets, T. C. Bond, G. R. Carmichael et al., "An inventory of gaseous and primary aerosol emissions in Asia in the year 2000," *Journal of Geophysical Research*, vol. 108, no. 21, pp. 1–23, 2003.

[18] T. Ohara, H. Akimoto, J. Kurokawa et al., "An Asian emission inventory of anthropogenic emission sources for the period 1980–2020," *Atmospheric Chemistry and Physics Discussion*, vol. 7, no. 16, pp. 6843–6902, 2007.

[19] X. Zhao, X. Y. Yan, and Z. Q. Xiong, "Spatial and temporal variation of inorganic nitrogen wet deposition to the Yangzi River Delta Region, China," *Water, Air, and Soil Pollution*, vol. 203, no. 1–4, pp. 277–298, 2009.

[20] F. S. Zhang, J. Q. Wang, W. F. Zhang et al., "Nutrient use efficiencies of major cereal crops in China and measures for improvement," *Acta Pedologica Sinica*, vol. 45, no. 5, pp. 915–924, 2008, in Chinese.

[21] X. T. Ju, B. J. Gu, and Z. C. Cai, "Suggestions to mitigate haze by reducing agricultural ammonia emission," *Science and Technology Review*, vol. 35, no. 13, pp. 11-12, 2017.

[22] B. Gu, X. Ju, and J. Chang, "Integrated reactive nitrogen budgets and future trends in China," *Proceedings of the National Academy of Sciences*, vol. 112, no. 28, pp. 8792–8797, 2015.

[23] W. Wang, W. Zhang, and S. Hong, "Geographical distribution of SO_2 and NO_x emission intensities and trends in China," *China Environmental Science*, vol. 16, pp. 161–167, 1996.

[24] Q. R. Sun and M. R. Wang, "Ammonia emission and concentration in the atmosphere over China," *Chinese Journal of Atmospheric Sciences*, vol. 21, no. 5, pp. 590–598, 1997.

[25] A. R. Townsend, B. H. Braswell, E. A. Holland, and J. E. Penner, "Spatial and temporal patterns in terrestrial carbon storage due to deposition of fossil fuel nitrogen," *Ecological Applications,* vol. 6, no. 3, pp. 806–814, 1996.

[26] Y. B. Chang, "Major environmental changes since 1950 and the onset of accelerated eutrophication in Taihu Lake, China," *Acta Palaeontologica Sinica*, vol. 35, no. 2, pp. 155–174, 1995, in Chinese.

[27] X. L. Dai, P. Q. Ye, L. Qian, and T. Song, "Changes in nitrogen and phosphorus concentrations in Lake Taihu, 1985-2015," *Journal of Lake Sciences*, vol. 28, no. 5, pp. 935–943, 2016.

[28] Y. Z. Song, B. Q. Qin, L. Y. Yang, and W. P. Hu, "Primary estimation of atmospheric wet deposition of nitrogen to aquatic ecosystem of Lake Taihu," *Journal of Lake Sciences,* vol. 17, no. 3, pp. 226–230, 2005.

[29] C. M. Li, S. G. Zhang, and W. P. Yao, "Study on agricultural non-point source pollution load of Taihu Lake Basin in Suzhou," *Research of Soil and Water Conservation*, vol. 23, no. 3, pp. 354–359, 2016.

[30] Y. P. Huang, *Water Environment and its Pollution Control in Taihu Lake*, Beijing Science Press, Beijing, China, 2001.

[31] H. J. Zhang and F. Chen, "Non-point pollution statistics and control measures in Taihu Basin," *Water Resources Protection,* vol. 26, no. 3, pp. 87–90, 2010.

[32] X. H. Wan and H. Q. Wang, "Analysis of agricultural surface source pollution and control measures in Lake Taihu basin of Jiangsu province," *Agro-Environment and Development,* vol. 25, no. 3, pp. 69–71, 2008.

[33] Y. Y. Shen, D. L. Hu, and Q. L. Jiang, "Characteristic of nitrogen and phosphorus load in the past three decades in the northwest of Lake Taihu basin based on the SWAT model," *Resources and Environment in the Yangtze Basin*, vol. 26, no. 6, pp. 902–914, 2017.

[34] Jiangsu Province Statistics Bureau, *Statistical Yearbook of Jiangsu Province from 2000 to 2016*, Jiangsu province Statistics Bureau, Jiangsu, China, 2016.

[35] Anhui Province Statistics Bureau, *Statistical Yearbook of Anhui Province from 2000 to 2016*, Anhui Province Statistics Bureau, Anhui, China, 2016.

[36] Zhejiang Province Statistics Bureau, *Statistical Yearbook of Zhejiang Province from 2000 to 2016*, Zhejiang Province Statistics Bureau, Zhejiang, China, 2016.

[37] Shanghai Statistics Bureau, *Statistical Yearbook of Shanghai from 2000 to 2016*, Shanghai Statistics Bureau, Shanghai, China, 2016.

[38] Y. L. Wu, H. Xu, G. J. Yang, G. W. Zhu, and B. Q. Qin, "Progress in nitrogen pollution research in Lake Taihu," *Journal of Lake Sciences*, vol. 26, no. 1, pp. 19–28, 2014.

[39] B. Gao, X. Y. Yan, X. S. Jiang, and C. P. Ti, "Research progress in estimation of agricultural sources pollution of the Lake Taihu region," *Journal of Lake Sciences*, vol. 26, no. 6, pp. 822–828, 2014.

[40] C. P. Ti, Y. Q. Xia, J. J. Pan, B. J. Gu, and X. Y. Yan, "Nitrogen budget and surface water nitrogen load in Changshu a case study in the Taihu Lake region of China," *Nutrient Cycling in Agroecosystems*, vol. 91, no. 1, pp. 55–66, 2011.

[41] Jiangsu Provincial Academy of Environmental Science, *Technical Specification of Water Environment Comprehensive Treatment Plan about the Main Rivers in Lake Tailu Basin*, Jiangsu Provincial Academy of Environmental Science, Jiangsu, China, 2008.

[42] R. R. Yan, J. Y. Chao, L. Zhang, Y. X. Cui, and W. Zhuang, "Research on the load of industrial pollution in the Taihu Lake Basin in Jiangsu Province," *China Rural Water and Hydropower*, vol. 3, pp. 39–43, 2012.

[43] R. G. Li, Y. L. Xia, A. Z. Wu, and Y. S. Qian, "Pollutants sources and their discharging amount in Taihu Lake area of Jiangsu Province," *Journal of Lake Science*, vol. 12, no. 2, pp. 147–153, 2000.

[44] Z. Liu, W. X. Li, Y. M. Zhang et al., "Estimation of non-point source pollution load in Taihu Lake Basin," *Journal of Ecology and Rural Environment*, vol. 26, no. 1, pp. 45–48, 2000.

[45] H. Zhang, H. P. Li, X. Y. Li, and Z. F. Li, "Temporal changes of nitrogen balance and their driving factors in typical agricultural area of Lake Taihu Basin," *Chinese Journal of Soil Science*, vol. 45, no. 5, pp. 1119–1129, 2014.

[46] N. Zhang, Y. H. Wang, Y. Qiu et al., "Quantification and environmental effects of waste nitrogen in crop-livestock-household system of Suzhou city," *Soils*, vol. 49, no. 5, pp. 926–934, 2017.

[47] X. Z. Wang, J. G. Zhu, R. Gao, and C. J. H. Bao, "Dynamics and ecological significance of nitrogen wet-deposition in Taihu Lake region—taking Changshu agro-ecological experiment station as an example," *Chinese Journal of Applied Ecology*, vol. 15, no. 9, pp. 1616–1620, 2004.

[48] Y. Zhang, X. J. Liu, A. Fangmeier, K. T. W. Goulding, and F. S. Zhang, "Nitrogen inputs and isotopes in precipitation in the North China Plain," *Atmospheric Environment*, vol. 42, no. 7, pp. 1436–1448, 2008.

[49] D. N. Zheng, X. S. Wang, S. D. Xie, L. Duan, and D. S. Chen, "Simulation of atmospheric nitrogen deposition in China in 2010," *China Environmental Science*, vol. 34, no. 5, pp. 1089–1097, 2014.

[50] Y. X. Xie, Z. Q. Xiong, G. X. Xing, X. Y. Yan, and S. L. Shi, "Source of nitrogen in wet deposition to a rice agroecosystem at Tai lake region," *Atmospheric Environment*, vol. 42, no. 21, pp. 5182–5192, 2008.

[51] Y. Wang, N. K. Liu, and J. F. Wang, "Study on atmospheric

deposition of nitrogen and phosphorus in Taihu Lake," *Environmental Science and Management*, vol. 40, no. 5, pp. 103–105, 2011.

[52] J. F. Wang, K. J. Zhou, and X. Q. Wang, "Atmospheric nitrogen and phosphorous deposition in Hangjiahu area," *China Environmental Science*, vol. 35, no. 9, pp. 2754–2763, 2015.

[53] X. M. Ye, J. M. Hao, and L. Duan, "On critical loads of nutrient nitrogen deposition for some major lakes in China," *Environmental Pollution and Control*, vol. 24, no. 1, pp. 54–58, 2002.

9

Response Surface Methodology for Biodiesel Production Using Calcium Methoxide Catalyst Assisted with Tetrahydrofuran as Cosolvent

Nichaonn Chumuang[1] and Vittaya Punsuvon[1,2,3]

[1]Department of Chemistry, Faculty of Science, Kasetsart University, Bangkok 10900, Thailand
[2]Center of Excellence-Oil Palm, Kasetsart University, Bangkok 10900, Thailand
[3]Center for Advance Studies in Tropical Natural Resource, National Research University, Kasetsart University, Bangkok 10900, Thailand

Correspondence should be addressed to Vittaya Punsuvon; fscivit@ku.ac.th

Academic Editor: Raj K. Gupta

The present study was performed to optimize a heterogeneous calcium methoxide ($Ca(OCH_3)_2$) catalyzed transesterification process assisted with tetrahydrofuran (THF) as a cosolvent for biodiesel production from waste cooking oil. Response surface methodology (RSM) with a 5-level-4-factor central composite design was applied to investigate the effect of experimental factors on the percentage of fatty acid methyl ester (FAME) conversion. A quadratic model with an analysis of variance obtained from the RSM is suggested for the prediction of FAME conversion and reveals that 99.43% of the observed variation is explained by the model. The optimum conditions obtained from the RSM were 2.83 wt% of catalyst concentration, 11.6 : 1 methanol-to-oil molar ratio, 100.14 min of reaction time, and 8.65% v/v of THF in methanol concentration. Under these conditions, the properties of the produced biodiesel satisfied the standard requirement. THF as cosolvent successfully decreased the catalyst concentration, methanol-to-oil molar ratio, and reaction time when compared with biodiesel production without cosolvent. The results are encouraging for the application of $Ca(OCH_3)_2$ assisted with THF as a cosolvent for environmentally friendly and sustainable biodiesel production.

1. Introduction

Currently, heterogeneous catalysts have been developed for use in a transesterification reaction for biodiesel production due to the problems of homogeneous catalysts in terms of water treatment and the nonreusability of the catalyst. Although heterogeneous catalysts have many advantages [1], their disadvantages are also many such as a high oil-to-alcohol molar ratio, high temperature, and a long reaction time [2]. The addition of a cosolvent such as tetrahydrofuran (THF) to the reaction medium is an alternative way to enhance the reaction rate, as well as increasing the solubility and mass transfer between the oil and methanol [3]. THF is favorable because it can dissolve organic compounds on the hydrophobic site and bind water or alcohol on the hydrophilic site [4]. In addition, THF is a nonhazardous and unreactive chemical with a low boiling point (67°C), and it can be distilled with methanol and recycled at the end of the reaction process. In the previous work [5], THF was used as a cosolvent to accelerate the biodiesel production using a calcium oxide (CaO) catalyst but no experiments have been reported on using THF as a cosolvent with $Ca(OCH_3)_2$ catalyst. Thus, our experiment represents the first report of such a study. Response surface methodology (RSM) has been applied to analyze research involving a complex variable process. RSM employs multiple regression and correlation analyses to assess the effects of two or more independent factor on the dependent variables. Its principal advantage is in reducing the number of experimental runs required to generate sufficient information for a statistically acceptable result. RSM has been successfully applied in the study and optimization of biodiesel production from various feed stocks [6, 7].

In this study, we select $Ca(OCH_3)_2$ as heterogeneous catalyst in WCO biodiesel production because $Ca(OCH_3)_2$ showed higher activity and very low solubility compared to CaO catalyst in the transesterification of rapeseed oil [8]. Calcium methoxide ($Ca(OCH_3)_2$) was prepared from quick lime and its properties were analyzed using X-ray diffraction (XRD) and attenuated total reflection Fourier transform infrared (ATR-FTIR) spectroscopy. The synthesized $Ca(OCH_3)_2$ was also tested as a heterogeneous catalyst assisted with THF in the transesterification of waste cooking oil (WCO). RSM was utilized for process optimization. The reusability of $Ca(OCH_3)_2$ in biodiesel production was also studied.

2. Materials and Methods

2.1. Materials. The WCO in this research was obtained from the Vehicle and Building Station, Kasetsart University, Bangkok, Thailand. The fatty acid composition in the WCO was 1.11 wt% myristic acid, 35.92 wt% palmitic acid, 4.65 wt% stearic acid, 46.02 wt% oleic acid, and 12.30 wt% linoleic acid. The calculated average molecular weight of the WCO based on the fatty acid composition was 855 g/mol and its free fatty acid content was 0.7 wt%. Quick lime powder was supplied by Suthagun Co., Ltd. (Thailand). Analytical grade methanol and n-heptane were purchased from Merck (Germany). Standard chromatographic grade fatty acid methyl esters (FAME) were purchased from Sigma-Aldrich (Switzerland).

2.2. Catalyst Preparation. $Ca(OCH_3)_2$ was prepared following the work of Suwanthai et al. [6]. Briefly, the quick lime powder was heated in a furnace at 700°C for 2 h. Then, 5 g of calcined quick lime powder was placed in a three-necked flask with 150 ml of methanol. The reaction of calcined quick lime and methanol was performed at 65°C for 2 h with vigorous stirring. After that, the residual $Ca(OCH_3)_2$ was collected by filtration and washed several times with methanol. Finally, the residual $Ca(OCH_3)_2$ was dried in an oven at 105°C for 1 h and kept in a desiccator until used.

2.3. Catalyst Characterization. The XRD analysis was carried out using a D8 Advance Bruker diffractometer (USA) with Cu k_α radiation to scan a 2θ range from 5° to 40°. The surface functional groups of the $Ca(OCH_3)_2$ catalyst were determined using ATR-FTIR on a Bruker Equinox 55 FTIR spectrometer (USA).

2.4. Transesterification Process. The transesterification reactions were carried out in a three-necked flask equipped with a reflux condenser and a thermometer on a magnetic stirrer heater at 65°C and 750 rpm. The $Ca(OCH_3)_2$ was added immediately to the flask when the reactants (oil and methanol) reached the required temperature. After the reaction was complete, the products were separated using centrifugation. The top layer consisted of biodiesel and the bottom layer contained a mixture of glycerol and $Ca(OCH_3)_2$.

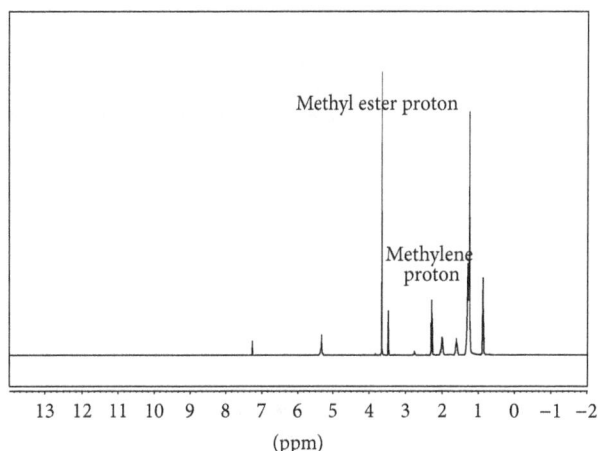

FIGURE 1: ^{1}H-NMR spectrum of WCO biodiesel.

The excess methanol contained in the biodiesel was further removed at 105°C in an oven. In the obtained biodiesel the % FAME conversion was investigated using proton nuclear magnetic resonance spectroscopy (^{1}H-NMR) [9].

2.5. Experimental Design and Statistical Analysis. RSM based on a central composite design (CCD) of experiments was used to optimize the biodiesel production process from the WCO and to investigate the influence of different transesterification process variables on the % FAME conversion. At five levels of independent variables ranging from −2 to +2, 30 experimental runs were carried out with the four independent variables: methanol-to-oil-molar ratio (A); catalyst concentration (B); reaction time (C); and cosolvent in methanol concentration (D). In addition, the 30 runs included 16 factorial points, 8 axial points, and 6 replicates at the center point to determine the experimental error for this study. The obtained experimental data was analyzed using a second-order polynomial (see (1)) to find the relationship between the independent variables and the % FAME conversion [6, 7].

$$Y = b_0 + \sum_{i=1}^{k} b_i x_i + \sum_{i=1}^{k} b_{ii} x_i^2 + \sum_{i=1}^{k} \sum_{j=1}^{k} b_{ij} x_i x_j, \qquad (1)$$

where Y is the response (% FAME conversion), b_0 is the intercept, b_i, b_{ii}, and b_{ij} are the linear, quadratic, and interactive coefficients, respectively, k is the number of factors, and x_i and x_j are the independent variables under study.

Statistical analysis of the model was performed to evaluate the analysis of variance (ANOVA) and Design-Expert software (State Ease Inc., Minneapolis, MN, USA) was used to design the experiments and carry out the regression and graphical analysis of the data [6, 7].

2.6. Proton Nuclear Magnetic Resonance Spectroscopy (^{1}H-NMR). Figure 1 shows the ^{1}H-NMR spectra of the WCO biodiesel that ^{1}H-NMR was performed for monitoring the transesterification reaction in the form of % FAME conversion.

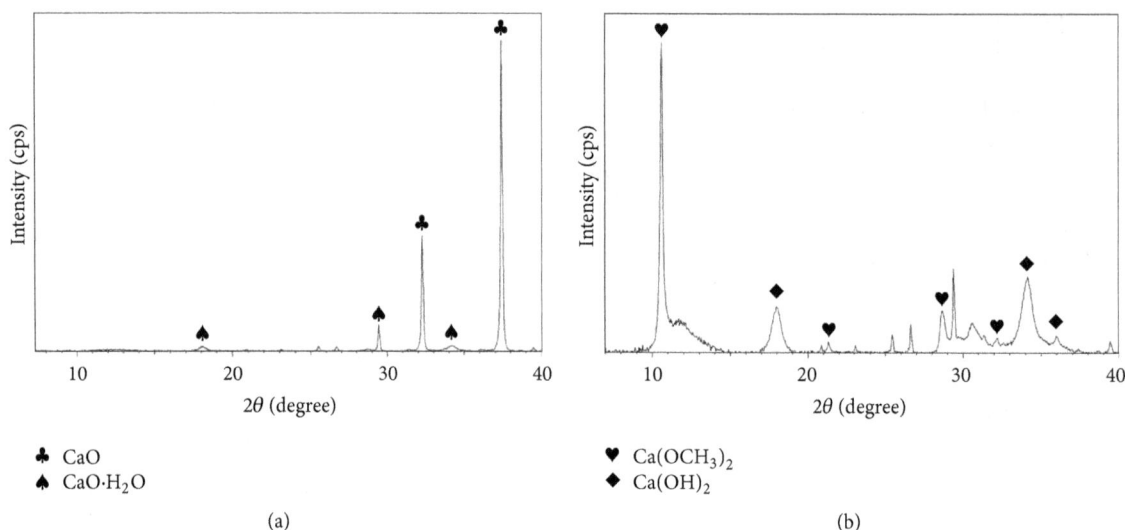

FIGURE 2: XRD patterns. (a) Calcined quick lime powder. (b) $Ca(OCH_3)_2$.

The % FAME conversion was analyzed following Knothe [10]. Briefly explained, the chemical shift at 3.6 ppm represented the methyl ester protons and at 2.3 ppm represented the methylene protons (α-CH_2). An equation to calculate the % FAME conversion is shown in [9, 10]

$$C = 100 \times \left\{ \frac{2A_{ME}}{3A_{\alpha\text{-}CH_2}} \right\}, \tag{2}$$

where C is the percentage of FAME conversion, A_{ME} is the integration value of the protons of the methyl esters and $A_{\alpha\text{-}CH_2}$ is the integration value of the methylene protons.

2.7. Physicochemical Characterization of Produced Biodiesel. The purified biodiesel obtained from transesterification was tested to evaluate its fuel properties using the recommended standard method: kinetic viscosity at 40°C (ASTM D445), density at 15°C (EN ISO 3675), acid value (ASTM D 664), water and sediment (ASTM D2709), and fatty acid methyl ester purity (EN 141003). All properties were analyzed in duplicate and reported as the average value.

3. Results and Discussion

3.1. Catalyst Characterization. In this study, the XRD results were compared with the standard diffraction pattern in the data base of the International Center of Diffraction Data (ICDD). Figure 2(a) shows the XRD patterns of the calcium oxide (CaO) obtained from the calcined quick lime power. The diffraction peaks at 2θ of 32.28° and 37.58° were attributed to the CaO (ICDD file number 00-001-1160) and three diffraction peaks at 2θ of 18.12°C, 28.73°C, and 34.23°C were attributed to calcium oxide hydrate (CaO·H_2O) (ICDD file number 00-002-0969). Figure 2(b) shows the XRD of $Ca(OCH_3)_2$ catalysts from calcined quick lime powder after reacting with methanol under reflux conditions. The four diffraction peaks that appeared at 2θ of 10.62°, 21.31°,

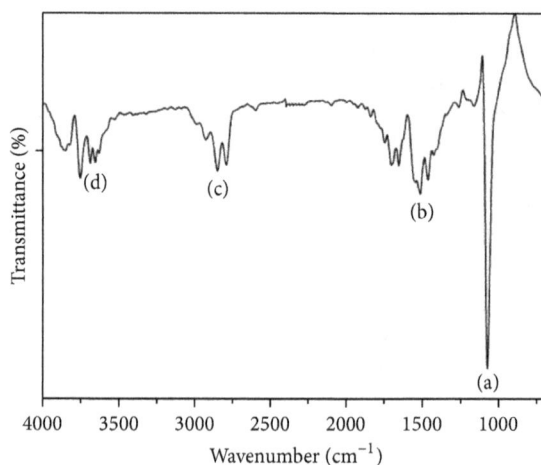

FIGURE 3: ATR-FTIR spectrum of $Ca(OCH_3)_2$.

28.69, and 32.24° were assigned to the characteristic peak of $Ca(OCH_3)_2$ (ICDD file number 00-031-1574) and ICDD file number 01-070-5492 for $Ca(OH)_2$ [9]. In order to confirm the functional group on the $Ca(OCH_3)_2$ catalyst, the ATR-FTIR spectrum is shown in Figure 3. The distinct peak at 1072.40. cm^{-1} (a) is assigned to the -C-O bond stretching vibration. A series of peaks at 1462.02 cm^{-1} (b), 2841.10 cm^{-1} (c), and 3645.40 cm^{-1} (d) are the -C-H bending vibration, -CH_3 stretching vibration, and -OH stretching vibration, respectively.

3.2. Optimization of Reaction Conditions by RSM. RSM was employed to evaluate the relations between the response (% FAME conversion) and the four reaction variables. The coded and uncoded independent variables for reaction experiment parameter were designed as shown in Table 1. Thirty experiments were performed in a randomized order. The results

TABLE 1: Reaction condition variables and levels for CCD.

Reaction condition variables	Symbol code	Range and levels				
		−2	−1	0	+1	+2
Methanol-to-oil molar ratio	A	6	8	10	12	14
Catalyst concentration (wt%)	B	1.5	2	2.5	3	3.5
Reaction time (min)	C	30	60	90	120	150
THF in methanol (% v/v)	D	4.5	6	7.5	9	10.5

TABLE 2: Experimental design with observed and predicted values from transesterification of WCO.

Run number	A: methanol-to-oil molar ratio	B: catalyst concentration (wt%)	C: reaction time (min)	D: THF concentration (v/v%)	Observed FAME (%)	Predicted FAME (%)
1	8 (−1)	2 (−1)	60 (−1)	6 (−1)	71.17	68.98
2	12 (+1)	2 (−1)	60 (−1)	6 (−1)	71.43	71.99
3	8 (−1)	3 (+1)	60 (−1)	6 (−1)	84.85	86.73
4	12 (+1)	3 (+1)	60 (−1)	6 (−1)	91.94	92.26
5	8 (−1)	2 (−1)	120 (+1)	6 (−1)	73.83	73.79
6	12 (+1)	2 (−1)	120 (+1)	6 (−1)	78.71	77.14
7	8 (−1)	3 (+1)	120 (+1)	6 (−1)	93.25	94.02
8	12 (+1)	3 (+1)	120 (+1)	6 (−1)	95.87	99.88
9	8 (−1)	2 (−1)	60 (−1)	9 (+1)	83.44	80.06
10	12 (+1)	2 (−1)	60 (−1)	9 (+1)	84.3	81.59
11	8 (−1)	3 (+1)	60 (−1)	9 (+1)	89.76	89.38
12	12 (+1)	3 (+1)	60 (−1)	9 (+1)	92.74	93.41
13	8 (−1)	2 (−1)	120 (+1)	9 (+1)	84.56	82.30
14	12 (+1)	2 (−1)	120 (+1)	9 (+1)	85.4	84.16
15	8 (−1)	3 (+1)	120 (+1)	9 (+1)	94.02	94.09
16	12 (+1)	3 (+1)	120 (+1)	9 (+1)	98.22	98.46
17	6 (−2)	2.5 (0)	90 (0)	7.5 (0)	81.03	83.15
18	14 (+2)	2.5 (0)	90 (0)	7.5 (0)	91.34	90.54
19	10 (0)	1.5 (−2)	90 (0)	7.5 (0)	50.57	56.33
20	10 (0)	3.5 (+2)	90 (0)	7.5 (0)	92.83	88.38
21	10 (0)	2.5 (0)	30 (−2)	7.5 (0)	81.72	83.68
22	10 (0)	2.5 (0)	150 (+2)	7.5 (0)	94.19	93.54
23	10 (0)	2.5 (0)	90 (0)	4.5 (−2)	91.94	89.41
24	10 (0)	2.5 (0)	90 (0)	10.5 (+2)	95.24	99.08
25	10 (0)	2.5 (0)	90 (0)	7.5 (0)	97.75	95.02
26	10 (0)	2.5 (0)	90 (0)	7.5 (0)	95.61	95.02
27	10	2.5	90	7.5	93.33	95.02
28	10	2.5	90	7.5	96.1	95.02
29	10	2.5	90	7.5	93.16	95.02
30	10	2.5	90	7.5	94.19	95.02

for each point base on the CCD experimental plans are shown in Table 2. The response obtained from the regression analysis was correlated with the four independent variables using second-order polynomial equation (see (3)). The % FAME conversion obtained at the design points of different reaction conditions is shown in Table 2. The observed values varied between 50.57% at 1.50% catalyst concentration, 10 : 1 methanol-to-oil molar ratio, 90 min reaction time, and 7.5% THF in methanol concentration and 98.22% at 3% catalyst concentration, 12 : 1 methanol-to-oil molar ratio, 120 min reaction time, and 9% THF in methanol concentration.

The Design-Expert software was used to determine and evaluate the coefficients of the full regression model equation and their statistical significance. The quadratic regression model used to predict the % FAME conversion is shown in

$$
\begin{aligned}
Y = &-231.274 + 10.387A + 140.443B + 0.393C \\
&+ 12.464D + 0.628AB + 0.001AC - 0.124AD \\
&+ 0.041BC - 2.811BD - 0.014CD - 0.511A^2 \\
&- 22.666B^2 - 0.002C^2 - 0.086D^2,
\end{aligned}
\tag{3}
$$

where Y is the response variable of % FAME conversion and $A, B, C,$ and D are the actual values of the predicted methanol-to-oil molar ratio, catalyst concentration, reaction time, and THF in methanol concentration, respectively.

TABLE 3: ANOVA for the response surface quadratic model.

Source of variation	Sum of squares	df[a]	Mean square	F-value	P value[b]	Significant at 5% level
Model	2957.24	14	211.23	20.65	<0.0001	Yes
A	81.96	1	81.96	8.01	0.0127	Yes
B	1541.28	1	1541.28	150.69	<0.0001	Yes
C	145.88	1	145.88	14.26	0.0018	Yes
D	140.12	1	140.12	13.70	0.0021	Yes
AB	6.31	1	6.31	0.62	0.4443	No
AC	0.11	1	0.11	0.011	0.9174	No
AD	2.23	1	2.23	0.22	0.6474	No
BC	6.14	1	6.14	0.60	0.4506	No
BD	71.11	1	71.11	6.95	0.0187	Yes
CD	6.64	1	6.64	0.65	0.4329	No
A^2	114.74	1	114.74	11.22	0.0044	Yes
B^2	880.73	1	880.73	86.11	<0.0001	Yes
C^2	70.46	1	70.46	6.89	0.0191	Yes
D^2	1.03	1	1.03	0.10	0.7550	No
Residual	153.42	15	10.23			
Lack of fit	137.45	10	13.74	4.30	0.0604	No
Pure error	15.97	5	3.19			
Cor total	3110.66	29				

[a]df = degree of freedom; [b]$P > 0.05$ is not significantly different at the 5% level.

The obtained data were evaluated using analysis of variance (ANOVA) for fitting a quadratic response surface model by the least squares method and to assess the quality of the fit. The significance of each coefficient parameter was determined by probability value (P value) as shown in Table 3. At the 95% confidence level, $F_{\text{model}} = 20.65$, the P value less than 0.05 clearly indicated that the high significance of the fitted model and is showing the reliability of the regression model for predicting the % FAME conversion [6, 11]. Furthermore, the variables of A, B, C, D, BD, A^2, B^2, and C^2 were found to be significant at the 95% confidence level according to the computed high F-value and the P values at the 5% level.

Thus, these statistical tests indicated that the selected model is satisfactory for predicting the % FAME conversion within the range of the experiment variables studies. The P value of the lack of fit was 0.0604, which reveals that it was not significant. Therefore, the number of experiments was sufficient to study the effect of the variable factors on % FAME conversion [6, 12]. The suitability of the model was tested using the determination coefficient (R^2). The high value of R^2 (0.9507) indicates that the fitted model can be used to predict reasonably precise outcome [6].

Figure 4 represents the actual results obtained from the experiments versus the predicted data by empirical model. The values of the adjusted determination coefficient (R^2_{adj}) and the determination coefficient (R^2) were 0.9046 and 0.9507, respectively. The high value of both coefficients justifies an excellent correlation between the independent variables and supports a high significance of the model. Meanwhile, the coefficient of variation was 3.66%. The relatively low value of the coefficient of variation reveals better reliability for this fitted model [6, 13].

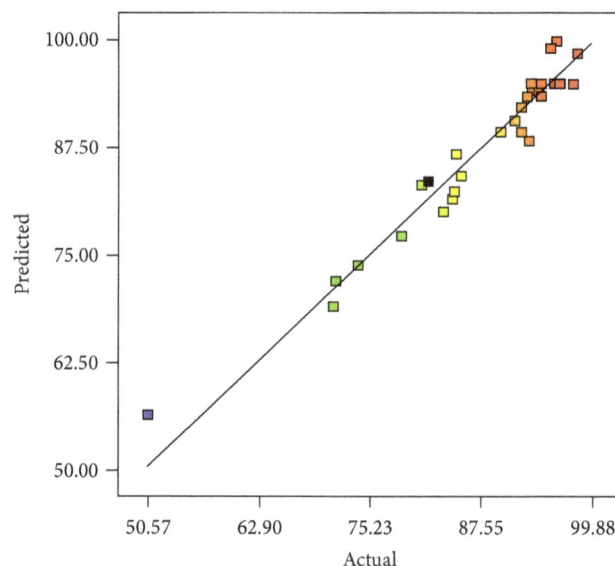

FIGURE 4: Plot of predicted % FAME conversion versus actual % FAME conversion.

The effects of the process variables on the FAME conversion were studied by plotting three-dimensional surface curves against any two independent variables while keeping the other variables at their central (0) level [14]. The 3D curves of the response from the effect of independent variables are shown in Figures 5(a)–5(f).

Figure 5(a) illustrates the effect of catalyst concentration and THF in methanol concentration on the % FAME conversion at 90 min reaction time and a 10 : 1 methanol-to-oil molar ratio. The result reveals that the % FAME

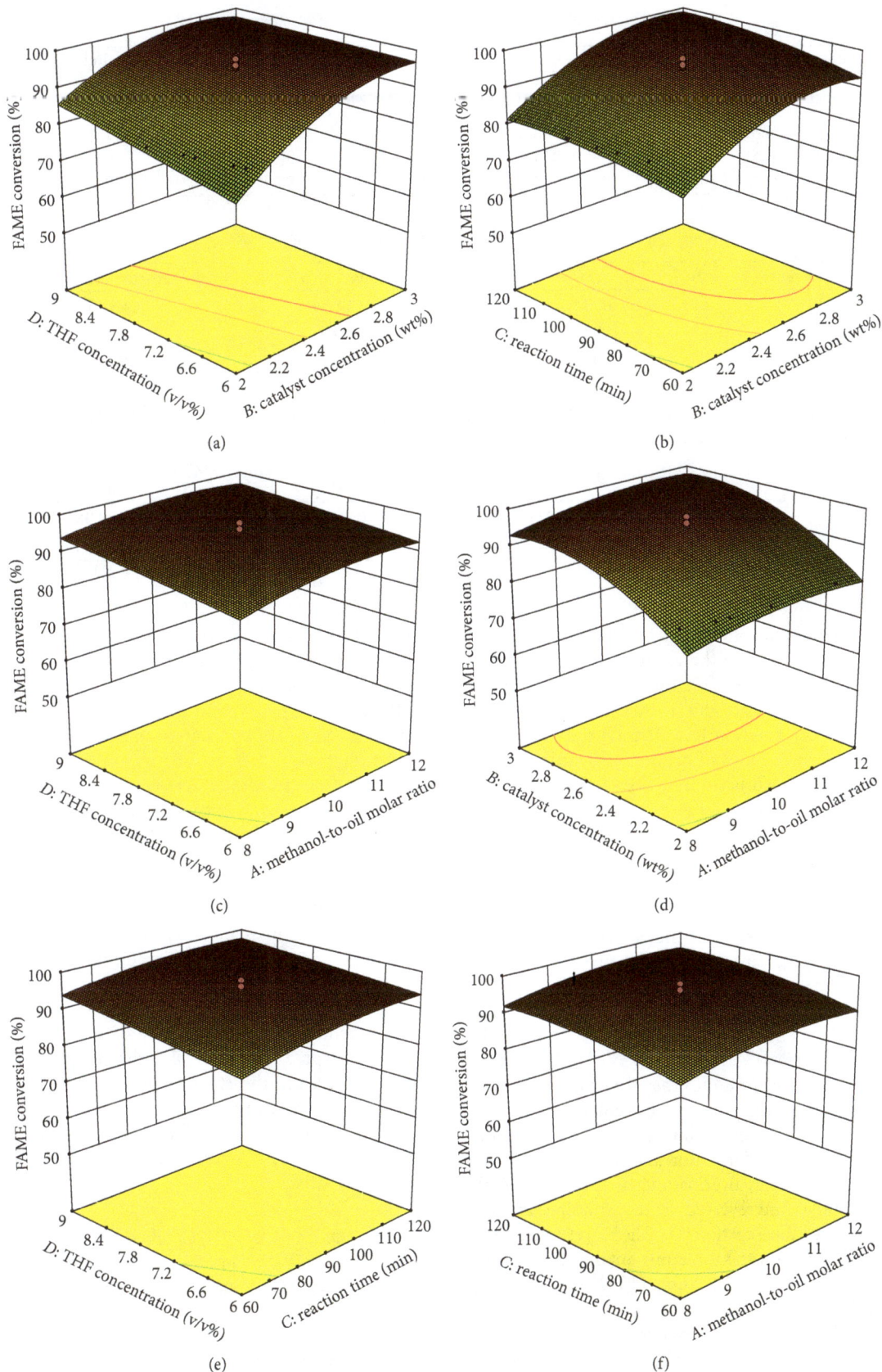

FIGURE 5: Response surface plots elucidating the effects of (a) THF in methanol concentration and catalyst concentration; (b) reaction time and catalyst concentration; (c) THF in methanol concentration and methanol-to-oil molar ratio; (d) catalyst concentration and methanol-to-oil molar ratio; (e) THF in methanol concentration and reaction time; and (f) reaction time and methanol-to-oil molar ratio.

TABLE 4: Numerical optimization of the reaction conditions using RSM.

Solution number	A	B	C	D	% FAME conversion	Desirability	
1	11.60	2.83	100.14	8.65	99.43	0.999	Selected
2	11.61	2.82	100.14	8.65	99.42	0.998	
3	11.61	2.83	100.14	8.63	99.42	0.998	
4	11.61	2.82	100.14	8.64	99.42	0.997	
5	11.58	2.83	100.14	8.34	99.39	0.974	

TABLE 5: Optimum reaction conditions and validation test.

Exp. number	A	B	C	D	Observed FAME (%)	Predicted FAME (%)	Error
1	11.60	2.83	100.14	8.65	98.70	99.43	0.73
2	11.60	2.83	100.14	8.65	99.57	99.43	0.12
3	11.60	2.83	100.14	8.65	99.12	99.43	0.31

TABLE 6: Optimum conditions using $Ca(OCH_3)_2$ with and without THF as cosolvent for WCO biodiesel production.

Method	Temperature (°C)	Methanol-to-oil molar ratio	$Ca(OCH_3)_2$ Catalyst (wt%)	Time (min)	FAME	
					Conversion	Purity (%)
Without THF	65°C	12:1	3	180	99.06	97.00
With THF	65°C	11.60:1	2.83	100.14	99.43	97.13

conversion increases with increasing catalyst concentration and THF concentration. For the amount of catalyst, there is a significant positive effect on the transesterification of vegetable oil to methyl ester due to the number of active sites available for the reaction [15]. The 3D response curve indicates that the interaction of catalyst concentration and THF concentration is significant in the reaction.

Figure 5(b) exhibits the effect of the catalyst concentration and reaction time at 10:1 methanol-to-oil molar ratio and 7.5% THF concentration. The % FAME conversion increases with increasing catalyst concentration and reaction time. Then, there is a slight decrease when the reaction period is too long due to the influence of the reversible reaction in transesterification [16]. The 3D response curve reveals that there is no significant interaction effect between the catalyst concentration and reaction time on the % FAME conversion.

Figure 5(c) shows the effect of the methanol-to-oil molar ratio and THF concentration at a 90 min reaction time and 2.5% catalyst concentration. The % FAME conversion increases with the increasing methanol-to-oil molar ratio and THF concentration. In general, a high molar ratio results in a higher rate of methyl ester formation and ensures completion of the reaction. However, overloading of methanol would inactivate the catalyst and reversed the reaction since transesterification is a reversible reaction [17]. The 3D response curve under this condition indicates no significant interaction effect between methanol-to-oil molar ratio and THF concentration on the % FAME conversion.

Figures 5(d)–5(f) illustrate the effect of the methanol-to-oil molar ratio and catalyst concentration, the effect of reaction time and THF concentration, and the effect of methanol-to-oil-molar ratio and reaction time, respectively. The interaction effect of two independent variables was studied by keeping the other variables at their central (0) level. The

results demonstrate that the % FAME conversion increased with an increase in all variables. The 3D response curves in Figures 5(d)–5(f) show no significant interaction effect of the methanol-to-oil molar ratio and catalyst concentration, reaction time, and THF concentration including the methanol-to-oil molar ratio and reaction time on % FAME conversion.

The optimal transesterification conditions were predicted by applying numerical optimization with the Design-Expert software using RSM. The results are shown in Table 4. The optimal conditions for the maximum value of % FAME conversion are as follows: 11.60:1 methanol-to-oil molar ratio, 2.83 wt% of catalyst, 100.14 min of reaction time, and 8.65% of THF in methanol concentration. Experiments were also conducted to verify the accuracy of the predicted model, and the experiment at the selected optimal conditions was performed with three replicates to confirm the experimental results as shown in Table 5. The predicted conversion value of 99.43 was approximately equal to the average observed value of 99.13. Therefore, the experimental (obtained) value showed acceptable agreement with the predicted values. The errors values between the predicted and the observed results were less than 1% FAME conversion indicating that the regression model was satisfactory.

3.3. Comparison of WCO Biodiesel Production Using $Ca(OCH_3)_2$ and without THF as Cosolvent.
In the authors' previous work (Chumuang and Punsuvon, 2016) [18], the production of WCO biodiesel using $Ca(OCH_3)_2$ without THF as cosolvent was studied and the results of the optimum conditions were compared with THF as a cosolvent as shown in Table 6.

For production without THF as cosolvent, the optimum conditions were 12:1 methanol-to-oil-molar ratio, 3% $Ca(OCH_3)_2$ catalyst concentration, 180 min reaction time,

TABLE 7: Optimum condition of difference catalyst assisted with THF as cosolvent for biodiesel production.

Method	Temperature (°C)	Methanol-to-oil molar ratio	Amount of catalyst (wt%)	THF in methanol (%v/v)	Time (min)	FAME Purity (%)
Present work	65°C	11.60 : 1	2.83	8.65	100.14	97.13
Reference work	65°C	12 : 1	5.00	10.00	90.00	98.5 ± 1.5

TABLE 8: Biodiesel properties.

Parameters	Testing method	Specification	WCO biodiesel
Viscosity at 40°C (cSt)	ASTM D 445	1.9–6.0	4.78
Density at 15°C (g/cm^3)	EN ISO 3675	0.86–0.90	0.89
Acid value (mg KOH/g)	ASTM D 664	0.80 max	0.53
Water and sediment (v%)	ASTM D 2709	0.05 max	<0.05
Methyl ester purity (wt%)	EN 14103	≥96.5	97.13

and 65°C reaction temperature resulting in 99.06% FAME conversion and 97.00% FAME purity. On the other hand biodiesel production with THF as a cosolvent was able to decrease the methanol-to-oil molar ratio (12 : 1 to 11.60 : 1), $Ca(OCH_3)_2$ catalyst concentration (3 to 2.83 wt%), and reaction time (180 to 100.14 min) while still maintaining the % FAME conversion and % FAME purity at nearly the same levels. The results indicated that THF can accelerate the catalyzed reaction by improving the mixing between the methanol, WCO, and $Ca(OCH_3)_2$ catalyst. Thus, our discovery can solve the problem of phase separation between hydrophilic methanol, hydrophobic oil, and a solid catalyst that is generally accepted to be major problem in using a heterogeneous catalyst in biodiesel production [5].

3.4. Comparison of Biodiesel Production between $Ca(OCH_3)_2$ Catalyst Assisted with THF and CaO Catalyst Assisted with THF as Cosolvent. In the work [5], the transesterification for palm oil biodiesel production was studied using CaO catalyst assisted with THF as cosolvent. The optimum condition between the reference work and our present work is shown in Table 7.

For the reference work, the optimum conditions were 65°C reaction temperature, 12 : 1 methanol-to-oil molar ratio, 5% wt CaO catalyst concentration, 10% v/v THF in methanol concentration, and 90 min reaction time resulting in 98.5 ± 1.5% FAME purity. On the other hand, our present work using $Ca(OCH_3)_2$ assisted with THF as cosolvent was able to decrease the methanol-to-oil molar ratio (12 : 1 to 11.60 : 1), amount of catalyst (5 to 2.83 wt), and THF in methanol concentration (10 to 8.65% v/v), while still maintaining the % FAME purity at nearly the same levels (97.13 and 98.5 ± 1.5%) on the same reaction temperature (65°C). The reaction time was only one variable of our present work that had longer time than the reference work (100.14 and 90 min). Thus, the result indicated that both $Ca(OCH_3)_2$ and CaO assisted with THF as cosolvent could produce both biodiesels with high purity of FAME but $Ca(OCH_3)_2$ catalyst was better than CaO catalyst in terms of catalyst concentration, methanol-to-oil

molar ratio, and THF in methanol concentration except reaction time.

3.5. Biodiesel Properties. The biodiesel properties were tested following the biodiesel standards of the USA (ASTM) and Europe (EN), as exhibited in Table 8.

While the result on viscosity result (at 40°C) of this WCO biodiesel is slightly lower (4.78 cst), the value is still within the range of ASTM D445 (1.9–6.0 cst). The density of WCO biodiesel falls in the range of the EN ISO 3675 specifications (0.86–0.90 g/cm^2). The acid value of this biodiesel was also found to be 0.53 mg KOH/g which is within the range of ASTM D664 specifications (≤0.80 mg KOH/g). The amounts of water and sediment produced in the biodiesel were less than 0.05% by volume which is in the range of the ASTM D2709 specification (≤0.05% v). The methyl ester purity was determined using the GC method and it was found to be 97.13% which falls in the range of the EN 14103 specification (≥96.5%).

3.6. Reusability of the Catalyst. One of the most important advantages of employing a heterogeneous catalyst is its reusability. $Ca(OCH_3)_2$ was separated from the reaction mixture by centrifugation followed by washing with hexane and methanol to remove the adsorbed stains. Then, $Ca(OCH_3)_2$ was collected by filtration and finally dried overnight at 105°C in an oven. The dried catalyst was further reused under the obtained optimal transesterification condition for biodiesel production. Figure 6 shows the relationship between the number of reused times and the % FAME conversion and indicates that a high % FAME (higher than 80%) was still obtained with five times of reuse.

4. Conclusion

RSM was applied to the transesterification reaction between WCO and methanol using a $Ca(OCH_3)_2$ catalyst assisted with THF as a cosolvent. The significant merit of THF as cosolvent is the short reaction time, low concentration of

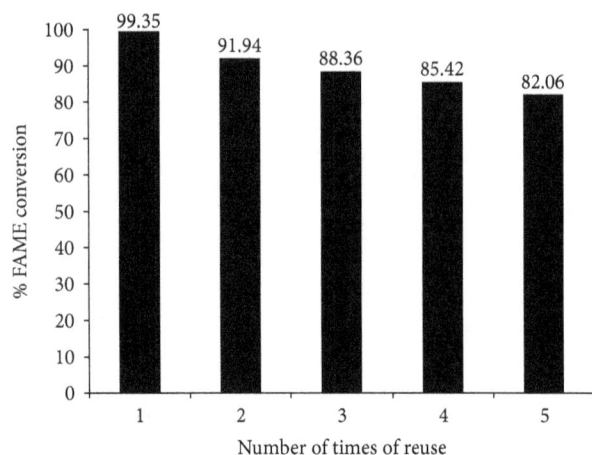

FIGURE 6: Ca(OCH$_3$)$_2$ catalyst reusability testing.

the catalyst, and the low value of the methanol-to-oil-molar ratio compared to the reaction without THF as a cosolvent. The synthesized Ca(OCH$_3$)$_2$ provides high catalytic activity for transesterification in terms of % FAME conversion and % FAME purity and reusability. These results indicate that Ca(OCH$_3$)$_2$ and THF as a cosolvent are both capable of improving the transesterification of WCO for biodiesel production.

Conflicts of Interest

The authors declare that there are no conflicts of interest regarding the publication of this paper.

Acknowledgments

This work was supported by the Higher Education Research Promotion and National Research University Project of Thailand, Office of the Higher Education Commission, and the Department of Chemistry, Faculty of Science, Kasetsart University, Bangkok, Thailand.

References

[1] Z. Helwani, M. R. Othman, N. Aziz, W. J. N. Fernando, and J. Kim, "Technologies for production of biodiesel focusing on green catalytic techniques: a review," *Fuel Processing Technology*, vol. 90, no. 12, pp. 1502–1514, 2009.

[2] R. Jothiramalingam and M. K. Wang, "Review of recent developments in solid acid, base, and enzyme catalysts (heterogeneous) for biodiesel production via transesterification," *Industrial and Engineering Chemistry Research*, vol. 48, no. 13, pp. 6162–6172, 2009.

[3] M. K. Lam and K. T. Lee, "Accelerating transesterification reaction with biodiesel as co-solvent: a case study for solid acid sulfated tin oxide catalyst," *Fuel*, vol. 89, no. 12, pp. 3866–3870, 2010.

[4] D. G. B. Boocock, S. K. Konar, V. Mao, and H. Sidi, "Fast one-phase oil-rich processes for the preparation of vegetable oil methyl esters," *Biomass and Bioenergy*, vol. 11, no. 1, pp. 43–50, 1996.

[5] W. Roschat, T. Siritanon, T. Kaewpuang, B. Yoosuk, and V. Promarak, "Economical and green biodiesel production process using river snail shells-derived heterogeneous catalyst and co-solvent method," *Bioresource Technology*, vol. 209, pp. 343–350, 2016.

[6] W. Suwanthai, V. Punsuvon, and P. Vaithanomsat, "Optimization of biodiesel production from a calcium methoxide catalyst using a statistical model," *Korean Journal of Chemical Engineering*, vol. 33, no. 1, pp. 90–98, 2016.

[7] W. Suwanthai and V. Punsuvon, "Optimization of transesterification reaction for biodiesel production from refined palm oil using calcined quick lime catalyst," *Asian Journal of Chemistry*, vol. 28, no. 2, pp. 423–428, 2016.

[8] S. Gryglewicz, "Rapeseed oil methyl esters preparation using heterogeneous catalysts," *Bioresource Technology*, vol. 70, no. 3, pp. 249–253, 1999.

[9] N. Chumuang and V. Punsuvon, "Application of calcium methoxide as solid base catalyst for biodiesel production from waste cooking oil," *Key Engineering Materials*, vol. 723, pp. 594–598, 1999.

[10] G. Knothe, "Analyzing biodiesel: standards and other methods," *Journal of the American Oil Chemists' Society*, vol. 83, no. 10, pp. 823–833, 2006.

[11] K. T. Lee, A. M. Mohtar, N. F. Zainudin, S. Bhatia, and A. R. Mohamed, "Optimum conditions for preparation of flue gas desulfurization absorbent from rice husk ash," *Fuel*, vol. 84, no. 2-3, pp. 143–151, 2005.

[12] D. C. Montgomery, *Design and Analysis of Experimentss*, Wiley, New York, NY, USA, 5th edition, 2001.

[13] X. Yuan, J. Liu, G. Zeng, J. Shi, J. Tong, and G. Huang, "Optimization of conversion of waste rapeseed oil with high FFA to biodiesel using response surface methodology," *Renewable Energy*, vol. 33, no. 7, pp. 1678–1684, 2008.

[14] D. O. Onukwuli, L. N. Emembolu, C. N. Ude, S. O. Aliozo, and M. C. Menkiti, "Optimization of biodiesel production from refined cotton seed oil and its characterization," *Egyptian Journal of Petroleum*, vol. 26, no. 1, pp. 103–110, 2016.

[15] G. Arzamendi, I. Campo, E. Arguinarena, M. Sanchez, M. Montes, and L. M. Gandia, "Synthesis of biodiesel with heterogeneous NaOH/alumina catalysts: comparison with homogeneous NaOH," *Chemical Engineering Journal*, vol. 134, pp. 123–130, 2007.

[16] C. Samart, P. Sreetongkittikul, and C. Sookman, "Heterogeneous catalysis of transesterification of soybean oil using KI/mesoporous silica," *Fuel Processing Technology*, vol. 90, no. 7-8, pp. 922–925, 2009.

[17] W. N. N. Wan Omar and N. A. Saidina Amin, "Optimization of heterogeneous biodiesel production from waste cooking palm oil via response surface methodology," *Biomass and Bioenergy*, vol. 35, no. 3, pp. 1329–1338, 2011.

[18] N. Chumuang and V. Punsuvon, "Synthesis of calcium methoxide for using as catalyst in biodiesel production from waste cooking oil," in *Proceedings of the Pure and Applied Chemistry International Conference (PACCON '16)*, pp. 242–247, 2016.

10

Adsorption Equilibrium and Kinetics of the Removal of Ammoniacal Nitrogen by Zeolite X/Activated Carbon Composite Synthesized from Elutrilithe

Yong Zhang,[1] Feng Yu,[2] Wenping Cheng,[2] Jiancheng Wang,[3] and Juanjuan Ma[1]

[1]*College of Water Resource Science and Engineering, Taiyuan University of Technology, Taiyuan 030024, China*
[2]*College of Chemistry and Chemical Engineering, Taiyuan University of Technology, Taiyuan 030024, China*
[3]*State Key Laboratory Breeding Base of Coal Science and Technology Co-Founded by Shanxi Province and the Ministry of Science and Technology, Taiyuan University of Technology, Taiyuan 030024, China*

Correspondence should be addressed to Feng Yu; yufeng@tyut.edu.cn and Juanjuan Ma; mjjsxty@163.com

Academic Editor: Mu. Naushad

Zeolite X/activated carbon composite material (X/AC) was prepared from elutrilithe, by a process consisting of carbonization, activation, and subsequent hydrothermal transformation of aluminosilicate in alkaline solution, which was used for the removal of ammoniacal nitrogen from aqueous solutions. Adsorption kinetics, equilibrium, and thermodynamic were studied and fitted by various models. The adsorption kinetics is best depicted by pseudosecond-order model, and the adsorption isotherm fits the Freundlich and Redlich-Peterson model. This explains the ammoniacal nitrogen adsorption onto X/AC which was chemical adsorption in nature. Thermodynamic properties such as ΔG, ΔH, and ΔS were determined for the ammoniacal nitrogen adsorption, and the positive enthalpy confirmed that the adsorption process was endothermic. It can be inferred that ammoniacal nitrogen removal by X/AC composite is attributed to the ion exchange ability of zeolite X. Further, as a novel sorbent, this material has the potential application in removing ammoniacal nitrogen coexisting with other organic compounds from industrial wastewater.

1. Introduction

Water pollution, such as heavy metals, dyes, organic, and inorganic pollutants, is a serious problem for the human being with the rapid urbanization, industrialization, and technological innovations in various disciplines. Many researchers have developed different methods and materials to remove the contaminants [1–7]. Among them, high concentration of ammonium ions is the a major pollutant and is harmful to animal and human health and also attacks the water plumbing systems [8]. The removal of ammoniacal nitrogen has attracted great attention in wastewater treatment.

For the water treatment, several methods such as chemical precipitation, biological processes, ion exchange, and adsorption have been taken and applied to the removal of ammoniacal nitrogen from wastewater [9–12]. Among these technologies, biological processes as one of the most widely used technologies are effective for wastewater with low concentration ammonium ions but require complicated configurations and process routing. However, the biological processes are usually helpless in dealing with the solution containing high concentration ammonium ions. So, the drive to remove the ammoniacal nitrogen with high concentration has motivated a significant increase in research activities. Compared with the other methods, ion exchange is more attractive owing to the advantages of simple operation and high effectiveness [13, 14].

Due to the excellent ion exchange ability and high surface area, natural zeolites [15, 16] and synthetic zeolites [17, 18] are employed to remove the ammoniacal nitrogen from aqueous solution by the ion exchange method. Karadag et al. [19] and Huang et al. [20] demonstrated that both the natural Turkish clinoptilolite and Chinese zeolite had strong ability to remove ammonium from aqueous solutions. In recent years, many researchers have investigated ammoniacal nitrogen removal from aqueous solutions by using eco-friendly material, such as the zeolites derived from agricultural waste, fly ash, or industrial waste. For example, Yusof et al. [21] reported that zeolite Y synthesized from rice husk ash waste was found to have higher adsorption capacity than mordenite for ammonium removal, and the equilibrium isotherm proved its monolayer adsorption. Mishra and Tiwari [22] found that zeolite 13X originated from Indian fly ash had a good sorption property for metal ions at acidic pH. However, the collection and processing of these raw materials are the major engineering problems to be solved in the case of commercializing those waste-originated zeolites.

Elutrilithe, an unusable solid waste, is widely discharged and piled up outside the coal mines in China. Moreover, the number is a steady increase of about 130 million tons each year [23, 24]. So much solid waste not only occupies a large area of farmland but also causes ecological damage, such as air pollution and water pollution. However, elutrilithe is a kaolinite-rich gangue containing aluminosilicate, according to the chemical composition of the raw material, and zeolites/activated carbon composites can be obtained, by a process consisting of carbonization, activation, and subsequent hydrothermal transformation of aluminosilicate in alkaline solution [25]. This composite material has the combination of the adsorptive properties of zeolite and activated carbon, and its applications have emerged in the wastewater treatment industries [26]. In our previous work, phenol adsorption on X/AC composite material has been studied, and this material showed an excellent adsorption capacity attributed to the existence of activated carbon [27]. However, the removal of ammoniacal nitrogen by this composite has not been reported.

So, this work aims to realize the value of zeolite X/activated carbon composite synthesized from solid waste on the removal of ammoniacal nitrogen from aqueous solutions. The optimal values of pH, temperature, and initial concentration were used for the batch experiments. In addition, the kinetic and equilibrium behaviors of the composite were investigated and several adsorption models such as Langmuir, Freundlich, and Redlich-Peterson were adopted to fit the adsorption isotherm. The adsorption kinetic rates were calculated to evaluate the possible adsorption mechanisms. The material has a significant potential in removing ammoniacal nitrogen coexisting with phenolic compounds from wastewater.

2. Materials and Methods

2.1. Preparation of Composite Materials.
The preparation of zeolite X/activated carbon composite was based on the procedure reported by Ma et al. [25]. The following is a typical synthesis example: the mixture of elutrilithe and 35 wt.% of pitch was used as starting material and extruded into cylinders (3.0 mm × 6.0 mm). The extrudate was carbonized by N_2 and then activated using CO_2 at 850°C for 24 h [27, 28]. After that, zeolite 13X was formed by hydrothermally treatment in NaOH solution at 65°C for 12 h, followed by 90°C for

24 h under stirring. Thus, the zeolite 13X/activated carbon composite was obtained and named as X/AC, in which the contents of SiO_2, Al_2O_3, and carbon were 29%, 20%, and 18%, respectively.

2.2. Characterization of Material.
The X-ray diffraction (XRD) patterns of the material were recorded on Shimadzu XRD-6000 with Cu Kα radiation at the 2θ of 20°–70°, in steps of 8°. Nitrogen adsorption and desorption isotherms were measured at −196°C on a Quantachrome analyzer. Before the measurement, the samples were evacuated for 3 h at 300°C. The surface area was calculated by Brunauer-Emmett-Teller (BET) formula and the pore volume was estimated at the relative pressure of 0.98. The pore size distribution was derived from the adsorption branch of the isotherm, using the density functional theory (DFT) method. The morphology was analyzed by a Hitachi S-4800 scanning electron microscope.

2.3. Batch Adsorption Experiments.
The adsorption isotherms of ammoniacal nitrogen were obtained by batch experiments at different temperatures (30, 35, and 40°C). For each experiment, 25 mL of ammoniacal nitrogen solution with different initial concentrations (C_0) and 6 g/L of X/AC adsorbent were mixed in a flask. The solution pH was adjusted to 6.5 by the addition of NaOH (0.1 mol/L) or HCl (0.1 mol/L). The mixture was shaken at 150 rpm for 20 h in a temperature-controlled shaker to ensure equilibrium. Finally, the adsorbent was filtered and the residual concentration of ammoniacal nitrogen was analyzed by Walter [29]. Adsorption kinetics was carried out with the same procedure at 25°C, the solution pH value was 6.5, and the initial ammoniacal nitrogen concentrations were 69, 122, 280, 460, and 500 mg/L. After different time intervals, the adsorbent was filtered and the residual concentrations were analyzed.

2.4. Adsorption Isotherm.
The equilibrium adsorption amount of ammoniacal nitrogen, q_e (mg/g), was calculated by the following formula:

$$q_e = \frac{(C_0 - C_e) \cdot V}{m}, \tag{1}$$

where C_0 is the initial concentration of the solution; C_e is the concentration at equilibrium; m is the mass of adsorbent. The equilibrium adsorption data was fitted by Langmuir, Freundlich, and Redlich-Peterson models.

2.4.1. Langmuir Adsorption Isotherm.
The Langmuir isotherm assumes that the adsorption takes place at specific homogeneous sites on the surface of the adsorbent and form a monomolecular adsorbed layer, which can be expressed as the following equation [30]:

$$q_e = \frac{q_m b C_e}{1 + b C_e}, \tag{2}$$

where q_m (mg/g) is the maximum adsorption capacity of the adsorbent and b (L/mg) is the Langmuir adsorption equilibrium constant.

2.4.2. Freundlich Adsorption Isotherm. The Freundlich isotherm provides an empirical isotherm, which assumes that nonideal adsorption takes place on a heterogeneous surface with different adsorption energy and characters [31]:

$$q_e = K_F C_e^{(1/n)}, \tag{3}$$

where K_F (mg/g)(L/mg) and n are Freundlich constants, related to adsorption capacity and adsorption intensity, respectively.

2.4.3. Redlich-Peterson (R-P) Adsorption Isotherm. R-P isotherm proposed by Redlich and Peterson [32] is a combined form of Langmuir and Freundlich expressions. It can be used for predicting homogenous and heterogeneous adsorption systems. The equation is as follows:

$$q_e = \frac{K_R C_e}{1 + a_R C_e^{\beta}}, \tag{4}$$

where K_R (L/g) and α_R (L/mg)$^{\beta}$ are the adsorption R-P constants and β is the exponent and ranges between 0 and 1. When $\beta = 0$, the R-P equation reduces to Henry's equation which is a linear isotherm and to the Langmuir isotherm for $\beta = 1$. For high adsorbate concentration, the R-P equation reduces to the Freundlich isotherm.

2.5. Adsorption Thermodynamics. The thermodynamic parameters, including change in Gibbs free energy (ΔG), enthalpy (ΔH), and entropy (ΔS), were determined by using following equations and represented as

$$\Delta G^0 = -RT \ln K_D,$$

$$K_D = \frac{q_e}{C_e}, \tag{5}$$

$$\ln K_D = \frac{\Delta S^0}{R} - \frac{\Delta H^0}{RT},$$

where K_D is the adsorption equilibrium constant, ΔG^0 was given from the classical Van't Hoff equation, and ΔH^0 and ΔS^0 were calculated from the slope and of $\ln K_D$ against $1/T$. R is the universal gas constant (8.314 J/mol) and T is the adsorption temperature (K).

2.6. Adsorption Kinetics

2.6.1. Pseudofirst-Order Model. The pseudofirst-order model is depicted as follows [33]:

$$\frac{dq_t}{dt} = k_1 (q_e - q_t). \tag{6}$$

When integrated under the boundary conditions $t = 0$, $q = 0$, and $t = t$, $q = q_t$, the equation becomes

$$\ln (q_e - q_t) = \ln q_e - k_1 t, \tag{7}$$

where k_1 is the pseudofirst-order rate constant and q_e and q_t are the adsorption capacity of the adsorbent at equilibrium and at time t, respectively.

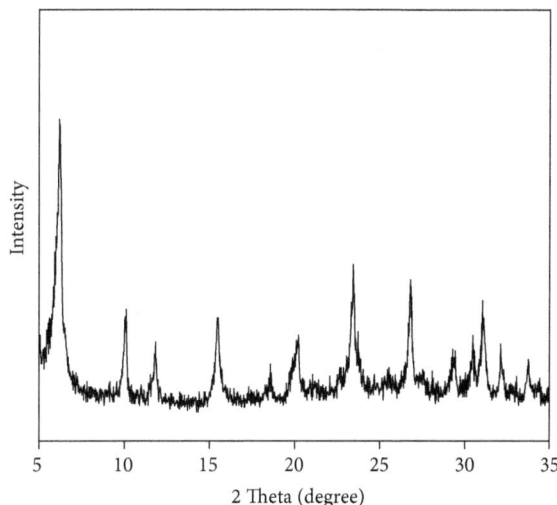

FIGURE 1: XRD patterns of the composite material.

2.6.2. Pseudosecond-Order Model. The pseudosecond-order model can be expressed as follows [34]:

$$\frac{dq_t}{dt} = k_2 (q_e - q_t)^2. \tag{8}$$

The linearized-integrated form of the equation is as follows:

$$\frac{t}{q_t} = \frac{1}{k_2 q_e^2} + \frac{t}{q_e}, \tag{9}$$

where k_2 is the pseudosecond-order rate constant.

3. Results and Discussion

3.1. Adsorbent Characterization. The XRD patterns of the samples are demonstrated in Figure 1. A well-crystallized X-ray diffraction pattern of typical zeolite X is found in the composite X/AC, which is in agreement with [35]. It is indicated that the zeolite X/activated carbon composite is successfully prepared from the waste raw materials.

N$_2$ adsorption-desorption isotherms of the composite are shown in Figure 2. As seen from Figure 2, the N$_2$ adsorption-desorption isotherms of X/AC composite exhibit both type I and IV isotherms, corresponding to hierarchical porosity ranging from micropore, mesopore, to macropore. The specific BET surface area and the total pore volume are 888 m^2/g and 0.63 cm^3/g, respectively. This result is much higher than the untreated elutrilithe, attributing to the formation of zeolite 13X from the aluminosilicate in the raw material by the hydrothermal crystallization.

The scanning electron microscopy (SEM) images of the samples are given in Figure 3. As demonstrated from Figure 3, the prepared X/AC material has the features of 13X and activated carbon. From the magnifying image of Figure 3(b), octahedral structure of 13X and rough structure of activated carbon coexisted, which confirms the 13X zeolite has been successfully prepared, and the crystal aggregates have been

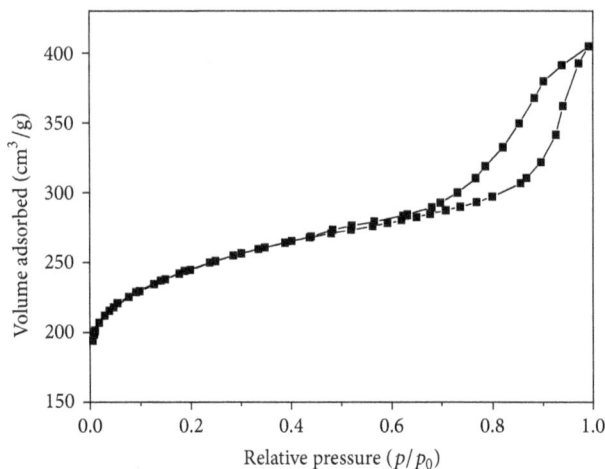

FIGURE 2: N_2 adsorption-desorption isotherms of the composite material.

covered by activated carbon. Furthermore, the surface of activated carbon in X/AC is looser and more porous compared with the raw material.

3.2. Effect of the Solution pH. The adsorption capacities of ammoniacal nitrogen in the pH range of 3.2–8.5 were performed and given in Figure 4. The maximum adsorption amount of ammoniacal nitrogen was achieved when the experiment was operated at pH 6.5. The pH has an important effect on ammoniacal nitrogen removal since it can impact the character of ammonium ion. When the pH value is higher than 7, the adsorption capacities of ammoniacal nitrogen decrease, because the ammonium ion is transformed to nonionized forms of ammonia gas, which is unfavorable for adsorption on X/AC composite [36–38]. When the pH is lower, ammonium ions compete with hydrogen ions on the adsorption sites. Hence, in this study, the pH of 6.5 is selected as the optimum pH on ammoniacal nitrogen adsorption.

3.3. Adsorption Isotherms. The composite was made into particles with the size of 20~60 mesh as adsorbent for ammonium. In a 100 mL beaker, the composite was added to 25 mL NH_4Cl solution with a ration of 6 g/L. The pH value of the solution was adjusted to 6.5, and the time for adsorption is 12 h. The thermodynamics and kinetics of the adsorption were studied by evaluating the effect of adsorption time and initial concentration.

Adsorption temperature and initial concentration affected the adsorption significantly. The concentration at equilibrium point and the rate of adsorption were affected by the initial concentration and temperature. The adsorption isotherms at the temperatures of 30, 35, and 40°C were shown in Figure 5. With the same initial concentration, the adsorption increased when the temperature was increased. At the concentration of 151.73 mg/L, the uptake of ammonium was 15.55, 17.44, and 19.02 mg/g, respectively, when the temperature was 30, 35, and 40°C. The removal of ammonium from solution by the composite of X/AC originated from the ion exchange by zeolite. The effectiveness and

efficiency of this material are close to the fly ash and 13X reported by Zheng et al. [7] and Zhang et al. [18]. This result demonstrates that ammoniacal nitrogen removal by this new composite synthesized from elutrilithe is feasible. Moreover, the process of ion exchange [39] is endothermic, which explains the result that the adsorption of ammoniacal nitrogen increased as the temperature increased. At the same temperature, the uptake increased when the concentration rose. With the same adsorption time, the solution with higher concentration could result in bigger difference of concentration between that in the solution and that in the adsorbent, which offered higher driving force for the ion exchange, and increased the efficiency of adsorption.

Three different models (Langmuir, Freundlich, and Redlich-Peterson (R-P)) were applied to fit the adsorption isotherms. The isotherm parameters, the values of the correlation coefficient R^2, and the statistical error RSME are summarized in Table 1. The values of R^2 of Langmuir model in the range of 0.9221–0.9571 are relatively low, which cannot describe the experimental data accurately. Moreover, the composite of X/AC does not have homogeneous surface, and the adsorption of ammonium is not in a monolayer. So Langmuir model does not apply in this case. The values of R^2 are higher for Freundlich and R-P model than that of Langmuir model, indicating that Freundlich and R-P model give the better fitting in the adsorption of ammoniacal nitrogen on X/AC. The Freundlich constant K increased with the increasing of the temperature, implying that adsorption of ammoniacal nitrogen is endothermic, Freundlich constant n was less than 1 revealing that the surface of X/AC is heterogeneous. Also, R-P model works in the situation of a wide range of concentration, explaining solid surface adsorption is heterogeneous. Further, the calculation of statistical error RMSE was also performed. The RMSE results indicate that the fitted data of Freundlich and Redlich-Peterson model are close to the actual value, it is superior to the Langmuir model, and the results are in a good agreement with the result of R^2.

3.4. Adsorption Thermodynamics. The adsorption equilibrium isotherms of ammoniacal nitrogen can be described better by R-P model. The values of K_D are obtained from R-P adsorption isotherm, according to literature [27]. In a study of the adsorption process in environmental engineering, Gibbs free energy (ΔG^0), enthalpy change (ΔH^0), and entropy change (ΔS^0) are normally evaluated to judge whether an adsorption of an adsorbate on an adsorbent can happen spontaneously or not. These parameters can be calculated from the graph of $\ln K_D$ versus $1/T$, which are listed in Figure 6 and Table 2. The free energy changes (ΔG^0) obtained were −36.81, −37.41, and −38.02 kJ/mol at 30, 35, and 40°C, respectively. When the value of ΔG^0 is negative, adsorption can happen by itself. On the other hand, when the value of ΔG^0 is positive, adsorption cannot happen spontaneously. The negative values of ΔG^0 indicate the spontaneous nature of ammonium uptake by the X/AC composite. The enthalpy change (ΔH^0) of adsorption was obtained as 24.44 kJ/mol. The positive value of ΔH^0 means the adsorption process is an endothermic nature [26]. This is in agreement with the expected higher

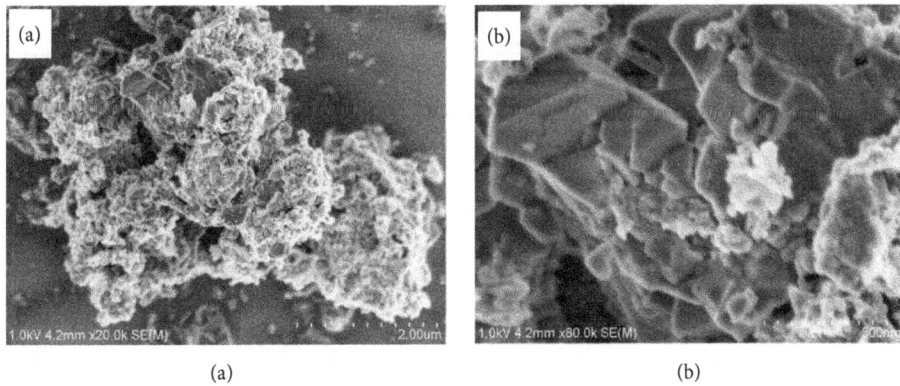

(a)   (b)

FIGURE 3: SEM images of the composite material.

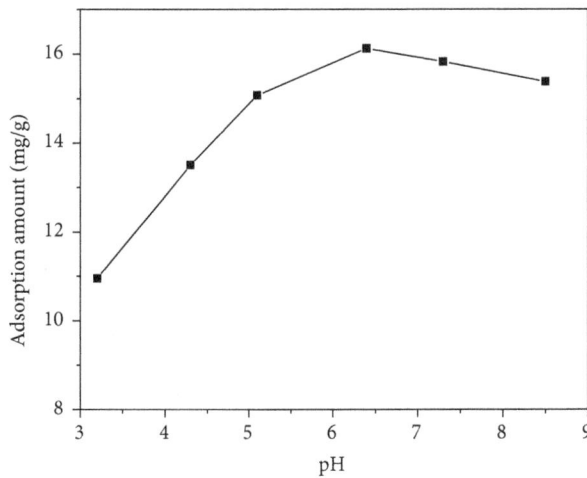

FIGURE 4: Effect of initial pH on ammoniacal nitrogen adsorption on X/AC composite at 298 K.

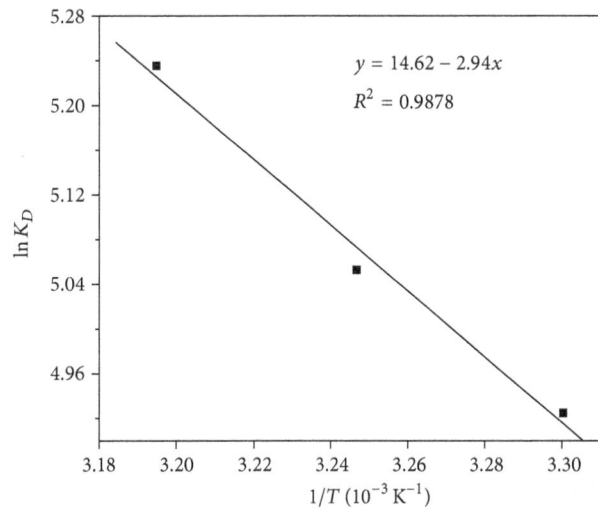

FIGURE 6: Plot of $\ln K_D$ versus $1/T$ for ammoniacal nitrogen adsorption on X/AC for thermodynamic parameters.

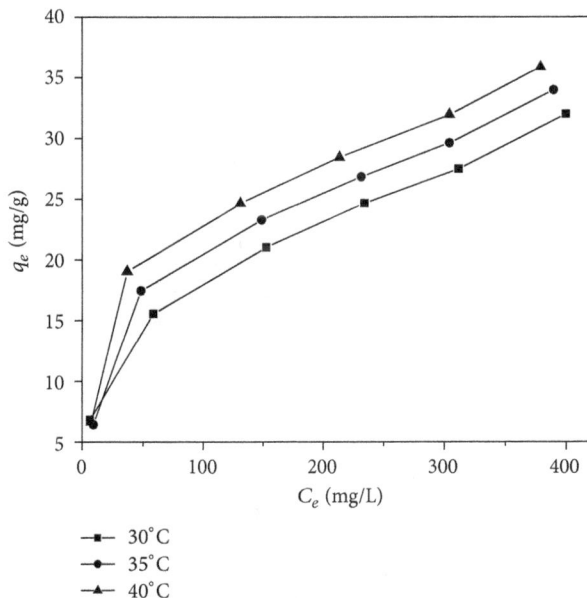

FIGURE 5: Adsorption isotherms of ammoniacal nitrogen on X/AC composite.

negative values of ΔG^0 at higher temperatures for endothermic adsorption. The entropy change (ΔS^0) was calculated as 0.1215 kJ/(mol·K). The positive value of ΔS^0 indicates the randomness at the solid/solution interface is related to the degree of freedom [19].

Generally, the values of ΔH^0 are 0~−20 KJ/mol and −80~−400 KJ/mol related to physical adsorption and chemical adsorption, respectively. The result in this work showed that the adsorption of ammoniacal nitrogen on the X/AC composite includes some chemisorption. The adsorption of ammoniacal nitrogen on the composite is attributed to ion exchange process. The values of ΔH^0 are 2.1~20.9 kJ/mol and 80~200 kJ/mol indicating physical adsorption and chemical adsorption. The value of ΔH^0 in Table 3 is higher than 20.9, indicating that adsorption of ammoniacal nitrogen on the composite is chemical adsorption. This result was in well agreement with ΔG^0.

3.5. Adsorption Kinetics.

In a typical adsorption experiment, 0.15 g of the composite was added in 25 mL ammonium

TABLE 1: Isotherm parameters of ammoniacal nitrogen adsorption on X/AC at different temperatures.

Isotherm model	Isotherm constants	Ammonium		
		30°C	35°C	40°C
Langmuir	q_m (mg/g)	35.04	35.50	35.91
	b_1 (L/mg)	0.013	0.018	0.026
	R^2	0.9221	0.9571	0.9512
	RMSE	1.76	1.96	2.29
Freundlich	K_F ((mg/g) (L/mg)$^{1/n}$)	3.23	3.56	4.69
	n	0.38	0.37	0.34
	R^2	0.9936	0.9805	0.9681
	RMSE	0.17	1.11	1.34
R-P	K_R (L/g)	$2.85 * 10^5$	1.83	$1.21 * 10^4$
	α_R (L/mg)$^\beta$	$8.84 * 10^4$	0.28	$2.6 * 10^3$
	β	0.62	0.73	0.67
	R^2	0.9936	0.9867	0.9681
	RMSE	0.19	0.91	1.64

TABLE 2: Thermodynamic parameters of ammoniacal nitrogen adsorption on X/AC.

Temperature/°C	Ammoniacal nitrogen		
	ΔG^0 (KJ/mol)	ΔH^0 (KJ/mol)	ΔS^0 (J/(mol·K))
30	−36.81		
35	−37.41	24.44	121.55
40	−38.02		

solution in 100 mL beaker, at the temperature of 25°C. After adsorption, the solution was separated from the adsorbent by centrifugation, and the concentration was analyzed. The figure of adsorption versus time is shown in Figure 7.

As shown in Figure 7, within 30 min, the adsorption speed increased while the concentration of ammonium increased. The higher the concentration of ammonium is, the bigger the difference is between the concentration in solution and that in the adsorbent, which offers high driving force for ion exchange in the composite and speeds up the adsorption of ammonium. The adsorption speed slowed down until getting equilibrium from 30 to 120 min. Figure 7 showed the uptakes of ammonium increased from 7.74 to 34.27 mg/g when the initial concentration of ammonium increased from 69 mg/L to 500 mg/L. The adsorption got equilibrium because ion exchange reached equilibrium, at which point ammonium could not be removed anymore. The kinetic results are fitted by pseudofirst-order kinetic model and pseudosecond-order kinetic model, as shown in Figures 7(b) and 7(c) and Table 3.

A curve of $\ln(q_e - q_t)$ versus t was fitted by the pseudofirst-order kinetic model and was shown in Figure 7(b). A curve of t/q_t and t was fitted by the pseudosecond-order kinetic model and was shown in Figure 7(c). The correlation coefficients (R^2) are used to describe the applicability of the adsorption kinetics model. The R^2 values for the pseudofirst-order model are the lowest among the used models. The R^2 values are 0.8405, 0.9517, 0.6940, 0.7841, and 0.6982, respectively. Moreover, the value of $q_{e(\exp)}$ differs significantly from that of $q_{e(cal)}$, which indicates the pseudofirst-order model does not work

for the adsorption of ammonium by the X/AC model. It has been reported that the pseudofirst-order model fits better the adsorption in the early stage, but not the whole adsorption process.

However, the R^2 values for the pseudosecond-order kinetic model are higher than 0.999 for five different initial concentrations, and the values of q_e(exp) are very close to that of q_e(cal), which indicates the pseudosecond-order kinetic model fit better the adsorption of ammonium by X/AC than the pseudofirst order. On the other hand, the q_e(cal.) value for the pseudofirst-order model is lower than the experimental adsorption capacity of ammoniacal nitrogen (q_e(exp.)), but the q_e(cal.) value for the pseudosecond-order model is in agreement with q_e(exp.) values. Generally, the pseudosecond-order model is proper to the adsorption kinetics.

In a water bath the temperatures were controlled at 298, 303, and 308 K; the adsorbent was added to the ammonium solution with a concentration of 65.50 mg/L and adsorbent to solution ratio of 6 g/L. The result is shown in Figure 8(a). The value of K was calculated from the pseudosecond-order kinetic model. From Arrhenius equation, $\ln K = -Ea/RT + \ln A$. The curve of $\ln K$ versus $1/T$ is shown in Figure 8(b). From the slope of the linear fitting, the active energy of adsorption Ea was calculated to be 47.74 KJ/mol. The activation energy of adsorption Ea for physical adsorption is 5~40 KJ/mol, while the activation energy of adsorption Ea for chemical adsorption is 40~800 KJ/mol. Thus, the adsorption of ammoniacal nitrogen on the composite is chemical adsorption.

4. Conclusions

In this work, it is demonstrated that the zeolite X/activated carbon composite originated from elutrilithe is an effective adsorbent for the removal of ammoniacal nitrogen. The adsorption equilibrium, thermodynamic, and kinetics parameters for the adsorption process have been investigated. Compared with Langmuir adsorption isotherm, the equilibrium adsorption data were better described by Freundlich

TABLE 3: Kinetics parameters for ammoniacal nitrogen adsorption on X/AC.

C_0 (mg/L)	$q_{e(exp)}$ (mg/g)	Pseudofirst order			Pseudosecond order		
		$q_{e(cal)}$ (mg/g)	K_1 (min^{-1})	R^2	$q_{e(cal)}$ mg/g	K_2 (min^{-1})	R^2
69	7.52	1.45	$2.2 * 10^{-2}$	0.8405	7.78	$6.55 * 10^{-2}$	0.9997
122	14.45	1.98	$3.6 * 10^{-2}$	0.9517	14.56	$6.56 * 10^{-2}$	0.9999
280	30.59	1.35	$6.1 * 10^{-3}$	0.6940	30.81	$4.41 * 10^{-2}$	0.9999
460	32.96	1.47	$2.0 * 10^{-2}$	0.7841	32.95	$7.23 * 10^{-2}$	0.9999
500	34.43	1.64	$2.0 * 10^{-2}$	0.6982	34.36	$5.76 * 10^{-2}$	0.9999

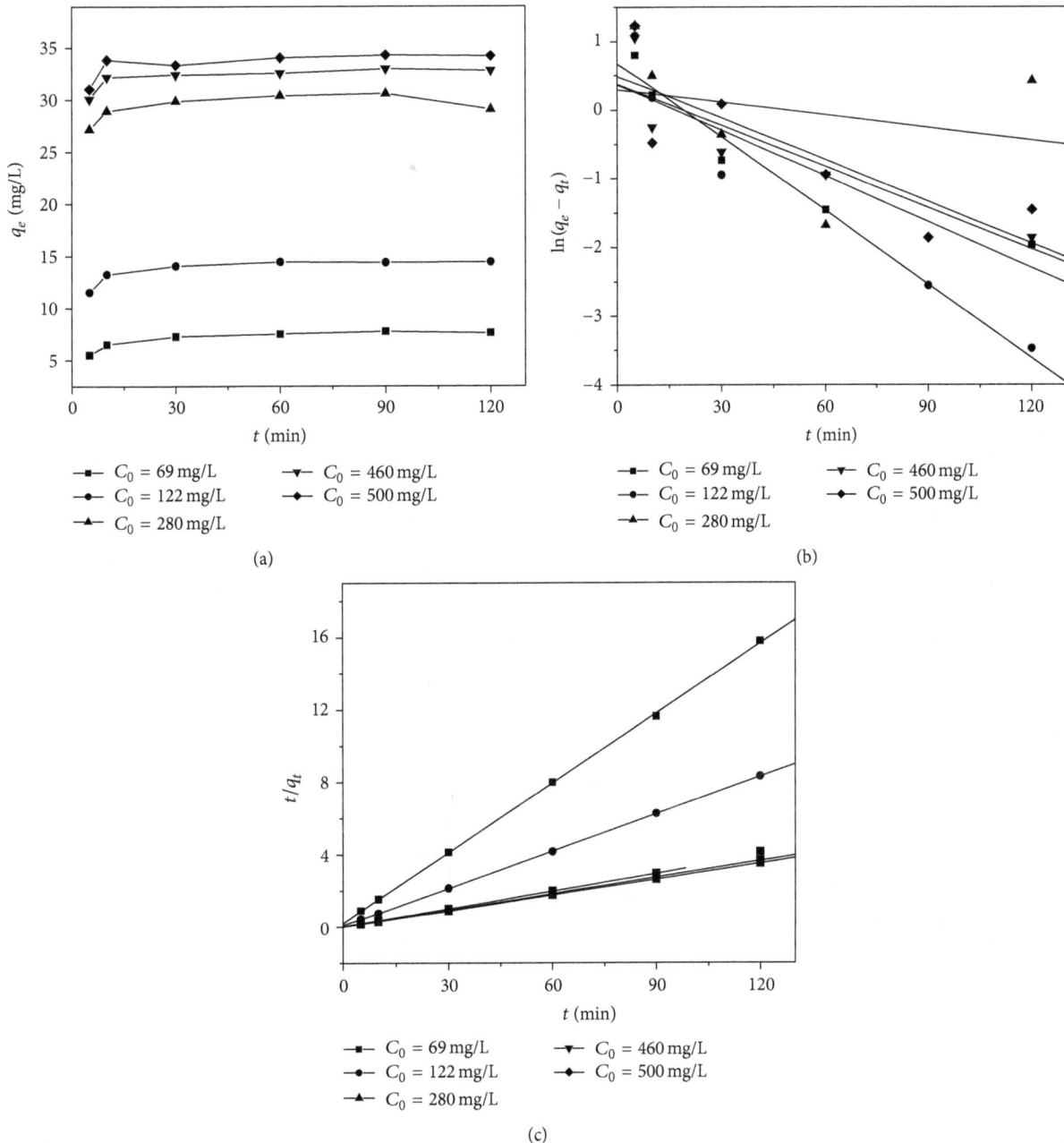

(a)

(b)

(c)

FIGURE 7: Adsorption kinetics of ammonium from aqueous solution: (a) effect of contact time on ammonium adsorption, (b) pseudofirst-order kinetic model, and (c) pseudosecond-order kinetic model.

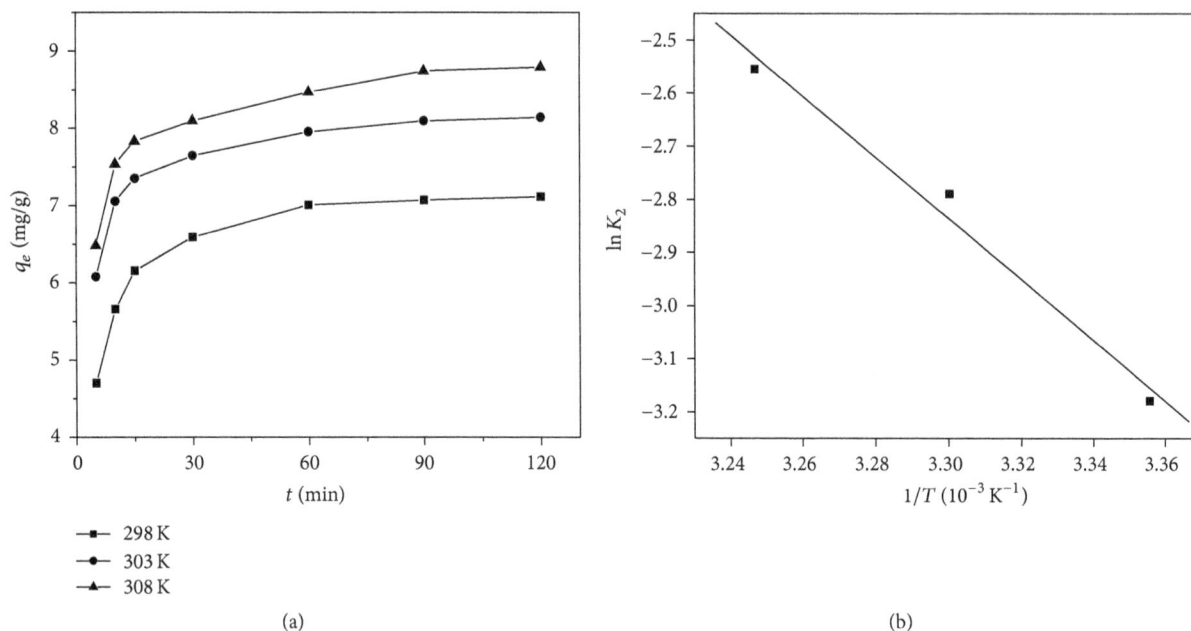

FIGURE 8: The activation energy of ammonia adsorption on X/AC (a) and plot of $\ln k$ versus $1/T$ (b).

models and the Redlich-Peterson. The thermodynamic properties of ammoniacal nitrogen adsorption concluded that the process was spontaneous and endothermic process by the adsorption of X/AC. The adsorption kinetics is best depicted by the pseudosecond-order model, indicating the adsorption process is chemisorption. This material has a significant potential in removing ammoniacal nitrogen coexisting with other organic compounds from industrial wastewater.

Conflicts of Interest

The authors declare that there are no conflicts of interest regarding the publication of this paper.

Acknowledgments

The authors gratefully appreciate the financial support from the National Science Foundation of China (nos. 51204120, 51579168, and 51409184), Scientific and Technological Project of Shanxi Province (no. 20140311016-6), and Natural Science Foundation of Shanxi (no. 2014021014-1). Also, the authors gratefully acknowledge Professor J. Ma and R. Li for their usefully discussion.

References

[1] D. Pathania, G. Sharma, and R. Thakur, "Pectin @ zirconium (IV) silicophosphate nanocomposite ion exchanger: photo catalysis, heavy metal separation and antibacterial activity," Chemical Engineering Journal, vol. 267, pp. 235–244, 2015.

[2] G. Sharma, M. Naushad, D. Pathania, A. Mittal, and G. E. El-desoky, "Modification of Hibiscus cannabinus fiber by graft copolymerization: application for dye removal," Desalination and Water Treatment, vol. 54, no. 11, pp. 3114–3121, 2015.

[3] S. Karthikeyan, M. Anil Kumar, P. Maharaja, T. Partheeban, J. Sridevi, and G. Sekaran, "Process optimization for the treatment of pharmaceutical wastewater catalyzed by poly sulpha sponge," Journal of the Taiwan Institute of Chemical Engineers, vol. 45, no. 4, pp. 1739–1747, 2014.

[4] A. Mittal, M. Naushad, G. Sharma, Z. A. Alothman, S. M. Wabaidur, and M. Alam, "Fabrication of MWCNTs/ThO$_2$ nanocomposite and its adsorption behavior for the removal of Pb(II) metal from aqueous medium," Desalination and Water Treatment, vol. 57, no. 46, pp. 21863–21869, 2016.

[5] T. Vidhyadevi, A. Murugesan, S. S. Kalaivani et al., "Optimization of the process parameters for the removal of reactive yellow dye by the low cost Setaria verticillata carbon using response surface methodology: thermodynamic, kinetic, and equilibrium studies," Environmental Progress & Sustainable Energy, vol. 33, no. 3, pp. 855–865, 2014.

[6] D. Pathania, D. Gupta, A. H. Al-Muhtaseb et al., "Photocatalytic degradation of highly toxic dyes using chitosan-g-poly(acrylamide)/ZnS in presence of solar irradiation," Journal of Photochemistry and Photobiology A: Chemistry, vol. 329, pp. 61–68, 2016.

[7] H. Zheng, L. Han, H. Ma et al., "Adsorption characteristics of ammonium ion by zeolite 13X," Journal of Hazardous Materials, vol. 158, no. 2-3, pp. 577–584, 2008.

[8] D. J. Randall and T. K. N. Tsui, "Ammonia toxicity in fish," Marine Pollution Bulletin, vol. 45, no. 1-12, pp. 17–23, 2002.

[9] N. Öztürk and T. E. Bektaş, "Nitrate removal from aqueous solution by adsorption onto various materials," Journal of Hazardous Materials, vol. 112, no. 1-2, pp. 155–162, 2004.

[10] L. A. Schipper and M. Vojvodić-Vuković, "Five years of nitrate removal, denitrification and carbon dynamics in a denitrification wall," Water Research, vol. 35, no. 14, pp. 3473–3477, 2001.

[11] K. Abe, A. Imamaki, and M. Hirano, "Removal of nitrate, nitrite, ammonium and phosphate ions from water by the aerial microalga Trentepohlia aurea," *Journal of Applied Phycology*, vol. 14, no. 2, pp. 129–134, 2002.

[12] B.-U. Bae, Y.-H. Jung, W.-W. Han, and H.-S. Shin, "Improved brine recycling during nitrate removal using ion exchange," *Water Research*, vol. 36, no. 13, pp. 3330–3340, 2002.

[13] D. Wu, B. Zhang, C. Li, Z. Zhang, and H. Kong, "Simultaneous removal of ammonium and phosphate by zeolite synthesized from fly ash as influenced by salt treatment," *Journal of Environmental Sciences*, vol. 19, no. 5, pp. 300–306, 2007.

[14] M. Sprynskyy, M. Lebedynets, A. P. Terzyk, P. Kowalczyk, J. Namieśnik, and B. Buszewski, "Ammonium sorption from aqueous solutions by the natural zeolite Transcarpathian clinoptilolite studied under dynamic conditions," *Journal of Colloid & Interface Science*, vol. 284, no. 2, pp. 408–415, 2005.

[15] M. Rožić, Š. Cerjan-Stefanović, S. Kurajica, V. Vančina, and E. Hodžić, "Ammoniacal nitrogen removal from water by treatment with clays and zeolites," *Water Research*, vol. 34, no. 14, pp. 3675–3681, 2000.

[16] J.-Y. Jung, Y.-C. Chung, H.-S. Shin, and D.-H. Son, "Enhanced ammonia nitrogen removal using consistent biological regeneration and ammonium exchange of zeolite in modified SBR process," *Water Research*, vol. 38, no. 2, pp. 347–354, 2004.

[17] Y. Takami, N. Murayama, K. Ogawa, H. Yamamoto, and J. Shibata, "Water purification property of zeolite synthesized from coal fly ash," *Shigen-to-Sozai*, vol. 116, no. 9, pp. 789–794, 2000.

[18] M. Zhang, H. Zhang, D. Xu et al., "Removal of ammonium from aqueous solutions using zeolite synthesized from fly ash by a fusion method," *Desalination*, vol. 271, no. 1–3, pp. 111–121, 2011.

[19] D. Karadag, Y. Koc, M. Turan, and B. Armagan, "Removal of ammonium ion from aqueous solution using natural Turkish clinoptilolite," *Journal of Hazardous Materials*, vol. 136, no. 3, pp. 604–609, 2006.

[20] H. M. Huang, X. M. Xiao, B. Yan, and L. P. Yang, "Ammonium removal from aqueous solutions by using natural Chinese (Chende) zeolite as adsorbent," *Journal of Hazardous Materials*, vol. 175, no. 1-3, pp. 247–252, 2010.

[21] A. M. Yusof, L. K. Keat, Z. Ibrahim, Z. A. Majid, and N. A. Nizam, "Kinetic and equilibrium studies of the removal of ammonium ions from aqueous solution by rice husk ash-synthesized zeolite Y and powdered and granulated forms of mordenite," *Journal of Hazardous Materials*, vol. 174, no. 1-3, pp. 380–385, 2010.

[22] T. Mishra and S. K. Tiwari, "Studies on sorption properties of zeolite derived from Indian fly ash," *Journal of Hazardous Materials*, vol. 137, no. 1, pp. 299–303, 2006.

[23] Z. Li, X. Cui, J. Ma, W. Chen, W. Gao, and R. Li, "Preparation of granular X-type zeolite/activated carbon composite from elutrilithe by adding pitch and solid SiO_2," *Materials Chemistry and Physics*, vol. 147, no. 3, pp. 1003–1008, 2014.

[24] L. Yang and Z. Wang, "Research and application on renewable resources coal gangue," *China Resources Comprehensive Utilization*, vol. 25, no. 3, pp. 15–16, 2007.

[25] J. Ma, J. Tan, X. Du, and R. Li, "Effects of preparation parameters on the textural features of a granular zeolite/activated carbon composite material synthesized from elutrilithe and pitch," *Microporous and Mesoporous Materials*, vol. 132, no. 3, pp. 458–463, 2010.

[26] K. Y. Foo and B. H. Hameed, "The environmental applications of activated carbon/zeolite composite materials," *Advances in Colloid and Interface Science*, vol. 162, no. 1-2, pp. 22–28, 2011.

[27] W. P. Cheng, W. Gao, X. Cui, J. H. Ma, and R. F. Li, "Phenol adsorption equilibrium and kinetics on zeolite X/activated carbon composite," *Journal of the Taiwan Institute of Chemical Engineers*, vol. 62, pp. 192–198, 2016.

[28] J. Ma, C. Si, Y. Li, and R. Li, "CO_2 adsorption on zeolite X/activated carbon composites," *Adsorption*, vol. 18, no. 5-6, pp. 503–510, 2012.

[29] W. G. Walter, "APHA standard methods for the examination of water and wastewater," *American Journal of Public Health & The Nations Health*, 1961.

[30] I. Langmuir, "The adsorption of gases on plane surfaces of glass, mica and platinum," *The Journal of the American Chemical Society*, vol. 40, no. 9, pp. 1361–1403, 1918.

[31] H. Freundlich, "Über die adsorption in lösungen," *Zeitschrift für Physikalische Chemie*, vol. 62, no. 5, pp. 121–125, 1906.

[32] O. Redlich and D. L. Peterson, "A useful adsorption isotherm," *Journal of Physical Chemistry*, vol. 63, no. 6, p. 1024, 1959.

[33] S. Lagergren, "About the theory of so-called adsorption of soluble substances," *Kungliga Svenska Vetenskapsakademiens Handlingar*, vol. 24, no. 4, pp. 1–39, 1898.

[34] Y. S. Ho and G. McKay, "Kinetic models for the sorption of dye from aqueous solution by wood," *Process Safety and Environmental Protection*, vol. 76, no. 2, pp. 183–191, 1998.

[35] J.-M. Lv, Y.-L. Ma, X. Chang, and S.-B. Fan, "Removal and removing mechanism of tetracycline residue from aqueous solution by using Cu-13X," *Chemical Engineering Journal*, vol. 273, pp. 247–253, 2015.

[36] Y. Zhao, B. Zhang, X. Zhang, J. Wang, J. Liu, and R. Chen, "Preparation of highly ordered cubic NaA zeolite from halloysite mineral for adsorption of ammonium ions," *Journal of Hazardous Materials*, vol. 178, no. 1–3, pp. 658–664, 2010.

[37] E. Marañón, M. Ulmanu, Y. Fernández, I. Anger, and L. Castrillón, "Removal of ammonium from aqueous solutions with volcanic tuff," *Journal of Hazardous Materials*, vol. 137, no. 3, pp. 1402–1409, 2006.

[38] K. Emerson, R. C. Russo, R. E. Lund, and R. V. Thurston, "Aqueous ammonia equilibrium calculations: effect of pH and temperature," *Journal De L'office Des Recherches Sur Les Pêcheries Du Canada*, vol. 32, no. 12, pp. 2379–2383, 2011.

[39] N. Widiastuti, H. Wu, H. M. Ang, and D. Zhang, "Removal of ammonium from greywater using natural zeolite," *Desalination*, vol. 277, no. 1–3, pp. 15–23, 2011.

11

Preparation of Sulfur-Free Exfoliated Graphite by a Two-Step Intercalation Process and Its Application for Adsorption of Oils

Jun He, Laizhou Song, Hongxia Yang, Xiaohui Ren, and Lifei Xing

School of Environmental and Chemical Engineering, Yanshan University, Qinhuangdao 066004, China

Correspondence should be addressed to Laizhou Song; songlz@ysu.edu.cn

Academic Editor: Carola Esposito Corcione

The sulfur-free exfoliated graphite (EG) was prepared by a two-step chemical oxidation process, using natural flake graphite (NFG) as the precursor. The first chemical intercalation process was carried out at a temperature of 30°C for 50 min, with the optimum addition of NFG, potassium permanganate, and perchloric acid in a weight ratio of 1 : 0.4 : 10.56. Then, in the secondary intercalation step, dipotassium phosphate was employed as the intercalating agent to further increase the exfoliated volume (EV) of EG. NFG, graphite intercalation compound (GIC), and EG were characterized by scanning electron microscope (SEM), energy dispersive spectrometer (EDS), X-ray diffractometer (XRD), Fourier transform infrared spectrometer (FTIR), BET surface area, and porosity analyzer. Also, the uptakes of crude oil, diesel oil, and gasoline by EG were determined. Results show that perchloric acid and hydrogen phosphate are validated to enter into the interlayer of graphite flake. The obtained EG possesses a large exfoliated volume (EV) and has an excellent affinity to oils; thus, the material has rapid adsorption rates and high adsorption capacities for crude oil, diesel oil, and gasoline.

1. Introduction

Although the petroleum consumption has resulted in a great deal of damage to natural environment [1], it is still extensively employed as one of the major fossil fuels. The increasing need of petroleum promotes the transport of this fossil; consequently, oil spill accidents have occurred and posed a serious threat to organisms and human beings [2]. Thus, for the purpose of alleviating the pollution for ocean, rivers, and lakes in the emergency accident caused by oil spill, it is urgent to dispose oil pollutants using the high efficient technology. Generally, the adsorption technique can be exactly competent for the disposal of spilled oils, so suction felt, suction linoleum, foam materials, and some porous polymers have always been employed as the adsorbents. Unfortunately, these adsorbents have an undesirable affinity to oils, thereby weakening the absorption efficiencies of them [3, 4]. In order to alleviate the pollution caused by spilled oils, the exploitation of novel adsorbents with a high affinity to oil is very imperative. Compared with the above-mentioned adsorbents, exfoliated graphite (EG) has received tremendous attention in

dealing with marine oil pollution due to its large adsorption capacity of oil and the noticeable disposal efficiency [5–8].

EG as a sort of versatile material is applied in various fields such as packing [9], sealing material [10], fire retardant [11], oil absorbing material [8], and electrode [12], due to its peculiar properties of flexibility, lubricity, and adsorption. There are three techniques applied to prepare the precursor of EG, namely, graphite intercalation compound (GIC)—gas phase intercalation, chemical oxidation, and electrochemical intercalation [13]. Among these three techniques, chemical oxidation [14] and electrochemical intercalation methods [15] are commonly used. Compared with electrochemical intercalation method, chemical oxidation method owns the merits of more convenience, lower cost, and higher stability of the products. EG with an exfoliated volume (EV) of 200–300 mL/g can be obtained via the conventionally chemical oxidation process using concentrated sulfuric acid and concentrated nitric acid as the oxidants and intercalating reagents [16, 17]. However, this process always releases sulfur dioxide and nitrogen oxide into the air. In this regard, the fabrication process of EG with the friendly environmental characteristics should be developed [18, 19].

The aim of this research is to develop a fabrication process of EG without using concentrated sulfuric acid and nitric acid as oxidants and intercalating reagents and then evaluate its adsorption efficiency of oils. In this study, a two-step chemical oxidation process was proposed; herein, reagents of perchloric acid ($HClO_4$) and dipotassium phosphate (KH_2PO_4) were employed as intercalating agents in the first and second steps, respectively. Structures, crystallinities, components, and chemical groups of natural flake graphite (NFG), GIC, and EG were characterized by scanning electron microscopy (SEM), X-ray diffraction (XRD), energy dispersive X-ray analysis (EDS), and Fourier transform infrared spectroscopy (FTIR) techniques. Influences of intercalation temperature, reaction time, and additions of potassium permanganate ($KMnO_4$) and $HClO_4$ on EV of EG were analyzed. Also, the uptakes of crude oil, diesel oil, and gasoline by EG were determined.

2. Materials and Methods

2.1. Materials. The NFG with an average size of 500 μm was purchased from Qingdao Tianheda Graphite Co., Ltd. (Shandong, China), and the weight content of carbon is 99.5 wt.%. Analytical reagents of $HClO_4$ (70~72 wt.%), $KMnO_4$ (99.5 wt.%), and KH_2PO_4 (99 wt.%) were provided by Jingchun Scientific Co., Ltd. (Shanghai, China); crude oil, diesel oil, and gasoline were obtained from China National Petroleum Corporation.

2.2. Preparations of GIC and EG. GIC was prepared by a chemical oxidation process as follows. Under the continuous stirring condition, 3 g of NFG and the amount of $KMnO_4$ (0.3–2.1 g) were added slowly into a glass beaker, where the amount of $HClO_4$ (9–27 mL) was previously added in this beaker. The addition of $KMnO_4$ should be controlled discreetly to prevent the temperature of the mixture exceeding 80°C. After the one-step intercalation reaction, 50 mL of saturated KH_2PO_4 solution was added to the mixture, and the oxidation process continued for 1 h at a temperature of 40°C. Then, the mixed solution was filtered and the collected GIC was cleaned with distilled water until the pH value of the washing effluent was neutral. Lastly, the cleaned GIC was dewatered by a suction filtration process followed by drying in an oven at 65°C for 24 h.

The preparation process of EG was as follows: 1 g of GIC was first added to a crucible, and then the crucible was placed into a muffle furnace and heated with a temperature of 950°C for 15 s. During the heating treatment, the laminar structure of GIC sample was exfoliated, and EG sample was finally obtained.

2.3. Characterization. The morphologies of NFG, GIC, and EG were characterized by a Hitachi S-4800 scanning electron microscope (SEM, Hitachi, Japan); the chemical components of GIC and EG were determined by an energy dispersive spectrometer (EDS) equipped into the above-mentioned microscope. The crystalline structures of NFG, GIC, and EG were measured using a Rigaku D/max 2500 PC X-ray diffractometer (XRD, Rigaku, Japan) with Cu Kα radiation (λ = 1.5418 Å) at 40 KV and 30 mA. The X-ray diffraction patterns were recorded in the range of 2θ = 10–90°. After being blended with KBr, a Nicolet IS10 Fourier transform infrared (FTIR) spectrometer (Nicolet, America) was employed to detect chemical groups of the obtained GIC and EG samples; FTIR data were recorded in the range of 4000–500 cm^{-1}. BET surface area, pore volume, and average pore diameter of the fabricated EG were measured by mercury intrusion porosimetry using an Autopore IV 9500 analyzer (Micromeritics, USA).

The exfoliated volume (EV) was determined by exfoliating 1.0 g of GIC in a temperature of 950°C for 15 s and its volume was measured using a graduated cylinder. Then the volume was recorded, and this datum was considered to be EV value. In order to ensure the accuracy, measurements were carried out in triplicate, and the average value was adopted.

2.4. Adsorption Rate and Capacity of EG for Oils. Crude oil, diesel oil, and gasoline as the oil candidates were selected as the target pollutants to measure the adsorption rate and capacity of EG. Three portions of EG samples with the same weight were placed in different beakers; then the excessive target oil was added and the solution was stirred at a temperature of 25°C. The mixture of EG and oil was separated by 35-numbered mesh net. After the different adsorption time, the collected EG sample was statically placed to make the excessive oil completely dropping down. The adsorption time was set at 1, 2, 4, 5, 6, 8, 10, 20, and 30 min. The adsorption capacity in different time was obtained from the weight increase before and after adsorption. The adsorption capacity of oil at certain time was calculated via

$$Q = \frac{M - m}{m}, \quad (1)$$

where Q is the adsorption capacity of EG at certain time (g/g); M and m (g) are the weights of EG before and after the oil adsorption. In order to ensure the accuracy, measurements were carried out in triplicate, and the average value was adopted.

The regeneration ratio of EG (the ratio between the readsorbed oil amount and the initial adsorbed value) was assessed with 1 g addition of EG. EG and oil were separated through air pump filtration process; then the regenerated EG was used for oil adsorption; this process was repeated three times. The regeneration ratio is calculated by

$$\frac{\text{regeneration ratio } (n)}{\%} = \frac{Q_{n+1}}{Q_n} \times 100\%, \quad (2)$$

where n is the times of cyclic regeneration.

3. Results and Discussion

3.1. Influence of Intercalation Temperature. The effect of intercalation temperature on EV was presented in Figure 1. As shown in Figure 1, EV of EG increases with the enhancement

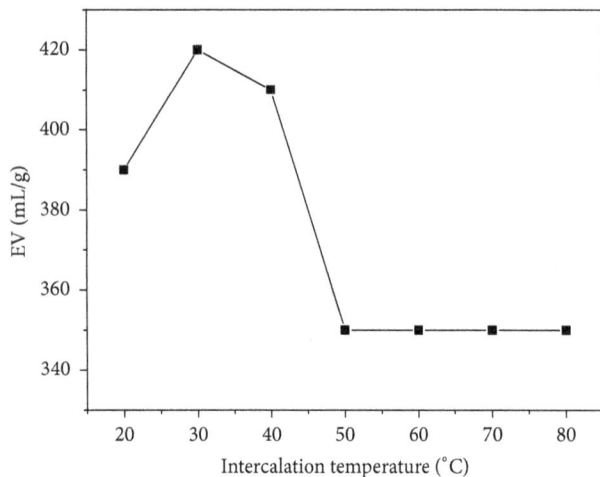

FIGURE 1: Effect of intercalation temperature on EV.

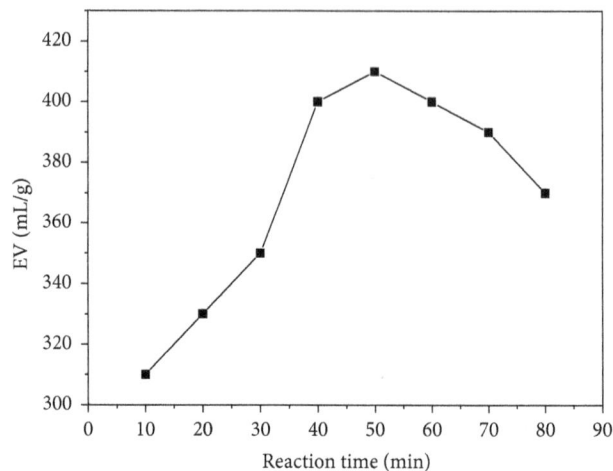

FIGURE 2: Effect of reaction time on EV.

in intercalation temperature, and then it reduces when the temperature was higher than 30°C. Thus, the temperature of 30°C can be identified as the optimum intercalation temperature. As the reaction time was 60 min and the mass ratio of NFG, $KMnO_4$, and $HClO_4$ was 1 : 0.4 : 10.56, the value of EV reaches the maximum value (420 mL/g).

It is well known that the preparation of GIC via the chemical oxidation technique is an exothermic process, so the low temperature can be favorable to the intercalating reaction. But a high reaction temperature (no more than 30°C) might be needed at the beginning, which will be valuable for the intercalating reagent entering into the laminar spacing of NFG. However, the intercalating temperature higher than 30°C will accelerate the volatilization of $HClO_4$. As a result, the amount of this agent entering the graphite interlayer reduces, thus resulting in a low EV.

3.2. Influence of Reaction Time. When the intercalating temperature and the mass ratio of NFG, $KMnO_4$, and $HClO_4$ were kept at 30°C and 1 : 0.4 : 10.56, respectively, the effect of reaction time on EV was analyzed and the results are shown in Figure 2. The value of EV increases from 310 to 420 mL/g as the reaction time elapsed from 10 to 50 min, and then it slightly decreases as time extends. When the reaction time was less than 50 min, the intercalating reaction cannot complete and the extended time will be helpful in destroying the edge layer of graphite by the oxidant. Thus, the prolonged time will guarantee more intercalating reagents entering into the laminar layer of graphite, which is beneficial to the achievement of high EV. This oxidation process is almost finished as the intercalating time is 50 min; then EV will decrease when the time was further prolonged because intercalating reagents will run away as time goes on. Therefore, the optimum reaction time of 50 min was ascertained for the preparation of EG.

3.3. Influence of $KMnO_4$ Addition. EV of EG increases significantly with the increase in $KMnO_4$ addition (from 0.1 to 0.4 g/g NFG) (Figure 3); then the EV change can be ignored

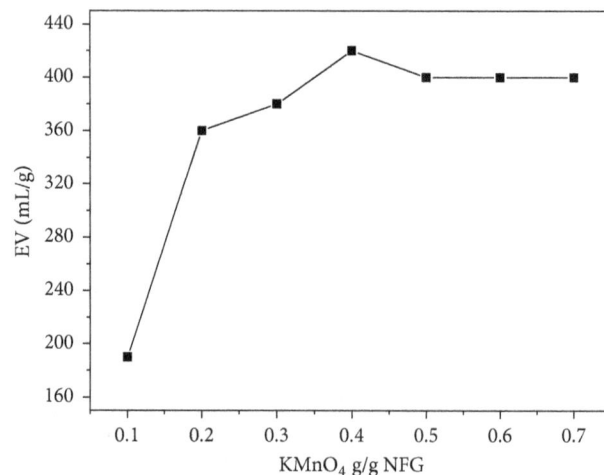

FIGURE 3: Effect of $KMnO_4$ addition on EV.

as the addition of $KMnO_4$ ranges from 0.5 to 0.7 g/g NFG. The maximum EV with a value of 420 mL/g can be obtained with $KMnO_4$ addition of 0.4 g/g NFG. During the intercalating reaction, the concentration of $KMnO_4$ in $HClO_4$ could hardly maintain saturated when its addition is below 0.4; as a consequence, NFG cannot be oxidized sufficiently, and the value of EV will be low. However, the excessive dosage of $KMnO_4$ (>0.4 g/g NFG) will erode the edge layers of NFG and destroy its lamellar structure. The intercalating agents will not be readily inserted into interlayer spacing, thereby resulting in the decrease in EV (<420 mL/g). The optical addition in weight ratio of NFG and $KMnO_4$ is validated as 1 : 0.4.

3.4. Influence of $HClO_4$ Addition. The influence of $HClO_4$ concentration on EV was also investigated and the result is depicted in Figure 4. Firstly, EV considerably increases with the enhancement in addition of $HClO_4$ (5.28~10.56 g/g NFG); but when the dosage exceeded 10.56 g/g NFG, EV decreases slightly and stays at a stable value of 390 mL/g. At a low addition of $HClO_4$, despite the help of $KMnO_4$ to destroy

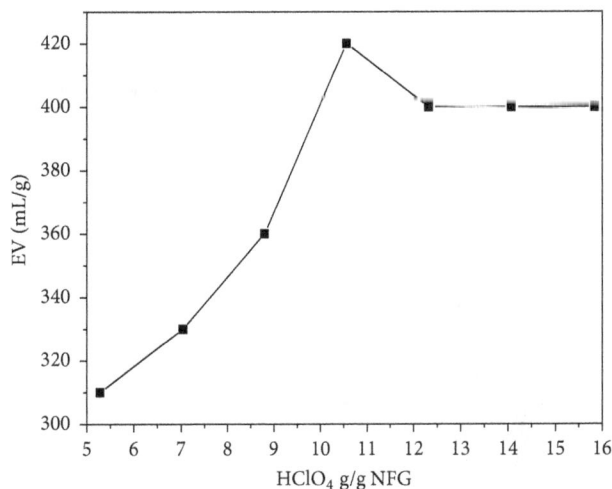

FIGURE 4: Effect of $HClO_4$ addition on EV.

TABLE 1: The atomic ratios of GIC and EG.

Atom	Atomic ratios of a	Atomic ratios of b
C	31.88	94.87
O	57.87	5.06
P	0.59	—
Cl	8.03	0.05
K	0.61	—
Mn	1.01	0.02

the lamellar structure of NFG, the amount of $HClO_4$ was not enough to guarantee that the interlayer spacing of graphite is filled as most possible. On the contrary, at the higher addition of $HClO_4$ (>10.56 g/g NFG), the excessive acid will dilute the concentration of $KMnO_4$, thereby weakening the oxidation potential of this oxidant; this is unfavorable for the enhancement in EV. The optimal weight ratio between NFG and $HClO_4$ is 1:10.56; with these additions of the above two reagents, the measured EV value is 420 mL/g.

In summary, the optimal conditions for the preparation of EG with EV value of 420 mL/g are as follows: mass ratio of NFG : $KMnO_4$: $HClO_4$ = 1 : 0.4 : 10.56; the intercalating temperature 30°C, and the reaction time 50 min. At the aforementioned preparation condition, the EV value of the prepared EG is significantly higher than those reported in literatures [20–22].

3.5. Characterization

3.5.1. SEM Analysis. Morphologies of NFG, GIC, and EG under the previously described conditions were characterized, and SEM micrographs of them are presented in Figure 5. It is clear that the interlayer spacing of NFG is compact and much smaller than that of GIC (Figure 5(a)). The change in interlayer spacing between these two samples can be due to the intercalation and exfoliation of NFG. After the oxidation treatment, the interlayer spacing of NFG is enlarged and its boundary layers are crimped (Figure 5(b)). The worm-like porous structure can be seen from the surface morphology of EG (Figures 5(c)–5(f)), and the pore size ranges from several microns to hundreds of microns. The slit-shaped gaps between the graphite platelets can also be identified; this can be due to the fact that the compounds in interlayers of GIC were decomposed and some micropores are formed during the exfoliation process. The presence of the micropores will be of benefit for guaranteeing an excellent adsorption property of the prepared EG.

3.5.2. EDS Analysis. The chemical components of GIC and EG were determined, and measured EDS spectra are displayed in Figure 6. The chloride (Cl) and phosphorus (P) elements can be identified on the surfaces of tested samples. For both GIC and EG, the determined elements (atomic weight) are shown in Table 1. Atomic weights of Cl and P in GIC are 8.03% and 0.95%, respectively. The existence of Cl and P may derive from the additions of $HClO_4$ and KH_2PO_4. Thus, it can be deduced that both $HClO_4$ and KH_2PO_4 were inserted into the interlayer of NFG; this process is a prerequisite for the preparation of EG. But the content of Cl in EG decreases to 0.05% and P cannot be detected. After the exfoliation treatment at 950°C, the inserted agents were decomposed and the worm-like EG was obtained.

3.5.3. XRD Analysis. The tested XRD patterns of NFG, GIC, and EG are shown in Figure 7. The peak appearing at $2\theta = 26.42°$ (Figure 7(a)) with a d-spacing of 3.37 Å for NFG is observed. For the sample of GIC, this peak with a d-spacing of 3.50 Å shifts to 25.40° (Figure 7(b)). This can be due to the fact that $KMnO_4$ can easily oxidize the graphite layers, and the repulsion interactions between the layers increase, thereby resulting in the increase in the interlayer spacing. As a result, $HClO_4$ migrates into the graphite layers and expands the lamellar spacing along the c-axis direction, which leads to the peak intensity of GIC lower than that of NFG. The similarity of these two peaks mentioned above evidences the insufficient oxidation reaction. As for the XRD pattern of GIC, another peak at $2\theta = 30.02°$ (Figure 7(b)) can be observed, suggesting that NFG is successfully intercalated by $HClO_4$. Also, for the XRD pattern of GIC, it is easy to find that the reflection peak at $2\theta = 54.56°$ is divided into two peaks ($2\theta = 51.88°$ and 56.94°). This change can also be attributed to the oxidation reaction. Two diffraction peaks ($2\theta = 26.44°$ and 54.54°) for EG are observed in Figure 7(c), but no significant difference in their locations can be identified between EG and NFG [21]. The crystallinity of EG is still analogous to that of NFG; between them, however, the discernable difference in diffraction intensities of the above peaks can be attested. After the exfoliation treatment at 950°C, the inserted agents will be decomposed to gaseous compounds and they escape from the edges of the particles, leaving spaces between the layers. Thus, the crystal structure of EG was partially destroyed, so the decrease in crystallinity for EG samples is validated [23].

3.5.4. Analysis of FTIR Spectra. The determined FTIR spectra of GIC and EG are shown in Figure 8. The peaks at 3450

TABLE 2: BET results of EG.

Sample	BET surface area (m^2/g)	Total pore volume (mL/g)	Average pore diameter (μm)
NFG	0.7	—	—
GIC	0.8	—	—
EG	245	37.5	22.1

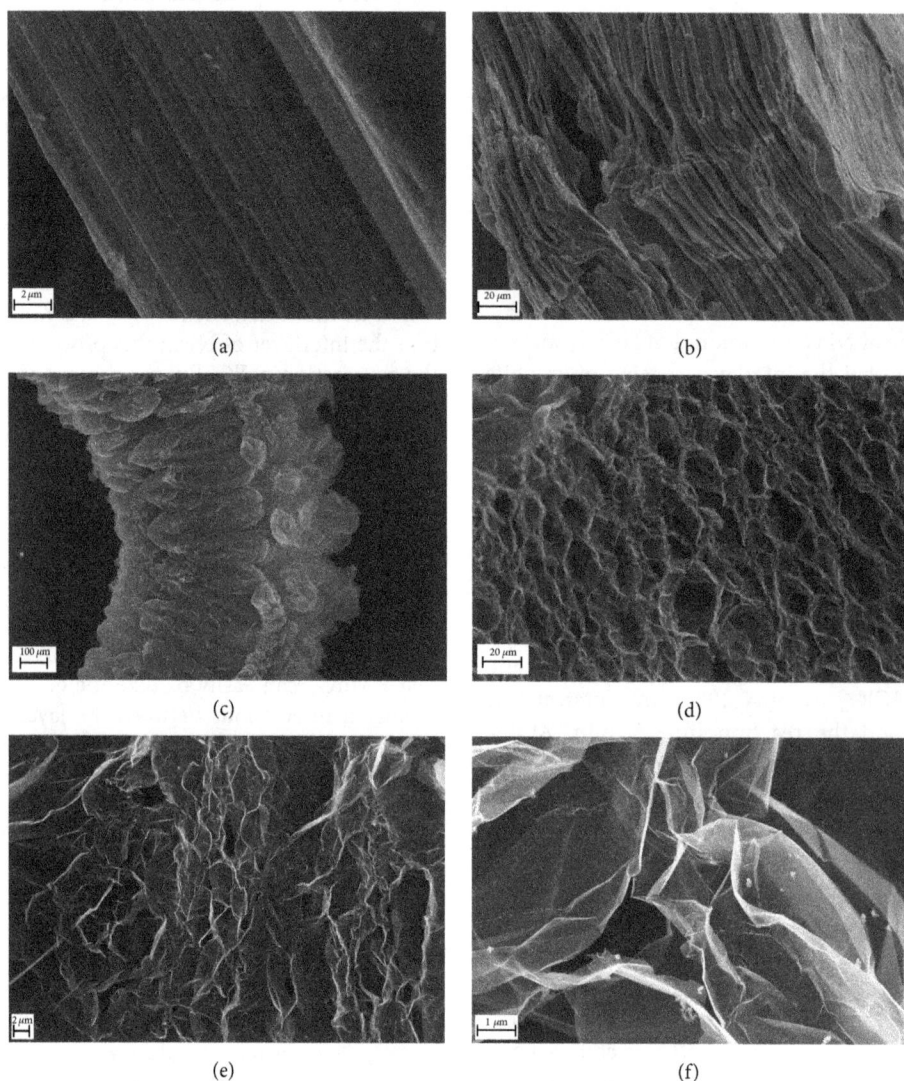

FIGURE 5: SEM images of (a) sectional morphology of natural flake graphite, (b) sectional morphology of GIC, and (c–f) surface morphology of EG: the magnification of each image is listed as follows: (a) ×5000, (b) ×500, (c) ×100, (d) ×500, (e) ×2000, and (f) ×10000.

and $1630\,cm^{-1}$ (for GIC) and those at 3444 and $1624^{-1}\,cm$ (for EG) are derived from adsorbed water [22]. The peak of GIC at $1397\,cm^{-1}$ is ascribed to hydroxyl groups because of the oxidation of NFG [24]. The presence of two peaks of GIC (625 and $1112\,cm^{-1}$) suggests the existence of ClO_4^{-}, and other peaks (1121–$1136\,cm^{-1}$) can be assigned to $H_2PO_4^{-}$. Therefore, reagents of $HClO_4$ and KH_2PO_4 are further ascertained to enter into graphite layers, being helpful in the preparation of EG.

3.5.5. Analysis of BET and Pore Structure. The surface area of NFG and GIC is low because they are bulk materials with an average size of $500\,\mu$m and less of porosity. BET surface area, pore volume, and pore diameter for EG sample were measured, and results are summarized in Table 2. The results show the feature of EG sample, that is, high surface area and micron pore structure. The surface area of the prepared EG was calculated to be $245\,m^2/g$, which is much higher than that of exfoliated graphite synthesized by microwave irradiation [8]. It is

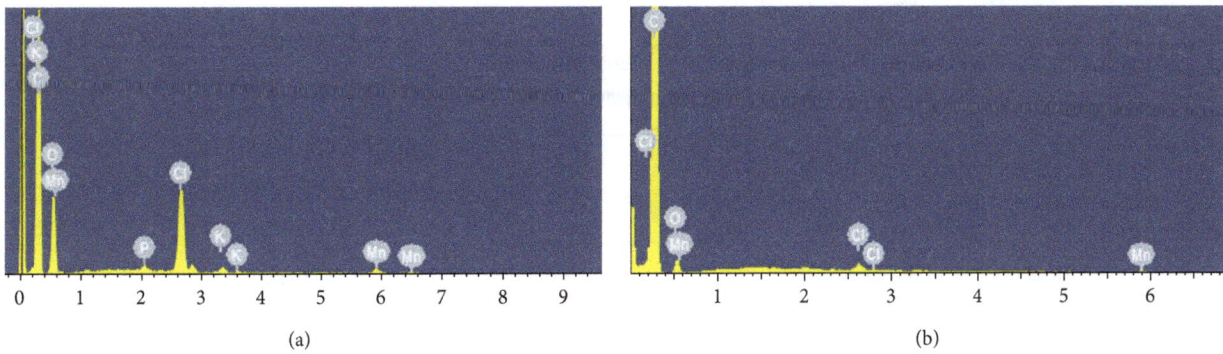

FIGURE 6: EDS spectra of (a) GIC and (b) EG.

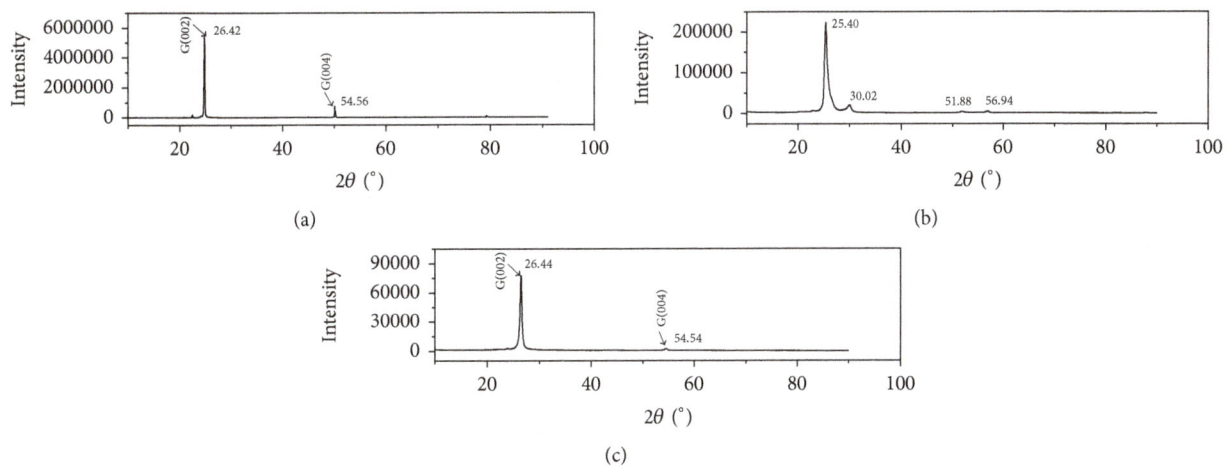

FIGURE 7: XRD patterns of (a) NFG, (b) GIC, and (c) EG.

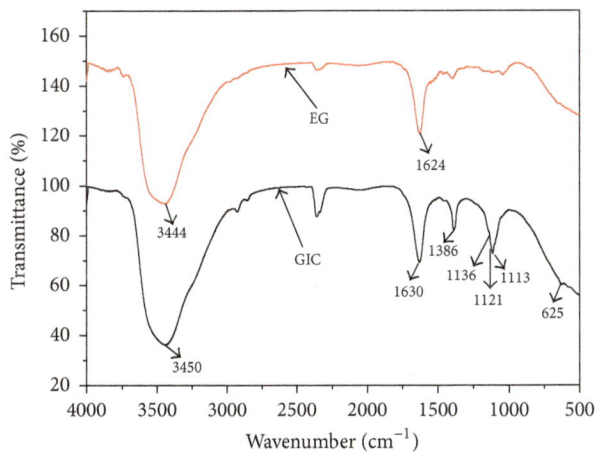

FIGURE 8: FTIR spectra of GIC and EG.

worth mentioning that the micrometer level of average pore diameter of EG will be advantageous to significantly enhance the adsorption performance toward oils [25, 26].

3.6. Adsorption Capacity and Adsorption Rate of EG for Crude Oil, Diesel Oil, and Gasoline. Figure 9 displays dynamic

adsorption behaviors of EG for crude oil, diesel oil, and gasoline. The maximum adsorption capacities are 123.3 g/g for crude oil, 76.5 g/g for diesel oil, and 61.4 g/g for gasoline. The adsorption rate is another important factor during adsorption process. The time to reach the equilibrium sorption capacity for each oil is 5, 2, and 2 min, respectively, reflecting rapid adsorption rates for these oils. These results indicate that the as-prepared EG will be competent for the application in oil spill accidents. Herein, the adsorption characteristics of EG toward the oil will be comprehensively investigated in the further research.

3.7. Cyclic Regeneration of EG. Table 3 exhibits cyclic regeneration abilities of EG after adsorbing the aforesaid three oils. The adsorption capacity of EG for each oil decreases slightly after regeneration, confirming that EG can be repeatedly applied as an effective adsorbent after the separation process by air pump filtration.

3.8. Mechanism of Adsorption. In the process of GIC sample preparation, $HClO_4$ is used as both an intercalation agent and oxidizing agent. The strong acidity, volatility, and oxidability of $HClO_4$ can lead to a high-temperature expansion and the formation of microns pores for EG sample, guaranteeing its

TABLE 3: Cyclic regeneration of EG for the three kinds of oils.

Categories of oils	Adsorption capacity number 1/g	Adsorption capacity number 2/g	Adsorption capacity number 3/g	Regeneration ratio (1)/%	Regeneration ratio (2)/%	Average regeneration ratio/%
Crude oil	45.6	40.3	34.5	88.38	85.61	86.99
Diesel oil	27.8	23.6	19.4	84.89	82.20	83.29
Gasoline	23.5	19.8	15.6	84.25	78.79	81.52

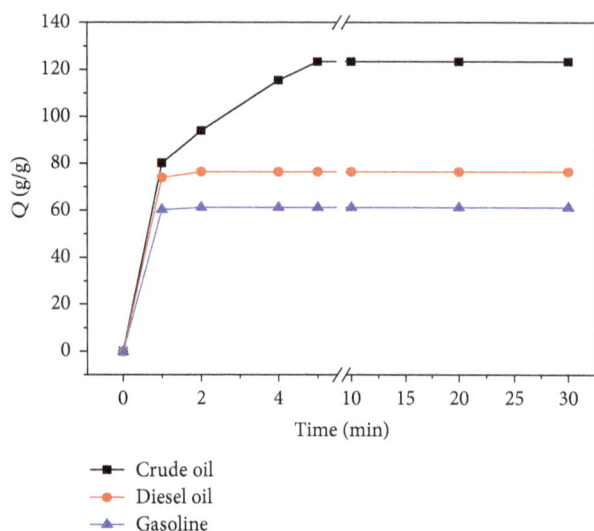

FIGURE 9: Dynamic adsorption curves of EG for crude oil, diesel oil, and gasoline.

excellent adsorption properties toward oils. In addition, high BET surface area and large pore volume of EG will also ensure rapid adsorption rates and high adsorption capacities.

4. Conclusions

In this study, a facile two-step intercalation method is proposed to prepare the sulfur-free EG, using $HClO_4$ and KH_2PO_4 as the inserting agent and $KMnO_4$ employed as the oxidant. The oxidant can easily open the graphite layers and the inserting agent will move into the interlayer spacing to obtain the GIC sample. The optimum conditions for the preparation of GIC are as follows: the mass ratio of NFG, $KMnO_4$, and $HClO_4$ is $1 : 0.4 : 10.56$; the reaction temperature and time are kept at $30°C$ and 50 min. EV of the prepared EG is 420 mL/g. The worm-like porous morphology, high BET surface area, and large pore volume of EG guarantee the excellent adsorption property toward oils. The maximum adsorption capacities of EG for crude oil, diesel oil, and gasoline are 123.3, 76.5, and 61.4 g/g, respectively.

Conflicts of Interest

The authors declare that there are no conflicts of interest regarding the publication of this paper.

Acknowledgments

This research was funded by Hebei Provincial Science and Technology Support Program of China (Grant no. 15273613).

References

[1] M. Toyoda and M. Inagaki, "Sorption and recovery of heavy oils by using exfoliated graphite," *Spill Science and Technology Bulletin*, vol. 8, no. 5-6, pp. 467–474, 2003.

[2] G. Wang, Q. Sun, Y. Zhang, J. Fan, and L. Ma, "Sorption and regeneration of magnetic exfoliated graphite as a new sorbent for oil pollution," *Desalination*, vol. 263, no. 1-3, pp. 183–188, 2010.

[3] A. Li, R. Lin, C. Lin et al., "An environment–friendly and multi–functional absorbent from chitosan for organic pollutants and heavy metal ion," *Carbohydrate Polymers*, vol. 148, pp. 272–280, 2016.

[4] S. O. Alayande, E. O. Dare, F. O. G. Olorundare, D. Nkosi, T. A. M. Msagati, and B. B. Mamba, "Superoleophillic electrospun polystrene/exofoliated graphite fibre for selective removal of crude oil from water," *Physics and Chemistry of the Earth*, vol. 92, pp. 3–6, 2016.

[5] C. Liu, Z. Chen, X. Cheng, Z. Wang, and X. Duan, "Preparation and structure analysis of expanded graphite–based composites made by phosphoric acid activation," *Journal of Porous Materials*, vol. 17, no. 4, pp. 425–428, 2010.

[6] V. G. Makotchenko, E. D. Grayfer, A. S. Nazarov, S.-J. Kim, and V. E. Fedorov, "The synthesis and properties of highly exfoliated graphites from fluorinated graphite intercalation compounds," *Carbon*, vol. 49, no. 10, pp. 3233–3241, 2011.

[7] B. Tryba, J. Przepiórski, and A. W. Morawski, "Influence of chemically prepared H_2SO_4-graphite intercalation compound (GIC) precursor on parameters of exfoliated graphite (EG) for oil sorption from water," *Carbon*, vol. 41, no. 10, pp. 2013–2016, 2003.

[8] N. Sykam and K. K. Kar, "Rapid synthesis of exfoliated graphite by microwave irradiation and oil sorption studies," *Materials Letters*, vol. 117, no. 15, pp. 150–152, 2014.

[9] B. Debelak and K. Lafdi, "Use of exfoliated graphite filler to enhance polymer physical properties," *Carbon*, vol. 45, no. 9, pp. 1727–1734, 2007.

[10] N. E. Sorokina, A. V. Redchitz, S. G. Ionov, and V. V. Avdeev, "Different exfoliated graphite as a base of sealing materials," *Journal of Physics and Chemistry of Solids*, vol. 67, no. 5-6, pp. 1202–1204, 2006.

[11] B. Dittrich, K.-A. Wartig, D. Hofmann, R. Mülhaupt, and B. Schartel, "Flame retardancy through carbon nanomaterials: carbon black, multiwall nanotubes, expanded graphite, multi–layer

graphene and graphene in polypropylene," *Polymer Degradation and Stability*, vol. 98, no. 8, pp. 1495–1505, 2013.

[12] M. Kujawski, J. D. Pearse, and E. Smela, "Elastomers filled with exfoliated graphite as compliant electrodes," *Carbon*, vol. 48, no. 9, pp. 2409–2417, 2010.

[13] X. Van Heerden and H. Badenhorst, "The influence of three different intercalation techniques on the microstructure of exfoliated graphite," *Carbon*, vol. 88, article 9752, pp. 173–184, 2015.

[14] Y. Chen, R. Luo, S. Li et al., "Preparation of highly–expandable graphite using waste liquid propellants of nitric-27S as one of intercalating agents," *Carbon*, vol. 50, no. 5, p. 2063, 2012.

[15] H. M. A. Asghar, S. N. Hussain, H. Sattar, N. W. Brown, and E. P. L. Roberts, "Environmentally friendly preparation of exfoliated graphite," *Journal of Industrial and Engineering Chemistry*, vol. 20, no. 4, pp. 1936–1941, 2014.

[16] M. Inagaki, R. Tashiro, Y.-I. Washino, and M. Toyoda, "Exfoliation process of graphite via intercalation compounds with sulfuric acid," *Journal of Physics and Chemistry of Solids*, vol. 65, no. 2-3, pp. 133–137, 2004.

[17] R. Bissessur and S. F. Scully, "Intercalation of solid polymer electrolytes into graphite oxide," *Solid State Ionics*, vol. 178, no. 11-12, pp. 877–882, 2007.

[18] Jihui-Li, Huifang-Da, Qian-Liu, and Shufen-Liu, "Preparation of sulfur-free expanded graphite with 320 μm mesh of flake graphite," *Materials Letters*, vol. 60, no. 29-30, pp. 3927–3930, 2006.

[19] J. Li, J. Li, and M. Li, "Ultrasound irradiation prepare sulfur–free and lower exfoliate–temperature expandable graphite," *Materials Letters*, vol. 62, no. 14, pp. 2047–2049, 2008.

[20] Y.-P. Chen, S.-Y. Li, R.-Y. Luo, X.-M. Lu, and X.-J. Wang, "Optimization of initial redox potential in the preparation of expandable graphite by chemical oxidation," *New Carbon Materials*, vol. 28, no. 6, pp. 435–441, 2013.

[21] X. H. Wei, L. Liu, J. X. Zhang, J. L. Shi, and Q. G. Guo, "HClO4-graphite intercalation compound and its thermally exfoliated graphite," *Materials Letters*, vol. 63, no. 18-19, pp. 1618–1620, 2009.

[22] Z. Ying, X. Lin, Y. Qi, and J. Luo, "Preparation and characterization of low–temperature expandable graphite," *Materials Research Bulletin*, vol. 43, no. 10, pp. 2677–2686, 2008.

[23] K.-C. Tsai, H.-C. Kuan, H.-W. Chou, C.-F. Kuan, C.-H. Chen, and C.-L. Chiang, "Preparation of expandable graphite using a hydrothermal method and flame–retardant properties of its halogen–free flame–retardant HDPE composites," *Journal of Polymer Research*, vol. 18, no. 4, pp. 483–488, 2011.

[24] S. Stankovich, R. D. Piner, S. T. Nguyen, and R. S. Ruoff, "Synthesis and exfoliation of isocyanate–treated graphene oxide nanoplatelets," *Carbon*, vol. 44, no. 15, pp. 3342–3347, 2006.

[25] S. J. Yang, J. H. Kang, H. Jung, T. Kim, and C. R. Park, "Preparation of a freestanding, macroporous reduced graphene oxide film as an efficient and recyclable sorbent for oils and organic solvents," *Journal of Materials Chemistry A*, vol. 1, no. 33, pp. 9427–9432, 2013.

[26] S. Kabiri, D. N. H. Tran, T. Altalhi, and D. Losic, "Outstanding adsorption performance of graphene–carbon nanotube aerogels for continuous oil removal," *Carbon*, vol. 80, no. 1, pp. 523–533, 2014.

12

Purification, Characterization, and Time-Dependent Adsorption Studies of Ghanaian Muscovite Clay

Samuel Tetteh ⓘ,[1] Andrews Quashie,[2] and Michael Akrofi Anang ⓘ[1]

[1]*Chemistry Department, School of Physical Sciences, College of Agriculture and Natural Sciences, University of Cape Coast, Cape Coast, Ghana*
[2]*Institute of Industrial Research, Council for Scientific and Industrial Research, Accra, Ghana*

Correspondence should be addressed to Samuel Tetteh; stoshgh2001@yahoo.com

Academic Editor: Amit Bhatnagar

Three clay samples (E1, E2, and C1) extracted from different parts of Ghana have been purified by sedimentation. The samples were further characterized by powder X-ray diffraction (PXRD), Fourier transform infrared spectroscopy (FT-IR), scanning electron microscopy (SEM), cation exchange capacity (CEC), and point of zero charge (pH_{pzc}). PXRD and FT-IR data revealed the samples to be predominantly muscovite clay with percentages ranging from 82.71 to 91.33%. The surfaces were mostly cationic with pH_{pzc} ranging from 5.58 to 6.40. Morphological studies by SEM confirmed the crystalline nature of the surfaces which is suitable for adsorption studies. Time-dependent adsorption studies show that C1 is a good candidate for the adsorption of chlorophenols, methyl orange, and Eriochrome Black T.

1. Introduction

Phenolic compounds are classified among the main raw materials used in industry and photochemical research [1–4]. Phenols are widely used in the production of materials such as pharmaceuticals, dyes, cosmetics, epoxy resins, and adhesives as well as photocells for various applications [5, 6]. They are the main contaminants in wastewater generated from petroleum refineries, dye- and phenol-producing industries, and general households which use phenol-based disinfectants [7]. Phenols also enter the environment through the use of phenoxy herbicides such as 2,4-dichlorophenoxyacetic acid or 4-chloro-2-methylphenoxyacetic acid and also phenolic biocides like pentachlorophenol [8]. These compounds are classified as noxious pollutants because of their toxicity even at low concentrations. The toxicity of phenols and their derivatives include histopathological changes, mutagenicity, and carcinogenicity [9]. Stringent US Environmental Protection Agency (USEPA) regulations have therefore been called for the lowering of phenol content in wastewater to less than 1 mg/L [10].

Several methods have been employed for the removal of phenolics from aqueous solutions. These techniques can be classified as biological, chemical, or physical [11]. Biodegradation methods, including microbial degradation, fungal decolonization, and bioremediation systems, have been employed in the treatment of industrial effluents, as microorganisms such as yeast, bacteria, and fungi are able to catalyze the breakdown of several organic pollutants into harmless products. However, if not controlled, these microorganisms could contaminate other systems leading to the outbreak of diseases [12] and emission of toxic products [13]. On the other hand, chemical methods such as coagulation, precipitation, flocculation with iron(II)/calcium hydroxide, electrokinetic coagulation, and conventional oxidation methods involving oxidizing agents such as hydrogen peroxide, ozone, and $KMnO_4$ have also been explored for the degradation of phenolics in wastewater [14–16]. Recently, advanced oxidation processes like the Fenton process have also been used [17, 18]. The Fenton process and related reactions usually involve reactions of peroxides (particularly hydrogen peroxide) with Fe(II) ions to form reactive oxygen species such as the hydroxyl radical

(HO) which initiates a series of chain reactions that catalyze the degradation of various organic compounds [19]. This method has been effectively used in wastewater treatment processes for the removal of hazardous organics from wastewater [20]. These methods are not environmentally friendly because the initiated free radical reactions are very difficult to terminate and could affect aquatic life when the effluents are discharged into water bodies. They are also expensive and commercially unattractive.

Several physical methods such as membrane-filtration processes and adsorption techniques are also used in the treatment of wastewater [21]. Adsorption processes involving solid adsorbents are widely used to remove certain pollutants from industrial effluents before final disposal into water bodies. Amongst the solid adsorbents used commercially, activated carbon is widely employed for the removal of a wide variety of phenolic compounds from wastewater because of its large capacity to adsorb phenols [22, 23]. This can be attributed to its structural characteristics and high porosity which gives it a large surface area [23]. However, there are several challenges which come with the production and use of activated carbon. These include the destruction of vegetation and the generation of harmful gases such as carbon, CO, and NOx which are not environmentally friendly. The regeneration of saturated carbon is also expensive and results in the loss of adsorbent [11].

Natural clay mineral, an environmentally friendly and relatively cheap adsorbent, has been explored for the removal of phenols and dyes from water and wastewater [24]; however, industrial scale application is still limited as a result of lack of specificity of the type of dye for a given clay. Some dyes adsorb strongly onto certain clays, whereas others adsorb poorly. According to Kurniawan et al., the use of chemicals for the surface modification of clay materials improves its adsorption characteristics [25], but these chemicals usually pose serious problems to the environment and usually require expensive waste treatment systems to remove the excess chemicals. Untreated clay samples which have been successfully applied in the adsorption of dyes from wastewater include bentonite [26–29], kaolinite [30], attapulgite [31], montmorillonite [32, 33], and alunite [34]. Extensive research have shown that the point of zero charge (pH_{pzc}) significantly influence the adsorption properties of the type of clay [35]. It determines the surface charge as well as the cation exchange capacity of the clay material. In this work, we report the purification and characterization of three muscovite clay samples extracted from the eastern and central regions of Ghana for the adsorption of chloro-substituted phenols and representative dyes from aqueous solutions. Extensive review of related literature has shown that there is no information available in this regard.

2. Materials and Methods

2.1. Sample Treatment. Raw clay samples E1 and E2 were collected from the eastern region, while C1 was obtained from the central region of Ghana. The samples were then transferred to clean polyethylene bags in the laboratory for further analyses. In the lab, foreign materials such as stones,

roots, and dead insects were separated, and the samples were ground using porcelain mortar and pestle. 1.00 kg of each clay sample was suspended in 10 L of distilled water and homogenized using a mechanical stirrer. The suspension was then left to sediment for 24 hours after which the fine clay particles settled due to gravitation. The clear supernatant and the upper gritty layer were then discarded. The process was repeated for three more times to obtain a pure clay sample for the work. The purified clay samples were air dried and subsequently oven dried at 105°C for 24 hours. They were then calcined at 500°C for 4 hours to ensure the breakdown of volatile organic compounds and carbonaceous materials.

2.2. Characterization. The cation exchange capacity (CEC) of the clay samples was determined by the ammonium acetate method [36]. The point of zero charge (PZC) was also determined by the salt addition method [35, 37] in the pH range 4–9.

Powder X-ray diffraction patterns of the samples were obtained with a Panalytical Empyrean Powder X-ray Diffractometer ($CuK\alpha1 = 1.541$ Å, $CuK\alpha2 = 1.544$ Å, $CuK\beta = 1.392$ Å, step scan size of 0.03° from 5.015° to 69.965° at 25°C). The samples were prepared by pressing composite samples with a flat surface down the sample cells. This ensured that the samples were densely packed and randomly oriented. The diffraction patterns were drawn from the data using Spectragryph [38]. The composition of the samples was determined using the MAUD software [39, 40].

The morphology of the samples was studied using images generated by JEOL JSM-6390LV Scanning Electron Microscope (SEM). A single layer of each sample was mounted on a stub using a double-sided adhesive carbon sheets. The magnification used was 500 for samples C1, 250 for E2, and 1200 for E1.

Fourier transform infrared spectroscopy (FT-IR) was carried out on a PerkinElmer UATR Two FT-IR spectrometer over a spectral region of 4000–400 cm^{-1} with a resolution of 1 cm^{-1}. The samples were analyzed in the form of powders without further dilutions.

2.3. Adsorption Experiment. Stock solutions of the respective adsorbates were prepared in methanol (100 mg/L). Working solutions were then prepared by appropriately diluting the stock solution with distilled water to get the adsorbates into aqueous solutions since the chlorophenols are insoluble in water (but the methanolic solution is soluble in water). For the batch adsorption studies, 0.5 g of each clay sample was added to 50 mL of aqueous solution of the adsorbate (5 mg/L). The mixture was stirred with a magnetic stirrer at 300 rpm. Exactly 5 mL of the samples was withdrawn after two hours, filtered, and the absorbance measured with a Shimadzu T70 UV-Vis spectrometer at predetermined wavelengths of maximum absorption (λ_{max}) (phenol = 280 nm, 2-chlorophenol = 275 nm, 3-chlorophenol = 300 nm, 4-chlorophenol = 260 nm, methyl orange = 440 nm, and Eriochrome Black T = 200 nm). All the experiments were carried out at 303 K in a temperature-controlled water bath.

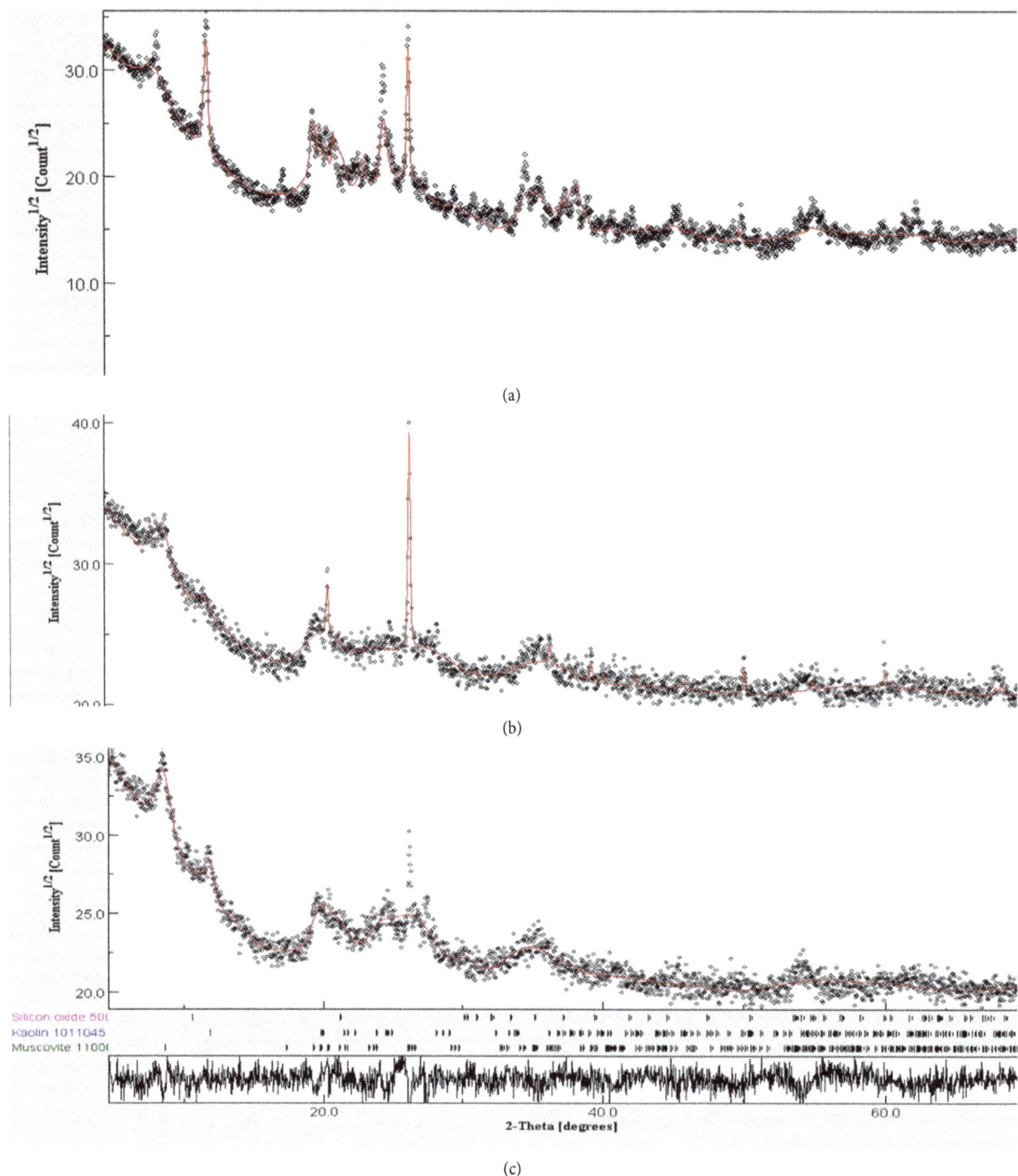

FIGURE 1: Powder X-ray diffraction pattern of the clay samples. (a) C1, (b) E1, and (c) E2.

3. Results and Discussion

Figure 1 shows the diffractograms of the X-ray diffraction pattern as well as the curve fitting patterns of the clay samples. This was done to elucidate the composition of the samples. All the patterns display the characteristic peak at $2\theta = 26.57°$ associated with silicon dioxide. Other conspicuous peaks include $2\theta = 8.83°$ for muscovite and $2\theta = 12.37°$ for kaolinite.

As shown in Table 1, muscovite was highest in all the samples with proportions of 82.71, 89.84, and 91.33% for C1, E1, and E2, respectively, followed by kaolinite and silicon dioxide.

Typical kaolinite and muscovite vibrational frequencies were observed in all the samples. These include conspicuous sharp doublets at 3694 and 3622 cm^{-1} in C1. These peaks were however weaker in E1 and E2. These bands arise as a result of interlayer O-H stretches within the clay structure

TABLE 1: Peak position and percent composition of the clay samples.

Sample	Peak position (2θ)	Intensity (counts)	(%) Muscovite	(%) Kaolinite	(%) Silicon dioxide
C1	26.62	1136	82.71	14.18	3.11
	8.85	1058			
	12.34	1251			
E1	26.62	1654	89.84	4.67	5.49
	8.58	1066			
	12.03	762			
E2	26.62	900	91.33	7.11	1.56
	8.94	1231			
	12.25	836			

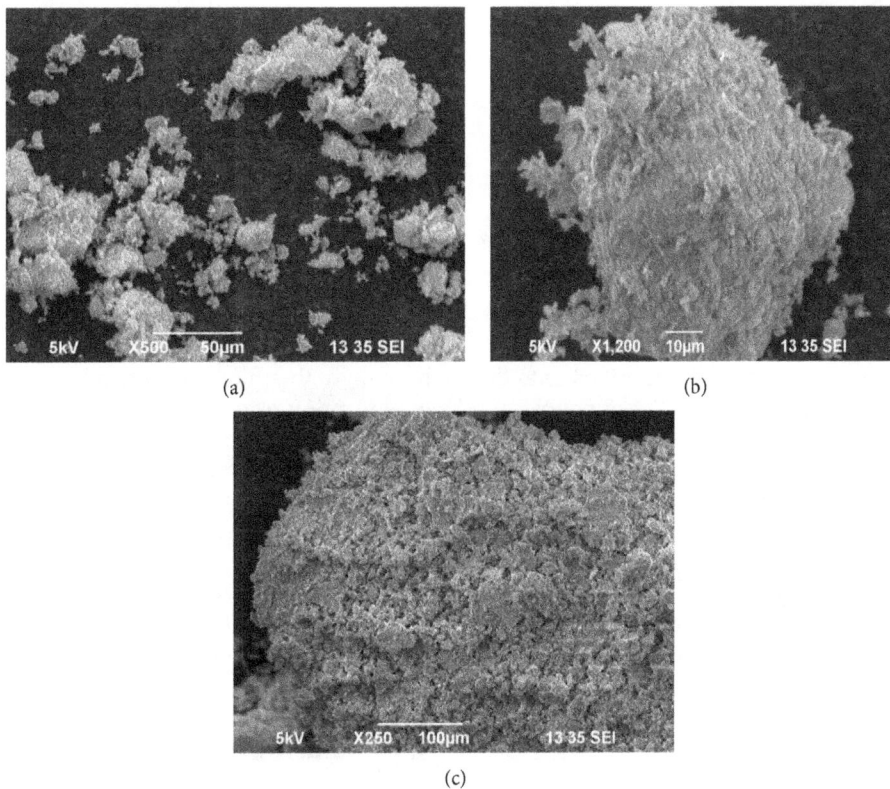

(a)

(b)

(c)

FIGURE 2: SEM micrographs of the clay samples. (a) E1, (b) C1, and (c) E2.

[41–43]. Broad peaks at $3404 \, cm^{-1}$ in E1 and E2 can be assigned to stretching of the hydroxyl group in adsorbed water molecules [44]. Strong O-H deformation bands of kaolinites were also observed at 912 and $911 \, cm^{-1}$ with typical Si-O stretching vibrations around $1002–1029 \, cm^{-1}$ in all the samples [41]. Si-O-Al stretching bands were also identified at 550, 667, and $531 \, cm^{-1}$ in E2, E1, and C1, respectively. According to Vaculikova and Plevova, vibrational peaks at 3622, 3436, 1026, 910, and $525 \, cm^{-1}$ are characteristic of muscovite clays with a confirmatory peak at 3698 attributable to kaolinites [42]. Therefore, the results of FT-IR data are important in the characterization of clay minerals. The surface topography of the samples was further studied by scanning electron microscopy.

Figure 2 shows the crystalline structure and orientation of the particle sizes making up the clay materials at X500

magnification for samples C1, X250 for E2, and X1200 for E1. These micrographs show some clay platelets of varying sizes arranged in various patterns.

In the determination of the point of zero charge, $NaNO_3$ was added to the clay suspensions to maintain a constant ionic strength. The changes in pH at fixed ionic strength were then measured as shown in Figure 3. The structural properties of clay minerals make them suitable as charge carriers and form the basis of cation exchange and the swelling ability of these minerals. The charge usually exists as structural or surface charge. The structural charge exists in-between the octahedral sheets and is permanent while the surface charge depends on the hydrolysis of Si-OH and Al-OH bonds along the surface of the clay structure and is pH dependent. Therefore, the net charge can either be positive or negative. The point of zero charge (pH_{pzc}) is the pH of the

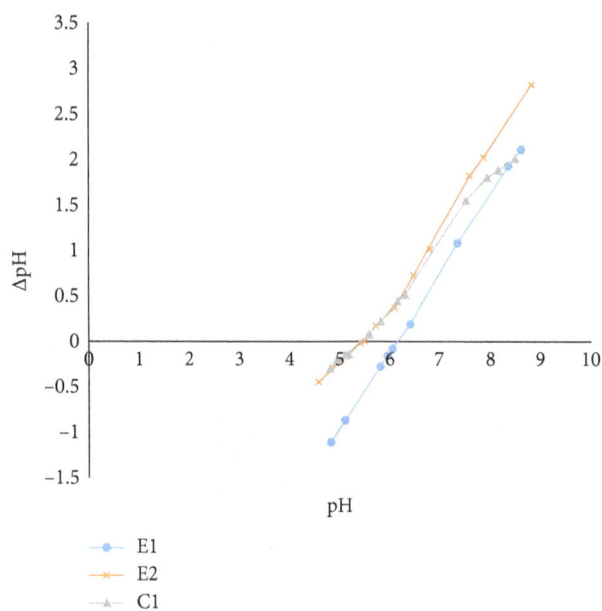

FIGURE 3: Point of zero charge (pH$_{pzc}$) of the respective clay samples.

TABLE 2: Concentration (cmol/kg) of exchangeable ions in the clay samples.

ion	Sample		
	C1	E1	E2
Ca^{2+}	0.880	13.204	2.600
Mg^{2+}	0.480	2.841	1.560
Na$^+$	0.314	0.377	0.377
K$^+$	0.064	0.077	0.077

clay suspension at which the net surface charge is zero. It determines how easily a pollutant can be adsorbed onto the surface of the clay. From Figure 3, the pH$_{pzc}$ of samples C1 and E2 was 5.6, whereas that of E1 was 6.4. According to Mahmood et al., pH$_{pzc}$ values are often divergent and no sample has a fixed pH$_{pzc}$ value [35]. Similar reports by Schindler et al. [45] and Weiland and Stumm [46] show that the pH$_{pzc}$ for kaolin varies from 4.0 to 7.5 and that for illite clay from 2.5 to 8.0. These differences have been attributed to the differences in the origin of the minerals and the methods used [35]. Nonetheless, the surface charge gives an indication of the type of pollutant that can be adsorbed onto the surface of the clay mineral [47]. The values recorded in Figure 3 show that the surface charge of the clay samples is generally positive and can adsorb negative species.

Cation exchange capacity is a measure of the clay mineral to substitute unfixed cations in the interlayer and surface layers for cations in solution. It expresses the ability of clay to adsorb pollutants such as heavy metals, phenols, dyes, and biocides.

Most common cations found in exchangeable positions include Ca^{2+}, Mg^{2+}, Na$^+$, and K$^+$. From Table 2, Ca^{2+} and Mg^{2+} were the dominant exchangeable ions in all the clay samples with K$^+$ being the least.

3.1. Adsorption Studies.

The pH$_{pzc}$ is an important factor that determines the surface active centers and the types of adsorbates that a clay sample can adsorb. The pH$_{pzc}$ values shown in Figure 3 illustrate the cationic nature of the clay surfaces. The lower the pH$_{pzc}$, the more cationic the surface layer of the clay, making it susceptible to adsorb anionic pollutants [48].

Figure 4 shows the time-dependent adsorption kinetics of phenols on the different clay samples. Generally, the clay

samples were very efficient in removing 4-chlorophenol. The high adsorption capacity of C1 is attributable to the high content (14.18%) of kaolinite as shown in Table 1. According to Dogan et al. [49], the 1:1 aluminosilicate structure of kaolinite clay carries a small negative charge as a result of the isomorphous substitution of Si^{4+} by Al^{3+}, leaving a slight negative charge for each substitution [49, 50]. The presence of this slight negative charge increases the electron density on the clay surface thereby facilitating the interaction between the clay surface and the phenolic hydrogen [51]. Phenols however appear to be amphoteric with the phenolic –OH acting as a hydrogen bond donor/acceptor in its interactions. This property of phenols makes the adsorption complex under different experimental conditions [52]. The slopes of the curves were however similar indicating that the position of the chloro-substituent has little effect on the rate of adsorption of the phenols onto the clay sample [53].

The adsorption kinetics of Eriochrome Black T (a phenolic dye) and methyl orange (an anionic dye) are shown in Figure 5.

It is evident that methyl orange is better adsorbed than Eriochrome Black T. This can partly be attributed to the cationic surface charge of the clay which has higher affinity for the anionic methyl orange. Similar studies by Nishimura et al. showed the adsorption propensity of dodecylammonium amine surfactants at muscovite-water interfaces [54].

4. Conclusion

Three clay samples (E1, E2, and C1) extracted from different parts of Ghana have been purified and characterized. Powder X-ray diffraction data shows that the samples were predominantly muscovite with the following compositions: E1 (89.84% muscovite, 4.67% kaolinite, and 5.59% SiO$_2$), E2 (91.33% muscovite, 7.11% kaolinite, and 1.56% SiO$_2$), and C1 (82.71% muscovite, 14.18% kaolinite, and 3.11% SiO$_2$). The composition was confirmed by comparing with characteristic peaks in the FT-IR spectrum of standard clay samples. Surface characterization by scanning electron microscopy revealed crystalline morphology of the samples which are suitable for adsorption studies. The cation exchange capacity was determined by the ammonium acetate method, and Ca^{2+} and Mg^{2+} were found as the most exchangeable cations. The surface was further characterized by the salt addition method to determine the point of zero charge (pH$_{pzc}$). The surfaces were found to be slightly cationic with pH$_{pzc}$ 5.58–6.40. Time-dependent adsorption studies show that C1 is a good candidate for the adsorption of chlorophenols, methyl orange, and Eriochrome Black T.

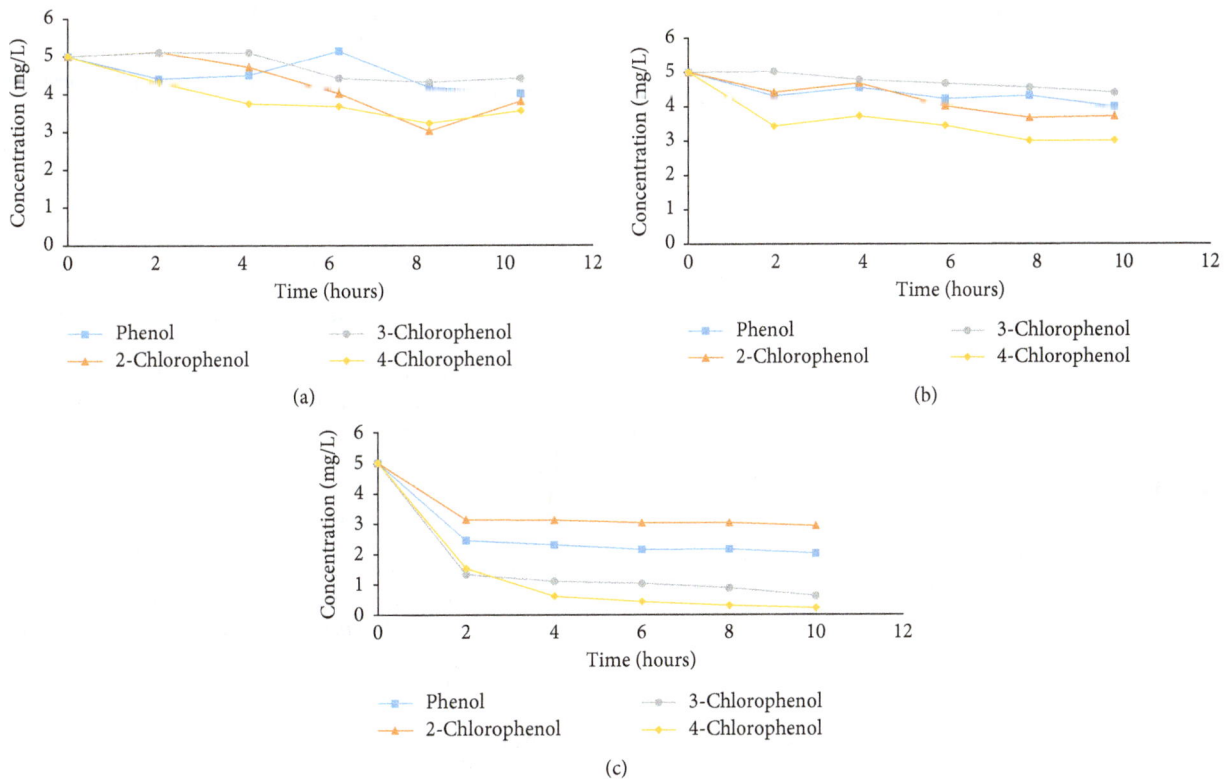

FIGURE 4: Time-dependent adsorption kinetics of phenols on the clay samples. (a) E1, (b) E2, and (c) C1.

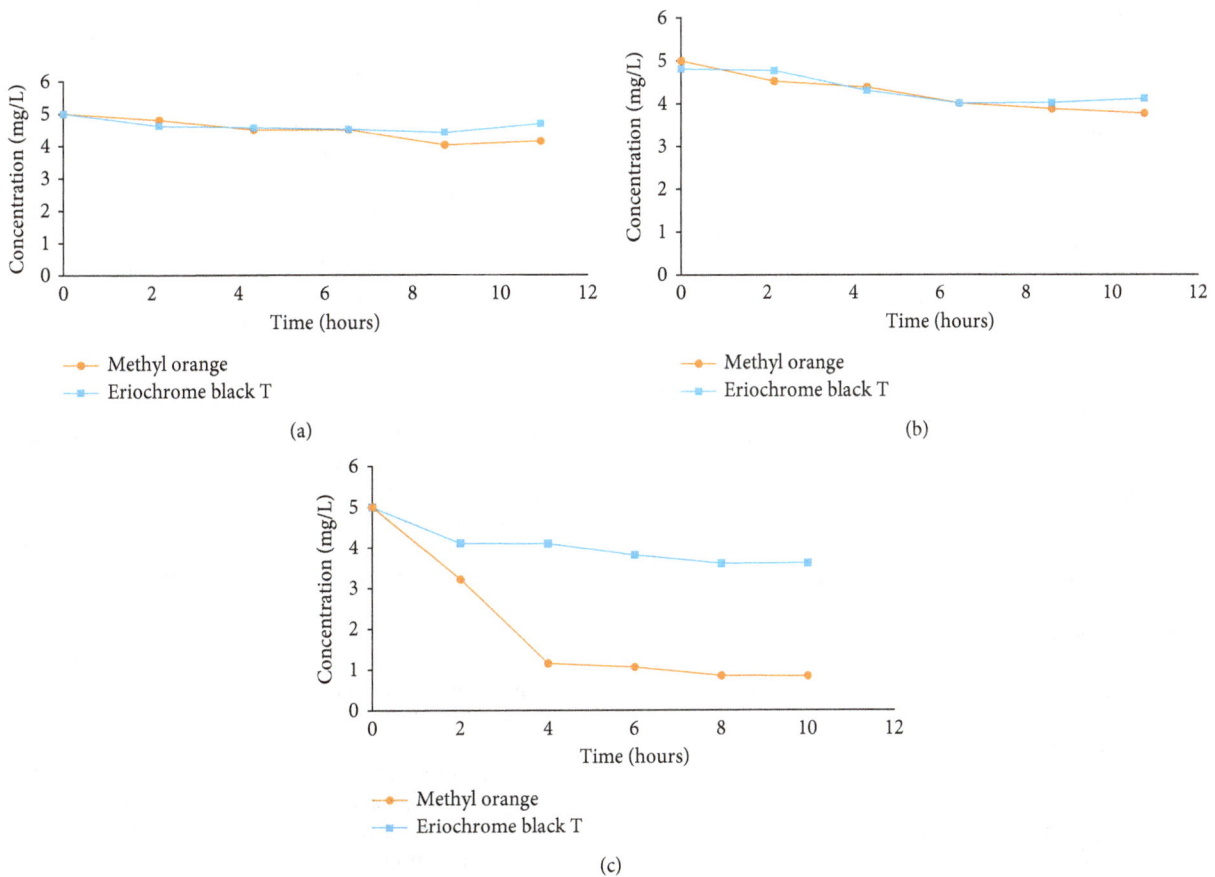

FIGURE 5: Time-dependent adsorption kinetics of Eriochrome Black T and methyl orange on the clay samples. (a) E1, (b) E2, and (c) C1.

Conflicts of Interest

The authors declare that there are no conflicts of interest regarding the preparation and publication of this manuscript.

Acknowledgments

The authors are grateful to Professor V. P. Y. Gadzepko for providing clay samples for this work.

References

[1] J. Eriksson, S. Rahm, N. Green, Å. Bergman, and E. Jakobsson, "Photochemical transformations of tetrabromobisphenol A and related phenols in water," *Chemosphere*, vol. 54, no. 1, pp. 117–126, 2004.

[2] R. Vinu and G. Madras, "Kinetics of simultaneous photocatalytic degradation of phenolic compounds and reduction of metal ions with nano-TiO_2," *Environmental Science & Technology*, vol. 42, no. 3, pp. 913–919, 2007.

[3] F. A. Tomás-Barberán and J. C. Espin, "Phenolic compounds and related enzymes as determinants of quality in fruits and vegetables," *Journal of the Science of Food and Agriculture*, vol. 81, no. 9, pp. 853–876, 2001.

[4] P. S. Murthy and M. M. Naidu, "Recovery of phenolic antioxidants and functional compounds from coffee industry by-products," *Food and Bioprocess Technology*, vol. 5, no. 3, pp. 897–903, 2012.

[5] K. Lee, K. Yoon, J. Kim, J. Bae, J. Yang, and S. Hong, "Effect of novolac phenol and oligomeric aryl phosphate mixtures on flame retardance enhancement of ABS," *Polymer Degradation and Stability*, vol. 81, no. 1, pp. 173–179, 2003.

[6] N. Dieme, "Influence of a semiconductor gap's energy on the electrical parameters of a parallel vertical junction photocell," *Energy and Power Engineering*, vol. 7, no. 5, pp. 203–208, 2015.

[7] M. Kitis, "Disinfection of wastewater with peracetic acid: a review," *Environment International*, vol. 30, no. 1, pp. 47–55, 2004.

[8] Y. Liu, B. Wen, and X. Q. Shan, "Determination of pentachlorophenol in wastewater irrigated soils and incubated earthworms," *Talanta*, vol. 69, no. 5, pp. 1254–1259, 2006.

[9] J. Michałowicz and W. Duda, "Phenols—sources and toxicity," *Polish Journal of Environmental Studies*, vol. 16, no. 3, pp. 347–362, 2007.

[10] H. B. Senturk, D. Ozdes, A. Gundogdu, C. Duran, and M. Soylak, "Removal of phenol from aqueous solutions by adsorption onto organomodified Tirebolu bentonite: equilibrium, kinetic and thermodynamic study," *Journal of Hazardous Materials*, vol. 172, no. 1, pp. 353–362, 2009.

[11] M. Ahmaruzzaman, "Adsorption of phenolic compounds on low-cost adsorbents: a review," *Advances in colloid and Interface science*, vol. 143, no. 1-2, pp. 48–67, 2008.

[12] M. Bucheli-Witschel and T. Egli, "Environmental fate and microbial degradation of aminopolycarboxylic acids," *FEMS Microbiology Reviews*, vol. 25, no. 1, pp. 69–106, 2001.

[13] A. Maleki, A. Mahvi, F. Vaezi, and N. R. Nabizadeh, "Ultrasonic degradation of phenol and determination of the oxidation by-products toxicity," *Iranian Association of Environmental Health*, vol. 2, no. 3, pp. 201–206, 2005.

[14] B. Bolto and J. Gregory, "Organic polyelectrolytes in water treatment," *Water Research*, vol. 41, no. 11, pp. 2301–2324, 2007.

[15] S. K. Dentel, "Coagulant control in water treatment," *Critical Reviews in Environmental Science*, vol. 21, no. 1, pp. 41–135, 1991.

[16] C. Y. Teh, P. M. Budiman, K. P. Y. Shak, and T. Y. Wu, "Recent advancement of coagulation–flocculation and its application in wastewater treatment," *Industrial & Engineering Chemistry Research*, vol. 55, no. 16, pp. 4363–4389, 2016.

[17] V. Kavitha and K. Palanivelu, "The role of ferrous ion in Fenton and photo-Fenton processes for the degradation of phenol," *Chemosphere*, vol. 55, no. 9, pp. 1235–1243, 2004.

[18] B. Iurascu, I. Siminiceanu, D. Vione, M. Vicente, and A. Gil, "Phenol degradation in water through a heterogeneous photo-Fenton process catalyzed by Fe-treated laponite," *Water Research*, vol. 43, no. 5, pp. 1313–1322, 2009.

[19] R. F. Pontes, J. E. Moraes, A. Machulek Jr., and J. M. Pinto, "A mechanistic kinetic model for phenol degradation by the Fenton process," *Journal of Hazardous Materials*, vol. 176, no. 1-3, pp. 402–413, 2010.

[20] G. C. Yang and Y. W. Long, "Removal and degradation of phenol in a saturated flow by in-situ electrokinetic remediation and fenton-like process," *Journal of Hazardous Materials*, vol. 69, no. 3, pp. 259–271, 1999.

[21] I. S. Chang and S. N. Kim, "Wastewater treatment using membrane filtration—effect of biosolids concentration on cake resistance," *Process Biochemistry*, vol. 40, no. 3-4, pp. 1307–1314, 2005.

[22] A. Dąbrowski, P. Podkościelny, Z. Hubicki, and M. Barczak, "Adsorption of phenolic compounds by activated carbon—a critical review," *Chemosphere*, vol. 58, no. 8, pp. 1049–1070, 2005.

[23] I. Tan, A. Ahmad, and B. Hameed, "Adsorption of basic dye on high-surface-area activated carbon prepared from coconut husk: Equilibrium, kinetic and thermodynamic studies," *Journal of Hazardous Materials*, vol. 154, no. 1-3, pp. 337–346, 2008.

[24] A. Gürses, Ç. Doğar, M. Yalçın, M. Açıkyıldız, R. Bayrak, and S. Karaca, "The adsorption kinetics of the cationic dye, methylene blue, onto clay," *Journal of Hazardous Materials*, vol. 131, no. 1-3, pp. 217–228, 2006.

[25] A. Kurniawan, H. Sutiono, Y. H. Ju et al., "Utilization of rarasaponin natural surfactant for organo-bentonite preparation: application for methylene blue removal from aqueous effluent," *Microporous and Mesoporous Materials*, vol. 142, no. 1, pp. 184–193, 2011.

[26] M. Hajjaji and H. El Arfaoui, "Adsorption of methylene blue and zinc ions on raw and acid-activated bentonite from Morocco," *Applied Clay Science*, vol. 46, no. 4, pp. 418–421, 2009.

[27] A. Tabak, N. Baltas, B. Afsin, M. Emirik, B. Caglar, and E. Eren, "Adsorption of reactive red 120 from aqueous solutions by cetylpyridinium-bentonite," *Journal of Chemical Technology & Biotechnology*, vol. 85, no. 9, pp. 1199–1207, 2010.

[28] S. F. P. Yesi, Y. H. Ju, F. E. Soetaredjo, and S. Ismadji, "Adsorption of acid blue 129 from aqueous solutions onto raw and surfactant-modified bentonite: application of temperature-dependent forms of adsorption isotherms," *Adsorption Science & Technology*, vol. 28, no. 10, pp. 847–868, 2010.

[29] S. Bouzid, A. Khenifi, K. Bennabou, R. Trujillano, M. Vicente, and Z. Derriche, "Removal of orange II by phosphonium-modified Algerian bentonites," *Chemical Engineering Communications*, vol. 202, no. 4, pp. 520–533, 2015.

[30] F. Priscila, Z. M. Magriotis, M. A. Rossi, R. F. Resende, and C. A. Nunes, "Optimization by response surface methodology of the adsorption of coomassie blue dye on natural and acid-treated clays," *Journal of Environmental Management*, vol. 130, pp. 417–428, 2013.

[31] H. Chen and J. Zhao, "Adsorption study for removal of Congo red anionic dye using organo-attapulgite," *Adsorption*, vol. 15, no. 4, pp. 381–389, 2009.

[32] D. Chen, J. Chen, X. Luan, H. Ji, and Z. Xia, "Characterization of anion–cationic surfactants modified montmorillonite and its application for the removal of methyl orange," *Chemical Engineering Journal*, vol. 171, no. 3, pp. 1150–1158, 2011.

[33] S. Yang, M. Gao, and Z. Luo, "Adsorption of 2-Naphthol on the organo-montmorillonites modified by Gemini surfactants with different spacers," *Chemical Engineering Journal*, vol. 256, pp. 39–50, 2014.

[34] S. Tunali Akar, T. Alp, and D. Yilmazer, "Enhanced adsorption of acid Red 88 by an excellent adsorbent prepared from alunite," *Journal of Chemical Technology & Biotechnology*, vol. 88, no. 2, pp. 293–304, 2013.

[35] T. Mahmood, M. T. Saddique, A. Naeem, P. Westerhoff, S. Mustafa, and A. Alum, "Comparison of different methods for the point of zero charge determination of NiO," *Industrial & Engineering Chemistry Research*, vol. 50, no. 17, pp. 10017–10023, 2011.

[36] K. P. Kitsopoulos, "Cation-exchange capacity (CEC) of zeolitic volcaniclastic materials: applicability of the ammonium acetate saturation (AMAS) method," *Clays and Clay Minerals*, vol. 47, no. 6, pp. 688–696, 1999.

[37] M. Kosmulski, "pH-dependent surface charging and points of zero charge. IV. Update and new approach," *Journal of Colloid and Interface Science*, vol. 337, no. 2, pp. 439–448, 2009.

[38] F. Menges, *Spectragryph-Optical Spectroscopy Software, Version 1.0. 7*, Universität Konstanz, Obersdorf, Germany, 2017.

[39] L. Lutterotti, M. Bortolotti, G. Ischia, I. Lonardelli, and H. Wenk, "Rietveld texture analysis from diffraction images," *Zeitschrift für Kristallographie Supplements*, vol. 2007, no. 26, pp. 125–130, 2007.

[40] S. Matthies, L. Lutteroti, and H. Wenk, "Advances in texture analysis from diffraction spectra," *Journal of Applied Crystallography*, vol. 30, no. 1, pp. 31–42, 1997.

[41] P. S. Nayak and B. Singh, "Instrumental characterization of clay by XRF, XRD and FTIR," *Bulletin of Materials Science*, vol. 30, no. 3, pp. 235–238, 2007.

[42] L. Vaculikova and E. Plevova, "Identification of clay minerals and micas in sedimentary rocks," *Acta Geodynamica et Geomaterialia*, vol. 2, no. 2, p. 163, 2005.

[43] Z. Chen, C. Huang, S. Liu, Y. Zhang, and K. Gong, "Synthesis, characterization and properties of clay-polyacrylate hybrid materials," *Journal of Applied Polymer Science*, vol. 75, no. 6, pp. 796–801, 2000.

[44] W. Rahmalia, J. F. Fabre, T. Usman, and Z. Mouloungui, "Adsorption characteristics of bixin on acid-and alkali-treated kaolinite in aprotic solvents," *Bioinorganic Chemistry and Applications*, vol. 2018, pp. 1–9, 2018.

[45] P. Schindler, P. Liechti, and J. Westall, "Adsorption of copper, cadmium and lead from aqueous solution to the kaolinite/water interface," *Journal of Agricultural Science*, vol. 35, pp. 219–230, 1987.

[46] E. Wieland and W. Stumm, "Dissolution kinetics of kaolinite in acidic aqueous solutions at 25 C," *Geochimica et Cosmochimica Acta*, vol. 56, no. 9, pp. 3339–3355, 1992.

[47] R. S. Bowman, "Applications of surfactant-modified zeolites to environmental remediation," *Microporous and Mesoporous Materials*, vol. 61, no. 1–3, pp. 43–56, 2003.

[48] M. T. Yagub, T. K. Sen, S. Afroze, and H. M. Ang, "Dye and its removal from aqueous solution by adsorption: a review," *Advances in Colloid and Interface Science*, vol. 209, pp. 172–184, 2014.

[49] M. Doğan, M. H. Karaoğlu, and M. Alkan, "Adsorption kinetics of maxilon yellow 4GL and maxilon red GRL dyes on kaolinite," *Journal of Hazardous Materials*, vol. 165, no. 1–3, pp. 1142–1151, 2009.

[50] A. Tehrani-Bagha, H. Nikkar, N. Mahmoodi, M. Markazi, and F. Menger, "The sorption of cationic dyes onto kaolin: Kinetic, isotherm and thermodynamic studies," *Desalination*, vol. 266, no. 1–3, pp. 274–280, 2011.

[51] Y. B. Acar, H. Li, and R. J. Gale, "Phenol removal from kaolinite by electrokinetics," *Journal of Geotechnical Engineering*, vol. 118, no. 11, pp. 1837–1852, 1992.

[52] U. F. Alkaram, A. A. Mukhlis, and A. H. Al-Dujaili, "The removal of phenol from aqueous solutions by adsorption using surfactant-modified bentonite and kaolinite," *Journal of Hazardous Materials*, vol. 169, no. 1–3, pp. 324–332, 2009.

[53] M. Boufatit, H. Ait-Amar, and W. McWhinnie, "Development of an Algerian material montmorillonite clay. adsorption of phenol, 2-dichlorophenol and 2,4,6-trichlorophenol from aqueous solutions onto montmorillonite exchanged with transition metal complexes," *Desalination*, vol. 206, no. 1–3, pp. 394–406, 2007.

[54] S. Nishimura, P. J. Scales, S. Biggs, and T. W. Healy, "An electrokinetic study of the adsorption of dodecyl ammonium amine surfactants at the muscovite mica– water interface," *Langmuir*, vol. 16, no. 2, pp. 690–694, 2000.

13

Fluoride Variations in Rivers on the Slopes of Mount Meru in Tanzania

Aldo J. Kitalika ⓘ, Revocatus L. Machunda, Hans C. Komakech, and Karoli N. Njau

Department of Water and Environmental Science and Engineering, Nelson Mandela African Institute of Science and Technology, P.O. Box 447, Tengeru, Arusha, Tanzania

Correspondence should be addressed to Aldo J. Kitalika; kitalikaa@nm-aist.ac.tz

Academic Editor: Maurizio Barbieri

This study reports the variations of fluoride ions in rivers on the slopes of Mount Meru in the northern part of Tanzania. More than 318 water samples were collected from Temi, Nduruma, Tengeru, and Maji ya Chai Rivers in both wet (mid-March and April) and dry (August) seasons. The samples were analyzed for fluoride levels using Ion Selective Electrode (ISE). The minimum and maximum average fluoride levels in the wet season were 0.24 ± 0.03 mg/l and 65.20 ± 0.03 mg/l, respectively, whereas the average lowest and highest levels in the dry season were 1.02 ± 0.02 mg/l and 69.01 ± 0.03 mg/l, respectively. Tengeru River had the lowest fluoride levels in both seasons, whereas Maji ya Chai recorded the highest fluoride levels in both seasons. The headwater of all rivers with the exception of Maji ya Chai met the World Health Organization's (WHO) maximum acceptable fluoride levels of 1.50 mg/l and the downstream environment qualified for Tanzania Bureau of Standards (TBS) maximum permissible fluoride concentration in drinking water of 4.00 mg/l. Also, the laboratory experiments showed that fluoride containing rocks exposed to pH above 7.6 display high leaching of F^- in solution which gradually increased with the increase in pH, indicating that dissolution of fluoride in water is a function of pH.

1. Introduction

The understanding of fluoride distribution in Tanzanian rivers is of great importance since majority of the Tanzanian population obtain their domestic freshwater from rivers, springs, and lakes. It is reported that 30% of these are water sources with fluoride concentration exceeding 1.5 mg/l [1]. Despite the fact that fluoride has health benefits, consumption above the optimal level is unhealthy. The WHO and TBS recommend that the healthy intake of fluoride in water should not exceed 1.5 mg/l and 4.0 mg/l, respectively [2, 3]. Excessive consumption of fluoride has been shown to cause crippling skeletal fluorosis due to the reaction of F and Ca in the bones; thus, it is extremely reactive in biological systems, thus affecting the enzymes and the whole organism as well [4, 5]. In Tanzania, fluorides are distributed in the regions of Arusha, Moshi, Singida, and Shinyanga, with a severely affected area being on the foothills of Mount Meru and Kilimanjaro [6, 7]. Fluoride-rich waters are associated with sediments of marine origin in mountainous areas and

volcanic, granitic, and gneissic rocks [8]. Being the case in Tanzania, the problem occurs both in the rift valley zones in the northern and southwestern part of the country associated with volcanic activity and in the crystalline basement complex of the central plateau [9]. Enrichment of fluoride minerals in water occurs through evaporation, weathering of volcanic rocks, and geothermal solutions in the rift valley system, as well as dissolution from saline rocks associated with fluoride [9, 10].

Fluorine is the most electronegative element with the electronegativity value of 3.98 on the Pauling Scale; thus, it is very reactive [11]. Therefore, this property makes the element exist in different forms of mineral salts in the environment rather than in its pure form [12]. The fluoride containing minerals are grouped into fluorides, phosphates, silicates, and mica [13]. In Tanzania, fluorapatite, fluorite, topaz, phlogopite, and lepidolite predominate. All these minerals are water-insoluble and hence their ability to release fluoride ions in surface and groundwater depends on the conditions which favor their solubility such as high temperature and pH. Therefore, fluorides enter surface water by leaching (being the main cause) and surface runoff from fertilizers containing phosphates, industrial emissions, and effluents [14]. The average dissolved fluoride content in major rivers of the world is between 0.01 mg/l and 0.02 mg/l, whereas in lakes it is below 0.5 mg/l [3], but in Tanzania the concentrations are above the mentioned values in the vulnerable regions [14].

Previous studies carried out to establish sources of high fluoride concentrations and distribution in the environment on the slopes of Mount Meru have mostly put emphasis on groundwater. In surface water, studies reported the fluoride levels of 12-13 mg/l, 21–26 mg/l, 61–65 mg/l, and 690 mg/l for Maji ya Chai and Engare Nanyuki Rivers, pond water, and Lake Momela, respectively [9, 14–18]. The contaminated areas have shown health implications to some children and adults living around the foot of Mount Meru [5, 15]. Further studies on fluoride levels were carried out in groundwaters within the same area and found a concentration of up to 68 mg/l which was highly associated with the alkaline volcanism and high pH [19, 20]. The vulnerability of alkaline soil for fluoride dissolution in soil has recently been associated with the presence of bicarbonate ions (HCO_3^-), which accelerates the alkalinity and fluoride availability [21]. Since surface/river water is a contribution of groundwater discharge and precipitation, fluorides in water are mainly from leaching of rocks from groundwater and fluoride containing stuff.

Despite the above facts, the general trends of fluorides in rivers after interaction with different environments of this study area have been little studied and mapped, thus raising the importance of this study. Thus, this work was conducted to investigate the spatial distribution of fluorides in rivers after such interactions have occurred together with tracing their point sources of contamination, if any. Studies in these rivers are of profound importance since water from these rivers is used in various domestic activities including cooking and drinking in Arusha City.

2. Materials and Methods

2.1. Description of the Study Area. The study area involved four rivers, namely, Temi and Nduruma which lie within the Arusha City and Tengeru and Maji ya Chai which lie within the Meru District. The four rivers originate from a common subcatchment of foothills of Mount Meru lying from the eastern part to the southwest of the mountain (Figure 1). The rivers run downstream from the mountain to the southeast. Natural vegetation is typically tropical forest to savannah. The topography of the study region is dominated by the Mount Meru volcanic cone of Pleistocene to recent origin. The local climate of the area is temperate Afro-Alpine, with an annual precipitation of 450 mm [22] and mean minimum and maximum daily temperature of 20.6°C and 28.5°C, respectively. Rainfall is irregularly distributed between a main wet season from February to mid-May (contributing 70% of the annual precipitation) and a minor one from September to November which provides much of the remainder giving the mean annual rainfall of 535.3 mm. The remaining months are

effectively dry, although occasional showers do occur during this period [23, 24]. The four rivers contribute to the Pangani basin subcatchment feeding the Pangani River. The study area was divided into three regions depending on the river and land development, namely, pristine (headwater) ($3°15'00''S$ to $3°20'00''S$), middle ($3°20'00''S$ to $3°25'00''S$), and flood plain ($3°25'00''S$ to $3°35'00''S$). The catchment area for the rivers is considered as pristine (headwater) which is characterized by artificial and natural forest conservation, the middle area of the river consists of mixed peasant agriculture and human settlement, and the floodplain is the downstream area of the river characterized by bare land, intensive grazing, large-scale agriculture, and serious flooding in wet season.

2.2. Sampling. The GPS predetermined sampling points were identified based on confluence, accessibility, and preestablished monitoring stations. Two-liter water samples were collected in each point downstream from the source of each river (Figure 1). One-liter water sample was used for chemical parameter measurement and the second liter for fluoride and nutrients measurement. Sampling was done thrice in one-week interval during the wet season (mid-March to early April) and dry season (August) in 2015. In each season, 159 representative samples were collected for analysis.

2.3. Chemicals. Analytical-grade reagents from Sigma-Aldrich (Merck) for the preparation of TISAB II and TISAB IV were prepared from glacial acetic acid, NaCl, NaOH, HCl, tris(hydroxymethyl) amino methane, and sodium tartrate dihydrate, respectively. Also, the analytical grade of Ion Electrolyte Reference Filling Solution (P/N 51344750) was obtained from the same company.

2.4. Extraction, Pretreatment, and Cleanup of Water Samples. Samples for major ions measurements were not pretreated with any chemical except for the samples with solids and other organic debris that were filtered using a 0.45 μm filter before measurement. Samples for $\delta^{18}O$ and δ^2H analysis were kept cooled at 4°C before and on transport to the Stable Isotope laboratory at Waterloo University in Canada.

2.5. Detection Limit of the Instrument and Calibration. The calibrated detection limit of the instrument was reconfirmed by measuring in triplicate the serially diluted primary standard solution with the lowest concentration of 0.02 mg/l, which was the calibrated minimum detection limit (MDL), and the maximum instrument detection limit was calibrated at 100 mg/l. Therefore, any undetected concentration measurement of a highly diluted sample was regarded to be below 0.02 mg/l and thus it was below the detection limit (BDL) of the instrument and for the same purpose similar measurements in double distilled water used for rinsing apparatus, its cleaners and TISAB were all considered as fluoride-free matrices.

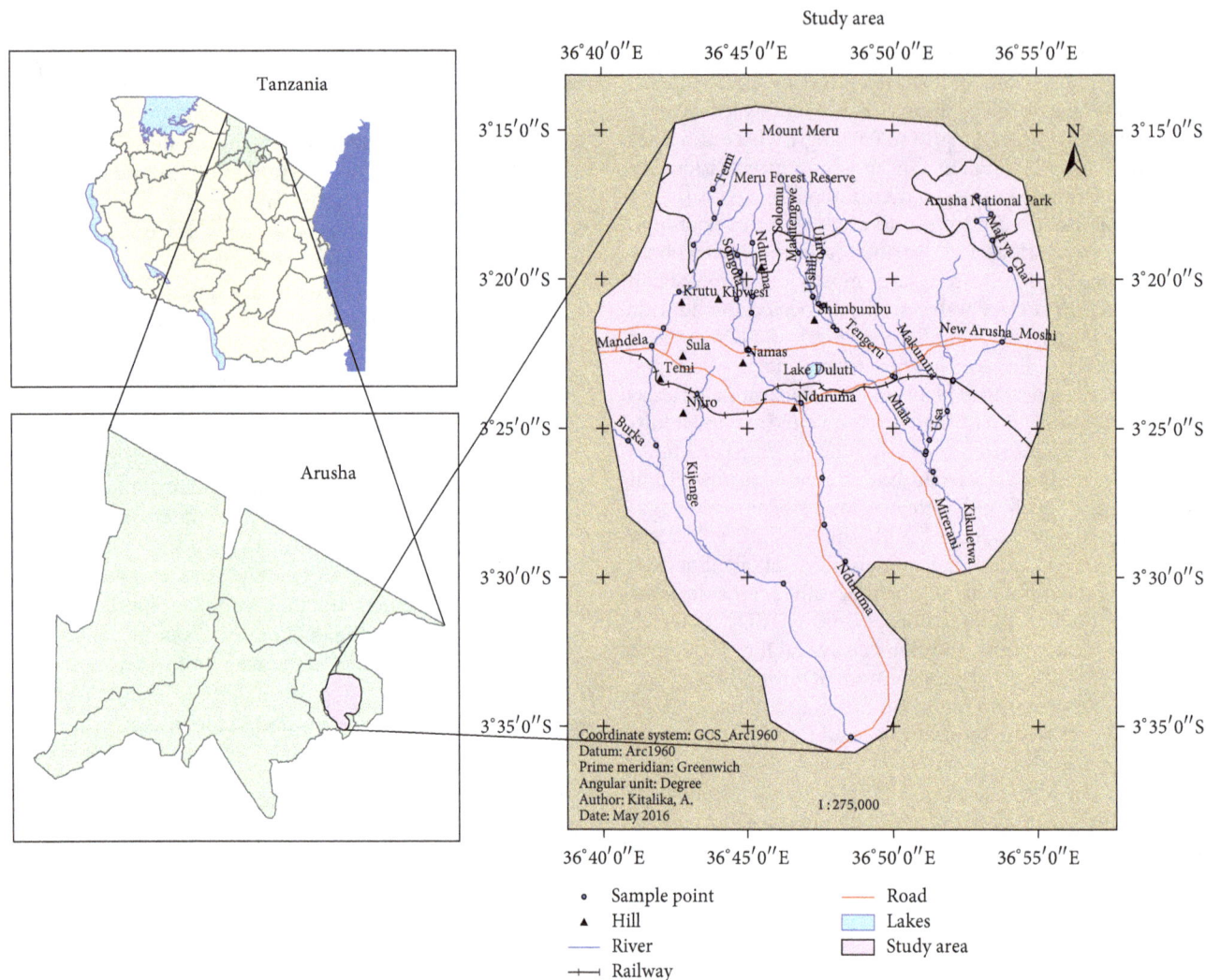

FIGURE 1: Location of the study area.

2.6. Analysis and Confirmation. The sample's physicochemical parameters (temperature, pH, conductivity, velocity, and TDS) were measured in situ from the sampling points. Quantification of free fluoride ions was done by mixing equal volumes of Total Ionic Strength Adjustment Buffer (TISAB) to provide a constant background ionic strength, decomplex fluoride ions, and adjust the solution pH. Thorough mixing was done using a magnetic stirrer. Measurement was done using the Ion Selective Electrode (ISE) from Mettler Toledo, perfectION™, with a Bayonet Neill-Concelman (BNC) connector, P/N 51344775. The electrode was immersed in serially diluted fluoride primary standard solution of 0.1 mg/l and 10 mg/l from Sigma-Aldrich (Merck) for calibration, followed by measurement of the mixture, and the steady readout was recorded. The electrode was filled with Ion Electrolyte Reference Filling Solution (P/N 51344750) to maintain its maximum sensitivity. Analysis for $\delta^{18}O$ and $\delta^{2}H$ stable isotopes was done using a modern technology of Los Gatos Research Laser processes analyzer with Integrated Cavity Output Spectroscopy (LGR-ICOS™) machine using the Vienna Standard Mean Ocean Water

(VSMOW) with a standard error of ±0.8‰. The experimental rock identification was done in the Geology Department of the University of Dar es Salaam (UDSM) using the X-ray powder diffraction (XRD) machine. The geostatistical analyses for the spatial distribution of fluoride and other related parameters were done using ArcGIS software, version 10.1, in the GIS Laboratory of the Nelson Mandela African Institute of Science and Technology, Arusha, Tanzania.

2.7. Analytical Quality Assurance. Double-distilled water that was used for rinsing all the instruments and apparatus was tested to check whether it is fluoride-free before use. Instruments were rinsed thrice with double-distilled water to ensure no traces of fluoride outside the sample before another consecutive measurement was done. Thereafter, the glass electrode cleanliness was reconfirmed by immersing in a beaker containing distilled water, and its cleaners were measured for fluoride-free ions and thereafter TISAB solution was also measured for any traces of fluoride to assess its purity before measurement of the analyte was made. Also, the linearity of the electrode was monitored in every two samples' measurements by measuring a known concentration (0.1 mg/l and 10 mg/l in serial dilutions) primary standard fluoride solution.

2.8. Statistical Test. The strength of the linear relationships between fluoride levels with other physicochemical parameters was assessed by calculating Pearson's correlation coefficient (r) using Sigma Plot software, version 11. Also, Student's t-test at the stated p value was carried out to assess whether the fluoride levels between seasons are significantly different. Significant difference was considered when the p values were <0.05.

2.9. Geostatistical Mapping. Fluoride concentration mapping was done by graduated symbols proportional to values using ArcMap version 10.1. The fluoride concentration ranges in water were grouped based on WHO (0.0–1.5 mg/l) and TBS (0.0–4.0 mg/l) standards simply for the purpose of assessing the status of the rivers according to the mentioned standards [2, 3].

2.10. Identification and Laboratory Experiments of Fluoride Leaching Rocks. A laboratory experiment was set up to assess the effect of pH in dissolution of fluoride containing rocks at 25°C. Two rocks were collected in the catchment river banks of Nduruma and Tengeru Rivers and two more rocks were collected in Maji ya Chai River, whereby the first rock was collected from Maksoro (M1, area with low fluoride levels) and the second rock was collected from Jamera (M2, with high fluorides in water). The rock types were initially identified by X-ray powder diffraction (XRD) and confirmed to be feldspar-quartz extrusive volcanic igneous rocks. All rocks were separately ground at variable grain sizes (Table 5). A 2.00 g sample of each ground rock was mixed with 200 ml of deionized water and the mixture was constantly stirred with a magnetic stirrer on a hot plate at 150 rpm and 25°C for three consecutive days. Monitoring of fluoride concentrations was done every twelve hours on all days and the results were recorded.

3. Results and Discussion

The average physicochemical trends for the four rivers are shown in Tables 1 and 2. The fluoride levels in various points of the rivers are shown in Table 3. The isotopic signatures of $\delta^{18}O$ and $\delta^{2}H$ from the LGR-ICOS laser process analyzer of water in the four rivers in different points are shown in Table 4. The particle size distribution of the ground rock and its pH dependence for fluoride leaching results are shown in Tables 5 and 6, respectively.

3.1. General Trends

3.1.1. Isotopic characteristics of Water. Table 1 shows the different stable isotopic compositions of water on the slopes of Mount Meru. The oxygen and deuterium isotopic compositions of each river water on these slopes are controlled by

meteoric precipitation, evaporation, and groundwater. There is a large difference between water from precipitation and running water with progressive variation from the catchment to the downstream. The isotopic data show enrichment downstream with an increase in water temperature (Figure 2). Such trend shows a sign of evaporation which can be an important factor contributing to the slight increase of some dissolved salts in water. Despite such isotopic variations, the data differ slightly from the Global Meteoric Water Line (GMWL) (Figure 2). Also, similar variations are noted within the river from the catchment area to the downstream. Such isotopic variations are further experienced in each river from the catchment to the downstream which can be caused by the different environments the rivers pass and groundwater recharged mixing with surface water.

Also, more observations show that ^{2}H and ^{18}O enrichment is relatively higher in Maji ya Chai and Temi Rivers; thus, it is expected that the evaporation effect will be much pronounced in the two rivers compared to Nduruma and Tengeru Rivers.

3.1.2. Physicochemical Trends. The pH of all rivers was above 7 in both seasons with the lowest pH being in the catchment areas of all rivers. The lowest pH (7.12 ± 0.11) was measured at Tengeru catchment during the wet season whereas the highest value (9.90 ± 0.14) was measured in Nduruma River downstream during the dry season. Tengeru River showed the lowest pH in both seasons which averaged 7.12 ± 0.11 in the wet season and 7.70 ± 0.36 in the dry season. Maji ya Chai had the highest average pH in both seasons with the values of 8.03 ± 0.57 and 8.57 ± 0.52 for wet and dry seasons, respectively (Figure 3). The minimum water temperature (12.21°C) was measured in the catchments area of Nduruma River with its highest temperature of up to 25°C being measured in the dry season. The lowest average temperature of 17.01 ± 1.80°C was recorded at Temi River in the wet season while its highest temperature of 20.44 ± 4.22°C was recorded in the dry season. The temperature variations in all rivers were generally associated with canopy cover of the riparian environment and its elevation such that the low water temperature was measured in high canopy cover environment and vice versa for the high water temperature.

The lowest conductivity of 82 μS/cm was measured at the catchments of Temi River, an indication of less salt being dissolved in it, whereas the lowest average value of 179.33 ± 26.73 μS/cm was measured at Nduruma River in the wet season. The highest EC of 1722 μS/cm was measured in the downstream of Maji ya Chai River during the dry season with the highest average value of 1183.43 ± 47.54 μS/cm being in the same river.

3.1.3. Distribution of Major Cations and Anions in Rivers. The dissolved major cations and ions in the four rivers are shown in Tables 2 and 3. The data are further summarized in the Piper diagram as shown in Figure 3. For Temi River, neither ion predominated in both seasons; thus, the Ca-Mg or Na$^+$ and K$^+$ were all very low with SO$_4^-$, HCO$_3^-$, and Cl$^-$ also being low, resulting in low EC and TDS and hence soft water.

TABLE 1: Isotopic signatures of δ^{18}O and δ^{2}H from the LGR-ICOS laser process analyzer of water in the four rivers.

	Temi River		Nduruma River			Tengeru River			Maji ya Chai River		
Point name	δ^{2}H‰	δ^{18}O‰	Point name	δ^{2}H‰	δ^{18}O‰	Point name	δ^{2}H‰	δ^{18}O‰	Point name	δ^{2}H‰	δ^{18}O‰
Prec.	−28.97	−6.16	Prec.	−28.96	−6.25	Prec.	−28.35	−6.22	Prec.	−28.41	−6.21
Te1	−24.71	−5.08	N1	−28.54	−5.80	T3	−21.95	−4.84	M1	−21.68	−4.68
Te5	−22.31	−4.70	N6	−24.83	−5.22	T8	−21.36	−4.69	M3	−20.12	−4.58
Te6	−20.36	−4.45	N9	−20.59	−4.60	T9	−20.85	−4.57	M5	−19.11	−4.20
Te9	−17.42	−3.77	N10	−19.24	−4.25	T19	−19.70	−4.40	M6	−16.14	−3.62
Te10	−14.50	−3.01	N12	−18.04	−3.99	T21	−18.58	−4.24	M7	−14.85	−3.28

Note. Te: Temi River; N: Nduruma River; T: Tengeru River; M: Maji ya Chai River; Prec.: precipitation; δ^{2}H = $x \pm 0.2$‰ and δ^{18}O = $x \pm 0.8$‰, where x are isotopic values from the table.

TABLE 2: The average seasonal physico- and hydrochemical parameters of Temi and Nduruma River water.

Season	pH	T (°C)	Ca^{2+}	Mg^{2+}	Na^+	K^+	CO_3^{2-}	HCO_3^-	Cl^-	SO_4^{2-}	TDS	EC ($\mu S/cm$)
Temi River												
Wet												
Min.	7.32	14.95	2	2.43	0.76	0.21	13.71	6.18	0	0	88	176
Max.	8.07	20	114	23.085	43.33	12.23	212.71	352.18	77.475975	6.18	400	802
Av.	7.70 ± 0.26	17.01 ± 1.80	22.5 ± 34.17	7.96 ± 7.20	7.96 ± 12.98	2.42 ± 3.66	57.08 ± 64.34	69.58 ± 105.51	33.28 ± 22.69	4.38 ± 6.67	169.83 ± 127.50	339.75 ± 255.38
Dry												
Min.	7.37	14.95	3.66	2.16	1.38	0.41	9.78	11.40	0.11	0.06	41	82
Max.	9.01	25	40.51	21.84	15.30	4.58	117.91	126.00	13	2.38	698	1396
Av.	8.31 ± 0.67	20.44 ± 2.44	23.97 ± 8.72	12.60 ± 6.09	9.05 ± 3.29	2.68 ± 0.99	67.89 ± 26.17	74.74 ± 27.13	4.58 ± 8.21	0.92 ± 0.57	245.42 ± 240.99	491.5 ± 481.11
Nduruma River												
Wet												
Min.	7.63	12.21	2.00	1.21	0.70	0.22	6.57	6.25	15.00	0.00	71	142
Max.	9.04	20.48	10.00	10.86	3.78	1.11	43.57	31.26	44.99	9.10	108	217
Av.	7.98 ± 0.37	17.20 ± 2.24	4.57 ± 2.51	5.40 ± 3.19	1.74 ± 0.95	0.54 ± 0.26	21.25 ± 10.94	15.35 ± 8.41	28.82 ± 8.70	1.28 ± 2.66	90.67 ± 12.48	179.33 ± 26.73
Dry												
Min.	7.40	12.01	1.21	5.00	0.48	0.14	26.92	3.81	45.00	1.35	81	88
Max.	9.90	24.32	9.22	38.00	3.61	1.06	92.20	28.92	122.50	23.49	154	308
Av.	8.22 ± 0.72	19.00 ± 3.25	4.63 ± 2.62	20.00 ± 10.04	1.95 ± 0.95	0.58 ± 0.28	59.00 ± 15.00	15.28 ± 7.70	73.12 ± 21.18	3.96 ± 6.18	117.83 ± 24.14	223.42 ± 64.31

Note. All concentration measurements are in mg/l. Min.: minimum; Max.: maximum; Av.: average.

Table 3: The average seasonal physico- and hydrochemical parameters of Tengeru and Maji ya Chai River water.

Season	pH	T (°C)	Ca^{2+}	Mg^{2+}	Na^+	K^+	CO_3^{2-}	HCO_3^-	Cl^-	SO_4^{2-}	TDS	EC ($\mu S/cm$)
Tengeru River												
Wet												
Min.	7.12	15.00	0.48	1.21	0.62	0.33	10.14	7.27	2.50	0.00	37.00	72
Max.	7.58	19	8.20	12.18	3.62	1.94	42.46	39.13	24.99	10.00	260.00	518
Av.	7.34 ± 0.13	17.53 ± 1.12	2.35 ± 2.19	3.76 ± 1.21	1.48 ± 0.84	0.79 ± 0.79	17.74 ± 7.82	16.29 ± 9.24	13.79 ± 5.70	1.20 ± 2.28	110.81 ± 52.62	221.52 ± 105.06
Dry												
Min.	7.23	14.00	8.00	1.21	2.56	0.79	19.14	20.15	12.50	0.00	42.00	83.00
Max.	8.59	21.00	24.00	15.78	8.27	2.56	73.22	62.44	137.00	25.00	243.00	474.00
Av.	7.90 ± 0.36	18.04 ± 1.80	12.95 ± 4.41	5.88 ± 4.10	4.25 ± 1.51	1.32 ± 0.47	39.99 ± 16.17	33.95 ± 11.44	42.71 ± 25.96	7.89 ± 8.27	131.43 ± 50.75	262.33 ± 100.41
Maji ya Chai River												
Wet												
Min.	7.37	17.00	0.49	1.21	1.52	0.22	6.57	6.27	0.00	5.00	395.00	791.00
Max.	8.60	19.48	2.43	6.08	4.56	0.67	32.86	18.82	16.00	39.99	594.00	1187.00
Av.	8.03 ± 0.57	18.27 ± 0.82	1.50 ± 0.63	3.75 ± 1.58	2.95 ± 0.91	0.43 ± 0.13	20.46 ± 8.61	12.21 ± 3.77	7.63 ± 6.38	18.56 ± 13.68	477.14 ± 77.17	954.86 ± 153.96
Dry												
Min.	7.91	17.37	10.00	2.43	3.77	1.10	3.35	31.14	35.00	11.00	448	896
Max.	9.60	21	14.00	7.28	5.28	1.53	4.69	43.59	87.50	30.00	861	1722
Av.	8.57 ± 0.52	19.95 ± 1.35	11.25 ± 1.50	3.81 ± 1.63	4.33 ± 0.52	1.26 ± 0.15	3.85 ± 0.47	35.81 ± 4.32	58.23 ± 16.44	23.98 ± 7.24	591.86 ± 173.82	1183.43 ± 347.54

Note. All concentration measurements are in mg/l. Min.: minimum; Max.: maximum; Av.: average.

TABLE 4: Fluoride levels in Temi, Nduruma, Tengeru, and Maji ya Chai Rivers.

Latitude	Longitude	Point code	River	F⁻_wet (mg/l)	F⁻_dry (mg/l)
−3.318744	36.780114	T1	Tengeru	0.94	1.45
−3.343263	36.788113	T2	Tengeru	0.24	1.25
−3.318238	36.794051	T3	Tengeru	0.68	1.11
−3.343263	36.788113	T4	Tengeru	1.17	1.44
−3.343118	36.788072	T5	Tengeru	1.48	1.51
−3.347191	36.791335	T6	Tengeru	0.66	0.97
−3.343113	36.788064	T7	Tengeru	1.32	1.50
−3.361913	36.801558	T10	Tengeru	1.20	1.56
−3.387562	36.833290	T11	Tengeru	1.13	1.34
−3.388095	36.834787	T13	Tengeru	1.27	1.53
−3.390375	36.867836	T14	Tengeru	2.30	5.96
−3.389705	36.867828	T15	Tengeru	2.19	3.13
−3.431228	36.851936	T17	Tengeru	1.47	1.89
−3.441064	36.856352	T19	Tengeru	1.31	1.53
−3.431185	36.852054	T20	Tengeru	1.77	1.92
−3.445723	36.857519	T21	Tengeru	2.23	3.13
−3.286817	36.882092	M1	Maji ya Chai	20.10	26.10
−3.297030	36.890270	M2	Maji ya Chai	65.20	69.01
−3.300897	36.882034	M3	Maji ya Chai	14.50	17.70
−3.311596	36.890981	M4	Maji ya Chai	14.80	18.20
−3.327916	36.901406	M5	Maji ya Chai	13.80	18.00
−3.368379	36.896311	M6	Maji ya Chai	11.75	16.40
−3.389771	36.868054	M7	Maji ya Chai	10.17	15.70
−3.348050	36.794225	T8	Tengeru	0.92	1.14
−3.359968	36.799709	T9	Tengeru	1.02	1.28
−3.406998	36.864624	T16	Tengeru	2.03	2.64
−3.429589	36.852371	T18	Tengeru	1.10	1.40
−3.423187	36.854269	T12	Tengeru	0.94	1.19
−3.313160	36.753599	N1	Nduruma	0.84	1.33
−3.319881	36.744824	N3	Nduruma	1.60	1.78
−3.329416	36.746495	N4	Nduruma	1.13	1.34
−3.344450	36.744286	N7	Nduruma	0.92	1.02
−3.373071	36.750155	N8	Nduruma	1.59	2.68
−3.402712	36.781335	N9	Nduruma	1.59	2.16
−3.444499	36.793244	N10	Nduruma	1.67	2.43
−3.470623	36.794341	N11	Nduruma	1.82	0.00
−3.491149	36.806301	N12	Nduruma	1.71	0.00
−3.342963	36.753636	N2	Nduruma	0.81	1.20
−3.352160	36.752858	N5	Nduruma	1.39	1.69
−3.372934	36.751276	N6	Nduruma	2.16	2.90
−3.314420	36.719588	Te1	Temi	1.02	1.60
−3.340386	36.711055	Te5	Temi	1.40	1.87
−3.360737	36.701942	Te6	Temi	1.19	1.83
−3.397515	36.721275	Te7	Temi	0.61	0.94
−3.370857	36.695152	Te8	Temi	1.26	1.83
−3.426367	36.697539	Te9	Temi	1.41	2.15
−3.503508	36.770893	Te10	Temi	1.54	2.68
−3.589528	36.809305	Te11	Temi	1.36	3.38

TABLE 4: Continued.

Latitude	Longitude	Point code	River	F⁻_wet (mg/l)	F⁻_dry (mg/l)
−3.282948	36.730934	Te3	Temi	1.36	1.96
−3.290865	36.735081	Te2	Temi	1.02	1.27
−3.423636	36.681599	Te12	Temi	1.37	2.01
−3.299415	36.731638	Te4	Temi	1.27	1.39

TABLE 5: Particle size distribution for ground igneous rocks.

Diameter	Solomu		Nduruma		Maksoro		Jamera	
	Weight (g)	%	Weight (g)	%	Weight (g)	%	Weight (g)	%
≥2 mm	23.9911	17.38	19.5052	14.13	21.7323	15.75	23.5537	17.07
1.5 mm	14.7302	10.67	13.9703	10.12	10.9429	7.93	13.1351	9.52
1 mm	19.339	14.01	19.4628	14.10	17.9295	12.99	17.8441	12.93
0.35 mm	39.5516	28.66	37.0772	26.14	34.4105	24.94	40.1626	29.10
250 μm	5.7149	4.14	9.0026	6.89	9.201	6.67	9.1059	6.60
180 μm	8.9251	6.47	6.0108	4.36	8.5363	6.19	6.5971	4.78
125 μm	14.8566	10.77	15.9047	11.53	10.4439	7.57	17.8368	12.93
90 μm	4.1171	2.98	11.317	8.20	14.6206	10.59	5.0321	3.65
63 μm	3.9996	2.90	4.0081	2.90	7.2159	5.23	3.2776	2.38
45 μm	1.4958	1.08	0.9102	0.66	1.7899	1.30	0.6323	0.46
32 μm	0.5373	0.39	0.6629	0.48	0.6107	0.44	0.4069	0.29
20 μm	0.5885	0.43	0.4715	0.34	0.3875	0.28	0.379	0.27
Total	*137.847*	*99.89*	*137.803*	*99.86*	*137.821*	*99.87*	*137.963*	*99.97*

The total weight of the sample was 138.0027 g.

This trend was different in Nduruma River in both seasons, whereby Ca-Mg and CO_3^-, SO_4^-, HCO_3^-, and Cl^- were moderate but with low concentrations of Na^+ and K^+ and thus the water alkalinity in this river was mainly contributed from hydrolysis of CO_3^- and HCO_3^- which increase the levels of OH^- as shown in (1) and (2). Therefore, the higher the levels of CO_3^- and HCO_3^-, the higher the river pH.

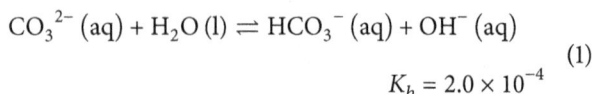

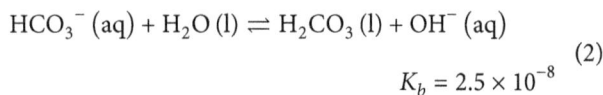

$$CO_3^{2-} (aq) + H_2O (l) \rightleftharpoons HCO_3^- (aq) + OH^- (aq)$$
$$K_b = 2.0 \times 10^{-4} \tag{1}$$

$$HCO_3^- (aq) + H_2O (l) \rightleftharpoons H_2CO_3 (l) + OH^- (aq)$$
$$K_b = 2.5 \times 10^{-8} \tag{2}$$

Therefore, the total carbon alkalinity (A_C) of water is practically contributed by hydrolysis of both CO_3^- and HCO_3^- [25].

$$A_C = [HCO_3^-] + [CO_3^-] \tag{3}$$

The increased levels of SO_4^- and Cl^- in the downstream can be attributed by the point source pollution from peasant horticultural farming in river banks where some practices use high amounts of fertilizers, herbicides, and pesticides containing such formulations. Generally, SO_4^- and Cl^- were relatively low, which is one of the characteristics for most natural water systems [6].

The levels of Ca-Mg in Tengeru River were higher in the dry season than in the wet season with the Na^+ and K^+ being low. The anions SO_4^{2-} and Cl^- were relatively increasing in the wet compared with the dry season, whereas CO_3^{2-} and HCO_3^- were low in the wet season compared to the dry season. A similar trend was shown in Maji ya Chai River.

3.1.4. Fluoride Ions Trends. The fluoride levels in various points of the respective rivers are indicated in Table 4. They are further summarized with geostatistical analysis and mapped as shown in Figures 4(a) and 4(b). The pH spatial distribution reflecting the pH dependency for fluoride levels in water is shown in Figures 5(a) and 5(b). The study shows pH to play a great role in the availability of fluoride ions in water. The lowest average fluoride level of 0.24 ± 0.02 mg/l was recorded during the wet season in Makitengwe stream which is a tributary subcatchment of Tengeru River. In the same season, the highest average fluoride level of 65.20 ± 0.26 mg/l was measured in Maji ya Chai River (Figure 4(a)). This river is located at Arusha National Park (ANP) and is used as one of the major drinking water sources for wildlife.

The lowest average fluoride level in the dry season was recorded in Kijenge stream in Temi River with an average level of 0.94 ± 0.07 mg/l, whereas the highest average level of 69.01 ± 0.21 mg/l was recorded in Maji ya Chai River (Figure 4(b)). The increase in fluoride levels in the dry season may be attributed to the absence of runoff to the river which may cause dilutions, and hence in that period the river flow is mainly from groundwater (old water) containing more fluoride ions. It should be noted that the groundwater mixing with surface water has enough time for interaction with

TABLE 6: The pH dependence of fluoride leaching from igneous rocks at 25°C.

Tengeru (Solomu)		Nduruma		Maksoro (M1)		Jamera (M2)	
pH	*[F⁻]	pH	*[F⁻]	pH	*[F⁻]	pH	*[F⁻]
3.72	0.26	3.82	0.28	3.85	0.43	3.75	1.28
4.20	0.29	4.17	0.31	4.32	0.49	4.00	1.44
5.34	0.33	5.44	0.72	5.63	0.46	4.05	1.77
6.95	1.37	6.37	0.97	6.76	0.64	4.33	2.15
7.22	2.83	7.47	1.74	7.49	0.68	4.6	2.15
8.51	3.86	8.82	2.73	8.03	1.21	5.69	2.17
9.72	3.95	9.49	2.53	9.17	3.71	6.24	2.28
10.21	3.95	10.73	2.53	10.6	5.95	7.07	2.79
11.94	3.94	10.99	2.53	11.3	6.20	8.25	9.05
12.64	3.94	12.01	2.53	12.34	26.60	9.48	10.3
13.08	3.94	13.37	2.53	13.17	26.85	10.85	18.75
13.55	3.94	-	-	13. 35	26.89	11.07	29.55
-	-	-	-	-	-	12.61	72.5
-	-	-	-	-	-	12.79	73.95
-	-	-	-	-	-	13.17	74.3

*[F⁻] in mg/l.

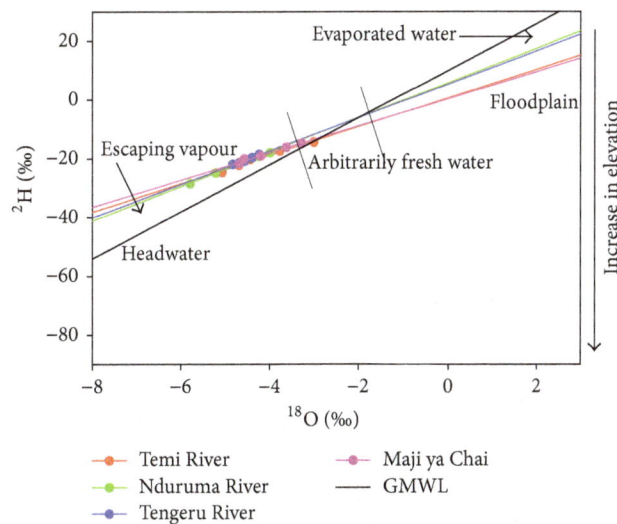

FIGURE 2: A plot of $\delta^2 H$ versus $\delta^{18} O$ indicating an increase in evaporation downstream.

fluoride containing rocks which allow dissolution of more fluoride ions in it.

The headwater environment of the three rivers (Temi, Nduruma, and Tengeru) showed low a fluoride level of <4.0 mg/l which is in line with the standards recommended by WHO and TBS except for the headwater environment of Maji ya Chai River which had extremely high fluoride levels of up to 69.0 mg ± 0.38 mg/l. Low fluoride levels in the three rivers (Temi, Nduruma, and Tengeru) in both seasons are mainly caused by low fluoride composition in the phonolite feldspar rocks which predominate at the upstream main catchment areas of these rivers, thus discharging low fluorides amount in water. Another general trend of interest is observed on the spatial distribution of fluoride at the headwater where fluorides were shown to increase from the southwestern part to the southeastern part (0.24 mg/l to 69.01 mg/l) with respect to Mount Meru. This trend can be explained by the fact that its headwater from the southwestern part is at a high altitude with low temperature, where the lower the temperature, the lower the dissolution rate of fluoride from rocks. Also, the aquifer lithology at such high altitude is predominantly phonolite which is characterized by low fluoride contents [19]. The headwater source in Maji ya Chai River is from lowland (foothills of Mount Meru) in the southeastern region of the mountain characterized by relatively high temperature incidence due to low canopy cover, and its aquifer lithology is basalt which is characterized by high fluoride levels (Table 6) [19]. All these features favor high fluoride contents in water of the river.

Seasonal hydrochemical variations in Mount Meru rivers

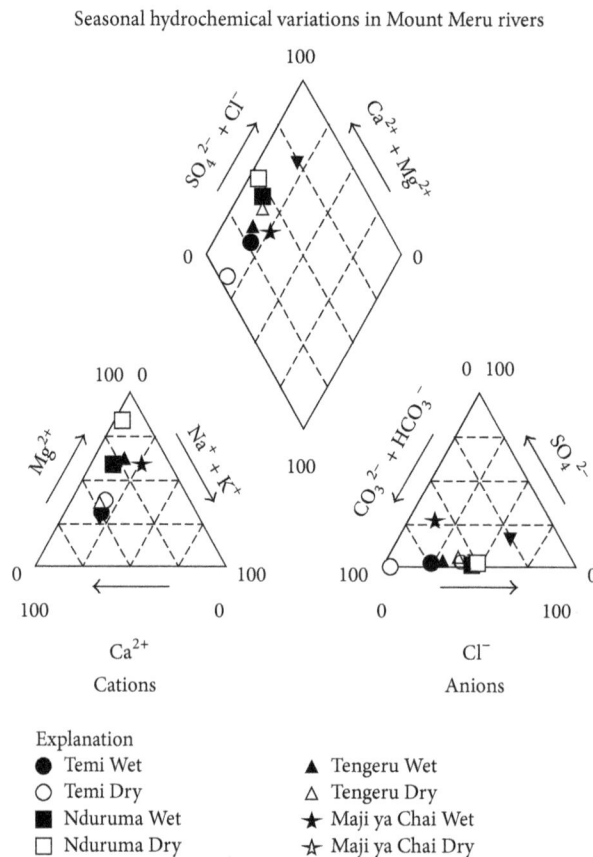

Explanation
● Temi Wet   ▲ Tengeru Wet
○ Temi Dry   △ Tengeru Dry
■ Nduruma Wet   ★ Maji ya Chai Wet
□ Nduruma Dry   ☆ Maji ya Chai Dry

FIGURE 3: Piper diagrams for distribution of major cations and anions in the four rivers.

The abovementioned trends are evidenced by the laboratory experiments for fluoride leaching from rocks collected at Nduruma, Tengeru (Solomu stream), Maksoro, and Jamera (M2) (Maji ya Chai) Rivers catchment banks as shown in Table 6. In this experiment, the rocks collected from the three rivers have been shown to release a low amount of fluoride ions at various pH with further negative response even at high alkaline environment which favors fluoride leaching. The low fluoride leaching in these rocks even at highly alkaline environment is evidence of low fluoride amounts contained in rocks. More experiments from the high fluoride containing rocks (collected from basalt rocks of Jamera) show high amounts of fluoride leaching which acquires its maximum levels in the alkaline environment with no further changes upon increase of its pH. At such point, complete fluoride leaching from the rocks is assumed (Table 6, Figure 7).

In addition to the above findings, it was observed that the average fluoride levels in the three rivers increase downstream up to 3.38 ± 0.16 mg/l at the floodplain in Temi River. Such observed trend does not happen at Maji ya Chai River where the fluoride levels were shown to decrease downstream. This is caused by dilutions from Ngurdoto and Shoripanga streams feeding to the main river in between with averaged fluoride concentration of 3.32 ± 0.26 mg/l and 4.21 ± 0.17 mg/l, respectively. Also, the extremely high difference in fluoride levels in the headwater environment of Maji ya Chai compared to other sampling areas of the same

river indicates the main point source pollution of fluoride in the river which has been identified at Jamera (Figure 4). The statistical tests showed good positive correlations (r) in the dry season between fluoride levels and pH and EC together with temperature simply because there were only groundwater (old water) recharges with the absence of runoff which may increase dilutions of river water at various rates.

3.2. Fluoride Distribution in Temi River. Fluoride levels in Temi River increased from the headwater downstream to the floodplain in both wet and dry seasons. The minimum average fluoride level was in the upstream headwater region of the river with average values of 1.02 ± 0.01 mg/l and 1.27 ± 0.07 mg/l during wet and dry seasons, respectively. The maximum level was measured at the floodplain with the average values of 1.54 ± 0.03 mg/l and 2.68 ± 0.02 mg/l in wet and dry seasons, respectively (Figures 4(a) and 4(b)). Furthermore, the headwater and middle regions of the river had fluoride levels lower than the WHO permissible levels for drinking water of 1.50 mg/l [3], whereas the remaining part had fluoride levels higher than WHO permissible standards but within the TBS maximum permissible levels of 4.0 mg/l in both seasons (Figures 4(a) and 4(b)) [2]. Higher fluoride levels in the floodplain can be explained by low water velocities which in turn increase the interaction time of water with basalt aquifers containing high fluoride element.

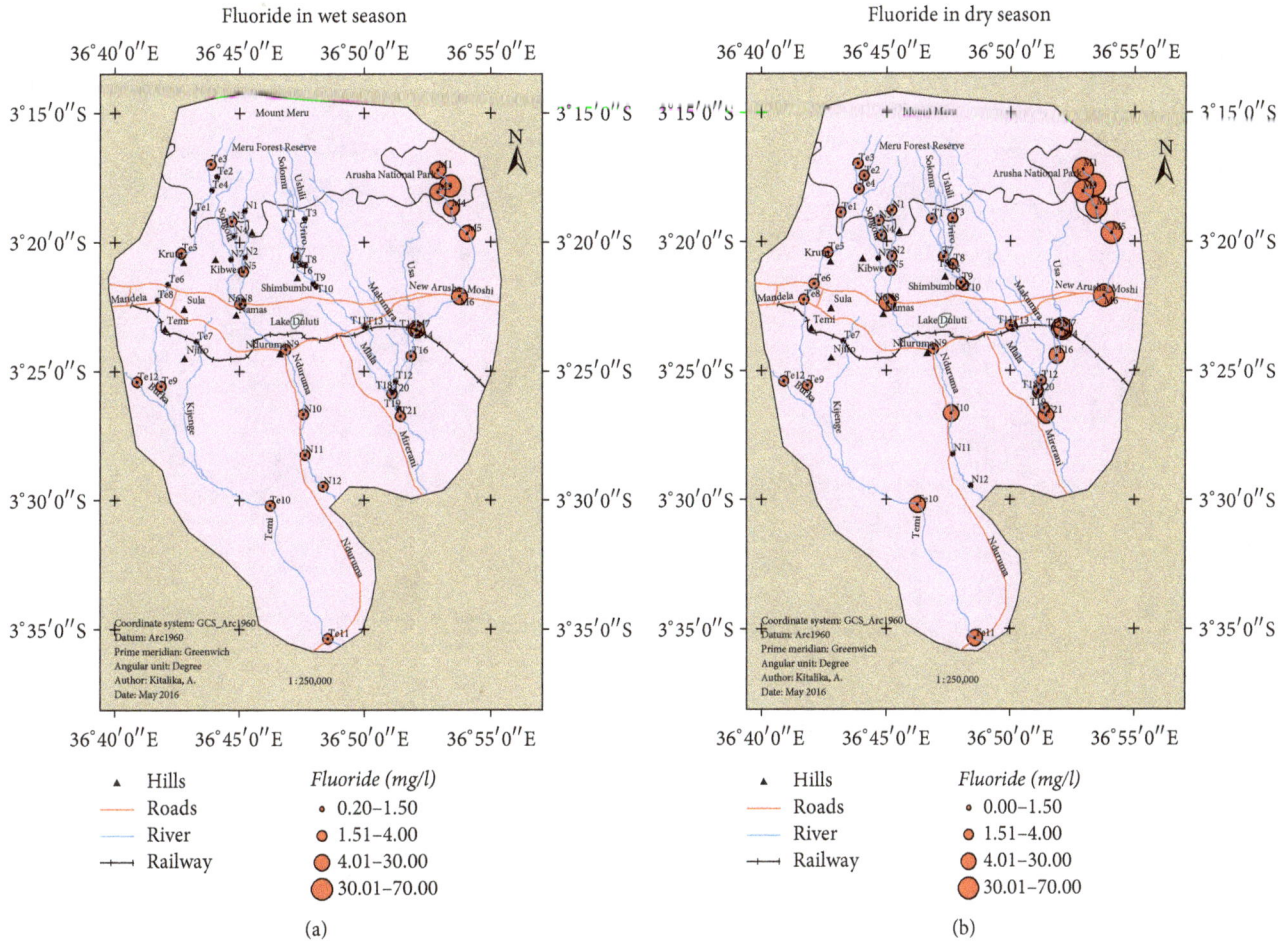

FIGURE 4: Fluoride variation in wet (a) and dry (b) seasons.

Together with this, the slightly alkaline condition in the floodplain favors the availability of fluorides in water as suggested by Saxena and Ahmed in their experiments [26].

In addition to the increase in fluoride levels downstream to the floodplain, the study shows a strong positive Pearson correlation of fluoride levels between wet and dry seasons ($r = 0.87$, $n = 13$, $p \leq 0.005$) which suggest common pollution sources between seasons (Figure 6(a)). Using the two-sample t-test for unequal variance, the means of fluoride levels were significantly different between seasons, and thus the fluoride levels were significantly higher in the dry season than in the wet season ($n = 13$, $p \leq 0.001$). The higher fluoride levels in the dry season can be explained by the river recharge from groundwater containing high fluoride levels as the major source without dilutions from runoff. Increase in pH showed a strong positive Pearson correlation with fluoride levels ($r = 0.9$, $n = 13$, $p \leq 0.0001$) (Figure 7(a)). Similar correlation trends have been shown by Saxena and Ahmad in their study on water-rock interaction on dissolution of fluoride containing feldspar rocks [26].

3.3. Fluoride Distribution in Nduruma River. Nduruma River had fluoride levels of 0.84 ± 0.05 mg/l to 2.16 ± 0.01 mg/l

and 1.02 ± 0.02 mg/l to 2.90 ± 0.05 mg/l during wet and dry seasons, respectively. The headwater and middle regions had fluoride levels within the WHO maximum permissible levels while the rest of the region was within the TBS standards (Figures 4(a) and 4(b)). Also, the fluoride concentrations in this river were shown to increase from the upstream to the floodplain, in favor of the increase of some its physicochemical properties including water temperature, pH, and EC. These good trends were much more pronounced in the dry season, depicting that the river water recharge was mainly from groundwater (old water) which had good interaction time with rocks containing salts. The very strong positive correlation in fluoride levels between wet and dry seasons ($r = 0.9$, $n = 12$, $p \leq 0.0002$) also suggests the common pollutant sources in the two seasons (Figure 6(b)).

The minimum and maximum average water temperature in the wet season were $12.01 \pm 0.06°C$ and $20.48 \pm 0.04°C$, respectively, whereas during the dry season, the minimum and maximum average water temperature were $12.61 \pm 0.03°C$ and $24.32 \pm 0.07°C$, respectively. While the pH trends in the wet season showed many irregularities due to various external factors such as surface runoffs, the minimum and maximum average pH ranges were from 7.63 ± 0.01 to 9.4

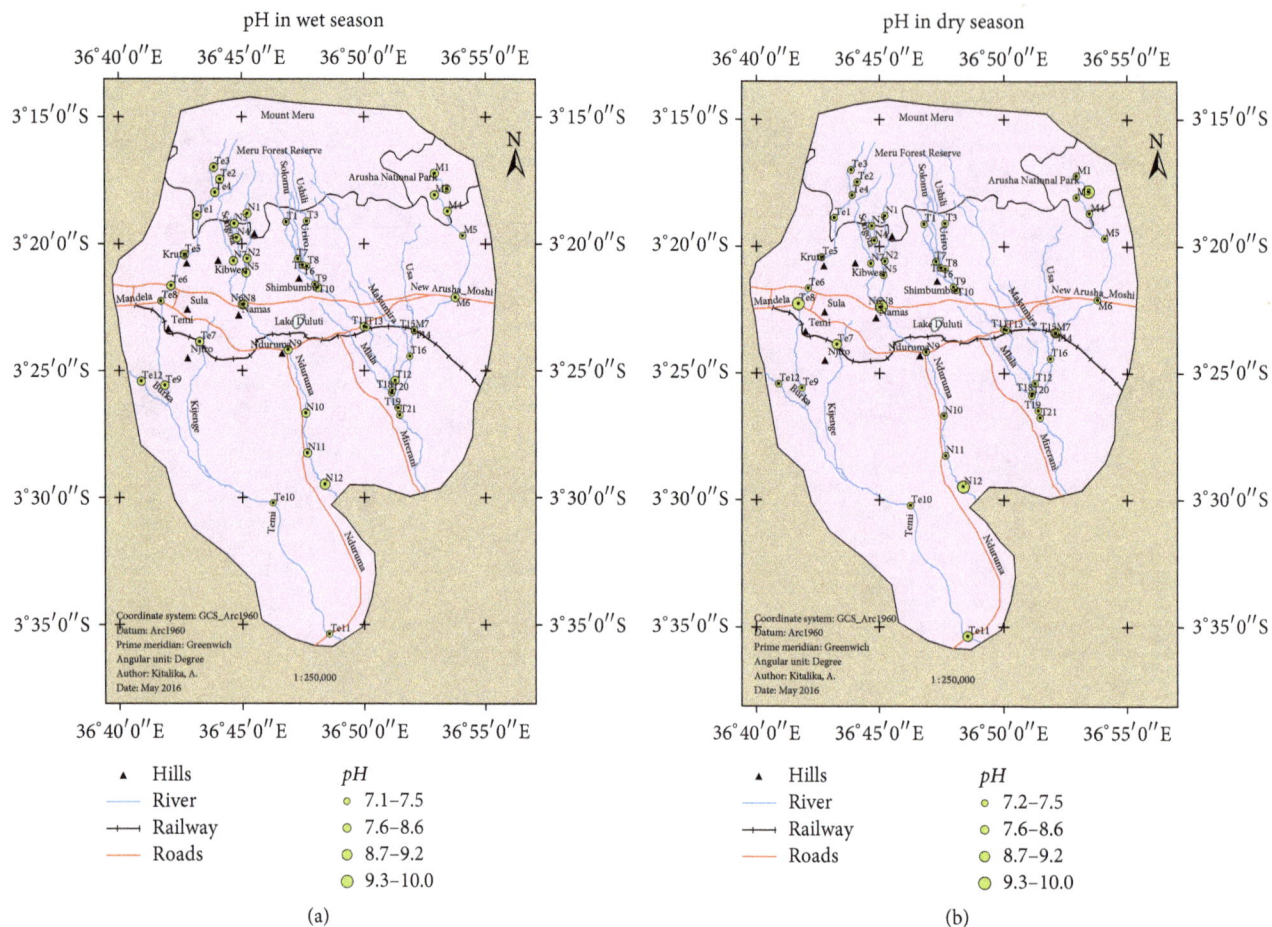

FIGURE 5: Water pH in wet (a) and dry (b) seasons.

± 0.03 and 7.4 ± 0.08 to 9.9 ± 0.01 for wet and dry seasons, respectively. A strong positive correlation between fluoride levels and pH in the dry season ($r = 0.93, n = 12, p \leq 0.02$) suggests that the increase in water pH increases the ability of dissolution of fluoride containing rocks in water (Figure 7(b)) [26], hence an increase in fluoride ions in water. Also, the mean fluoride levels between seasons were significantly not different ($p \leq 0.7, n = 12$), suggesting that seasonal dilution in this river had no significant impact on the fluoride availability in water.

3.4. Fluoride Distribution in Tengeru River. The fluoride levels in this river were shown to fluctuate as a result of dilution from inputs of several tributaries such as Ngare Sero and Malala which had very low fluoride levels. The headwater of this river showed lower values than the WHO standards while the floodplain showed higher values due to effects from Maji ya Chai River (Figures 4(a) and 4(b)). Despite such fluctuations, fluoride levels in this river increased when the confluence between Usa River and Maji ya Chai contributed its water to form the main Kikuletwa River. Higher levels in this downstream were caused by Maji ya Chai River which had the highest fluoride levels among the studied rivers. Despite their low levels, the wet season had lower levels than

the dry season such that the minimum average levels in this river were 0.24 ± 0.03 mg/l and 0.97 ± 0.01 mg/l whereas its average maximum levels were 2.23 ± 0.01 mg/l and 3.13 ± 0.03 mg/l for wet and dry seasons, respectively. In addition, the higher values in the floodplain did not exceed the TBS standards of 4.0 mg/l. Generally, this river showed the lowest fluoride levels in the study area. Water temperature in this river had slight changes due to high canopy cover in its riparian environment throughout, whereby the minimum temperatures were 14.00 ± 0.02°C and 15.00 ± 0.01°C with its maximum temperature being 18.71 ± 0.02°C and 21.00 ± 0.04°C for wet and dry seasons, respectively.

The statistical test showed a very strong positive correlation in fluoride levels between the two seasons ($r = 0.9, n = 21, p \leq 0.0002$), suggesting a common source of pollutant in the two seasons (Figure 6(c)). In addition to this, a very weak positive correlation ($r = 0.4, n = 21, p \leq 0.4$) was shown between fluoride levels and pH during the dry season, suggesting that the slight changes in pH had no significant effect on fluoride variations (Figure 7(c)). Together with these, their mean fluoride levels between seasons were shown to be significantly different ($n = 21, p = 0.02$), indicating that water from rainfall had a significant effect on the water quality changes and the contributing water in the dry season was mainly from groundwater recharge.

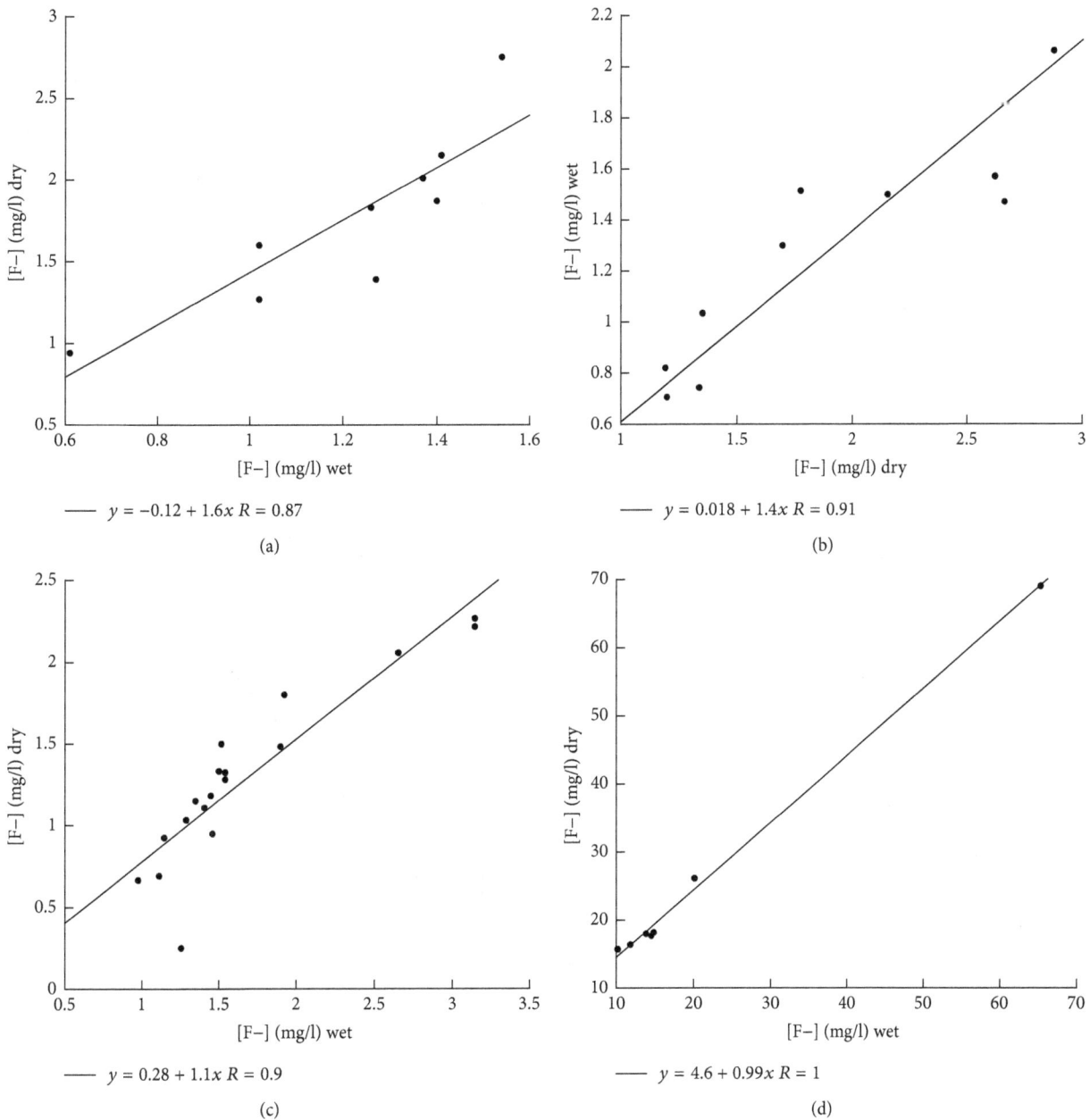

— $y = -0.12 + 1.6x$ $R = 0.87$

(a)

— $y = 0.018 + 1.4x$ $R = 0.91$

(b)

— $y = 0.28 + 1.1x$ $R = 0.9$

(c)

— $y = 4.6 + 0.99x$ $R = 1$

(d)

FIGURE 6: Correlation of fluoride between wet and dry seasons in (a) Temi, (b) Nduruma, (c) Tengeru, and (d) Maji ya Chai.

3.5. Fluoride Distribution in Maji Ya Chai River. During the wet season, the average fluoride levels were between 11.75 ± 0.70 mg/l and 65.20 ± 0.03 mg/l whereas the minimum average levels in the dry season ranged from 16.40 ± 0.05 mg/l to 69.01 ± 0.03 mg/l, both highest being recorded in the headwater at Jamera (M2). A similar concentration of 59 mg/l–68 mg/l fluoride was recorded in the past five years in the nearby hydrothermal spring feeding its water in Engare Nanyuki River which is a lowland river with respect to Maji ya Chai River [19]. These similarities may entail common fluoride containing rock in all rivers around the area. Interestingly, Jamera (M2) was discovered to be the main point source for fluoride ions in this river and

contains the highest fluoride concentration among others. This area is characterized by high fluoride levels in its river bank rocks (Table 6) and highest pH (9.6 ± 0.02) among all measurements recorded, creating a good alkaline environment which favors high fluoride dissolution from rocks compared to other areas. Also, this area is characterized by very low water velocity of 0.27 m/s which favors a good interaction time of water with fluoride containing rocks.

Low fluoride concentration in the downstream of Maji ya Chai River is due to dilutions from springs contributing water to the main river which have a low fluoride level range of 1.62 ± 0.01 mg/l and 3.01 ± 0.04 mg/l in the wet season and 3.32 ± 0.26 mg/l and 4.21 ± 0.17 mg/l in the dry seasons,

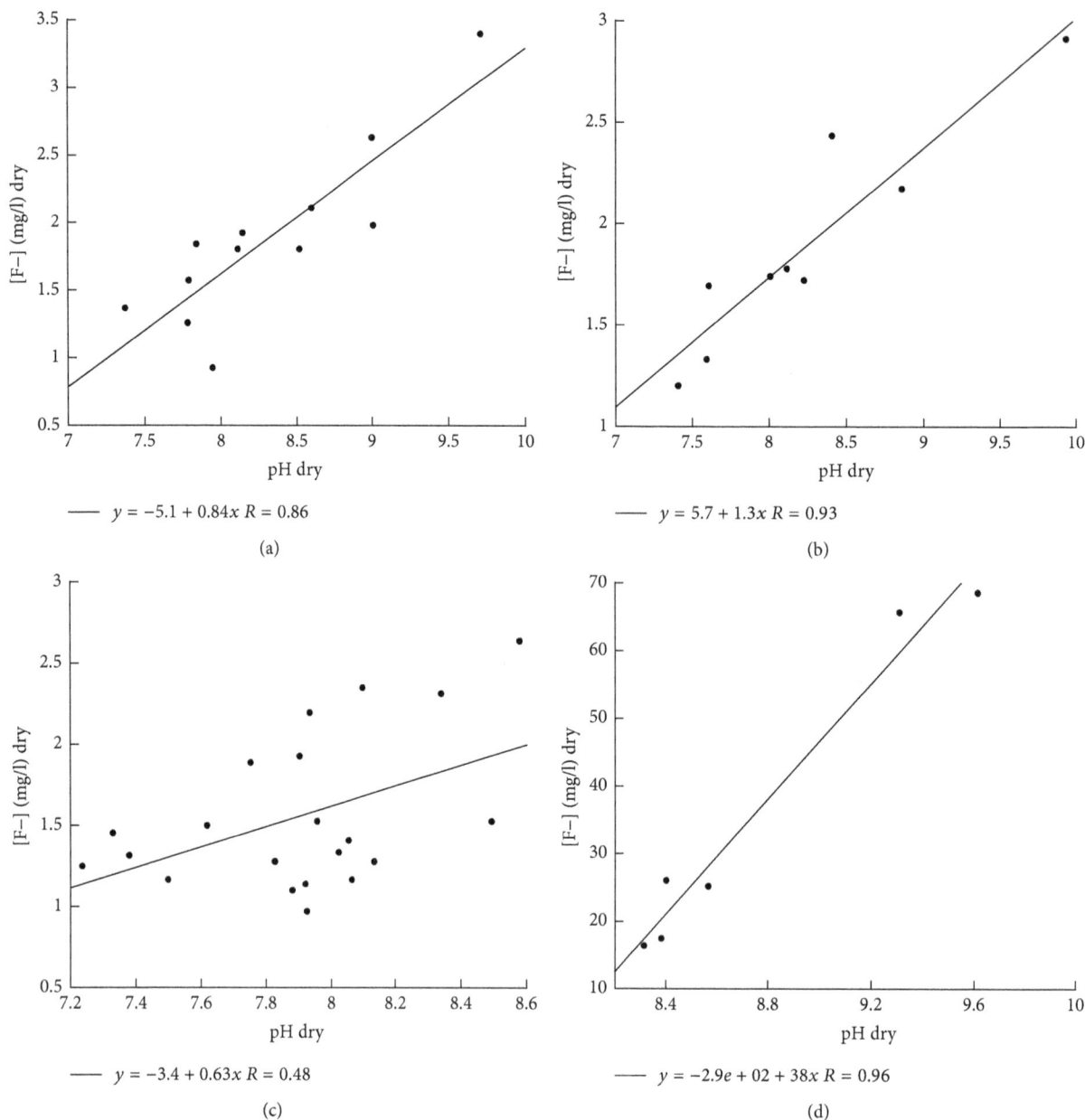

$$y = -5.1 + 0.84x \quad R = 0.86$$

(a)

$$y = 5.7 + 1.3x \quad R = 0.93$$

(b)

$$y = -3.4 + 0.63x \quad R = 0.48$$

(c)

$$y = -2.9e + 02 + 38x \quad R = 0.96$$

(d)

FIGURE 7: Correlation of fluoride with pH for the dry season in (a) Temi, (b) Nduruma, (c) Tengeru, and (d) Maji ya Chai.

namely, Ngurdoto and Shoripanga springs, respectively. The contributed values are quite lower than minimum levels observed in the two seasons of the main river. In addition, waters from this river had very high EC compared to other nearby rivers, indicating that the river contains more soluble salts. The area noted with the highest fluorides and pH also had the highest EC (1958 ± 0.7 μS/cm and 1187 ± 0.3 μS/cm) with their average minimum values being 896 ± 0.3 μS/cm and 791 ± 0.6 μS/cm in dry and wet seasons, respectively. This good pH dependence correlation was previously shown by Saxena and Ahmed in their experiment on pH rock interaction in dissolution of fluoride containing granitic rock, and it was found that the pH range of 7.6–8.6 favors the rock dissolution [26].

The comparison of fluoride levels between seasons indicated a very strong positive correlation between seasons ($r = 1$, $n = 7$, $p \leq 0.002$), indicating that the pollutant originates from a common source (Figure 6(d)). Also, a very strong positive correlation ($r = 0.9$, $n = 7$, $p \leq 0.005$) was observed between the fluoride levels and pH in the dry season, an indication that dissolution of fluoride rocks is a function of pH (Figure 7(d)) [20, 26].

3.6. The pH Dependence in Fluoride Leaching from Rocks. The plot of pH against [F$^-$] from Table 6 was done to assess the relationship between the two parameters. The study from Figure 8 shows three phases of rock leaching whereby *Phase I*

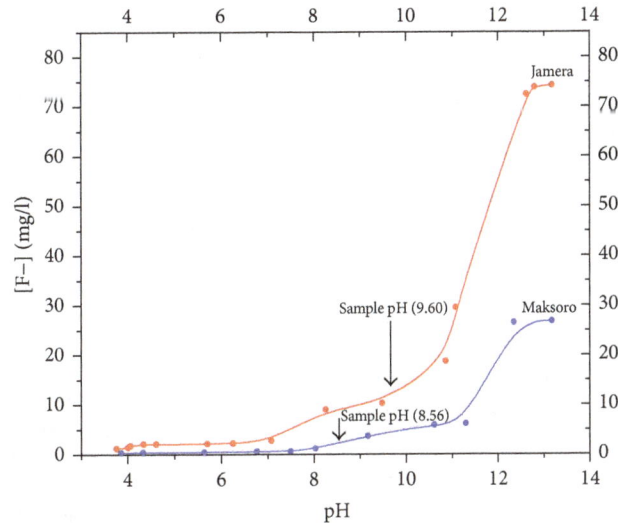

FIGURE 8: pH dependent fluoride leaching in feldspar-quartz igneous rock at 25°C.

involves pH below 7.6 where there is little response of fluoride leaching until neutral pH. *Phase II* involves the pH range of 7.6 to 11 where there is a good response to the increase of alkalinity (pH) with fluoride leaching. *Phase III* which starts at pH 11 shows a quick response in fluoride leaching in the rock. The pH of water samples ranged from 7.12 to 9.90 which is slightly above neutral to alkaline, thus all corresponding to *Phase II* in our experiment. These results are supported by Saxena and Ahmad in a similar study with granite rock which showed good response in pH 7.6–8.8 [26].

At low pH, slow fluoride leaching is expected from rocks in the environment. However, there were some water samples which had high pH with a lower fluoride concentration than expected. Such results depict an important message to scientists that the availability of fluoride ions in water depends mainly on the nature of rocks that they interact with, whether they contain fluoride or not, and also their response to leaching under favorable conditions. From this experiment, it was found that the amount of fluoride in rocks is another major limiting factor for the presence of this pollutant in water, whose availability is favored by alkaline environments.

3.6.1. The Role of OH⁻ in Fluoride Leaching from Rocks.

3.6.1. The Role of OH⁻ in Fluoride Leaching from Rocks. The alkaline environment associated with fluoride leaching occurs as a result of several reactions, including the reactions between fluorite (CaF_2) and HCO_3^- which release fluoride ions (see (4)) [20].

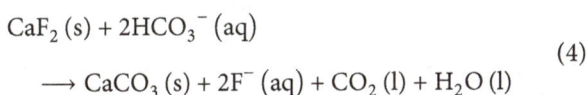

$$CaF_2 \,(s) + 2HCO_3^- \,(aq)$$
$$\longrightarrow CaCO_3 \,(s) + 2F^- \,(aq) + CO_2 \,(l) + H_2O \,(l) \tag{4}$$

Also, the carbonate ions present in the aqueous solution increase the alkalinity of water through reaction (1). When CO_2 (g) (from reaction (4) and atmospheric) is dissolved in water, it gives carbonic acid (reaction (5)), where in turn the acid decreases the pH of water due to increased H^+ (reaction

(6)). This process is very minimum due to the low solubility of CO_2 (g) in water and the resulting acid is weak.

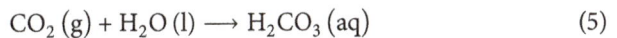

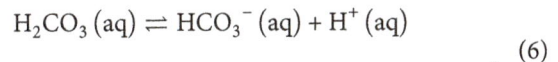

$$CO_2 \,(g) + H_2O \,(l) \longrightarrow H_2CO_3 \,(aq) \tag{5}$$

$$H_2CO_3 \,(aq) \rightleftharpoons HCO_3^- \,(aq) + H^+ \,(aq) \tag{6}$$

$$K_a = 2.5 \times 10^{-4}$$

The formed HCO_3^- from (6) accelerates further the dissolution of CaF_2 (s) rock to release more F^- (aq.). However, in this experiment, since the fluoride containing rock is a feldspar, at low pH (acidic), alumina surfaces can be hydrogenated into a neutrally charged species which also gives free aluminum ions (reaction (7)) [27].

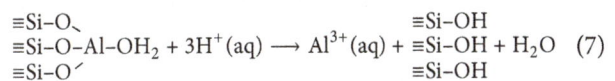

$$\equiv Si-O \diagdown \atop \equiv Si-O-Al-OH_2 + 3H^+ \,(aq) \longrightarrow Al^{3+} \,(aq) + {\equiv Si-OH \atop \equiv Si-OH} + H_2O \atop \equiv Si-O \diagup \qquad\qquad\qquad\qquad\qquad \equiv Si-OH \tag{7}$$

Thus, every one Al^{3+} will polarize three water molecules to give three H^+ which increase the acidity of water (see (8)). This reaction reduces the free fluoride ions in solution.

$$Al^{3+} \,(aq) + 3H_2O \,(l) \longrightarrow Al(OH)_3 \,(aq) + 3H^+ \,(aq) \tag{8}$$

Since CO_3^- did not react with CaF_2 previously, it was hydrolyzed in water to give more OH^- which is consumed by H^+ in (8), increasing the dissolution of feldspar surfaces containing fluoride, which in turn further exposes the new surfaces from feldspar to the reaction since more H^+ is consumed (*Phase II* responses). This decreases the resistance of igneous rock against weathering, exposing more free fluorides in water (*Phase III*).

Therefore, the different phases in the rock dissolution are caused by a mixture of reactions and nature of the rock; when these processes actively work in a particular environment, their collective effect becomes substantial.

4. Conclusions

The present study shows that the fluoride levels in all rivers were spatially distributed such that the levels in the pristine (headwater) regions of Temi, Nduruma, and Tengeru rivers were lower than what WHO and TBS recommended. The spatial variation of fluoride in rivers is a function of rock composition (the amount of fluoride present in it) and several other chemical reactions which lead to changes in the pH of the environment. Moreover, interaction of groundwater and surface water patterns and external inputs such as surface runoff, salinity, and climate change, which affects the water temperature, also govern its availability. Also, these findings show the importance of environmental conservation in water sources or catchment areas in which well conserved environments reduce the rates of several chemical reactions due to low water temperature such that an increase in water temperature alters the natural water composition due to the increased rate of chemical reactions. The availability of fluoride in water is highly affected by pH such that it is more favored in alkaline environments. However, nonconserved floodplains have shown higher levels of the pollutant as they are catalyzed by external factors such as high water temperature and high water-rock interaction time due to low water velocity and its lithology. An exception is observed in Maji ya Chai River where the pristine (headwater) environment showed elevated fluoride levels, which is basically caused by its high fluoride in fluoride bearing rocks in the river banks, in which its dissolution is supported by the high alkaline environment in the river. Also, this study is of great help to different authorities as it enlightens some of the best practices of the catchment area and its management which can also minimize the cost for water processing.

Conflicts of Interest

The authors declare that they have no conflicts of interest.

Acknowledgments

The authors would like to acknowledge the Tanzanian Government for their financial assistance and the VILR-OUS Project for providing transport and analytical instruments in this entire study.

References

[1] M. C. Lathman and P. Grech, "The effects of excessive fluoride intake.," *American Journal of Public Health*, vol. 57, no. 4, pp. 651–660, 1967.

[2] R. B. Cattell, "The scree test for the number of factors," *Multivariate Behavioral Research*, vol. 1, pp. 245–276, 1966.

[3] WHO, "Guidelines for Drinking Water Quality. Recommendations, 2004".

[4] WHO, "Fluoride in drinking-water," in *A Series of World Health Organization Monographs*, K. Bailey, J. Chilton, E. Dahi, M. Lennon, P. Jackson, and J. Fawell, Eds., IWA Publishers, Geneva, 2006.

[5] E. S. Johansen, *The effects of fluoride on human health in Eastern Rift Valley*, 2013.

[6] T. C. Davies, "Chemistry and pollution of natural waters in western Kenya," *Journal of African Earth Sciences*, vol. 23, no. 4, pp. 547–563, 1996.

[7] D. M. Deocampo, "Hydrogeochemistry in the Ngorongoro Crater, Tanzania, and implications for land use in a World Heritage Site," *Applied Geochemistry*, vol. 19, no. 5, pp. 755–767, 2004.

[8] K. F. Nair, F. Manji, and J. N. Gitonga, "The occurance and distribution of fluoride in ground waters of Kenya," *IAHS Publ*, vol. 144, pp. 75–86, 1984.

[9] J. T. Nanyaro, U. Aswathanarayana, J. S. Mungure, and P. W. Lahermo, "A geochemical model for the abnormal fluoride concentrations in waters in parts of northern Tanzania.," *Journal of African Earth Sciences*, vol. 2, no. 2, pp. 129–140, 1984.

[10] P. Kilham and R. E. Hecky, "Fluoride: geochemical and ecological significance in east african waters and sediments," *Limnology and Oceanography*, vol. 18, no. 6, pp. 932–945, 1973.

[11] L. P. Pauling, *The Nature of Chemical Bond and the Structure of Molecules and Crystals: an Introduction to Modern Structural Chemistry*, Cornell University Press, New York, NY, USA, 1960.

[12] J. D. Hem, *Study and Interpretation of the Chemical Characteristics of Natural Water*, Water Supply Paper 2254, Washington, DC, USA, 3rd edition, 1989.

[13] S. P. S. Teotia, M. Teotia, and R. K. Singh, "Hydro-geochemical aspects of endemic skeletal fluorosis in India - An epidemiologic study," *Fluoride*, vol. 14, no. 2, pp. 69–74, 1981.

[14] U. Aswathanarayana, P. Lahermo, E. Malisa, and J. T. Nanyaro, "High fluoride waters in an endemic fluorosis area," in *Proceedings of the International Symposium on Geochemistry in Health*, Royal Society, 1986.

[15] WHO, "Fluoride in drinking-water back ground document," *Multivariate Behavioral Research*, vol. 1, pp. 629–637, 1966.

[16] A. J. Mcharo, *The Occurrence and Possibilities of Fluoride Removal in Arusha Area, Tanzania*, Tampere University of Technology, 1986.

[17] J. S. M. Mungure, *A collective Study of Incidences of Fluorosis and Possible Fluoride Sources in Maji ya Chai in Arumeru District, Arusha Region, Tanzania*, University of Dar es Salaam, Tanzania, 1984.

[18] H. Mjengera, *Excess Fluoride in Portable Water in Tanzania and the Defloridation Technology with Emphasis on the Use of Polaluminium chloride and Magnesite*, Tampere University of Technology, 1988.

[19] G. Ghiglieri, R. Balia, G. Oggiano, and D. Pittalis, "Prospecting for safe (low fluoride) groundwater in the eastern african rift: the arumeru district (northern Tanzania)," *Hydrology and Earth System Sciences*, vol. 14, no. 6, pp. 1081–1091, 2010.

[20] G. Ghiglieri, D. Pittalis, G. Cerri, and G. Oggiano, "Hydrogeology and hydrogeochemistry of an alkaline volcanic area: The NE Mt. Meru slope (East African Rift-Northern Tanzania)," *Hydrology and Earth System Sciences*, vol. 16, no. 2, pp. 529–541, 2012.

[21] J. Mkungu, "Application of Soil Composition for Inferring Fluoride Variability in Volcanic Areas of Mt. Meru, Tanzania," *International Journal of Environmental Monitoring and Analysis*, vol. 2, no. 5, p. 231, 2014.

[22] R. J. Hijmans, S. E. Cameron, J. L. Parra, P. G. Jones, and A. Jarvis, "Very high resolution interpolated climate surfaces for global land areas," *International Journal of Climatology*, vol. 25, no. 15, pp. 1965–1978, 2005.

[23] UNDP, Arusha Region Water Master Plan, Final Report, 2000.

[24] F. Gea, *Analysis of Issues Connected with Water Supplying in Maasai Pasturelands Around the Village of Uwiro (Arusha, Tanzania)*, Graduation Thesis University of Torino, 2005.

[25] IAEA, *Environmental Isotopes in the Hydrological Cycle Principles and Applications Environmental*, vol. 1, International Atomic Energy Agency (IAEA) and United Nations Educational, Scientific and Cultural Organization, 2000.

[26] V. K. Saxena and S. Ahmed, "Dissolution of fluoride in groundwater: A water-rock interaction study," *Environmental Geology*, vol. 40, no. 9, pp. 1084–1087, 2001.

[27] J. V. Walther, "Comment and reply: Feldspar dissolution at 25°C and low pH," *American Journal of Science*, vol. 297, no. 10, pp. 1012–1032, 1997.

14

Studies on Characterization of Corn Cob for Application in a Gasification Process for Energy Production

Anthony I. Anukam,[1,2] Boniswa P. Goso,[1]
Omobola O. Okoh,[1] and Sampson N. Mamphweli[2]

[1]Department of Chemistry, University of Fort Hare, Private Bag X1314, Alice 5700, South Africa
[2]Fort Hare Institute of Technology, University of Fort Hare, Private Bag X1314, Alice 5700, South Africa

Correspondence should be addressed to Anthony I. Anukam; aanukam@ufh.ac.za

Academic Editor: Kaustubha Mohanty

Quintessential characteristics of corn cob were investigated in this study in order to determine its gasification potential. Results were interpreted in relation to gasification with reference to existing data from the literature. The results showed that the gasification of corn cob may experience some challenges related to ash fouling, slagging, and sintering effects that may be orchestrated by high ash content recorded for corn cob, which may contribute to increasing concentration of inorganic elements under high temperature gasification conditions, even though EDX analysis showed reduced concentration of these elements. The study also found that the weight percentages of other properties such as moisture, volatile matter, and fixed carbon contents of corn cob as well as its three major elemental components (C, H, and O) including its clearly exhibited fiber cells make corn cob a suitable feedstock for gasification. FTIR analysis revealed the existence of –OH, C–O, C–H, and C=C as the major functional group of atoms in the structure of corn cob that may facilitate formation of condensable and noncondensable liquid and gaseous products during gasification. TGA results indicated that complete thermal decomposition of corn cob occurs at temperatures close to 1000°C at a heating rate of 20°C/min.

1. Introduction

The quest for energy is expanding rapidly all around the world, resulting in increasing pressure on power generating systems in most countries, especially in developing countries such as South Africa. These have necessitated the use of alternative means of power generation that will not only ease the pressure on power generating systems but also have reduced consequential impact on the environment. Renewable source of energy is the most convenient alternative energy source to dwindling power generating capabilities imposed by rising energy demand.

South Africa is one of many developing countries in the world with quite a number of rural settlements that are associated with energy challenges because of the difficulties in extending the national electricity grid to these settlements. Therefore, the energy needs of these remote areas have to be met by off-grid technologies that are economically viable and sustainable. There are basically four different types of off-grid power generation methods, namely, solar, wind, and hydro- as well as dendro (biomass)-power generation. Because of the high cost associated with the first three, power generation from biomass happens to be the most convenient and yet efficient method of providing energy to remote settlements; however the importance of the type and availability of the biomass to be used as feedstock in the energy conversion processes cannot be overemphasized [1].

Corn cob (CC) is a biomass feedstock with direct potential as an energy resource that can be used in gasification systems for energy production. It has a number of advantages over other biomass feedstocks including its dense and uniform nature as well as its increased energy content and its low sulfur and nitrogen concentrations [2]. It is an agricultural residue that is generated from maize and remains part of the ear on which the kernels grow. In most established countries, CC is usually disposed and destroyed by fire on the farms to prepare for the next coming season. The dumping and burning of CC on the farms constitute gross air pollution. In South Africa, corn is a very important food for many people and it remains the most critical horticultural harvest for more than 70 million homestead families around the world [3, 4].

CC residues are sufficiently available in South Africa and are produced in large quantities by the maize industry, which happens to be one of the largest producers of agricultural residues in the country as approximately 11 million tons of CC is produced per year [5]. The conversion of CC into an energy carrier gas known as syngas through gasification is a viable alternative to electricity generation needed to meet the ever escalating energy demands of remote settlements [6, 7]. Before this can be achieved, there is a need to investigate the characteristics of CC that are relevant to gasification in order to accurately predict its performance during gasification since the operation of energy conversion systems has been quite compromised because of the wide variety of biomass origin that affects its composition and characteristics.

Ethanol production from lignocellulosic biomass is facing quite a number of challenges. Among these issues are getting rid of the lignin content of the biomass (delignification) as well as conversion of its cellulose and hemicellulose contents into fermentable sugars through fermentation [8]. The process of finding a way around these challenges is still under investigation. However, with gasification, any biomass material can be successfully converted into useable energy without the need for delignification. The syngas produced from gasification can be further processed into other chemicals via different reforming processes. It can also be converted into fuels through the Fischer Tropsch process. Another advantage of gasification as compared to other bioenergy generation technologies is its ability to utilize a wide variety of biomass feedstocks ranging from any agricultural or plant residue, industry organic by-products, or even municipal wastes, and hence gasification is considered a viable technique for producing energy from biomass feedstocks, which cannot be technically or economically fermented to ethanol.

Gasification occurs under a sequence of successive reactions that are mostly endothermic in nature. The following reactions take place during biomass gasification [9, 10].

Oxidation reaction is

$$C + O_2 \longrightarrow CO_2 \quad (+393\,\text{MJ/kg mole}) \quad (1)$$

Reduction reaction is

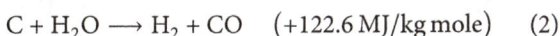

$$C + H_2O \longrightarrow H_2 + CO \quad (+122.6\,\text{MJ/kg mole}) \quad (2)$$

Water-gas shift reaction is

$$CO + H_2O \longleftrightarrow CO_2 + H_2 \quad (-41.2\,\text{MJ/kg mole}) \quad (3)$$

Methanation reaction is

$$C + 2H_2 \longleftrightarrow CH_4 \quad (+75\,\text{MJ/kg mole}) \quad (4)$$

Reaction (1) typically occurs in the combustion zone of the gasifier with a theoretical temperature of over 1200°C;

hence oxidation reactions of a gasification process are also referred to as combustion zone reactions. The second reaction (reaction (2)) represents the medium where the product of partial oxidation (i.e., products that were not fully combusted in reaction (1)) passes through, which is the reaction representing the red-hot bed of charcoal that is capable of reducing gas temperature because of the reaction's endothermic nature, while reaction (3) depicts the prominent water-gas shift reaction that is the major determinant of the yield and quality of the syngas produced from a gasification process. The fourth reaction (reaction (4)) is the hydrogasification reaction that is also known as the methanation reaction, which forms very minute levels of methane during gasification [9].

Gasifiers operate satisfactorily only within certain ranges of feedstock characteristics, as such knowledge of the characteristics of the feedstock to be used during gasification is required in order to predict its performance prior to gasification [11]. Analytical instruments such as CHNS analyzer, atomic absorption spectrometer (AAS), thermogravimetric analyzer (TGA), and scanning electron microscopic (SEM) analyzer are very useful instruments to determine the characteristics of biomass materials for the purpose of gasification. Previous studies from other researchers have used these instruments to determine the sintering characteristics and mineral transformation behaviour of corn cob ash (CCA) [7]. Kumar et al., 2008 [12], used TGA to study the thermal characteristics of corn stover (CS) as a gasification and pyrolysis feedstock, while Arun and Ramanan, 2016 [13], conducted experimental studies on the gasification of CC in a fixed-bed system after determining the characteristics of CC using a muffle furnace, an ultimate analyzer, and a bomb calorimeter to provide information on physical and chemical properties of the material as well as its energy content, respectively. In another study, Aboyade et al., 2013 [14], determined the nonisothermal thermokinetics of copyrolysis of a blend of two different biomass materials that included corn residue with coal. In addition to the physical, chemical, and thermal properties of CC, there is a lack of specific information on its characterization intended to reveal its surface and internal structural properties for the purpose of gasification. The objective of this study, therefore, is to establish the characteristics of CC relevant to gasification in a downdraft system and to interpret the information obtained from these characteristics in relation to gasification based on existing data from the literature.

The characteristics of CC in terms of proximate and ultimate analysis as well as in terms of energy value (heating value) reported by previous researchers for various applications different from gasification are presented in Tables 1 and 2, respectively.

The difference in the reported values in both Tables 1 and 2 may be essentially due to a combination of factors that includes the source of the CC, its handling, storage, and climatic conditions as well as soil type and texture where the corn was grown including the tendency of the corn plant to uptake nutrients from the soil. The ash content varied probably because of different methods of harvesting and the amount of nutrients (fertilizers) applied to the corn plant during growth. It is valuable from another overview that

TABLE 1: Proximate analysis of corn cob from previous studies (wt.%).

Author	MC	VM	AC	FC
[15]	4.6	79.9	1.8	13.7
[16]	11.74	72.33	10.67	4.97
[17]	—	78.7	0.9	16.2

TABLE 2: Ultimate analysis of corn cob from previous studies (wt.%).

Author	C	H	N	S	O	HHV (MJ/kg)
[15]	50.2	5.9	0.42	0.03	43.5	19.14
[16]	46.2	5.42	0.92	0.24	47.22	18.36
[17]	45.5	6.2	1.3	—	47.0	—

the inorganic compounds contained in CC with higher ash content have potential to be used as catalyst in thermal conversion systems [16].

2. Experimental

2.1. Sample Preparation. The corn cob (CC) used for this study was obtained from a local farm in Alice, in the Eastern Cape Province of South Africa. It was dried outdoors at an average temperature of about $30°C$ to lower its moisture content. This was followed by milling to a size required by the instruments that were used for analyses. The dried and milled CC was preserved in a desiccator prior to analyses.

2.2. Proximate Analysis. Information required for moisture, volatile matter, and ash as well as fixed carbon contents of CC was given by proximate analysis. These properties are relevant to the thermal conversion of any biomass material into energy [19]. The proximate analysis data of CC was obtained from the TGA plot presented in Figure 2 following a modified ASTM D 5142-04 standard test method [18, 20]. They were obtained according to the equations in Table 3.

Moisture content was determined by weight loss at temperatures close to $100°C$, while volatile matter content represented the mass evolved between the temperatures of 100 and $550°C$. After heating of the sample to about $1000°C$ during TGA, the remaining mass was considered as being ash and the fixed carbon content of CC was obtained by difference.

2.3. Ultimate Analysis. This analysis provided information on the elemental components of CC both in qualitative and in quantitative terms. A Thermo Quest CHNS elemental analyzer was used for this purpose. The proportion of carbon (C), hydrogen (H), sulfur (S), and nitrogen (N) were determined, while oxygen (O) was obtained by difference.

About 5 mg of milled CC was placed in a tin capsule that contained an oxidizer prior to combustion in a reaction at $1000°C$. This led to a violent reaction as the sample and tin capsule decomposed, creating a condition where all heat resistant substances became fully oxidized. The products obtained were made to pass through a high purity copper at

TABLE 3: Equation parameters used for proximate analysis determination from TGA curve [18].

Equation name	Parameters
Moisture content	$\left(\dfrac{[\text{Initial Mass}] - [\text{Moisture Mass}]}{[\text{Initial Mass}]}\right) \times 100$
Volatile matter content	$\left(\dfrac{[\text{Moisture Mass}] - [\text{Volatile Mass}]}{[\text{Initial Mass}]}\right) \times 100$
Ash	$\left(\dfrac{[\text{Ash Mass}]}{[\text{Initial Mass}]}\right) \times 100$
Fixed carbon	$100 - ([\text{Moisture}] + [\text{Volatile}] + [\text{Ash}])$

$500°C$ in order to rid the process of any oxygen that was not completely consumed during the combustion process. There is always a need to employ high purity substances during CHNS analyses for the purpose of oxidation and to remove unwanted materials that may interfere with analyses results [21]. Complete oxidation was ensured by using tungsten trioxide and copper downstream of the combustion chamber of the instrument. Combustion products such as carbon dioxide (CO_2), sulfur dioxide (SO_2), and nitrogen dioxide (NO_2) were obtained after the analysis, which were all separated by gas chromatography and the elements measured with a thermal conductivity detector.

The energy value, also known as heating value and reported in terms of higher heating value (HHV) of CC, was calculated from the mass fractions of the elemental components obtained from CHNS analysis, which was done according to [22]

$$HV\ (MJ/kg) = -1.3675 + 0.3137 \times C + 0.7009 \times H + 0.0318 \times O, \tag{5}$$

where HV is the heating value measured in MJ/Kg, while C, H, and O are the carbon, hydrogen, and oxygen contents of CC.

2.4. FTIR Analysis. The Fourier Transform Infrared (FTIR) spectroscopy also deals with quantitative and qualitative analysis of organic samples and recognizes chemical bonds in a molecule by generating an infrared retention range; the spectra generate a profile of the sample, a particular molecular fingerprint that can be utilized to screen and scan samples for a wide range of segments [23]. FTIR is an operative analytical instrument for distinguishing functional groups and characterizing covalent bonding data. In this study, it was used to determine the most reactive components of CC in terms of functional groups since the rate of gasification reactions depend on the chemically active group of components of the biomass used as feedstock [24].

About 0.5 mg of the sample was mixed with 0.25–0.50 of KBr and placed in the FTIR test holder. The sample was examined by a fully computerized Perkin Elmer FTIR system which produces the absorbance spectra that demonstrate the unique chemical bonds and the atomic structure of the sample material. This profile was in the form of an absorption spectrum that indicated peaks representing components in higher concentration. Absorbance peaks on the spectrum also indicated the functional groups. Different types of bonds and thus different functional groups absorbed infrared radiation of various wavelengths. Despite the fact that the analysis was performed in absorbance mode, it can be converted into a transmittance mode since they are just the reverse of each other. The analytical spectrum is then contrasted in a reference library program with cataloged spectra to identify components for unknown material using the cataloged spectra for known materials.

2.5. Thermogravimetric Analysis (TGA).

The thermal behaviour of biomass materials are usually measured by a thermogravimetric analyzer (TGA), which measures the percentage weight loss of the biomass as a function of temperature and the resulting thermogram has a peculiar shape for biomass materials [25]. In addition to studying the thermal behaviour of CC, this analysis was undertaken in order to establish the thermal parameters that would impact on the gasification of the material. It is worth noting that most TGA experiments are conducted under a chemically inactive environment (of which nitrogen or argon is often used) to show the effect of heat degradation that includes carbonization; oxygen is highly reactive and usually not recommended during analyses involving TGA because it reacts with sample components, leading to loss of original sample in the process [24].

A 7.81 mg of the sample was combusted in a SDT Q600 TGA instrument under a nitrogen atmosphere at a flow rate of 35 mL/min between 35 and 1000°C. Nitrogen was used to create a chemically inactive environment so as to prevent the TGA instrument from overheating. A heating rate of 20°C/min was used during TGA because this is characteristic of gasification systems using the downdraft gasifier [11].

2.6. SEM Analysis.

Scanning Electron Microscopy (SEM) is a high resolution imaging system with an extraordinary depth of field. It indicates topographical, structural, and elemental data at low magnifications up to 200,000x [26]. The utilization of SEM innovation is a priceless guide in distinguishing and portraying mineral and material stages together with surface components. SEM in this study was used for surface morphological view of the material to establish if CC is enough carbonaceous material that would be suitable for gasification using the downdraft gasification system.

The SEM analysis of CC was undertaken by a JEOL (JSM-6390) operating with accelerating voltage of 15 kV. The micrographs were generated at different magnifications (250–1000x) by a computer program. The data was collected over a selected area of the surface of the sample and a two-dimensional image was generated that displayed spatial variations in properties.

TABLE 4: Measured physical characteristics of corn cob.

Proximate analysis (wt.%)	
Moisture content	5.1
Volatile matter content	65.1
Ash content	8.5
Fixed carbon	21.3

3. Results and Discussion

In this section, the findings of this study are presented, discussed, and substantiated with reference to existing data from the literature.

3.1. Physical Characteristics of Corn Cob.

Gasifiers, among other factors, operate satisfactorily with regard to efficiency only within certain ranges of feedstock characteristics [11]. Table 4 shows the data obtained from the physical characterization of CC, which were obtained from the thermogravimetric plot of the sample presented in Figure 2, employing the equations presented in Table 3.

From Table 4, moisture content of CC was measured as 5.1%, which is quite different when compared to the values reported in the literature. It is lower than the value of 11.74% reported by Danish et al., 2015 [16], and higher than the value of 4.6% reported by Danje, 2011 [15], in Table 1. The difference in these values may be attributed to a number of reasons including the source of the CC and handling conditions. However, this value (5.1%) for moisture content is desirable for gasification to take place as materials with moisture content beyond 20% would create technical difficulties linked to poor combustion conditions within the gasification system and will inhibit immediate combustion of the material at the same time increasing its smoking propensity with a consequent reduction in gasification process efficiency [24]. It can also be noted that the CC used for this study is characterized by relatively high volatile matter content (65.1%), which was anticipated because of the organic nature of the material. The contents of volatile matter in biomass materials are usually high due to the organic nature of the biomass, which indicates the biomass potential to create huge amounts of inorganic vapours when used as feedstock in a gasification process; the higher the volatile matter content of biomass, the better its combustion and gasification rates because of the biomass yield upon carbonization [25]. The material is also characterized by high ash content that may also be attributed to a number of factors that include those previously given in Section 1. This high content of ash may not be favourable to gasification because of issues linked to sintering and slagging that may be experienced during gasification, which might also contribute to reduction in process efficiency. Biomass ash content greater than 6% is not desirable for gasification because it creates technical issues related to agglomeration, fouling, and sintering as well as slagging that may together reduce gasification efficiency; however, ash may exert some catalytic effect that may allow for cracking of higher molecular weight compounds such as

TABLE 5: Measured elemental components of corn cob.

Ultimate analysis (wt.%)	
C	44.4
H	5.6
N	0.43
S	1.3
O (by difference)	48.27

TABLE 6: Weight percentages of the metallic elemental components of corn cob.

Element	Composition (wt.%)
Al	0.31
K	1.53
Si	0.44
Na	1.32
Ca	0.11
Mg	0.42
Fe	0.06

tar into lighter ones for optimum gasification efficiency [27–29]. The fixed carbon content of CC was also found to be about 21%, which is high enough to allude that there will be increased formation of char during gasification as the relative proportions of the content of volatile matter and fixed carbon are related to the yields and composition of solid, liquid, and gaseous products formed during gasification [30].

3.2. Chemical Characteristics of Corn Cob. The ratio of the products formed during gasification of biomass is influenced not just by its physical characteristics but also by the chemical composition of the biomass fuel and the operating conditions of the gasifier [31]. The chemical properties of CC was studied in order to obtain information regarding the relative proportions of the major elemental components of the material and to predict the impact of these components on syngas quality and yield as well as on the environmental effects of gasifying CC. Table 5 shows the elemental components of CC as measured by the CHNS analyzer.

The data in Table 5 shows that CC is composed of three major elements with a higher proportion of oxygen than carbon. The higher oxygen proportion is the reason for the low energy value reported for CC in Section 3.5. However, this higher oxygen content implies increased thermal reactivity during gasification. Increased biomass oxygen content is an indication of increased thermal reactivity of biomass during thermochemical conversion processes; the gasification of biomass is centered on carbon conversion [11]. The content of hydrogen is in agreement with most findings in the literature and had positive contribution to the energy value of CC reported in Section 3.5 together with its content of carbon. Oxidation of carbon and hydrogen contents of biomass are usually initiated by exothermic reactions during gasification, forming CO_2 and H_2O, with the CO_2 emitted as a product of complete combustion [24]. The relatively low nitrogen and sulfur contents imply lower amounts of NH_3, HCN, and H_2S (which are environmentally harmful compounds) may be anticipated during gasification.

3.3. Metallic Elemental Components of Corn Cob. In addition to nonmetallic elemental components of biomass, there are also metallic elemental components such as Na, K, Mg, and Si that are especially responsible for the concentration of ash in biomass materials; in other words, the weight percentage of these metallic elements, to an extent, determines the overall weight percentage of ash contained in biomass as high concentration of these elements creates technical hitches such as fouling, sintering, and slagging because of volatilization of

the elements, which forms liquid slags on cooling when the biomass is used as feedstock in gasification processes [32]. Table 6 shows the weight percentages of the metallic elements contained in CC, which were obtained after analysis using a Thermo Scientific Model ICE 3500 Atomic Absorption Spectrometer (AAS) equipped with hollow cathode lamps.

It is quite obvious that the concentrations of the metallic elements are relatively low, implying that there may be little or no technical issues related to those previously mentioned when CC is used as feedstock in a gasification process. The reasons for the low concentration of these elements are the same as those given for the high content of ash reported in Table 4. These ash-forming elements are usually taken up by plants during growth; the elemental composition of biomass, especially with regard to the weight percentages of the ash-forming elements, has key impact on ash transformation sequences and sintering behaviours [7, 24]. Ash-forming elements are usually characterized by complex transformation reactions during biomass gasification, creating technical issues linked to those previously mentioned; however, reactions involving the oxides of calcium or magnesium with potassium silicates lead to formation of high-temperature-melting calcium-magnesium-potassium silicates that play significant roles in the reduction of sintering issues during gasification because of the limit in the formation of silicates that are rich in potassium [7, 24, 33, 34]. For fixed-bed gasification systems such as the downdraft system, ash-related sintering proceeds with the formation of slag as a consequence of certain factors like bridging, coalescence, and accumulation of the sintered ash residues on gasifier grates. The slag with large sizes cannot be transported out from the grate, which then interferes with the gasification process and reduces the performance of gasification appliances [35–37].

3.4. Reactive Components of Corn Cob. To gain a deeper understanding of the chemistry of CC and to provide a baseline for the prediction of its gasification performance, a diagnosis of the internal structure of the material is necessary. This diagnosis relates to analysis of the material's reactive components in terms of the functional groups present in its structure. The spectrum associated with the structure of CC and the indicated peaks relative to each functional group are presented in Figure 1. The absorbance at various wavenumbers corresponds to the functional groups.

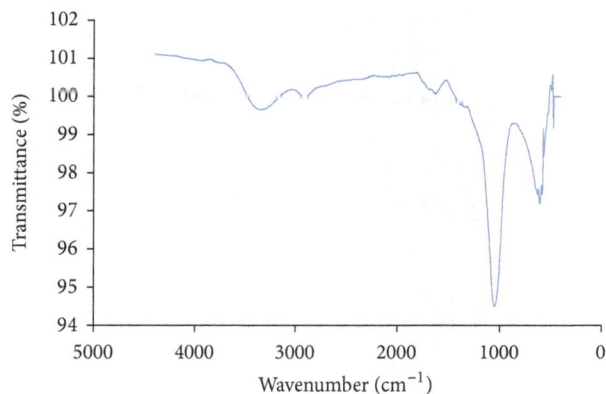

FIGURE 1: FTIR spectrum of corn cob.

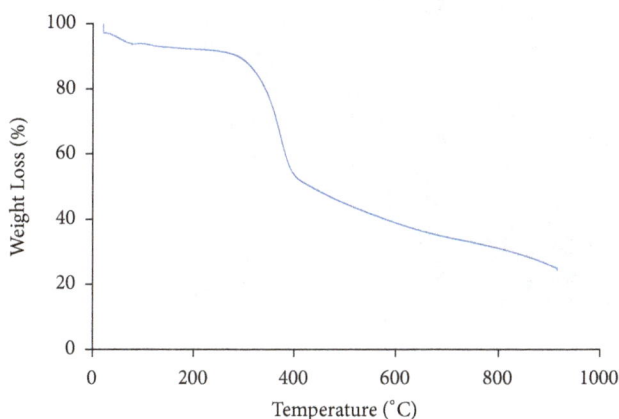

FIGURE 2: Thermogram resulting from the thermal analysis of corn cob.

TABLE 7: Functional groups present in the structure of corn cob.

Frequency range (cm^{-1})	Groups	Class of compounds
3303	O–H stretching	Alcohol, phenols
2844	C–H stretching	Alkanes
1589	C=C bending	Aromatic compounds
1029	C–O stretching	Alcohol, phenols & esters
582	C–H bending	Aromatic compounds

cellulose and hemicellulose, speeds up the rate of other reactions such as decarboxylation reactions that leads to the breakage of glycosidic bonds that consequently forms a series of less oxygen-containing compounds such as ethers, acids, and aldehydes and noncondensable gases such as CO and CO_2 [24, 38].

Plant photosynthesis is usually driven by energy from the sun that is usually stored in chemical bonds of the structural components of the plant, implying that an amount of energy would be required to break these bonds in order to harness the energy, which is mostly achieved through initiation of gasification reactions when the plant material is to be used as feedstock in a gasification process [11, 28].

3.5. Energy Value of Corn Cob. Plants convert energy from the sun into chemical energy that is stored in the structural components of the biomass by using CO_2 in the atmosphere [24]. The energy value of CC was determined to evaluate the amount of energy available for conversion, which is a very important property of biomass because conversion efficiency of a gasification process depends on it [11]. In this study, the energy value of CC was measured as 18.02 MJ/kg, a value that is in agreement with those reported by Danje, 2011 and Danish et al., 2015 [15, 16], in Table 2. It is therefore sufficient to allude that the energy value of CC measured in this study is in agreement with most findings in the literature.

3.6. Thermal Behaviour of Corn Cob. In order to better understand the gasification characteristics of CC, thermal analysis of the sample using an instrument relevant to gasification is necessary. This analysis is intended to establish the thermal behaviour of the sample under both high and low temperatures as well as determine the thermal parameters that would influence its gasification. Figure 2 shows the thermogram obtained from the thermogravimetric analysis of CC.

The plot in Figure 2 shows that as temperature increases there is a marked reduction in the weight of the sample. The plot also shows that the thermal degradation behaviour of CC is characterized by three different weight loss stages with the initial one at 94°C, which signifies the removal of moisture from the sample. A significant weight loss could be observed between 200 and 500°C and represents the second stage of the decomposition process of the sample. This may be attributed to the decomposition of basic organic components of CC such as cellulose, hemicellulose, and lignin; the decomposition of

It is quite obvious from Figure 1 that the peak at 3303 cm^{-1} corresponds to O–H stretching vibrations that indicates the presence of hydroxyl groups, while that near 2844 cm^{-1} depicts C–H stretching that corresponds to the presence of alkanes. 1000 cm^{-1} depicts C–O stretching, with the peak near 600 cm^{-1} showing characteristics of C–H bending. These functional groups represent the chemically active components of biomass that accelerates the rates of the gasification reactions presented in Section 1 [24].

Nonetheless, for better understanding of the functional groups common to the structure of CC, Table 7 presents the chemically active components related to the bonds of the atoms that make up the material and which take part during thermal conversion processes.

During gasification, the presence of the –OH group will initiate and accelerate the rate of condensation reactions created by dehydroxylation as a result of thermal decomposition of the cellulose content of the material caused by rising temperatures within the gasifier, while C–H presence due to alkanes is connected to the reactions leading to hemicellulose degradation [24]. The existence of the C=C group, which is an indication of the presence of alkenes, facilitates reactions leading to lignin decomposition; while the group C–O, which is assigned to carboxylic groups in

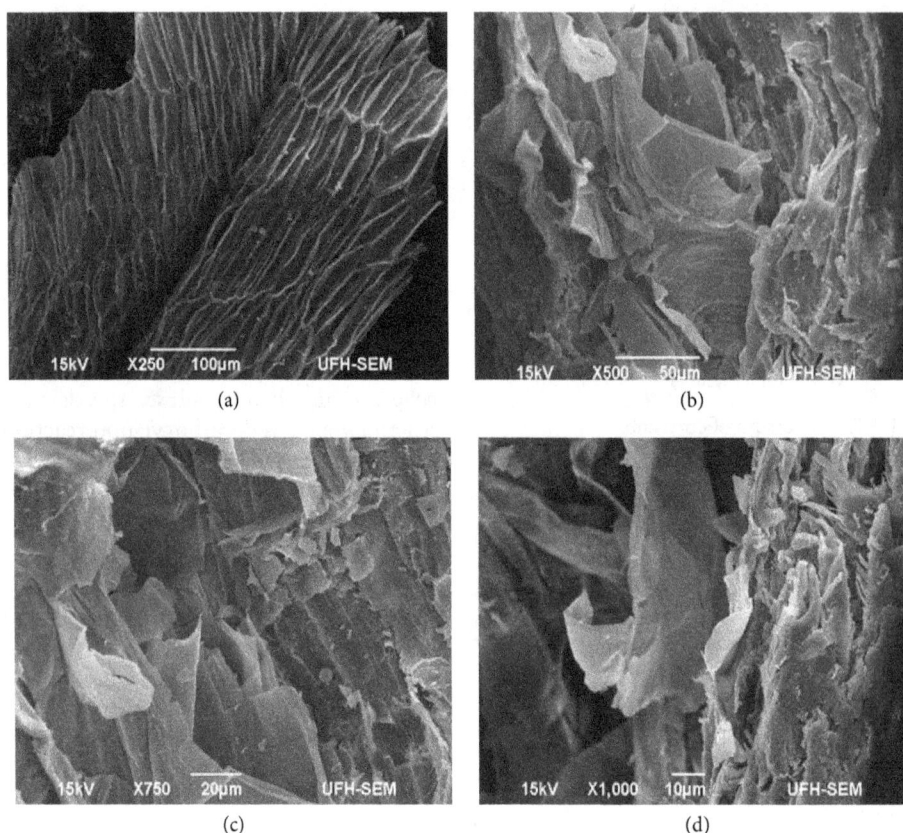

FIGURE 3: SEM images of corn cob obtained at different magnifications.

these components releases volatile gases such as CO_2 and CH_4 that are mainly formed due to the decomposition of hemicellulose between the temperatures of 190 and 320°C. This degradation temperature for hemicellulose implies less production of tar and char during gasification of CC [24]. The third stage of the thermal decomposition process of CC is indicated by cellulose and lignin degradation between 280 and 400°C for cellulose and between 320 and 450°C for lignin, with total combustion of the sample taking place as its weight is reduced in the process to give rise to decomposition of hydrocarbons. During gasification, cellulose and lignin degradation at higher temperatures depict the production of carbonized biomass as well as heavy organic and inorganic compounds [39, 40].

3.7. Microstructural Characteristics of Corn Cob. The surface structure of CC was examined with a scanning electron microscopic instrument that offered detailed information on imaging and surface composition of the sample. This provided a guide as to whether CC is enough carbonaceous material suitable for gasification in a downdraft gasifier. Figure 3 shows the SEM images of CC obtained at different magnifications. The images were magnified by a factor of 250 for better understanding and interpretation of the microstructural characteristics of the material.

As can be seen from the images in Figures 3(a)–3(d), the shapes are quite irregular and agglomerated. The sample is clearly seen to have no pores even at higher magnifications, an evidence of lack of pretreatment prior to analysis but it exhibits cells on the surface without much characterized structure. However, at 250 magnification (Figure 3(a)) there seem to be plenty of parallel lines that appear on the surface of the sample, which look like cells of residual pith that provides a pathway for the transportation of water and nutrients from the soil, but on increasing magnification to 500 (Figure 3(b)) these lines seem to disappear, showing more vascular bundles with not too conspicuous fragmented cells which indicates fibrous lignocellulosic nature of CC, which is a common feature of agricultural biomass residues [24]. At higher magnification (×750, Figure 3(c)), the vascular bundles are more pronounced with the fragmented cell structures more visible. The presence of the vascular bundles and cell structures are an indication of carbon-oriented structures which corroborates the carbon content data of CC presented in Table 5. These cell structures are also associated with the formation of pathways for the production of gaseous products; these features make CC amenable to high temperature gasification that connotes optimum efficiency [24, 39, 41]. As image magnification was increased to a maximum of 1000 (Figure 3(d)), more features were revealed including the size of the vascular bundles and their compact nature, which are important features used to understand the combustion behaviour of biomass materials [42].

4. Conclusions

In order to evaluate CC with regard to its gasification-related characteristics, a detailed assessment based on fuel analysis was performed. The experiments conducted and the results presented showed that CC is a biomass feedstock suitable for gasification due to its low moisture content and due to its low concentration of metallic elements. However, its high percentage of ash may create a bit of technical challenges that may lower gasification efficiency. Its low concentration of nitrogen and sulfur implies reduced emissions of NO_X and SO_2 during gasification; and its hydrogen concentration is high enough to initiate the water-gas shift reaction that is the dominant chemical reaction which forms the major portion of the syngas. The energy value analysis showed that CC contains a manageable amount of energy that can be converted into useful energy through gasification. The reactive components of CC were mostly oxygen-containing functional groups that may play important roles during gasification. The study also established that the thermal decomposition of CC began at temperatures below 100°C, with its complete degradation occurring at temperatures close to 1000°C, releasing enormous amount of gases, while SEM analysis revealed compacted vascular bundles and fiber tissues linked to carbon-orientation that are among the features of CC that may favour high temperature gasification.

The results obtained form a significant basis for the development of a gasification system that would be tailored to the demands of the characteristics of CC.

Even though the CC used for this study exhibited low concentration of metallic elements, an elevated weight percentage of these elements is anticipated when the material is used as feedstock in a gasification process. This is because of the high amount of ash recorded for CC. The weight percentage of the metallic elements of biomass increases with rising gasification temperature [24]. As such, further research is required on reduction of the weight percentage of CC ash content. This study did not involve the gasification of CC either via simulation or via experimental investigation of its gasification process. This is where challenges could be experienced with the use of CC. It is therefore recommended that research be undertaken on the gasification of CC in order to adequately establish the impact of fuel characteristics on gasification process efficiency. The reaction kinetics of the thermal decomposition of CC also require further studies.

Conflicts of Interest

The authors declare that they have no conflicts of interest.

Acknowledgments

The authors wish to acknowledge the financial support of the National Research Foundation of South Africa (NRF), the Govan Mbeki Research and Development Center (GMRDC), and the Chemistry Department of both the University of Fort Hare and the Fort Hare Institute of Technology (FHIT), for their technical assistance.

References

[1] D. Gunarathne, *Optimization of the performance of downdraft biomass gasifier installed at national engineering research and development (NERD) Centre of Sri Lanka [MSc thesis]*, KTH School of Industrial Engineering and Management, Sweden, 2012.

[2] Extension Farm Energy, "Corn cobs for biofuel production," http://articles.extension.org/pages/26619/corn-cobs-for-biofuel-production, 2016.

[3] J. T. Oladeji and C. C. Enweremadu, "A predictive model for the determination of some densification characteristics of corncob briquettes," in *Materials and Processes for Energy: Communicating Current Research and Technological Developments*, pp. 169–177, 2013.

[4] Y. Zhang, A. E. Ghaly, and B. Li, "Physical properties of corn residues," *American Journal of Biochemistry and Biotechnology*, vol. 8, no. 2, pp. 44–53, 2012.

[5] SA Department of Agriculture, "Maize production," http://nda.agric.za/publications, 2016.

[6] M. Y. Suberu, A. S. Mokhtar, and N. Bashir, "Potential capability of corn cob residue for small power generation in rural Nigeria," *ARPN Journal of Engineering and Applied Sciences*, vol. 7, no. 8, pp. 1037–1046, 2012.

[7] L. Wang, J. E. Hustad, and M. Grønli, "Sintering characteristics and mineral transformation behaviors of corn cob ashes," *Energy and Fuels*, vol. 26, no. 9, pp. 5905–5916, 2012.

[8] J. Lee, "Biological conversion of lignocellulosic biomass to ethanol," *Journal of Biotechnology*, vol. 56, no. 1, pp. 1–24, 1997.

[9] R. N. André, F. Pinto, C. Franco et al., "Fluidised bed co-gasification of coal and olive oil industry wastes," *Fuel*, vol. 84, no. 12-13, pp. 1635–1644, 2005.

[10] J. Fermoso, *Pressure co-gasification of coal and biomass for the production of hydrogen*, University of Oviedo, Spain, 2009.

[11] A. Anukam, S. Mamphweli, E. Meyer, and O. Okoh, "Computer simulation of the mass and energy balance during gasification of sugarcane bagasse," *Journal of Energy*, vol. 2014, Article ID 713054, 9 pages, 2014.

[12] A. Kumar, L. Wang, Y. A. Dzenis, D. D. Jones, and M. A. Hanna, "Thermogravimetric characterization of corn stover as gasification and pyrolysis feedstock," *Biomass and Bioenergy*, vol. 32, no. 5, pp. 460–467, 2008.

[13] K. Arun and M. V. Ramanan, "Experimental studies on gasification of corn cob in a fixed bed system," *Journal of Chemical and Pharmaceutical Research*, vol. 8, pp. 667–676, 2016.

[14] A. O. Aboyade, J. F. Görgens, M. Carrier, E. L. Meyer, and J. H. Knoetze, "Thermogravimetric study of the pyrolysis characteristics and kinetics of coal blends with corn and sugarcane residues," *Fuel Processing Technology*, vol. 106, pp. 310–320, 2013.

[15] S. Danje, *Fast pyrolysis of corn residues for energy production [dissertation, thesis]*, Stellenbosch University, 2011.

[16] M. Danish, M. Naqvi, U. Farooq, and S. Naqvi, "Characterization of South Asian agricultural residues for potential utilisation in future energy mix," *Energy Procedia*, vol. 75, pp. 2974–2980, 2015.

[17] J. Wannapeera, N. Worasuwannarak, and S. Pipatmanomoi, "Product yields and characteristics of rice husk, rice straw and corncob during fast pyrolysis in a drop-tube/fixed-bed reactor," *Songklanakarin Journal of Science and Technology*, vol. 30, no. 3, pp. 393–404, 2008.

[18] Leco Corporation, *Moisture, Volatile Matter, Ash, and Fixed Carbon Determination-Solid Fuel Characterization Measurements in Coke*, Organic Application Note, Form 203-821-381, LECO Corporation, St. Joseph, Mich, USA, 2010, http://www.leco.co.za/wp-content/uploads/2012/02/TGA701_COKE_203-821-381.pdf.

[19] P. Tanger, J. L. Field, C. E. Jahn, M. W. DeFoort, and J. E. Leach, "Biomass for thermochemical conversion: targets and challenges," *Frontiers in Plant Science*, vol. 4, article 218, 2013.

[20] ASTM Standards, *ASTM D 5142-04: Standard Test Method for Proximate Analysis of the Analysis Sample of Coal and Coke by Instrumental Procedures*, vol. 5, ASTM Standards, West Conshohocken, PA, USA, 2008.

[21] P. Elmer, *2400 Series II CHNS/O Elemental Analysis: Organic Elemental Analysis (2016)*, 2016, https://www.perkinelmer.com/labsolutions//resources/docs/BRO_2400_SeriesII_CHNSO_Elemental_Analysis.pdf.

[22] C. Sheng and J. L. T. Azevedo, "Estimating the higher heating value of biomass fuels from basic analysis data," *Biomass & Bioenergy*, vol. 28, no. 5, pp. 499–507, 2005.

[23] D. A. Skoog and J. J. Leary, *Principles of Instrumental Analysis*, Chapter 12, Harcourt Brace Jovanovich Philadelphia, Philadelphia, PA, USA, 1992.

[24] A. Anukam, S. Mamphweli, P. Reddy, and O. Okoh, "Characterization and the effect of lignocellulosic biomass value addition on gasification efficiency," *Energy Exploration and Exploitation*, pp. 1–16, 2016.

[25] B. M. Jenkins Jr. and T. Miles, "Combustion properties of biomass," in *Fuel Processing Technology*, T. L. Baxter, Ed., vol. 54, pp. 17–46, 1998.

[26] A. Abdolali, H. H. Ngo, W. Guo et al., "Characterization of a multi-metal binding biosorbent: chemical modification and desorption studies," *Bioresource Technology*, vol. 193, pp. 477–487, 2015.

[27] E. Gustafsson, *Characterization of Particulate Matter from Atmospheric Fluidized Bed Biomass Gasifiers [PhD. thesis]*, Linnaeus University, 2011.

[28] P. McKendry, "Energy production from biomass (part 3): gasification technologies," *Bioresource Technology*, vol. 83, no. 1, pp. 55–63, 2002.

[29] R. Fahmi, A. V. Bridgwater, I. Donnison, N. Yates, and J. M. Jones, "The effect of lignin and inorganic species in biomass on pyrolysis oil yields, quality and stability," *Fuel*, vol. 87, no. 7, pp. 1230–1240, 2008.

[30] J. S. Brar, K. Singh, J. Wang, and S. Kumar, "Co-gasification of coal biomass. A review," *International Journal of Forestry Research*, pp. 1–10, 2012.

[31] T. Chandrakant, "Biomass gasification-technology and utilisation," in *Humanity Development Library (2002)*, 2012, http://www.pssurvival.com.

[32] T. R. Miles, T. R. Miles Jr., L. L. Baxter, R. W. Bryers, B. M. Jenkins, and L. L. Oden, "Boiler deposits from firing biomass fuels," *Biomass and Bioenergy*, vol. 10, no. 2-3, pp. 125–138, 1996.

[33] D. Boström, N. Skoglund, A. Grimm et al., "Ash transformation chemistry during combustion of biomass," *Energy and Fuels*, vol. 26, no. 1, pp. 85–93, 2012.

[34] B.-M. Steenari, A. Lundberg, H. Pettersson, M. Wilewska-Bien, and D. Andersson, "Investigation of ash sintering during combustion of agricultural residues and the effect of additives," *Energy and Fuels*, vol. 23, no. 11, pp. 5655–5662, 2009.

[35] L. Wang, G. Skjevrak, J. E. Hustad, and M. G. Grønli, "Effects of sewage sludge and marble sludge addition on slag characteristics during wood waste pellets combustion," *Energy and Fuels*, vol. 25, no. 12, pp. 5775–5785, 2011.

[36] S. Xiong, J. Burvall, H. Örberg et al., "Slagging characteristics during combustion of corn stovers with and without kaolin and calcite," *Energy and Fuels*, vol. 22, no. 5, pp. 3465–3470, 2008.

[37] E. Lindström, M. Sandström, D. Boström, and M. Öhman, "Slagging characteristics during combustion of cereal grains rich in phosphorus," *Energy and Fuels*, vol. 21, no. 2, pp. 710–717, 2007.

[38] D. Chen, Z. Zheng, K. Fu, Z. Zeng, J. Wang, and M. Lu, "Torrefaction of biomass stalk and its effect on the yield and quality of pyrolysis products," *Fuel*, vol. 159, article no. 9381, pp. 27–32, 2015.

[39] M. Wilk, A. Magdziarz, and I. Kalemba, "Characterisation of renewable fuels' torrefaction process with different instrumental techniques," *Energy*, vol. 87, pp. 259–269, 2015.

[40] G. S. Miguel, M. P. Domínguez, M. Hernández, and F. Sanz-Pérez, "Characterization and potential applications of solid particles produced at a biomass gasification plant," *Biomass and Bioenergy*, vol. 47, pp. 134–144, 2012.

[41] S. Gaqa, S. Mamphweli, E. Meyer, and D. Katwire, "Synergistic evaluation of the biomass/coal blends for co-gasification purposes," *International Journal of Energy And Environment*, vol. 5, pp. 251–265, 2014.

[42] A. Anukam, S. Mamphweli, P. Reddy, O. Okoh, and E. Meyer, "An investigation into the impact of reaction temperature on various parameters during torrefaction of sugarcane bagasse relevant to gasification," *Journal of Chemistry*, vol. 2015, Article ID 235163, pp. 1–12, 2015.

15

Kinetics of Methane Hydrate Formation in an Aqueous Solution with and without Kinetic Promoter (SDS) by Spray Reactor

Yaqin Tian,[1,2,3] **Yugui Li,**[1,2,3] **Hongping An,**[1,3] **Jie Ren,**[1,3] **and Jianfeng Su**[1,2]

[1]*College of Materials Science and Engineering, Taiyuan University of Science and Technology, Taiyuan 030024, China*
[2]*Shanxi Provincial Key Laboratory of Metallurgical Device Design Theory and Technology,*
 Taiyuan University of Science and Technology, Taiyuan 030024, China
[3]*Collaborative Innovation Center of Taiyuan Heavy Machinery Equipment,*
 Taiyuan University of Science and Technology, Taiyuan 030024, China

Correspondence should be addressed to Yaqin Tian; tianyaqin203@163.com

Academic Editor: Barbara Gawdzik

Hydrate formation apparatus reported so far was mainly concentrated in stirred-tank batch environments. It was difficult to produce the high gas storage hydrate efficiently. Some nonstirred technology has been attracting more attention by researchers. This work proposed a new apparatus for hydrate formation by spraying water into a gaseous phase with a fine nozzle. It can get sufficient contact surface area for gas-liquid reaction. Methane hydrate formation experiments have been conducted using pure water and sodium dodecyl sulfate (SDS) aqueous solution for comparison at 277.15 K. The experiments were conducted at 7.0 and 6.0 MPa, respectively. Kinetics of methane hydrate formation have been investigated by methane consumption per mole of water and reaction rate. The mechanism of hydrate formation and kinetics property by spraying atomization were studied with the theory of crystal chemistry.

1. Introduction

Gas hydrates are nonstoichiometric crystal structures that can trap guest molecules in the well-defined host lattices built up from hydrogen-bonded water molecules. There are three distinct crystal types: structure I (sI), structure II (sII), and structure H (sH), which differ in sizes and shapes of the cages formed [1]. It has long been known that gas hydrates possess the remarkable capacity to compress a large volume of gas into a relatively small volume of hydrate where a commonly quoted number as $1\,m^3$ of hydrate, at standard conditions, can contain up to $170\,m^3$ of natural gas [2].

Recently, there have been numerous proposals to make use of gas hydrates in water desalination, CO_2 capture and sequestration [3–5], food industry, refrigeration, gas separation [6, 7], and gas storage and transportation [8]. However, gas hydrates based technology requires the ability to be formed in a rapid and predictable manner (i.e., kinetics) [9]. There are two approaches to overcome the slow kinetics of

hydrate formation: the first approach is via the use of kinetic promoters to enhance the kinetics [10] and the second approach is via the use of innovative gas/liquid contact modes [11].

During the gas hydrate formation methane is generally less soluble in water. In the case of no disturbance, few hydrates are generated only at the gas-liquid interface [12]. Although the ideal hydrate formation rate can be obtained by mechanical stirring, the mechanical reliability problems caused by high-pressure mixing and liquid viscosity increased in the later period of the formation of gas hydrate [13]. This will affect the efficiency of mechanical stirring and then reduce the gas storage rate of unit volume [14]. Therefore, a new efficient method for high gas storage hydrate formation is urgently needed. The hydrate formation is the physical and chemical reaction of the gas and liquid which is easy to generate gas hydrate in the gas-liquid interface. According to the principle of chemical reaction kinetics, the increase of the number of the interfaces is very important

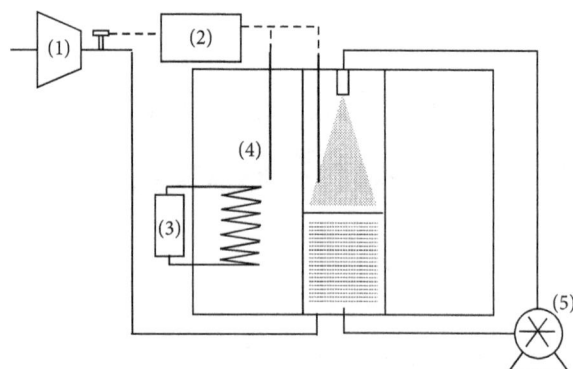

(1) Gas supply system   (4) Bath trough
(2) Data display gathering system   (5) Circulating pump
(3) Refrigeration system

FIGURE 1: Schematic diagram of forming hydrates in spray reactor.

for accelerating the reaction [15]. So, the spray method can produce high gas storage hydrate quickly. This method is efficient and simple and can meet the requirements of the experiment and production [16].

In this study, methane hydrate formation experiments have been conducted with pure water and aqueous solution of sodium dodecyl sulfate (SDS) at 277.15 K. Methane consumption per mole of water and reaction rate have been reported in the spray reactor with SDS aqueous solution. Kinetics of methane hydrate formation for pure water and SDS aqueous systems have been investigated at 7.0 and 6.0 MPa.

2. Experiment

The experimental apparatus was shown in Figure 1. The device consisted of a gas supply system, a control display system, a refrigeration system, and a circulating reaction system. In the gas supply system (1), the experimental gas flowed out of the cylinder through the valve to the buffer tank. Then the pressure of the gas was controlled at the desired fixed value by the pressure regulator. Finally, the gas entered the reactor for hydrate formation. Pressure and temperature of the reactor were controlled and recorded by the system of the control system (2). Refrigeration system (3) was kept at a constant temperature of bath by cooling medium through the coil. The water or solution in the reaction was directly added to the reactor before the reaction. In the reaction process, the solution was pumped to the top of the nozzle cycle by the circulating pump (5). The experimental data was automatically detected and recorded by the control display system (2).

2.1. Instrument and Equipment. The formation of gas hydrate is a high-pressure and low-temperature reaction. In the process of reaction, interface of gas-liquid needs to be disturbed, until the formation of ice crystal gas hydrate. In this paper, the effective volume of the reactor was 14 L; the maximum working pressure was 35 MPa. The reactor consists

of a thick-walled steel tube (diameter: 150 mm, thickness: 30 mm, and height: 800 mm) and two upper/lower flange covers. The form of circular micro spray was used in the agitation mode. In this way, the reaction system was mixed evenly and promoted the reaction. The low temperature of the reactor was provided by the cooling medium of the constant temperature bath. The whole reactor was placed in a low-temperature bath. Two temperature sensors were set in the reactor and the bath tank for measuring and controlling the reaction temperature. High-pressure gas from the gas bottle was released in a buffer tank (effective volume 20 L) before the hydrate formation for maintaining the stability of the entrained flow reactor and preventing dangerous gas reflux.

2.2. Measuring Instruments. The pressure of the reactor was measured by a pressure transducer. MPM480 explosion-proof type of pressure transmitter was selected from Shaanxi Mike sensor Co., Ltd. Its highest working pressure was 100 MPa, with an accuracy of ±0.25% FS. The temperature was measured by a sheathed platinum resistance thermometer (model WZPK2103). Then it was transformed into a digital signal by a platinum resistance temperature transmitter, which was transported to the controller to be recorded and displayed.

2.3. Experimental Materials. In the present analysis, deionized water and 99.9% methane gas (Beijing AP BAIF Gases Industry Co., Ltd., Beijing, China) were used. The surface-active agent used in the experiment was sodium dodecyl sulfate (SDS), and the purity was 99%. SDS solutions were prepared at 300 ppm (mg/kg) by the deionized water. All chemicals were used as received. All other chemicals were used without further purification.

2.4. Calculation. The moles of gas (methane) consumed in hydrate and rate of hydrate formation were calculated using the recorded pressure and temperature data during the period of experiment.

2.4.1. Moles of Gas Consumption. The consumption of moles of gas during hydrate formation has been calculated using

$$\Delta n_H = V_g \left(\frac{P_i}{Z_i RT} - \frac{P_t}{Z_t RT} \right), \tag{1}$$

where Δn_H is the moles of methane consumed at time, t, V_g is the volume of the gas inside the reactor, P_i is the pressure at the start of hydrate formation, Z_i is the compressibility factor at the start of hydrate formation, R is the ideal gas constant, T is the average temperature of the gas during hydrate formation experiment, P_t is the pressure of the reactor at t, and Z_t is the compressibility factor of the gas in the reactor at t. Z_i and Z_t are calculated using Pitzer correlation (Smith et al., 2001). The moles of methane consumed (Δn_H) per mole of water (n_w) are calculated using

$$NG_t = \Delta n_\downarrow \; (\text{mol/mol}) = \frac{\Delta n_H}{n_w}. \tag{2}$$

TABLE 1: Experimental parameters for hydrate formation.

Pressure/MPa	SDS/ppm	Reaction time/min	Gas storage mmol/mol	Reaction rate mmol/min
6	300	265	145.97	0.07
6	0	445	97.31	0.03
7	0	209	154.55	0.12

2.4.2. Rate of Hydrate Formation. The rate of methane hydrate formation has been calculated using a discrete forward difference method as shown in

$$\frac{dN_t}{dt} = \left(\frac{d\Delta n_H}{dt}\right)_t = \frac{\Delta n_{H,t+\Delta t} - \Delta n_{H,t}}{\Delta t}, \qquad (3)$$

where Δt is the time difference between two observations and the average rate of methane hydrate formation was calculated every 60 min and plotted.

2.5. Experimental Processes. First of all, the experimental gas was injected into the reactor with a water filled reactor until the remaining volume of the reaction was required. Then the gas supply booster pump started, and the freezing unit began to cool the reaction kettle. When the pressure and temperature were stable, the circulating pump was opened for the hydrate formation process. As the reaction was carried out, the consumption of the gas was automatically added by the booster pump to ensure that the pressure was constant during the reaction. When the pressure and temperature were constant, the pressure relief operation can be carried out. Finally, the formation of the hydrate was removed to determinate the gas storage and other physical and chemical indicators. The experiment data was automatically detected and recorded by the control display system.

In this paper, three experiments were conducted to determine the SDS and the pressure factor. The amount of water in the reaction was 2.0 L. Specific experimental conditions were shown in Table 1.

3. Experimental Results and Discussion

3.1. Consumption of Gas and Temperature. The hydrate formation process is an exothermic reaction and usually experiences three stages: induction stage, rapid reaction stage, and ending stage. If the hydrate formed in a short time, the reaction heat cannot be transferred in time which results in rapid rise of the temperature in the reactor.

Figure 2 shows the temperature and the moles of gas consumed per mole of water variation during methane hydrate formation process in the presence of 300 ppm SDS aqueous solution at 6.0 MPa. In the course of hydrate formation, the high-pressure reactor is a semiclosed system. In the reactor, methane is continuously added as a result of the hydrate formation process, in order to ensure the reaction under constant pressure. At 6.0 MPa, the temperature and gas consumption of the whole process of gas hydrate nucleation and crystal growth are shown in Figure 2. The three stages

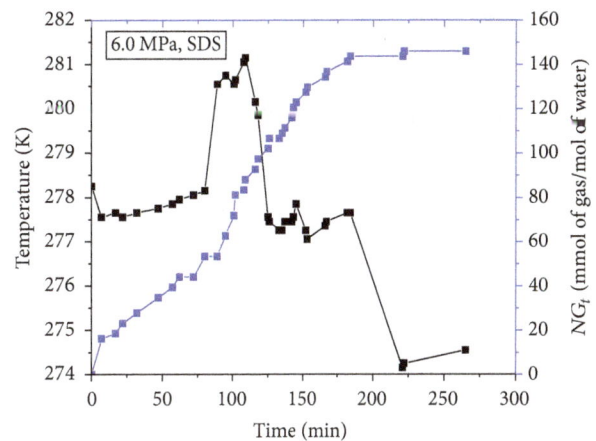

FIGURE 2: Hydrate formation in 300 ppm SDS solution at 6.0 MPa.

of the formation of the hydrate formation process are clearly shown in the figure, that is, induction stage, rapid reaction stage, and ending stage.

In the induction stage, the gas consumption growth rate is slow, and there was a basic linear rise. Temperature is basically kept constant, which shows that the reaction heat is released in time. It is obviously shown in Figure 2 that the temperatures rise by about 0.5 K after 80 min of the high-pressure gaseous methane entering into the cool cell. At the rapid reaction stage, the gas consumption has increased sharply. Due to the rapid reaction rate, the heat produced by the reaction cannot be released in time and the temperature in the reactor increased. The system temperatures rise by 3.3 K secondly between 80 and 180 min of the gas intake process being completed, which indicates that lots of methane hydrates form during this period. The gas consumption growth rate increased evidently. In the ending stage, the amount of gas consumption became smaller and gradually reached zero, the temperature dropped sharply, and the reaction stopped.

Figures 3 and 4 show the moles of gas consumed per mole of water and temperature during methane hydrate formation in the presence of SDS and pure water. As shown in Figures 3 and 4, the different aqueous solution and pressures of the systems are of similar morphology which still have the three stages. The experiments conducted at 300 ppm SDS aqueous solution show higher moles of gas consumption per mole of water as compared with pure water (see Figure 4) at 6.0 MPa. But the temperature change is relatively stable. The experiment conducted at 300 ppm of SDS aqueous solution at 7.0 MPa shows the highest moles of gas consumption per mole of water as compared with 300 ppm of SDS aqueous solutions and pure water (see Figures 3 and 4) at 6.0 MPa, while the change of temperature in pure water at 6.0 MPa (shown in Figure 4) is smoother than that of two other reaction conditions.

The comparison between Figures 3 and 4 can be seen: the amount of gas consumption in the pure water in the nucleation stage is smaller but the time is longer than that of SDS aqueous solution. In the rapid reaction stage, the rate of reaction is more rapid in SDS solution and the temperature of the

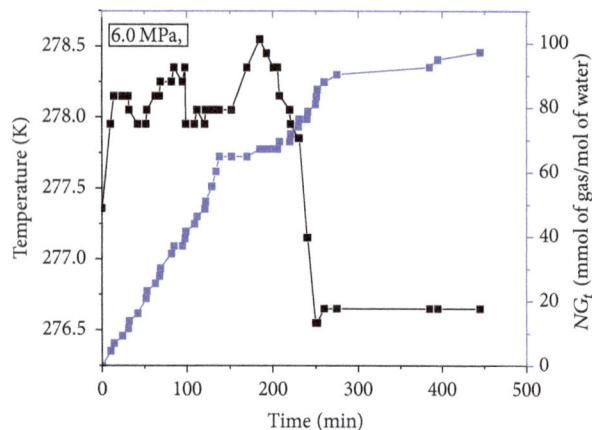

FIGURE 3: Hydrate formation in pure water at 6.0 MPa.

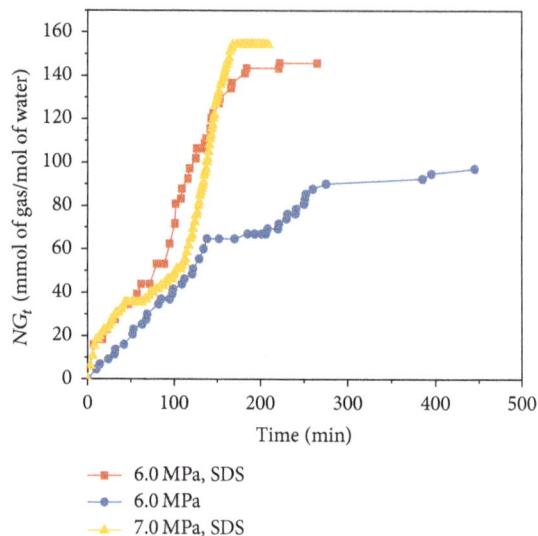

FIGURE 5: Methane consumed per mole of water in different condition.

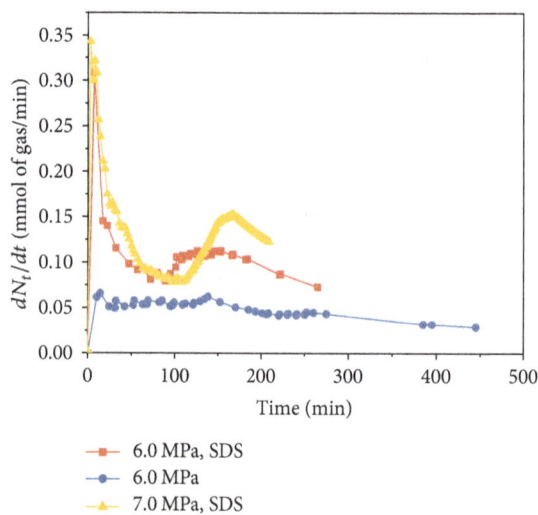

FIGURE 4: Hydrate formation in 300 ppm SDS solution at 7.0 MPa.

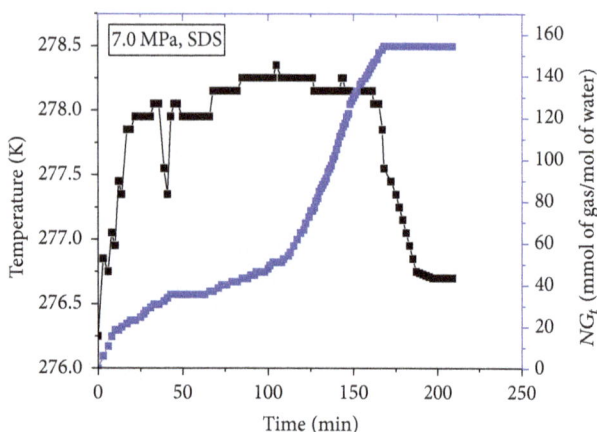

FIGURE 6: Rate of hydrate formation in different condition.

reactor increases to 281 K, while in pure water the temperature is relatively stable. From the whole process of the reaction of gas and pure water, nucleation induction stage is relatively long, but the rapid reaction stage is short and gas consumption is less. The surfactant SDS solution can provide more interface reactions, dominant heterogeneous nucleation, and fast reaction speed.

Figure 5 shows comparison of the gas consumptions for the selected best sets of hydrate system at 7.0 and 6.0 MPa. It has been observed that the hydrate system containing pure water with SDS (300 ppm) shows higher cumulative moles of gas consumption per mole of water as compared with the pure water systems. One of the distinct advantages of using SDS is that it lowers the induction time as compared to pure water by achieving faster nucleation. Through the experiment, it can be found that the contact area of gas and water can be greatly improved by using the spray atomization to generate hydrate. Therefore, the effect of the spray atomization method is better than that in the promotion of the rate of hydrate preparation.

3.2. Rate of Hydrate Formation. Figure 6 shows the rate of methane hydrate formation at 300 ppm SDS aqueous solution

in comparison with pure water at 7.0 and 6.0 MPa. It has been observed that, for the initial hours of the hydrate formation experiment, system with SDS maintained the highest rate of hydrate formation followed by pure water.

Overall, SDS aqueous solution has maintained relatively a high rate of hydrate formation as compared to pure water throughout the experiment. This indicates that the induction time using SDS has been reduced and the rate of hydrate formation has been favorable.

The hydrate formation rate is controlled by the pressure, temperature, and composition of the impact. In addition, the surface tension of the liquid has an important influence on the formation. Adding appropriate surfactants can greatly promote the crystallization reaction.

4. Conclusions

Gas hydrate formation is the process of gas-liquid reaction in the formation of gas-liquid phase interface. The process is affected by the formation conditions, heat transfer, and the intrinsic characteristics of the reaction. The reaction rate can be improved by the experiments and measurement, and the induction period of the reaction can be greatly reduced. There are two reasons for this: (1) the contact surface is increased, so the reaction is carried out; (2) the reaction heat is rapidly spreading out. Gas hydrate is proved to be synthesized by spray atomization, and the high efficiency can be realized by artificial means. Thus, gas storage, separation, transportation, and so on can be reached through the hydrate. As spray atomization factors of hydrate preparation, the thermodynamic process involves contact heat and mass transfer, induction time, and nucleation. The study of these factors will have an important role in the development of gas hydrate storage and transportation technology.

Conflicts of Interest

The authors declare that there are no conflicts of interest regarding the publication of this paper.

Acknowledgments

The financial supports were received from Research Project Supported by Shanxi Scholarship Council of China (no. 2016-095), Fund Program for the Scientific Activities of Selected Returned Overseas Professionals in Shanxi Province (no. 201697), Fund Program for the Scientific Activities of Selected Returned Overseas Professionals in MOHRSS (no. 2015192), and Doctoral Project of Taiyuan University of Science and Technology of Shanxi Province in China (Grant no. 20122003). Thanks are due to Collaborative Innovation Center of Taiyuan Heavy Machinery Equipment and Shanxi Provincial Key Laboratory of Metallurgical Device Design Theory and Technology.

References

[1] E. D. Sloan and C. A. Koh, *Clathrate Hydrates of Natural Gases*, CRC Press, Taylor & Francis Group, New York, NY, USA, 3rd edition, 2008.

[2] C. A. Koh, E. D. Sloan, and A. Sum, *Natural Gas Hydrates in Flow Assurance*, Gulf Professional Publishing, Burlington, NJ, USA, 2011.

[3] X.-S. Li, H. Zhan, C.-G. Xu, Z.-Y. Zeng, Q.-N. Lv, and K.-F. Yan, "Effects of tetrabutyl-(ammonium/phosphonium) salts on clathrate hydrate capture of CO_2 from simulated flue gas," *Energy and Fuels*, vol. 26, no. 4, pp. 2518–2527, 2012.

[4] M. Yang, W. Jing, J. Zhao, Z. Ling, and Y. Song, "Promotion of hydrate-based CO_2 capture from flue gas by additive mixtures (THF (tetrahydrofuran) + TBAB (tetra-n-butyl ammonium bromide))," *Energy*, vol. 106, pp. 546–553, 2016.

[5] P. Babu, R. Kumar, and P. Linga, "Pre-combustion capture of carbon dioxide in a fixed bed reactor using the clathrate hydrate process," *Energy*, vol. 50, no. 1, pp. 364–373, 2013.

[6] Y. Zhao, J. Zhao, D. Shi, Z. Feng, W. Liang, and D. Yang, "Micro-CT analysis of structural characteristics of natural gas hydrate in porous media during decomposition," *Journal of Natural Gas Science and Engineering*, vol. 31, pp. 139–148, 2016.

[7] J. Zhao, Y. Zhao, and W. Liang, "Hydrate-based gas separation for methane recovery from coal mine gas using tetrahydrofuran," *Energy Technology*, vol. 4, no. 7, pp. 864–869, 2016.

[8] D.-L. Zhong, Z. Li, Y.-Y. Lu, J.-L. Wang, and J. Yan, "Evaluation of CO_2 removal from a CO_2 + CH_4 gas mixture using gas hydrate formation in liquid water and THF solutions," *Applied Energy*, vol. 158, pp. 133–141, 2015.

[9] P. Linga and M. A. Clarke, "A review of reactor designs and materials employed for increasing the rate of gas hydrate formation," *Energy & Fuels*, vol. 31, no. 1, pp. 1–13, 2017.

[10] Y. Zhong and R. E. Rogers, "Surfactant effects on gas hydrate formation," *Chemical Engineering Science*, vol. 55, no. 19, pp. 4175–4187, 2000.

[11] K. Yoshikawa, K. Yuichi, and K. Takahiro, "Production method for hydrate and device for proceeding the same," United States Patent: 524753, 2000.

[12] D. Kashchiev and A. Firoozabadi, "Nucleation of gas hydrates," *Journal of Crystal Growth*, vol. 243, no. 3-4, pp. 476–489, 2002.

[13] J.-Y. Liu, J. Zhang, Y.-L. Liu, X.-H. Tan, and J. Zhang, "Experimental and modeling studies on the prediction of gas hydrate formation," *Journal of Chemistry*, vol. 2015, Article ID 198176, 5 pages, 2015.

[14] S. Chang-yu, C. Guang-jin, and G. Tian-min, "The status of the kinetics of hydrate nucleation," *Acta Petrolei Sinica*, vol. 22, no. 4, pp. 82–86, 2001 (Chinese).

[15] Y. H. Mori, "Estimating the thickness of hydrate films from their lateral growth rates: application of a simplified heat transfer model," *Journal of Crystal Growth*, vol. 223, no. 1-2, pp. 206–212, 2001.

[16] N. Gnanendran and R. Amin, "Modelling hydrate formation kinetics of a hydrate promoter-water-natural gas system in a semi-batch spray reactor," *Chemical Engineering Science*, vol. 59, no. 18, pp. 3849–3863, 2004.

16

The Internal Recycle Reactor Enhances Porous Calcium Silicate Hydrates to Recover Phosphorus from Aqueous Solutions

Wei Guan[1] and Shichao Tian[2]

[1]Chongqing Key Laboratory of Environmental Materials & Remediation Technologies, Chongqing University of Arts and Sciences, Chongqing 402160, China
[2]Shenzhen Environmental Science and New Energy Technology Engineering Laboratory, Tsinghua-Berkeley Shenzhen Institute, Shenzhen 518055, China

Correspondence should be addressed to Shichao Tian; 18812672619@163.com

Academic Editor: Nicolas Roche

In this experiment, the porous calcium silicate hydrates (P-CSHs) were prepared via a hydrothermal method and then modified by polyethylene glycol (PEG). The modified P-CSHs combined with an internal recycle reactor could successfully recover the phosphorus from electroplating wastewater. The modified P-CSHs were characterized by X-ray diffraction (XRD), N_2 adsorption-desorption isotherms, and Fourier transform infrared spectroscopy (FT-IR). After compared with different samples, the modified P-CSHs-PEG2000 sample had larger specific surface area of 87.48 m^2/g and higher pore volume of 0.33 cm^3/g, indicating a high capacity for phosphorus recovery. In the process of phosphorus recovery, the pH value of solution was increased to 9.5, which would enhance the recovery efficiency of phosphorus. The dissolution rate of Ca^{2+} from P-CSH-PEG2000 was fast, which was favorable for phosphorus precipitation and phosphorus recovery. The effects of initial concentration of phosphorus, P-CSHs-PEG2000 dosage, and stirring speed on phosphorus recovery were analyzed, so the optimal operation conditions for phosphorus recovery were obtained. The deposition was analyzed by XRD, N_2 adsorption-desorption, and SEM techniques; it was indicated that the pore volume and surface area of the P-CSHs-PEG2000 were significantly reduced, and the deposition on the surface of P-CSHs-PEG2000 was hydroxyapatite.

1. Introduction

In recent years, the electroplating industry is developing rapidly and the yield of electroplating wastewater is up to 4 billion m^3 each year [1]. Cyanide has been gradually replaced by phosphate in plating cleaning, descaling, and anticorrosion due to the lower toxicity of phosphate [2]. The total concentration of phosphorus in electroplating wastewater is higher than 200 mg/L. In the process of electroplating wastewater treatment, people focus on the removal of heavy metals and toxic anion and ignore the phosphorus resource recycling, causing huge loss of phosphorus resources. However, natural reserves of high-grade rock phosphorus are limited and dissipated on a global scale [3, 4]. Hence, considerable attention has been paid to phosphorus recovery for realizing closed-loop phosphorus recycling [5].

In recent years, various methods have been investigated to remove or recover phosphorus from wastewater, such as chemical precipitation [6], adsorption [7, 8], ion exchange [9], electrocoagulation [10], and membrane filtration [11]. Among these methods, chemical precipitation is difficult to be used to treat high concentration of phosphorus. In some areas, electrocoagulation and membrane filtration cannot be adopted due to the higher energy cost [12, 13]. Adsorption process is widely used for the recovery of phosphorus. However, with respect to the sustainable recycling of phosphorus, adsorption techniques are limited owing to the difficult regeneration of adsorbents and subsequent complex desorption process [14]. Therefore, it is urgent to develop a simple and beneficial method for phosphorus recovery.

Compared with other adsorbents, calcium silicate hydrates (CSHs) show unique characteristics for phosphorus

TABLE 1: Chemical components of carbide residue and white carbon black ((1) carbide residue; (2) white carbon black).

	Chemical composition/%									
	CaO	SiO_2	Al_2O_3	SO_2	MgO	Fe_2O_3	SrO	NaOH	CuO	H_2O
(1)	79.34	3.57	2.14	1.22	0.62	0.21	0.26	—	—	12.64
(2)	0.08	97.46	0.16	1.82	—	0.03	—	0.29	0.02	0.14

recovery. Firstly, the higher phosphorus adsorption capacity of CSHs is 137 mg/g [15], while the maximum adsorption capacity of phosphorus for natural calcium-rich sepiolite is 34 mg/g [16] and for ferrihydrite-modified diatomite it is 37 mg/g [17]. Secondly, CSHs are inexpensive and easily available materials compared with ferrihydrite-modified diatomite, lanthanum-modified bentonite clay, and zirconium-based oxides [18]. Thirdly, CSHs are nontoxic inorganic material made of carbide residue and white carbon black by hydrothermal synthesis method [19]. Therefore, CSHs are kinds of ideal materials for phosphorus recovery.

The internal recycle reactor for phosphorus recovery is designed with both inner and outer tubes, including mixing reaction zone, crystallization reaction zone, centrifugal solid-liquid separation zone, crystal return flow zone, effluent overflow area, and phosphorus collection area [20]. The crystals of porous calcium silicate hydrates can be circulated in the system, which is beneficial for effectively recovering the high concentration of phosphorus in the wastewater. The circulating crystal reactor can be used as phosphorus nuclei, which is conducive to recover low concentration of phosphorus.

In this experiment, calcium silicate hydrates were prepared with carbide residue and white carbon black. Then, it was modified by polyethylene glycol (PEG) to improve the performance of phosphorus recovery. Based on the fundamental principle of phosphorus recovery and technological characteristics of reactor, a fluidized bed of internal circulation for phosphorus crystallization was designed. The modified CSHs were characterized by X-ray diffraction (XRD), N_2 adsorption-desorption isotherms, and Fourier transform infrared spectroscopy (FT-IR). The effects of initial concentration of phosphorus, P-CSHs-PEG2000 dosage, and stirring speed on recovery efficiency of phosphorus were analyzed, and the products of phosphorus recovery in this system were confirmed.

2. Materials and Methods

2.1. Materials. Carbide residue was used as calcium sources and obtained from Chongqing Changshou Chemical Co. Ltd. White carbon black was used as Si sources and purchased from Chongqing Jianfeng Chemical Co. Ltd. Chemical constituents of carbide residue and white carbon black were shown in Table 1. The polyethylene glycol (PEG) regents were purchased from Chongqing Jianfeng Chemical Co. Ltd and the molecular weight of PEG was 200, 1000, 2000, 6000, and 20000, respectively. The phosphorus solution was prepared by adding KH_2PO_4 (analytical reagent, Chongqing Boyi Chemical reagent Co. Ltd.) and the initial concentration of phosphorus was 30 mg/L. All chemicals were used without

further purification and all solutions were prepared using Milli-Q water.

2.2. Preparation and Modification of Porous Calcium Silicate Hydrates

2.2.1. Preparation of Porous Calcium Silicate Hydrates. Preparation of porous calcium silicate hydrates was shown as follows: firstly, the mixture of carbide residue and white carbon black was prepared at the molar ratios of Ca/Si at 1.6 : 1. And the slurry was generated by dissolving the mixture with a liquid/solid ratio of 30. Secondly, the slurry hydrothermally reacted at 170°C for 6 hours in reaction still and then taken out when the temperature was reduced to room temperature. Thirdly, the slurry was filtrated and dried at 105°C for 2 hours. Then the final products were ground and filtrated through a sieve of 200 mesh. The prepared samples were porous calcium silicate hydrates (P-CSHs) [21].

2.2.2. Modification of Porous Calcium Silicate Hydrates. The modification of porous calcium silicate hydrates was shown as follows: firstly, the prepared porous calcium silicate hydrates (1.0 g) were dissolved and then added to polyethylene glycol regent with different molecular weights (200, 1000, 2000, 6000, and 20000). The solution was kept stirring at the rate of 60 r/min for 1 hour at water bath. Secondly, the slurry was filtrated and hydrothermally reacted at 500°C for 2 hours in muffle. After the sample cooled down to room temperature, the final products were ground and filtrated through a sieve of 200 mesh. The prepared samples were named as P-CSHs-PEG200, P-CSHs-PEG1000, P-CSHs-PEG2000, P-CSHs-PEG6000, and P-CSHs-PEG20000 [22].

2.3. Experimental Section. The experiments were performed in a glass reactor with 1.0 L potassium dihydrogen phosphate solutions and the initial concentration of KH_2PO_4 is in the range of 5~40 mg/L. The schematic diagram of internal recycle reactor was shown in Figure 1, and the operation of reactor was continuous inflow and indirect outflow. During all the experiments, the phosphorus content (%) in the solid product was calculated by

$$P = \frac{\left[\sum \left(c_0 - c_{ti}\right)\right] \times v}{w} \times 100\%, \tag{1}$$

where C_{ti} is the phosphorus concentration at a given time, mg/L; C_0 is the phosphorus concentration at the initial time, mg/L; w is the weight of solid product, mg; v is the volume of solution, L.

2.4. Analytical Method. The crystal structures and phase compositions of the samples were characterized by X-ray

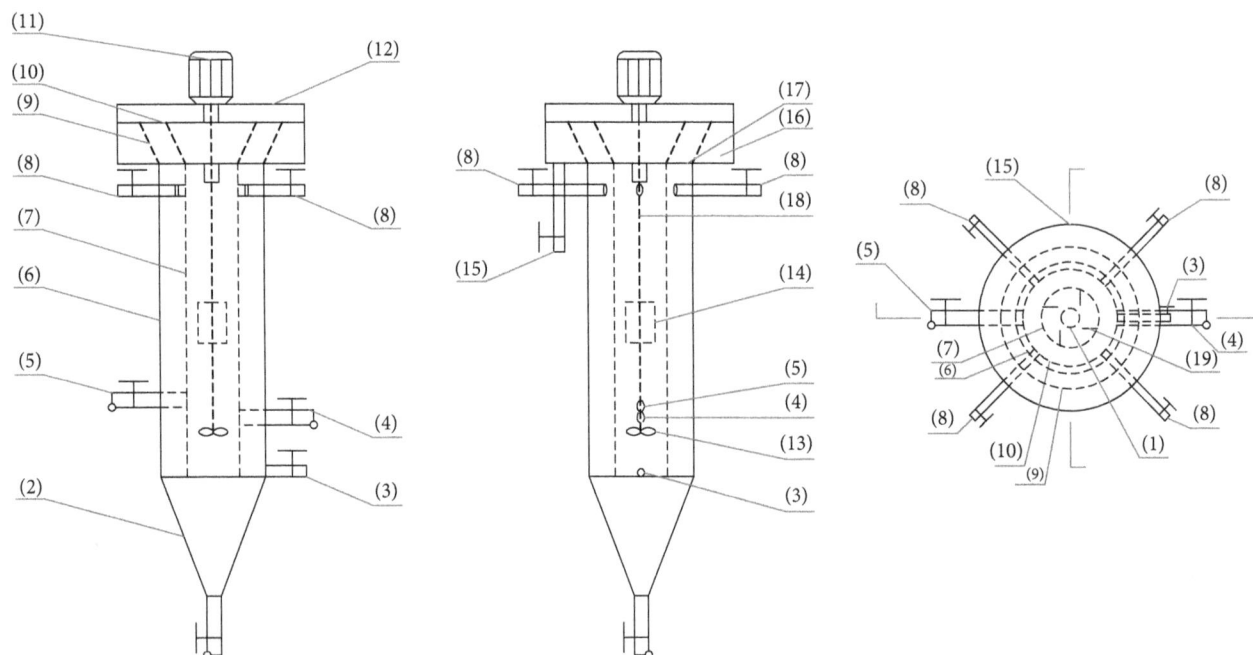

FIGURE 1: The schematic diagram of internal recycle reactor for phosphorus recovery. (1) Outlet of product; (2) collecting bowl for crystalline particle; (3) sampling port; (4) water inlet; (5) feed inlet; (6) outer tube; (7) inner tube; (8) reagent addition; (9) outer overflow nozzle; (10) internal overflow nozzle; (11) electrical machine; (12) mounting plate of electrical machine; (13) spiral stirrer; (14) rectangle stirrer; (15) water outlet; (16) effluent flume; (17) reflowing valve; (18) transmission shaft; (19) baffle of turbulent flow.

diffraction (XRD) on a D/Max IIIB X-ray powder diffractometer (Rigaku, Japan) with Cu Kα radiation. The specific surface area and pore-size distribution of the as-prepared samples were calculated by N$_2$ adsorption-desorption analysis (ASAP2020, Micromeritics, USA). The chemical bonds of samples were analyzed by the Fourier transform infrared spectroscopy (FT-IR) (Nexus 670). Morphology, microstructure, and size of products were measured by scanning electronic microscopy (SEM, JEM-6490) [23]. The concentration of dissolved Ca^{2+} ion was measured using a 710 series inductively coupled plasma optical emission spectrometer (ICP-OES, Agilent Technology, USA).

3. Results and Discussion

3.1. XRD Characterization of the Modified CSHs. The XRD spectra of porous calcium silicate hydrates modified by PEG with different molecular weights were shown in Figure 2. Before CSHs modification, the diffraction peaks with $2\theta = 8.40°$, $29.92°$, $31.78°$, $35.74°$, and $49.70°$ (JCPDS card No. 25-1464) are observed [24]. The characteristic peaks of the modified CSHs samples at $2\theta = 31.78°$ and $35.74°$ have disappeared obviously, and the strongest characteristic peaks at $2\theta = 29.92°$ have weakened. The addition of PEG results in distortion of the crystalline structure and reduction of the crystallinity of porous calcium silicate hydrates [25]. The diffraction peaks of porous calcium silicate hydrates with $2\theta = 8.40°$ are according to (002), which almost disappear after modified by PEG. The PEG reagent enters the interlayer of porous calcium

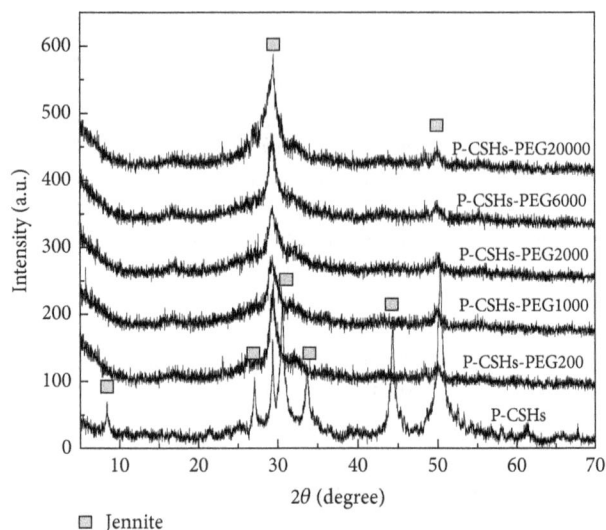

FIGURE 2: The XRD patterns of porous calcium silicate hydrates modified by PEG with different molecular weights.

silicate hydrates and increases the interlayer space of porous calcium silicate hydrates. The disappearing of diffraction peaks shows that the degree of crystallinity of porous calcium silicate hydrates is decreased significantly. The fitting of MDI jade5.0 indicates that the crystallinity of P-CSHs,

TABLE 2: The data of pore structure for different materials.

Samples	Pore volume/(cm^3/g)	Pore diameter/(nm)	S_{BET}/(m^2/g)
P-CSHs-PEG200	0.24	14.90	64.00
P-CSHs-PEG1000	0.24	14.85	65.74
P-CSHs-PEG2000	0.33	15.43	87.48
P-CSHs-PEG6000	0.20	10.28	77.95
P-CSHs-PEG20000	0.23	20.35	46.03
P-CSHs	0.29	26.37	43.97

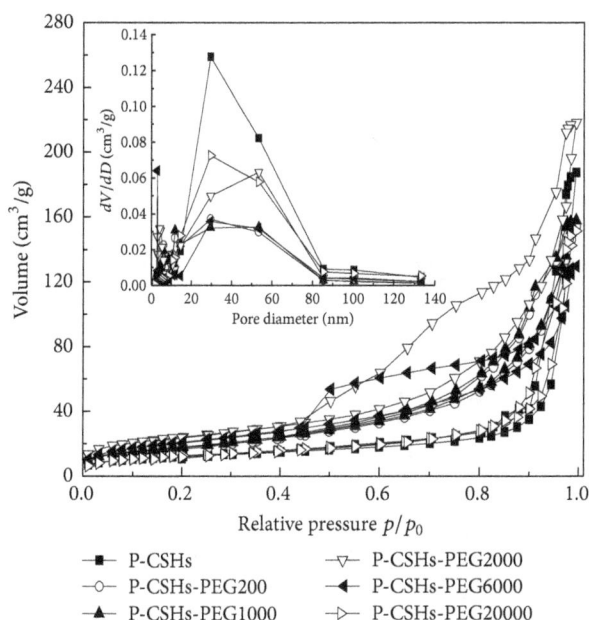

FIGURE 3: The N$_2$ adsorption-desorption isotherms and pore-size distribution curves for different materials.

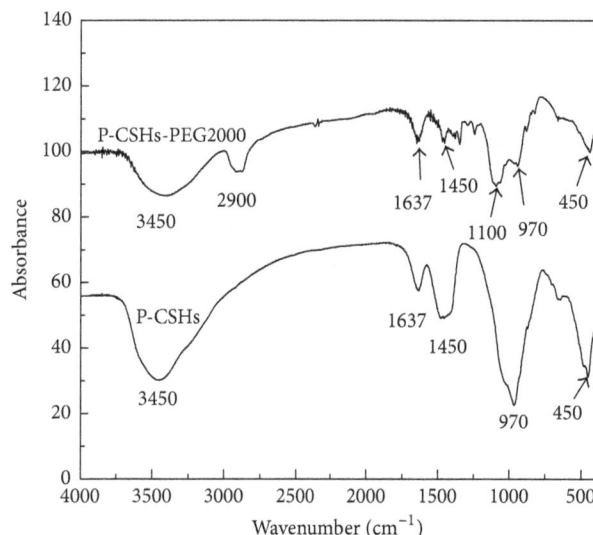

FIGURE 4: FT-IR spectra of calcium silicate hydrates before and after modification.

P-CSHs-PEG200, P-CSHs-PEG1000, P-CSHs-PEG2000, P-CSHs-PEG6000, and P-CSHs-PEG20000 is 43.37%, 26.89%, 24.9%, 16.94%, 18.24%, and 38.16%, respectively.

3.2. Nitrogen Adsorption-Desorption Analysis. The N$_2$ adsorption-desorption isotherms and pore-size distribution curves for different materials were shown in Figure 3. The phenomena of hysteresis loop in N$_2$ adsorption-desorption isotherms for different materials are caused by the condensation of pores [26]. The data of pore structure of various materials is shown in Table 2. The specific surface area and pore volume of porous calcium silicate hydrates are increased by the modification of PEG. However, when the molecular weight of PEG is too low or too high, the surface area and pore volume of porous calcium silicate hydrates are not obviously changed. Therefore, when the porous calcium silicate hydrates are modified by PEG at molecular weight of 2000, it has the largest specific surface area of 87.48 m^2/g and pore volume of 0.33 cm^3/g.

3.3. FT-IR Analysis. In order to investigate the chemical bonding of the samples, the calcium silicate hydrates before and after modification were analyzed by FT-IR. As shown in Figure 4, there are a series of absorption peaks at 3450 cm^{-1} (OH$^-$), 970 cm^{-1} (Si-O-Si), 1637 cm^{-1} (O-H), 1450 cm^{-1} (C-O-H), and 450 cm^{-1} (Si-O-Si) in FT-IR spectra of two samples [27]. The characteristic peaks are ascribed to the silicon oxygen tetrahedron structure of calcium silicate hydrates. However, the intensity of characteristic peaks for P-CSHs-PEG2000 sample is obviously weaker than calcium silicate hydrates, indicating that the order degree of calcium silicate hydrates is decreased with the addition of PEG. In addition, P-CSHs-PEG2000 samples exhibit a stretching vibration peak of C-H at 2900 cm^{-1} and a stretching vibration absorption peak of C-O at 1100 cm^{-1}. These characteristic peaks indicate that PEG is a polymer with -CH$_2$-CH$_2$-O- group in the main chain of molecules and belongs to nonionic dispersant [28]. The molecular structure consists of two parts: anchoring group and solvation chain. Before modification, the surface structure of calcium silicate hydrates is relatively dense and the molecular structure is closely arranged. When PEG is added to the slurry of calcium silicate hydrates, the polymer enters the defect site of the silicon oxygen tetrahedron chain of the calcium silicate hydrates or enters the interlayer. The anchoring groups are adsorbed on the surface of calcium silicate hydrates. And the solvation chains are expanded in

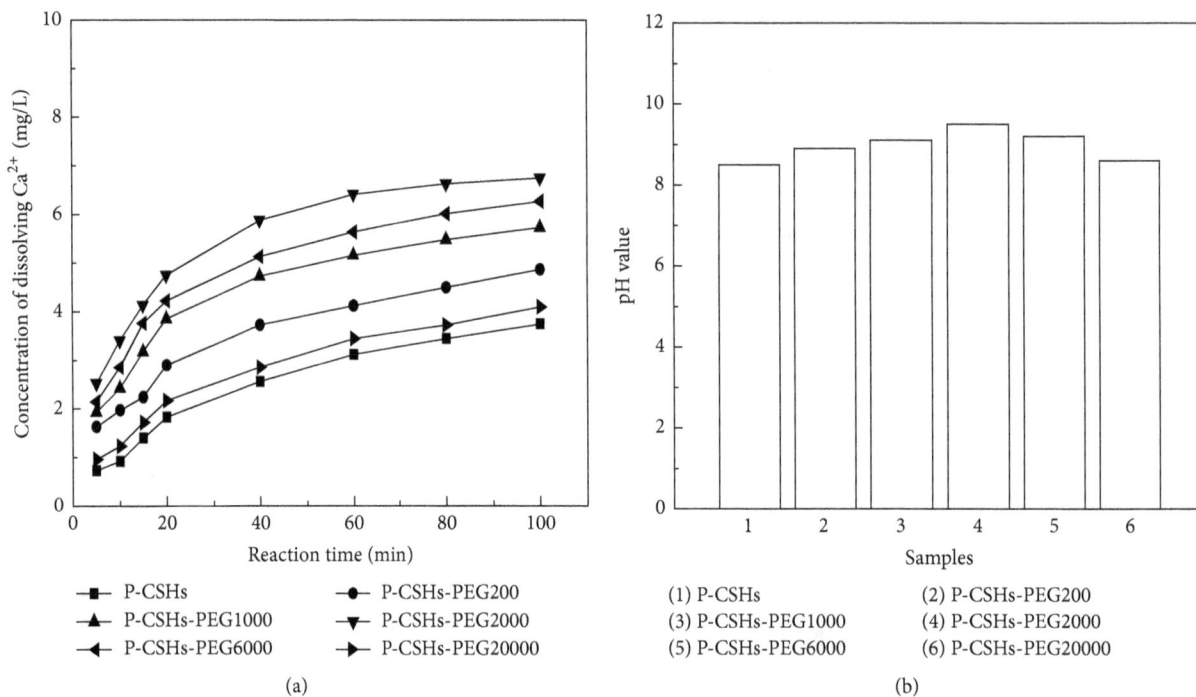

FIGURE 5: (a) The variation of dissolved Ca^{2+} concentration for different samples; (b) the variation of pH value in solution.

the medium to form a barrier layer, which prevents the aggregation of solid particles and achieves the function of steric hindrance. PEG is burned away by calcinations and generated more pore structure.

3.4. The Analysis of Dissolving Property for Calcium Silicate Hydrates.

The concentration of dissolved Ca^{2+} and the variation of pH for different samples were shown in Figure 5. The modified calcium silicate hydrates can increase the concentration of dissolved Ca^{2+} and pH of solution [29]. Comparing with the P-CSHs sample, the concentration of dissolved Ca^{2+} from the modified P-CSHs-PEG2000 has been increased from 3.75 mg/L to 6.75 mg/L when it reached the reaction equilibrium. However, when the molecular weight of PEG is too large or too small, the calcium solubility of the modified calcium silicate hydrates is not significantly increased. The concentration of dissolved Ca^{2+} for the P-CSHs-PEG20000 is only 4.10 mg/L (Figure 5(a)). The modified calcium silicate hydrates can increase the pH of solution [30]. When various materials are reached to the solubility equilibrium, the pH value of solution is basically maintained at 8.6~9.5. Among them, the P-CSHs-PEG2000 samples can increase the pH value of solution to 9.5 (Figure 5(b)). The high concentration of dissolved Ca^{2+} and high pH value of solution are beneficial for phosphorus recovery. Therefore, the P-CSHs-PEG2000 samples are used as the material for phosphorus recovery in internal recycle reactor. Therefore, the P-CSHs-PEG2000 samples are used as the material for phosphorus recovery in internal recycle reactor.

3.5. The Analysis of Operation Efficiency for Phosphorus Recovery

3.5.1. The Effect of P-CSHs-PEG2000 Dosage on Phosphorus Recovery System.

The effect of P-CSHs-PEG2000 dosage on phosphorus recovery system was investigated. While keeping parameters such as initial concentration of phosphorus and stirring speed fixed, the phosphorus recovery system was performed using different dosages of P-CSHs-PEG2000 (0.5, 1, 2, 3, and 4 g/L). As shown in Figure 6, when the dosages of P-CSHs-PEG2000 are changed in the range of 1~4 g/L, the residual concentration of phosphorus remained at 4~5 mg/L after 60 min reaction. When the dosage of P-CSHs-PEG2000 is too small, the residual concentration of phosphorus is increased significantly and remained at 9.44 mg/L. The results are different from the conventional phenomena of coagulation and sedimentation. Comparing the traditional coagulation and sedimentation process, when the initial concentration of phosphorus is the same, the more dosage of the reagent can increase the recovery efficiency of phosphorus. However, the principle of phosphorus recovery in this system is different from the traditional principles of coagulation and sedimentation. In conventional coagulation and sedimentation process, the recovery efficiency of phosphorus mainly depends on the alkalinity generated by adding excess chemicals, which results in the lower utilization of agent for phosphorus recovery. Considering the effective utilization of P-CSHs-PEG2000 samples, the optimum dosage of crystal for phosphorus recovery is 1.0 g/L.

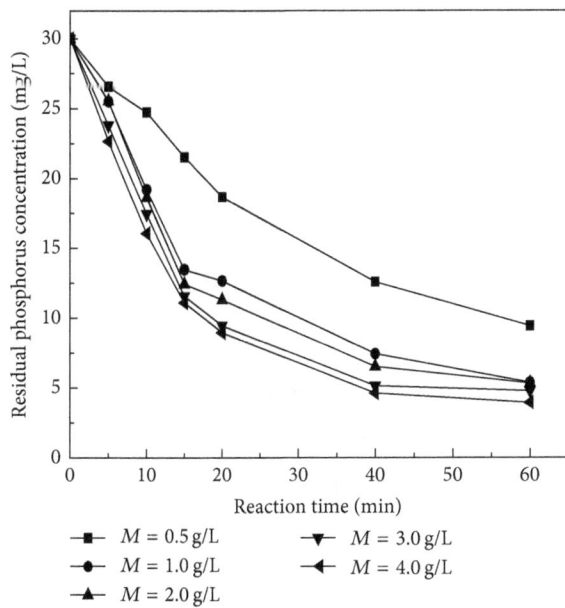

FIGURE 6: The effect of P-CSHs-PEG2000 dosage on phosphorus recovery system (initial concentration of phosphorus, 30 mg/L; initial pH, 7.0; stirring speed, 60 r/min).

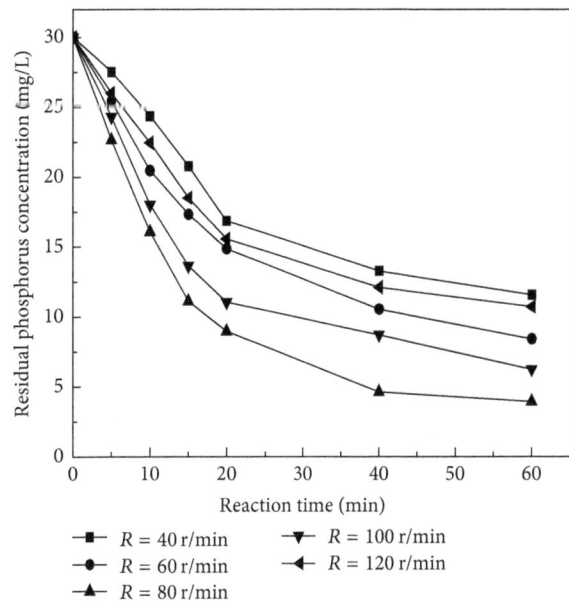

FIGURE 7: The effect of stirring speed on phosphorus recovery system (initial concentration of phosphorus, 30 mg/L; initial pH, 7.0; P-CSHs-PEG2000 dosage, 1.0 g/L).

3.5.2. The Effect of Stirring Speed on Phosphorus Recovery System.

The effect of stirring speed on phosphorus recovery system was investigated. While keeping parameters such as initial concentration of phosphorus and P-CSHs-PEG2000 dosage fixed, the phosphorus recovery system was performed using different stirring speed (40, 60, 80, 100, and 120 r/min). As shown in Figure 7, the stirring speed has a greater influence on residual concentration of phosphorus. When the stirring speed is 80 r/min, the residual concentration of phosphorus reaches the minimum (3.93 mg/L). However, when the stirring speeds are too higher or too lower, this will affect the operation efficiency and increase the residual concentration of phosphorus. In the process of phosphorus recovery, hydroxyapatites are generated on the surface of porous calcium silicate hydrates. The lower stirring speed cannot make the crystal fully contact with the sewage and the higher stirring speed cannot make the hydroxyapatites grow well on the surface of the crystal. Therefore, the best stirring speed for phosphorus recovery is 80 r/min.

3.5.3. The Effect of Initial Concentration of Phosphorus on Operation Efficiency.

The effect of initial concentration of phosphorus on operation efficiency was investigated. While keeping parameters such as P-CSHs-PEG2000 dosage and stirring speed fixed, the phosphorus recovery system was performed using different concentrations of phosphorus (40, 30, 20, 10, and 5 mg/L). As shown in Figure 8, the initial concentration of phosphorus can only determine the rate of reaction; however, it cannot change the chemical equilibrium. When the reaction equilibrium is reached, the residual phosphorus concentration is basically stable between 3.5~4.0 mg/L. With respect to phosphorus treatment, the effluent of phosphorus concentration is not up to the standard in

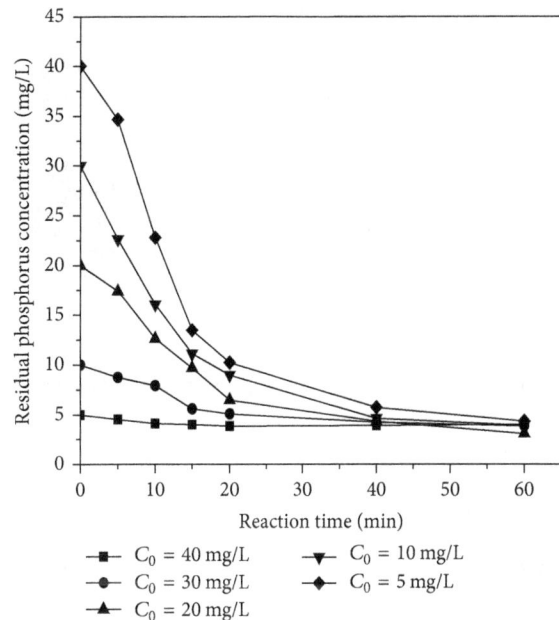

FIGURE 8: The effect of initial concentration of phosphorus on operation efficiency (initial pH, 7.0; P-CSHs-PEG2000 dosage, 1.0 g/L; stirring speed, 80 r/min).

this system. But from the point of phosphorus recovery, the increase of phosphorus concentration is beneficial to improve the phosphorus recovery efficiency. Therefore, the initial concentration of phosphorus for phosphorus recovery is 40 mg/L.

3.6. The Characterization of Deposition.

The residual concentration of phosphorus changed greatly in the four use

(a)

(b)

(c)

(d)

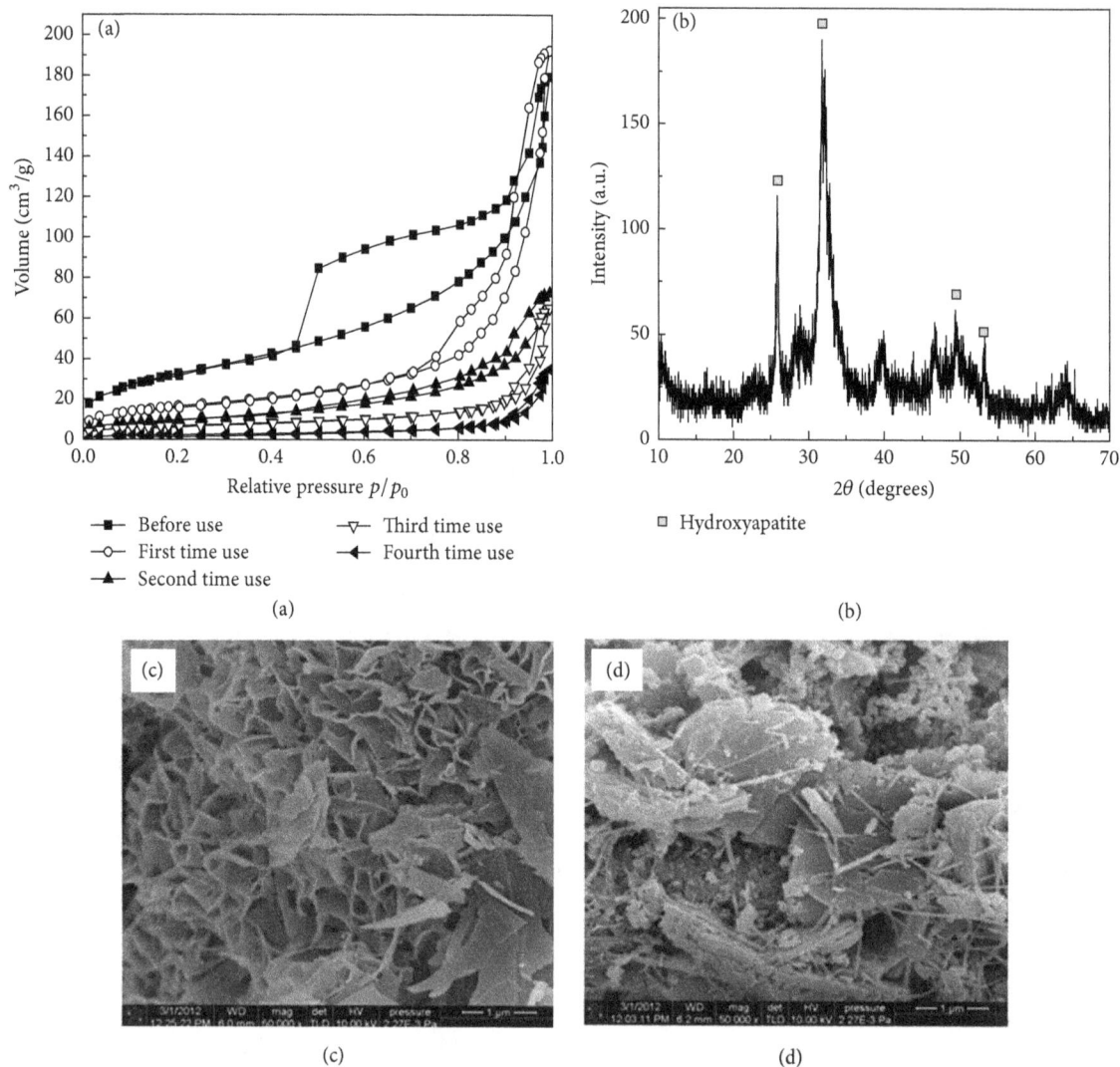

FIGURE 9: The characterization of deposition: (a) the N_2 adsorption-desorption isotherms for different times of using P-CSHs-PEG2000; (b) XRD patterns of deposit; (c) SEM images of P-CSHs-PEG2000 before use; (d) SEM images of P-CSHs-PEG2000 after four times use.

times of P-CSHs-PEG2000 samples. Therefore, the influence of use times on pore structure of P-CSHs-PEG2000 samples was investigated. As shown in Figure 9(a), the mesoporous adsorption mainly occurred in the middle pressure zone ($0.4 < p/p_0 < 0.9$). In this region, the phenomenon of hysteresis loop in various samples gradually disappeared and the adsorption curves are gradually declined with the increasing times of using calcium silicate hydrates. The data of pore volume and surface area about P-CSHs-PEG2000 samples after different use times are shown in Table 3.

With increasing the use times of P-CSHs-PEG2000 samples for phosphorus recovery, the pore volume and surface area are significantly reduced. The structural properties of P-CSHs-PEG2000 samples used for phosphorus recovery were further investigated by XRD in the range of 10–70°. As shown in Figure 9(b), the pronounced diffraction peaks in the XRD pattern at 2θ of 26.4°, 33.4°, 50.7°, and 77.52° corresponded to hydroxylapatite in accordance with the standard [31] and

TABLE 3: The data of pore structure for P-CSHs-PEG2000 samples after different use times.

Samples	Pore volume/(cm³/g)	S_{BET}/(m²/g)
Before use	0.33	87.48
First time use	0.25	55.00
Second time use	0.14	35.01
Third time use	0.10	14.90
Fourth time use	0.05	3.75

the characteristic peaks of porous calcium silicate hydrate disappeared basically. Indicating porous calcium silicate hydrate is mainly covered by hydroxyapatite and achieves phosphorus recovery. The surface variation of porous calcium silicate hydrate in the process of phosphorus recovery is analyzed by SEM. As shown in Figures 9(c)-9(d), before P-CSHs-PEG2000 samples are used for phosphorus recovery,

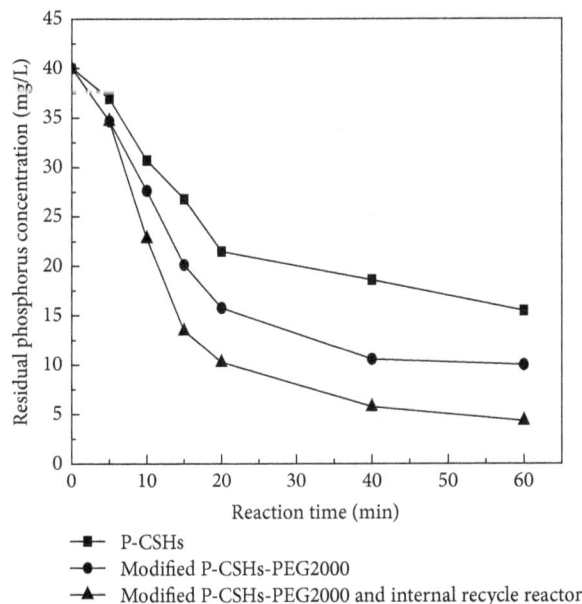

FIGURE 10: Comparing the three systems for phosphorus recovery (initial concentration of phosphorus, 40 mg/L; initial pH, 7.0; dosage of crystal, 1.0 g/L; stirring speed, 80 r/min).

it obviously has a porous structure [32]. After four times using P-CSHs-PEG2000 samples for phosphorus recovery, it has no obvious pore structure and a large amount of flocculent substance appears. According to the analysis of XRD, the deposition on the surface of calcium silicate hydrate is hydroxyapatite. The phosphorus content of the deposition was 14.43%, which is basically closed to the phosphorus content of phosphorus minerals (15%). Therefore, it provides a simple and beneficial method for phosphorus recovery from the electroplating wastewater.

3.7. The Process of Phosphorus Recovery in Internal Recycle Reactor. Under the optimum conditions, the large surface area of porous calcium silicate hydrate quickly makes the materials dissolve calcium and effectively maintain the high concentrations of Ca^{2+} and OH^-, which makes the pH of the solution kept at 8.5~9.5. Under this pH condition, the existence form of phosphorus is HPO_4^{2-}. Thus, high concentrations of Ca^{2+}, OH^-, and HPO_4^{2-} formed, which facilitates the formation of hydroxyapatite in a weak alkaline environment [33]. As shown in Figure 10, comparing the three systems (i.e., P-CSHs, modified P-CSHs-PEG2000, and the modified P-CSHs-PEG2000 combined with the internal recycle reactor) for phosphorus recovery, the internal recycle reactor enhances porous calcium silicate hydrates to recover phosphorus from electroplating wastewater.

The internal recycle reactor for phosphorus recovery is structured with an inner cylinder and an external cylinder, wherein the inner cylinder consisted of mixing reaction zone, crystallization reaction zone, and solid-liquid separation zone [34]. The external cylinder includes hydroxyapatite formation zone and crystallization settling zone. In addition, the top of the reactor is the effluent settling zone and the

bottom of the reactor is the crystal collecting zone. The design of double cylinder structure makes the reactor efficiently recover phosphorus. The operation way of reactor is simple and the operating cost is low. The functions of different reaction zones are introduced as follows: the mixed reaction zone makes the porous calcium silicate hydrate and phosphorus wastewater in full contact; the crystallization reaction zone makes the phosphate rapidly form hydroxyapatite; the solid-liquid separation zone makes the treated wastewater overflow from the top of the reactor and the porous calcium silicate hydrates get into the external cylinder under centrifugal force; the growth of hydroxyapatite is based on the principle of secondary nucleation, which makes the hydroxyapatite continually grow. The cyclic precipitation zone allows the hydroxyapatite to flow back to the inner cylinder and improves phosphorus recovery; the collection zone is used for deposition and collection products of phosphorus recovery [35]. Therefore, the design of double cylinder structure makes the reactor efficiently recover phosphorus.

4. Conclusion

The porous calcium silicate hydrates could be circulated in the internal recycle reactor, which could effectively recover high concentration of phosphorus from electroplating wastewater. Compared with different samples, the modified P-CSHs-PEG2000 sample had larger specific surface area and higher pore volume, which was beneficial for phosphorus recovery. Moreover, the dissolution rate of Ca^{2+} from P-CSH-PEG2000 was fast, which was favorable for phosphorus precipitation and phosphorus recovery. The effects of initial concentration of phosphorus, P-CSHs-PEG2000 dosage, and stirring speed on phosphorus recovery were analyzed, so the optimal operation conditions for phosphorus recovery were obtained. The deposition was analyzed by XRD, N_2 adsorption-desorption, and SEM techniques; it was indicated that the pore volume and surface area of the P-CSHs-PEG2000 were significantly reduced, and the deposition on the surface of P-CSHs-PEG2000 was hydroxyapatite. The main products in the process of phosphorus recovery were hydroxyapatite and the phosphorus content of the products was 14.43%, which is basically closed to the phosphorus content of phosphorus minerals (15%).

Conflicts of Interest

The authors declare that they have no conflicts of interest.

References

[1] F. V. Hackbarth, D. Maass, A. A. U. de Souza, V. J. P. Vilar, and S. M. A. G. U. de Souza, "Removal of hexavalent chromium from electroplating wastewaters using marine macroalga Pelvetia canaliculata as natural electron donor," *Chemical Engineering Journal*, vol. 290, pp. 477–489, 2016.

[2] Z. Rajabalizadeh and D. Seifzadeh, "Strontium phosphate conversion coating as an economical and environmentally-friendly pretreatment for electroless plating on AM60B magnesium alloy," *Surface and Coatings Technology*, vol. 304, pp. 450–458, 2016.

[3] N. P. Springer, "Physical, technical, and economic accessibility of resources and reserves need to be distinguished by grade: application to the case of phosphorus," *Science of the Total Environment*, vol. 577, pp. 319–328, 2017.

[4] Y. Wang, Y. Yu, H. Li, and C. Shen, "Comparison study of phosphorus adsorption on different waste solids: Fly ash, red mud and ferric–alum water treatment residues," *Journal of Environmental Sciences (China)*, vol. 50, pp. 79–86, 2016.

[5] K. Zhou, M. Barjenbruch, C. Kabbe, G. Inial, and C. Remy, "Phosphorus recovery from municipal and fertilizer wastewater: China's potential and perspective," *Journal of Environmental Sciences (China)*, vol. 52, pp. 151–159, 2017.

[6] M. Hupfer, K. Reitzel, A. Kleeberg, and J. Lewandowski, "Long-term efficiency of lake restoration by chemical phosphorus precipitation: scenario analysis with a phosphorus balance model," *Water Research*, vol. 97, pp. 153–161, 2016.

[7] C. Wan, S. Ding, C. Zhang et al., "Simultaneous recovery of nitrogen and phosphorus from sludge fermentation liquid by zeolite adsorption: mechanism and application," *Separation and Purification Technology*, vol. 180, pp. 1–12, 2017.

[8] Y. Zhang, W. Zhang, and B. Pan, "Struvite-based phosphorus recovery from the concentrated bioeffluent by using HFO nanocomposite adsorption: effect of solution chemistry," *Chemosphere*, vol. 141, pp. 227–234, 2015.

[9] K. Asoh and M. Ebihara, "Accurate determination of trace amounts of phosphorus in geological samples by inductively coupled plasma atomic emission spectrometry with ion-exchange separation," *Analytica Chimica Acta*, vol. 779, pp. 8–13, 2013.

[10] H. Huang, D. Zhang, Z. Zhao, P. Zhang, and F. Gao, "Comparison investigation on phosphate recovery from sludge anaerobic supernatant using the electrocoagulation process and chemical precipitation," *Journal of Cleaner Production*, vol. 141, pp. 429–438, 2017.

[11] E. R. Hall, A. Monti, and W. W. Mohn, "A comparison of bacterial populations in enhanced biological phosphorus removal processes using membrane filtration or gravity sedimentation for solids-liquid separation," *Water Research*, vol. 44, no. 9, pp. 2703–2714, 2010.

[12] L.-C. Hua, C. Huang, Y.-C. Su, T.-N. Nguyen, and P.-C. Chen, "Effects of electro-coagulation on fouling mitigation and sludge characteristics in a coagulation-assisted membrane bioreactor," *Journal of Membrane Science*, vol. 495, pp. 29–36, 2015.

[13] M. L. Gerardo, N. H. M. Aljohani, D. L. Oatley-Radcliffe, and R. W. Lovitt, "Moving towards sustainable resources: Recovery and fractionation of nutrients from dairy manure digestate using membranes," *Water Research*, vol. 80, pp. 80–89, 2015.

[14] Y.-J. Tu and C.-F. You, "Phosphorus adsorption onto green synthesized nano-bimetal ferrites: Equilibrium, kinetic and thermodynamic investigation," *Chemical Engineering Journal*, vol. 251, pp. 285–292, 2014.

[15] X. C. Chen, H. N. Kong, D. Y. Wu, X. Z. Wang, and Y. Y. Lin, "Phosphate removal and recovery through crystallization of hydroxyapatite using xonotlite as seed crystal," *Journal of Environmental Sciences*, vol. 21, no. 5, pp. 575–580, 2009.

[16] H. Yin, M. Kong, and C. Fan, "Batch investigations on P immobilization from wastewaters and sediment using natural calcium rich sepiolite as a reactive material," *Water Research*, vol. 47, no. 13, pp. 4247–4258, 2013.

[17] D. Xu, S. Ding, Q. Sun, J. Zhong, W. Wu, and F. Jia, "Evaluation of in situ capping with clean soils to control phosphate release from sediments," *Science of the Total Environment*, vol. 438, pp. 334–341, 2012.

[18] C. Li, H. Yu, S. Tabassum et al., "Effect of calcium silicate hydrates (CSH) on phosphorus immobilization and speciation in shallow lake sediment," *Chemical Engineering Journal*, vol. 317, pp. 844–853, 2017.

[19] W. Guan, F. Ji, D. Fang et al., "Porosity formation and enhanced solubility of calcium silicate hydrate in hydrothermal synthesis," *Ceramics International*, vol. 40, no. 1, pp. 1667–1674, 2014.

[20] Z. Liu, Q. Zhao, D.-J. Lee, and N. Yang, "Enhancing phosphorus recovery by a new internal recycle seeding MAP reactor," *Bioresource Technology*, vol. 99, no. 14, pp. 6488–6493, 2008.

[21] S. Wang, X. Peng, Z. Tao, L. Tang, and L. Zeng, "Influence of drying conditions on the contact-hardening behaviours of calcium silicate hydrate powder," *Construction and Building Materials*, vol. 136, pp. 465–473, 2017.

[22] O. Mendoza, C. Giraldo, S. S. Camargo, and J. I. Tobón, "Structural and nano-mechanical properties of Calcium Silicate Hydrate (C-S-H) formed from alite hydration in the presence of sodium and potassium hydroxide," *Cement and Concrete Research*, vol. 74, pp. 88–94, 2015.

[23] W. Guan and X. Zhao, "Fluoride recovery using porous calcium silicate hydrates via spontaneous Ca^{2+} and OH^- release," *Separation and Purification Technology*, vol. 165, pp. 71–77, 2016.

[24] K. Okano, S. Miyamaru, A. Kitao et al., "Amorphous calcium silicate hydrates and their possible mechanism for recovering phosphate from wastewater," *Separation and Purification Technology*, vol. 144, pp. 63–69, 2015.

[25] F. Pelisser, P. J. P. Gleize, and A. Mikowski, "Structure and micro-nanomechanical characterization of synthetic calcium-silicate-hydrate with Poly(Vinyl Alcohol)," *Cement and Concrete Composites*, vol. 48, pp. 1–8, 2014.

[26] J. Zhao, Y.-J. Zhu, J. Wu et al., "Chitosan-coated mesoporous microspheres of calcium silicate hydrate: environmentally friendly synthesis and application as a highly efficient adsorbent for heavy metal ions," *Journal of Colloid and Interface Science*, vol. 418, pp. 208–215, 2014.

[27] S. C. Mojumdar and L. Raki, "Preparation, thermal, spectral and microscopic studies of calcium silicate hydrate-poly(acrylic acid) nanocomposite materials," *Journal of Thermal Analysis and Calorimetry*, vol. 85, no. 1, pp. 99–105, 2006.

[28] X. Zhang, H. Liu, Z. Huang et al., "Preparation and characterization of the properties of polyethylene glycol @ Si3N4 nanowires as phase-change materials," *Chemical Engineering Journal*, vol. 301, pp. 229–237, 2016.

[29] T. Missana, M. García-Gutiérrez, M. Mingarro, and U. Alonso, "Analysis of barium retention mechanisms on calcium silicate hydrate phases," *Cement and Concrete Research*, vol. 93, pp. 8–16, 2017.

[30] A. A. Rouff, N. Ma, and A. B. Kustka, "Adsorption of arsenic with struvite and hydroxylapatite in phosphate-bearing solutions," *Chemosphere*, vol. 146, pp. 574–581, 2016.

[31] Y. Kim and Y. J. Lee, "Characterization of mercury sorption on hydroxylapatite: batch studies and microscopic evidence for adsorption," *Journal of Colloid and Interface Science*, vol. 430, pp. 193–199, 2014.

[32] C. Hu, Y. Han, Y. Gao, Y. Zhang, and Z. Li, "Property investigation of calcium-silicate-hydrate (C-S-H) gel in cementitious composites," *Materials Characterization*, vol. 95, pp. 129–139, 2014.

[33] K. Okano, M. Uemoto, J. Kagami et al., "Novel technique for phosphorus recovery from aqueous solutions using amorphous calcium silicate hydrates (A-CSHs)," *Water Research*, vol. 47, no. 7, pp. 2251–2259, 2013.

[34] Z. Deng, T. Wang, N. Zhang, and Z. Wang, "Gas holdup, bubble behavior and mass transfer in a 5m high internal-loop airlift reactor with non-Newtonian fluid," *Chemical Engineering Journal*, vol. 160, no. 2, pp. 729–737, 2010.

[35] H. Huang, J. Yang, and D. Li, "Recovery and removal of ammonia-nitrogen and phosphate from swine wastewater by internal recycling of struvite chlorination product," *Bioresource Technology*, vol. 172, pp. 253–259, 2014.

17

UV Photocatalytic Activity for Water Decomposition of $Sr_xBa_{1-x}Nb_2O_6$ Nanocrystals with Different Components and Morphologies

Guoqiang Han,[1,2] Shuguang Cao,[1] and Bo Lin[1]

[1]*School of Mechanical Engineering and Automation, Fuzhou University, Fuzhou, Fujian 350108, China*
[2]*Fujian Institute of Research on the Structure of Matter, Chinese Academy of Science, Fuzhou, Fujian 350002, China*

Correspondence should be addressed to Guoqiang Han; galehan@hotmail.com

Academic Editor: Roberto Comparelli

Strontium barium niobate $Sr_xBa_{1-x}Nb_2O_6$ (SBN) nanocrystals with different components (x = 0.2, 0.4, 0.6, and 0.8) were synthesized by Molten Salt Synthesis (MSS) method at various reaction temperatures (T = 950°C, 1000°C, 1050°C, and 1100°C). The SBN nanocrystals yielded through flux reactions possess different morphologies and sizes with a length of about ~100 nm~7 μm and a diameter of about ~200~500 nm. The Scanning Electron Microscopy (SEM) and X-ray Diffraction (XRD) techniques were used to study the compositions, structures, and morphologies of the nanocrystals. The absorption edges of the SBN nanocrystals are at a wavelength region of approximate 390 nm, which corresponds to band-gap energy of ~3.18 eV. The SBN nanocrystals with different sizes display different photocatalytic activity under ultraviolet light in decomposition of water. The SBN60 nanocrystals exhibit stable photocatalytic rates (~100~130 μmol of $H_2 \cdot g^{-1} \cdot h^{-1}$) for hydrogen production. The SBN nanocrystals can be a potential material in the application of photocatalysis and micro/nanooptical devices.

1. Introduction

The synthesis of nanostructures has attracted extensive attention in the past two decades due to their novel size-dependent properties. The SBN nanocrystals have a complex structure of tetragonal tungsten-bronze with NbO_6 octahedrons [1]. The strontium barium niobate $Sr_xBa_{1-x}Nb_2O_6$ (SBN) nanocrystals have exhibited a variety of ferroelectricity [2–4], photorefractive [5], electrooptic, nonlinear optical [1, 6], and thermoelectric properties [7] which are potentially important for many applications in various fields. The advantage of SBN nanocrystals is generally attractive for some device applications, such as thin film devices [2], optical sensors [8], and piezoelectric ceramic resonators [9]. In addition, SBN nanocrystals have lead-free composition which is concerned with environment, safety, and health. SBN nanocrystals with varied compositions can be fabricated into different sizes and complex shapes through low-cost and easy methods such as template-assisted synthesis and template-free synthesis, Czochralski method, dual-stage sintering method,

and low-temperature combustion synthesis process [10–13]. The conventional SBN ferroelectric thin films are usually crystallized at the temperature of about 700~800°C and the SBN nanocrystals can be synthesized at the temperature of about 1100°C [14–16].

The hydrogen generation through splitting of water has been a topic of severely popular research interest because it is considered to be a promising method of providing renewable energy. Hydrogen energy is increasingly becoming significant from the viewpoint of both solar energy employment and environment, particularly in the case of the exhaustion of fossil fuels. Furthermore, the combustion product of hydrogen is water which has less pollution to the environment than other organic fuels. Many materials have been used as catalysts for photochemical decomposition of water, such as $Na_2Ta_4O_{11}$, $Ba_5Nb_4O_{15}$, $MgO\text{-}ZrO_2$, and $Sr_2Nb_2O_7$ [17–21]. However, SBN nanocrystals have received very little attention in this area. The morphology of nano-crystal has an important effect on the performance of photocatalytic. As stimulated by the promising photocatalytic and optical application, the

(a)

(b)

(c)

(d)

FIGURE 1: SEM images of the flux-synthesized SBN nanocrystals with flux-to-reactant weight ratio of 2 : 1 at a reaction temperature of 1100°C for 3 h: (a) SBN20, (b) SBN40, (c) SBN60, and (d) SBN80.

synthesis of more efficient light-driven SBN nanocrystals with complex morphologies is a subject of considerable research interest.

2. Experimental

$Sr_xBa_{1-x}Nb_2O_6$ (SBN) nanocrystals with x = 0.2, 0.4, 0.6, and 0.8 were prepared from molten NaCl salts (hereafter abbreviated as SBN20, etc.). The starting powders of $SrCO_3$, $BaCO_3$, and Nb_2O_5 were premixed according to the desired Sr : Ba : Nb ratios in ethanol for 24 h. The mixtures were again mixed with NaCl for 8 h, with the weight ratio of oxides to salts being equal to 1 : 2. The final mixtures of powders and salts were heated in covered high alumina crucibles for Molten Salt Synthesis (MSS). The mixtures were heated at 950°C, 1000°C, 1050°C, and 1100°C, respectively for 3 h. The nanocrystals were collected after washing away the remaining salt with hot deionized water.

The SBN nanocrystals were characterized by powder X-Ray Diffraction (XRD, Ultima-IV, Rigaku, Japan), Scanning Electron Microscopy (SEM, JSM6700-F, JEOL, Japan), and UV-vis Diffraction Reflected Spectra (DRS, Lambda900, PerkinElmer, USA). The surface area measurements were performed with laser Particle Size and Zeta Potential Analyzer (PSZPA, BI-200SM, BROOKHAVEN, USA). The photocatalytic reaction was carried out in a gas-closed circulation

system. The sample of 0.05 g was dispersed in 50 ml of solution in an irradiation cell of volume 100 cm^3 quartz reactor and was illuminated with 400 W Hg lamp. The solution was stirred for 30 mins in the dark to obtain a good dispersion of the catalyst in water. The net balanced photocatalytic reaction (under irradiation) is $H_2O \rightarrow H_2$ + $1/2O_2$. The evolved H_2 and O_2 were separated with molecular sieve-5 Å column and the amount of gas was determined by using gas chromatography with thermal conductivity detector.

3. Results and Discussion

These SBN nanocrystals were found to have rod-like shapes with the length of ~100 nm~7 μm and the diameter of ~200~ 500 nm as shown in Figures 1 and 2. Figure 1 shows the SEM images of the SBN nanocrystals with various components (x = 0.2, 0.4, 0.6, and 0.8) at a reaction temperature of 1100°C for 3 h. The SBN20 nanocrystals possess the smallest size. Figure 2 indicates different morphologies of the SBN60 nanocrystals at different reaction temperatures (T = 950°C, 1000°C, 1050°C, and 1100°C) for 3 h. The grain size is determined by the annealing temperature and the SBN60 nanocrystals at a reaction temperature of 950°C get a smaller size. The XRD patterns for the nanocrystals obtained are shown in Figures 3 and 4. All peaks are assigned to SBN nanocrystals with a tetragonal tungsten-bronze (TB)

(a)

(b)

(c)

(d)

FIGURE 2: SEM images of the flux-synthesized SBN60 nanocrystals with flux-to-reactant weight ratio of 2 : 1 at different reaction temperatures for 3 h: (a) 950°C, (b) 1000°C, (c) 1050°C, and (d) 1100°C.

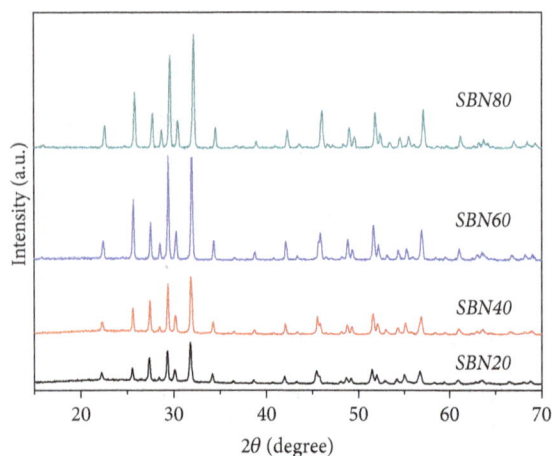

FIGURE 3: XRD patterns of the flux-synthesized SBN nanocrystals with flux-to-reactant weight ratio of 2 : 1 at a reaction temperature of 1100°C for 3 h.

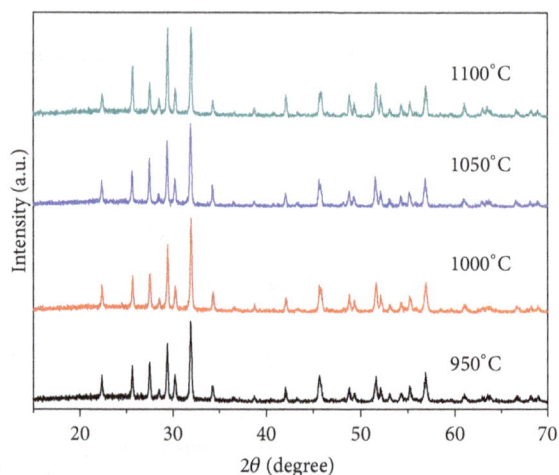

FIGURE 4: XRD patterns of the flux-synthesized SBN60 nanocrystals with flux-to-reactant weight ratio of 2 : 1 at different reaction temperatures for 3 h.

structure. The tested results confirm that all SBN nanocrystals are obtained randomly without having any impurity phase.

The SBN nanocrystals represent the similar absorption properties as observed from UV-vis DR spectra (Figure 5).

Figure 5(a) indicates SBN nanocrystals with different components ($x = 0.2, 0.4, 0.6$, and 0.8) possess similar UV-vis absorption properties. In pure SBN nanocrystals, the UV-vis near the absorption edge will create a pair of electron–hole excitons. The constraints of these excitons are

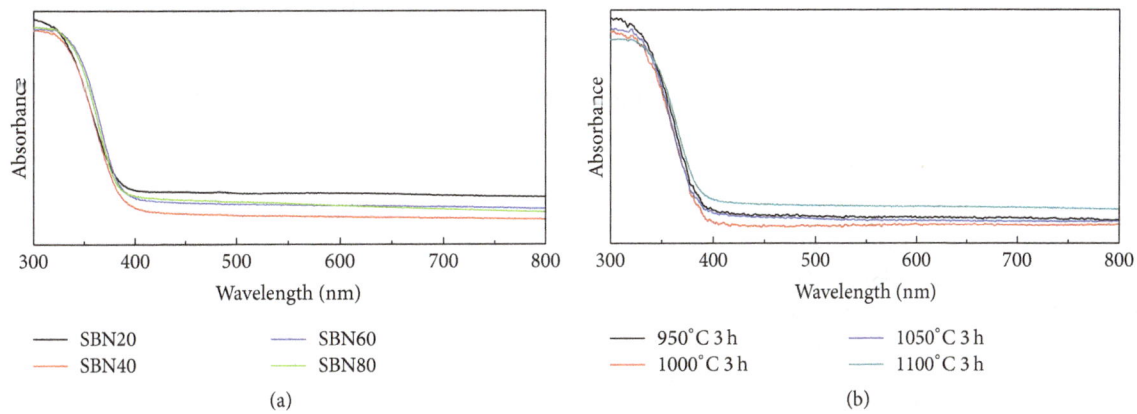

FIGURE 5: The UV-vis absorption spectra of the flux-synthesized SBN nanocrystals: (a) with different components; (b) at different reaction temperatures.

TABLE 1: Surface areas and photocatalytic rates for hydrogen production of the flux-prepared SBN nanocrystals.

Samples	Temperature/Dwelling time	Surface area [m^2/g]	Photocatalysis rate [μmol of H$_2 \cdot$g$^{-1} \cdot$h^{-1}]
SBN20	1100°C/3 h	3.8476	26.5
SBN40	1100°C/3 h	2.2028	56.2
SBN60	950°C/3 h	2.4400	122.35
SBN60	1000°C/3 h	2.4193	99.6
SBN60	1050°C/3 h	2.1549	102.1
SBN60	1100°C/3 h	1.7346	102.2
SBN80	1100°C/3 h	2.2127	79.55

weak and can be dissociated with the help of photons and phonons. The absorption edges of the SBN nanocrystals are at the wavelength region of approximate 390 nm and the band-gap of SBN crystal is about 3.18 eV at room temperature. So that UV-vis absorption properties are similar to different components. Simultaneously the reaction temperature of SBN nanocrystals has little effect on the UV-vis absorption nature as shown in Figure 5(b). Figure 5 also displays that the absorption edges of all kinds of SBN nanocrystals are at the same wavelength region of approximate 390 nm. The band-gap energy (E) can be calculated by the following equation [22]

$$E = \frac{1240}{\lambda},\tag{1}$$

where λ is the wavelength in nanometers [23]. The calculated band-gap energy (E) corresponding to the 390 nm light wave is ~3.18 eV.

In metal-oxide photocatalysts, the light-driven excitation of electrons into their conduction bands can be used to drive the water splitting reactions for the reduction and oxidation of water to hydrogen and oxygen, respectively. The photocatalytic rates of SBN nanocrystals vary obviously with different components (x = 20, 40, 60, and 80) and reaction temperatures (T = 950°C, 1000°C, 1050°C, and 1100°C) as shown in Table 1. When the temperature of the sample is 1100°C, SBN60 has the highest photocatalytic efficiency. Table 1 indicates that the SBN60 nanocrystals exhibit stable photocatalytic rates for

hydrogen production of ~100~130 μmol of H$_2 \cdot$g$^{-1} \cdot$h^{-1} and especially the photocatalytic rate of SBN60 nanocrystals at a reaction temperature of 950°C gets the peak (122.35 μmol of H$_2 \cdot$g$^{-1} \cdot$h^{-1}). The structure of the SBN is nonfilled [1], so the photocatalytic rates can be controlled by changing the content of metal ions Sr and Ba in a large range. And the photocatalytic rates of SBN can be significantly improved by adjusting the Sr/Ba ratio in a certain range [24]. In the SBN nano-crystal structure, different positions are occupied by the metal ion, Sr and Ba. In addition, the radii of Sr and Ba ions are different. Therefore, the change of the composition leads to the distortion about the structure of the SBN. So different components have different UV-vis absorption properties and SBN60 has the highest photocatalytic efficiency. Table 1 also indicates that the surface area of SBN nanocrystals has little effect on the material photocatalytic activity. The surface area indicates one gram of the sample can be expanded at a certain temperature. The SBN nanocrystals with small size and powdered morphology display the higher photocatalytic rate. Especially the photocatalytic rate of SBN60 nanocrystals with the surface area of 2.4400 m^2/g gets the peak (122.35 μmol of H$_2 \cdot$g$^{-1} \cdot$h^{-1}). This is because the nanocrystals with small size and powered morphology are uniformly dispersed in the photocatalytic solution. They can absorb more light energy. So the photocatalytic rates of SBN nanocrystals can be controlled through changing the component and morphology of SBN nanocrystals, which are synthesized by using different experimental parameters.

4. Conclusions

The SBN nanocrystals are synthesized as single-crystal particles with sizes of $\sim$100 nm$\sim$7 μm through changing component and reaction temperature. In order to gain some small nanocrystals with high surface area, the component and reaction temperature must be controlled properly. The absorption edges of the SBN nanocrystals are at the wavelength region of approximate 390 nm, which corresponds to band-gap energy of $\sim$3.18 eV. Under UV light, the SBN60 nanocrystals exhibit excellent photocatalytic rates ($\sim$100$\sim$130 μmol/g/h of $H_2 \cdot g^{-1} \cdot h^{-1}$) in UV photocatalytic activity for water decomposition. The SBN60 nanocrystals synthesized at a reaction temperature of $T = 950°C$ are found to possess highest photocatalytic rate (122.35 μmol of $H_2 \cdot g^{-1} \cdot h^{-1}$). The SBN nanocrystals with powdered morphology and small size display the higher photocatalytic rate. In a word, the water decomposition performance under UV mainly depends on the component and morphology of photocatalytic materials.

Conflicts of Interest

The authors declare that they have no conflicts of interest.

Acknowledgments

This work is financially supported by the National Natural Science Foundation of China (Grant no. 51205063), China Postdoctoral Science Foundation (Grant no. 2012M521281 and no. 2013T60643), and the Education Department Foundation of Fujian Province (Grant no. JA14032).

References

[1] W. C. Liu, A. D. Li, C. L. Mak, and K. H. Wongt, "Fabrication and electro-optic properties of ferroelectric nanocrystal/polymer composite films," *Journal of Physical Chemistry C*, vol. 112, no. 36, pp. 14202–14208, 2008.

[2] X. Y. Liu, Y. M. Liu, S. Takekawa, K. Kitamura, F. S. Ohuchi, and J. Y. Li, "Nanopolar structures and local ferroelectricity of Sr0.61Ba 0.39Nb2O6 relaxor crystal across Curie temperature by piezoresponse force microscopy," *Journal of Applied Physics*, vol. 106, no. 12, Article ID 124106, 2009.

[3] N. Ortega, P. Bhattacharya, and R. S. Katiyar, "Enhanced ferroelectric properties of multilayer SrBi2Ta2O9/SrBi2Nb 2O9 thin films for NVRAM applications," *Materials Science and Engineering B: Solid-State Materials for Advanced Technology*, vol. 130, no. 1-3, pp. 36–40, 2006.

[4] J. A. Bock, "Investigation of reduced $(Sr_xBa_{1-x})Nb_2O_6$ as a ferroelectric-based thermoelectric," *Materials Science and Engineering*, pp. 1–198, 2016.

[5] W. Horn, J. V. Bassewitz, and C. Denz, "Slow and fast light in photorefractive SBN:60," *Journal of Optics*, vol. 12, no. 10, Article ID 104011, 2010.

[6] J. Nuja, C. S. Suchand Sandeep, P. Reji, and K. Nandakumar, "Nonlinear optical properties of nanosized rare-earth-doped strontium barium niobate ceramics," *Spectroscopy Letters*, vol. 44, no. 5, pp. 334–339, 2011.

[7] S. Lee, R. H. T. Wilke, S. Trolier-Mckinstry, S. Zhang, and C. A. Randall, "SrxBa1-x Nb2 O6-δ Ferroelectric- thermoelectrics: Crystal anisotropy, conduction mechanism, and power factor," *Applied Physics Letters*, vol. 96, no. 3, Article ID 031910, 2010.

[8] T. Yamazawa, T. Honma, H. Suematsu, and T. Komatsu, "Synthesis, ferroelectric and electrooptic properties of transparent crystallized glasses with SrxBa1-xNb2O 6 nanocrystals," *Journal of the American Ceramic Society*, vol. 92, no. 12, pp. 2924–2930, 2009.

[9] A. Ando, M. Kimura, T. Minamikawa, and Y. Sakabe, "Layered piezoelectric ceramics for fine-tolerance resonator applications," *International Journal of Applied Ceramic Technology*, vol. 2, no. 1, pp. 33–44, 2005.

[10] Y. Yang, Y. Liu, B. Dou, X. Hao, and X. Liu, "Preparation and characterization of SBN50 ceramic powder by the combustion synthesis," *Ferroelectrics*, vol. 334, no. 1, pp. 97–103, 2006.

[11] P. M. Rorvik, T. Grande, and M. A. Einarsrud, "One-dimensional nanostructures of ferroelectric perovskites," *Advanced Materials*, vol. 23, no. 35, pp. 4007–4034, 2011.

[12] B. Yang, F. Li, J. P. Han et al., "Structural, dielectric and optical properties of barium strontium sodium niobate (Sr0.7Ba0.3)2NaNb5Oi 15 single crystals," *Journal of Physics D: Applied Physics*, vol. 37, no. 6, pp. 921–924, 2004.

[13] S.-I. Kang, J.-H. Lee, J.-J. Kim, H. Y. Lee, and S.-H. Cho, "Effect of sintering atmosphere on densification and dielectric characteristics in Sr0.5Ba0.5Nb2O6 ceramics," *Journal of the European Ceramic Society*, vol. 24, no. 6, pp. 1031–1035, 2004.

[14] C. M. Rouleau, G. E. Jellison Jr., and D. B. Beach, "Influence of MgO substrate miscut on domain structure of pulsed laser deposited SrxBa1-xNb2O6 as characterized by x-ray diffraction and spectroscopic ellipsometry," *Applied Physics Letters*, vol. 82, no. 18, pp. 2990–2992, 2003.

[15] I. S. Kim, S.-I. Kim, and Y. T. Kim, "Improvement of ferroelectric properties of Pt-SrBi2Nb2O9-SiO2-Si gate structure through oxygen plasma rapid thermal annealing," *Physica Status Solidi (A) Applied Research*, vol. 201, no. 1, pp. 125–129, 2004.

[16] X.-H. Wang, H. Gu, Q.-W. Huang, and M. Čeh, "Cation occupancy at the A1/A2 sites in strontium barium niobate microcrystals grown from molten NaCl and KCl salts," *Acta Materialia*, vol. 55, no. 16, pp. 5304–5309, 2007.

[17] A. Ratnamala, G. Suresh, V. D. Kumari, and M. Subrahmanyam, "Template synthesized nano-crystalline natrotantite: Preparation and photocatalytic activity for water decomposition," *Materials Chemistry & Physics*, vol. 110, no. 1, pp. 176–179, 2008.

[18] F. E. Osterloh, "Inorganic materials as catalysts for photochemical splitting of water," *Chemistry of Materials*, vol. 20, no. 1, pp. 35–54, 2008.

[19] A. Kudo, H. Kato, and S. Nakagawa, "Water splitting into H_2 and O_2 on new $Sr_2M_2O_7$ (M = Nb and Ta) photocatalysts with layered perovskite structures: factors affecting the photocatalytic activity," *Journal of Physical Chemistry*, vol. 104, no. 3, pp. 571–575, 2000.

[20] N. McLamb, P. P. Sahoo, L. Fuoco, and P. A. Maggard, "Flux growth of single-crystal $Na_2Ta_4O_{11}$ particles and their photocatalytic hydrogen production," *Crystal Growth & Design*, vol. 13, no. 6, pp. 2322–2326, 2013.

[21] F. Ciesielczyk, W. Szczekocka, K. Siwińska-Stefańska, A. Piasecki, and D. Paukszta, "Evaluation of the photocatalytic ability of a sol-gel-derived MgO-ZrO2 oxide material," *Open Chemistry*, vol. 15, no. 1, pp. 7–18, 2017.

[22] M. Pudukudy, A. Hetieqa, and Z. Yaakob, "Synthesis, character-ization and photocatalytic activity of annealingdependent quasi spherical and capsule like ZnO nanostructures," *Applied Surface Science*, vol. 319, no. 1, pp. 221–229, 2014.

[23] Z. Liu, W. Xu, J. Fang et al., "Decoration of BiOI quantum size nanoparticles with reduced graphene oxide in enhanced visible-light-driven photocatalytic studies," *Applied Surface Science*, vol. 259, pp. 441–447, 2012.

[24] C. Shekhar Pandey, J. Schreuer, M. Burianek, and M. Mühlberg, "Relaxor behavior of ferroelectric Ca0.22Sr0.12Ba 0.66Nb2O6," *Applied Physics Letters*, vol. 102, no. 2, Article ID 022903, 2013.

18

Water Pollution and Water Quality Assessment of Major Transboundary Rivers from Banat (Romania)

Andreea-Mihaela Dunca (iD)

Department of Geography, Faculty of Chemistry, Biology, Geography, West University of Timişoara, Blvd. V. Pârvan No. 4, Timişoara, 300223 Timiş, Romania

Correspondence should be addressed to Andreea-Mihaela Dunca; andreea.dunca@e-uvt.ro

Academic Editor: Narcis Duteanu

This study focuses on water resources management and shows the need to enforce the existing international bilateral agreements and to implement the Water Framework Directive of the European Union in order to improve the water quantity and quality received by a downstream country of a common watershed, like Timiş-Bega hydrographical basin, shared by two countries (Romania and Serbia). The spatial trend of water quality index *(WQI)* and its subindexes are important for determining the locations of major pollutant sources that contribute to water quality depletion in this basin. We compared the values of WQI obtained for 10 sections of the two most important rivers from Banat, which have a great importance for socioeconomic life in southwestern part of Romania and in northeastern part of Serbia. In order to assess the water quality, we calculated the *WQI* for a long period of time (2004–2014), taking into account the maximum, minimum, and the mean annual values of physical, chemical, and biological parameters (*DO, pH, BOD5, temperature, total P, N-NO2$^-$,* and *turbidity*). This article highlights the importance of using the water quality index which has not been sufficiently explored in Romania and for transboundary rivers and which is very useful in improving rivers water quality.

1. Introduction

The water quality from the rivers has a considerable importance for the reason that these water resources are generally used for multiple matters such as: drinking domestic and residential water supplies, agriculture (irrigation), hydroelectric power plants, transportation and infrastructure, tourism, recreation, and other human or economic ways to use water [1].

For a given river the water quality is the result of several interrelated parameters with a local and temporal variation which are influenced by the water flow rate during the year [2].

In the context of sustainable water management, many hydrological studies have been published around the world, which highlights the ecological role of water from the rivers. Moreover, there have been more researches based upon water quality evaluation [3–5]. This category of studies is related to the quality of watercourses which generally use many statistical and mathematical models.

Most of the studies related to the assessment of the water resources quality use several water quality indices among

the most important are water quality index (WQI), water pollution index (WPI), and river habitat survey (RHS) [6, 7].

Studies focusing on water quality of water bodies from Romanian territory and especially of major transboundary rivers from Banat hydrographical area are scarce, so this study has a great importance for the reason that it describes the suitability of surface water sources from this hydrographical area for human consumption being useful for communication of overall water quality information to the concerned citizens and policy makers.

To determine the locations of major polluting sources that contribute to water quality depletion in the Timiş-Bega hydrographical basin and its tributaries, an analysis has been made in order to evaluate the two largest waterways from Banat (Timiş and Bega transboundary rivers), using the water quality index (WQI) method, which is one of the most reliable indicators of the watercourses pollution and the most convenient way to express the water quality at the same time [8].

Timiş-Bega hydrographical system is located in the western part of Romania, overlapping the hydrographical basins

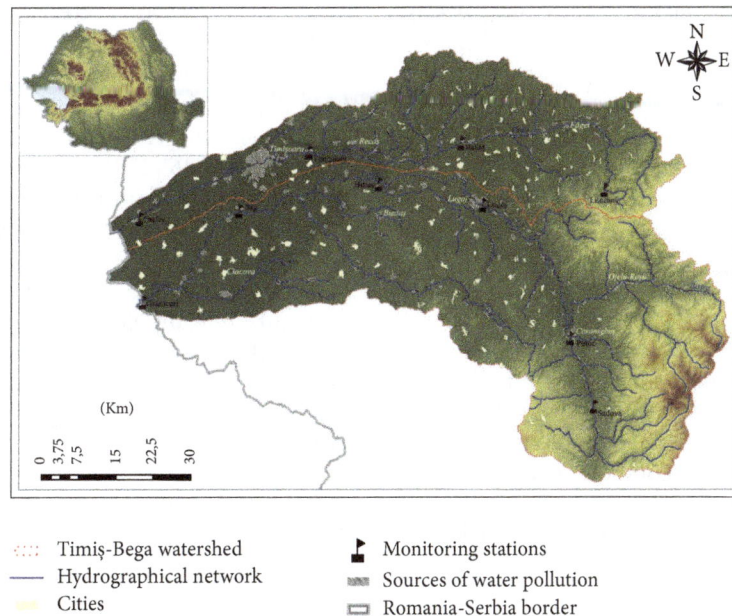

:::: Timiş-Bega watershed   ▐ Monitoring stations
— Hydrographical network   ▨ Sources of water pollution
   Cities   ▭ Romania-Serbia border

FIGURE 1: The geographical position of the Timiş-Bega hydrographical basin within Romania.

of Timiş and Bega rivers, named after the hydrotechnical works constructed in the two basins. These works are meant to ease the better management of the water resources within them, interconnecting the two rivers, through the Coştei-Chizătău Supply Channel and Topolovăţ-Hitiaş Discharge Channel (Figure 1).

Timiş River is the most important river of Banat historical region, springing from the crystalline massif of Semenic, under the Piatra Goznei Peak, from the approximate altitude of 1135 m, and it discharges on a total length of 249 km up to the confluence point with the Danube, located in the South of Pancevo locality, on the current territory of Serbia [9, as amended]. The river gathers its tributaries that spring from the Banat Mountains, Ţarcului Mountains, Poiana Ruscă Mountains, and, finally, the piedmont hills of Lugoj and Pogăniş, summing up a total length of the watercourses of approximately 462 km and a watershed surface of 5505 km^2 on the territory of Romania, representing approximately 2.31% from the total surface of Romania (238391 km^2) [10].

Bega River springs from Poiana Ruscă Mountains, under Padeş Peak (1359 m), from an altitude of 1150 m, discharging into Tisa River, near Titel locality, found on the current territory of Serbia where it collects its tributaries from Poiana Ruscă Mountains, Lipova Piedmont, the pied mountainous, and divagation field of Banat that are summing up a length of 176 km and a surface of 2675 km^2 on the territory of Romania (almost 1.12% from the total surface) [10].

Human activities in the basin of the two most important rivers from Banat have a great importance for socioeconomic life in southwestern part of Romania and in northeastern part of Serbia. Moreover, these human activities have an important influence in the geographical environment generally speaking with a particularity in what concerns the water resources, their quantity, and quality.

The problems involving the water resources management activity from Timiş-Bega hydrographical system consist of the assurance of the required water demand by the various social-economic objectives, the prevention of damaging effects of the waters, and the maintenance of a good environmental quality.

The water intakes from Timiş-Bega hydrographical system are providing the drinkable water supply or the use of water for industrial purposes, which can influence the river hydromorphological level, changing the features of the natural water discharge regime on their courses.

In Timiş-Bega hydrographical system the river monitoring activity started in the 19th century, when the achievement of drainage works of great amplitude was started, in the subsidence area of the Western Plain, where several swamps and frequent floods took place, and when several hydrotechnical works were performed, based on studies and projects, for which several observations and hydrological measurements where necessary.

According to the Water Framework Directive of European Union (WFD 2000), Timiş-Bega watershed has been selected several watercourses, well-delimited in the territory, for the operational monitoring of the surface and ground waters and for the determination of water quality status, as follows: 14 surface water bodies found in natural status, 12 surface water bodies which are heavily modified and artificial, 3 surface reservoirs, and 8 monitored ground water bodies [10].

2. Materials and Methods

The water quality index (WQI) is a numeric expression used to evaluate the quality of a given water body meant to be easily understood by managers from many countries [11].

TABLE 1: Index value intervals and the corresponding quality category [17].

Value intervals (%)	Water quality status
90–100	Excellent
70–90	Good
50–70	Medium
25–50	Bad
0–25	Very bad

In order to calculate the water quality index, Horton proposed in 1965 the first formula which takes into account all parameters necessary for determining the quality of the surface waters and which reflects the composite influence of different parameters important for the assessment and management of water quality [12, 13].

This index was for the first time used to highlight the physical-chemical changes that may occur during the year on the flowing water quality [14, 15].

Most often, the water quality index is used in the evaluation of surface water quality. This index incorporates data from multiple parameters into a mathematical equation that rates the quality of water bodies with numbers from 1 to 100 which can be separated in five classes, each class with a different quality state and with a different usage domain [13, 16].

One of the most computation formulas used to determine the water quality index can be noticed in the following arithmetic expression:

$$WQI = \frac{1}{100} \left(\sum_{i=1}^{9} q_i w_i \right)^2, \tag{1}$$

where i is the quality parameter, q_i is the registered value, and w_i is the rank of implication of the parameter in the computation formula [12].

Seven factors have been chosen, in order to rate this index; each of them is more important than others, so weighted mean is used to combine the values of these factors.

The classes of the water quality status obtained according to the quality intervals of the WQI are presented in Table 1.

WQI scores above 80 represent stations of "lowest concern" that generally meet state water quality standards, WQI scores between 40 and 80 indicate stations of "marginal concern," and WQI scores below 40 did not meet expectations and are of "highest concern" [18].

In order to obtain the WQI values a selection of the parameters has been made according to the Global Quality Classes established through the norms regarding the classification of surface water quality towards to the determination the ecological status of the water bodies.

Thereby, some of the most important parameters of water quality index have been taken into account with the impossibility of considering two of these parameters (total coliforms and turbidity), for the reason that, first of all, the total coliforms parameter is monitored only in the sections where the water is targeted for the potable use and secondly because turbidity was not considered from sampling stations analyzed.

The results have been further analyzed using current Romanian legislation (the Water Law number 107/1996, as amended and supplemented, the Law number 310/2004, and Order number 161/2006 approving the norms concerning the classification of surface water quality to determine the ecological status of water bodies) which complies with WFD 2000 [19–22].

This directive has been adopted by European Parliament and Council (Directive 2000/60/EC) on establishing a framework for European Community action in the field of water, and it contains for each parameter the limit values for corresponding chemical status of all five classes set, namely, very good (1st grade of quality), good condition (2nd grade of quality), moderate condition (3rd grade of quality), poor condition (4th grade of quality), and bad condition (5th grade of quality) [23].

Another step in calculating the values of water quality index from each sampling sections analyzed has been the one that brings all the measurement units at the same reference scale.

Determining the degree of involvement of the parameters has been accomplished in correspondence with the specific methodology, which takes into account the role of each analyzed parameter in defining the status of the water bodies and of the aquatic ecosystems [12]. Afterward the last step has been completed using an online calculator of the water quality index advanced by Mr. Brian Oram in 2010, according to the Field Manual for Water Quality Monitoring book [12, 17].

The computation of the WQI for two of the most important rivers from Banat (Timiş and Bega) has been performed taking into account the mean annual values of each quality parameter, which were registered at the six monitoring stations on the Timiş River (Sadova Veche, Potoc, Lugoj, Hitiaş, Şag, and Grăniceri) and at the four monitoring stations on the Bega River (Luncanii de Jos, Balinţ, Timişoara, and Otelec).

3. Results and Discussions

In order to assess the water quality of Timiş and Bega rivers the water quality index for a long period of time (2004–2014) has been calculated which has been applied also for 10 sampling sections, along the Timiş (6) and Bega (4) rivers, taking into account the maximum annual, the minimum annual, and the mean annual values of 7 following physical, chemical, and biological parameters: DO (oxygen saturation in percent), pH (in pH units), BOD5 (biochemical oxygen demand in mg O_2/L), temperature (°C), total P (total phosphorus in mg P/L), N-$NO2^-$ (total nitrates in mg N/L), and turbidity (mg/L), with units of measurement adapted according to International Union of Pure and Applied Chemistry (IUPAC).

The average values of physical (temperature and turbidity), chemical (pH, total phosphorus and nitrates), and biologic/organic (oxygen saturation and biochemical oxygen demand) parameters of water from Timiş and Bega rivers

TABLE 2: Water quality status and WQI values at sampling stations from the Timiş River.

| Water quality parameters (unit) | Water sampling station | | | | | | | | |
| | Sadova Veche | | | Potoc | | | Lugoj | | |
	Max.	Min.	Mean	Max.	Min.	Mean	Max.	Min.	Mean
DO (%)	125.9	56.1	84.5	128	57.7	85.7	163.8	54.1	98.1
pH (U pH)	7.9	7.1	7.5	7.9	7.2	7.6	9	6.8	7.6
BOD5 (mg O_2/L)	3	1.2	1.9	3	1.1	1.9	5.8	1.2	2.8
Temperature (°C)	17.3	1	7.9	17	1.9	8.6	24.5	1.5	12.9
Total P (mg P/L)	0.075	0.032	0.050	0.106	0.026	0.052	0.298	0.028	0.106
N-NO2⁻ (mg N/L)	0.021	0.004	0.010	0.037	0.008	0.017	0.041	0.004	0.016
Turbidity (mg/L)	37.2	18.7	26.1	41.4	17.9	27.1	198.3	4.6	37.5
Overall WQI	*79*	*83*	*86*	*78*	*84*	*86*	*58*	*81*	*83*
Class	II	II	II	II	II	II	III	II	II
Water quality status	Good	Good	Good	Good	Good	Good	Medium	Good	Good
Water quality concern	*Marginal concern*	*Lowest concern*		*Marginal concern*	*Lowest concern*		*Marginal concern*	*Lowest concern*	

Source. Data processed by the Banat Water Basin Administration (ABAB) Archives, Timişoara.

TABLE 3: Water quality status and WQI values at sampling stations from the Timiş River.

| Water quality parameters (unit) | Water sampling station | | | | | | | | |
| | Hitiaş | | | Şag | | | Grăniceri | | |
	Max.	Min.	Mean	Max.	Min.	Mean	Max.	Min.	Mean
DO (%)	130.9	40	79.1	147.7	51.2	89.3	144.7	47.4	86.7
pH (U pH)	7.7	6.8	7.3	7.7	7	7.4	8	7.2	7.6
BOD5 (mg O_2/L)	10.8	1.3	4.2	4.8	1.1	2.4	6	1.6	3.7
Temperature (°C)	23.6	2.7	12.6	23.6	2.1	12.1	23.3	3.1	12.7
Total P (mg P/L)	0.437	0.042	0.145	0.263	0.041	0.123	0.434	0.058	0.171
N-NO2⁻ (mg N/L)	0.078	0.009	0.030	0.042	0.007	0.021	0.046	0.006	0.021
Turbidity (mg/L)	291.4	8.9	68	185.2	6.2	44.9	124.9	11.5	46.8
Overall WQI	*66*	*76*	*79*	*66*	*81*	*83*	*62*	*79*	*81*
Class	III	II	II	III	II	II	III	II	II
Water quality status	Medium	Good	Good	Medium	Good	Good	Medium	Good	Good
Water quality concern	*Marginal concern*			*Marginal concern*	*Lowest concern*		*Marginal concern*		*Lowest concern*

Source. Data processed by the Banat Water Basin Administration (ABAB) Archives, Timişoara.

and the results of water quality data analysis are presented in Tables 2, 3, 4, and 5.

At Sadova Veche and Potoc monitoring stations, located on the upper course of Timiş River, the water quality status is good (70–90%), according to the average, maximum, and minimum annual values of the analyzed parameters during the period under review (2004–2014), which make these sampling stations fall into the "lowest concern" category (Table 2).

Downstream on the Timiş River beginning with Lugoj monitoring section until the border between Romania and Serbia the water quality is preserved in good condition according to the mean and minimum annual values. Only the maximum annual values decreased, which cause the medium status of water quality (50–70%) at all other sections (Lugoj,

Hitiaş, Şag, and Grăniceri), fitting them into the "marginal concern" category (Table 3).

The values of the water quality index from these stations correspond to the moderate class, which are influenced by the nutrients, respectively, by the high values of the nitrates from Timiş river water, as a result of the agricultural practices, municipal and industrial wastewaters, manure from farms, and so on.

The water quality of the Timiş River is influenced by many factors including the quantitative variation of biogenic and organic substances. All biogenic elements within the water bodies are the result of the decomposition process of the organic substances therefore the regime of the biogenic elements depends directly on the vital activity of the organisms from the rivers. Moreover this river is characterized

TABLE 4: Water quality status and WQI values at sampling stations from the Bega River.

Water quality parameters (unit)	Water sampling station					
	Luncanii de Jos			Balinț		
	Max.	Min.	Mean	Max.	Min.	Mean
DO (%)	146.3	60.6	97.6	154.1	48.3	89.1
pH (U pH)	8.3	7.6	8	8.1	7.2	7.7
BOD5 (mg O_2/L)	3.7	0.8	2	6.1	1.5	3.4
Temperature (°C)	19.8	2.2	10	23	1.7	11.4
Total P (mg P/L)	0.242	0.032	0.099	0.381	0.037	0.128
N-NO2$^-$ (mg N/L)	0.015	0.002	0.007	0.041	0.006	0.017
Turbidity (mg/L)	54.9	4.4	18.8	232.4	5.2	55.5
Overall WQI	66	84	85	61	80	82
Class	III	II	II	III	II	II
Water quality status	Medium	Good		Medium	Good	
Water quality concern	Marginal concern	Lowest concern		Marginal concern	Lowest concern	

Source. Data processed by the Banat Water Basin Administration (ABAB) Archives, Timișoara.

TABLE 5: Water quality status and WQI values at sampling stations from the Bega River.

Water quality parameters (unit)	Water sampling station					
	Timișoara			Otelec		
	Max.	Min.	Mean	Max.	Min.	Mean
DO (%)	148.9	44.9	85.6	136.9	23.8	61.9
pH (U pH)	7.8	7	7.4	7.7	6.9	7.3
BOD5 (mg O_2/L)	3.8	0.8	1.8	10.3	3.2	5.7
Temperature (°C)	23.9	1.4	12.2	24.7	1.5	13.1
Total P (mg P/L)	0.395	0.026	0.117	0.881	0.161	0.418
N-NO2$^-$ (mg N/L)	0.041	0.004	0.013	0.268	0.011	0.056
Turbidity (mg/L)	132.1	5.7	30.1	134.8	5.2	28.9
Overall WQI	65	80	84	64	69	68
Class	III	II	II	III	III	III
Water quality status	Med.	Good		Medium		
Water quality concern	Marginal concern	Lowest concern		Marginal concern		

Source. Data processed by the Banat Water Basin Administration (ABAB) Archives, Timișoara.

by the presence of several impurities in natural state with a composition which depends on the types of soils from the reception basin, waste water spills from different kind of users, and the dissolving capacity of the gases in the atmosphere [24, 25].

Within the water of unpolluted rivers, the concentration of nitrates often oscillates between the limits of a few tenths of mg/l. The main cause for the loading of the flowing waters with nitrates consists in the eviction of the urban waste waters [26]. This is the reason why the content in N-NO$_2^-$ of the river water is almost double at Lugoj station and the reason why the water quality is changing from good to a moderate status according to the maximum annual values.

Generally, the best water quality status from Bega River concerning average and minimum annual is centralized in the sections from the upper course, which falls into the "lowest concern" category (Tables 4 and 5).

Downstream from Timișoara until the Romanian Serbian border the water quality status is deteriorated according to the average annual, maximum annual, and minimum annual

values of the water quality index (50–70% – medium state), so the water quality of Bega River has a moderate status at the exit of our country, weaker compared with Timiș River, which causes Grăniceri station to have a "marginal concern" regarding water quality.

Water quality of the most important rivers from Timiș-Bega hydrographical basin is a result of human activity and demographic characteristics on one side and urbanization and industrialization on the other side. Discharging of untreated waste waters from industry, households, and pollution from agriculture (sewage water from rural localities, from animal farms and from industry) are the main causes of pollution on surface water resources and groundwater in this region [27, 28].

The human stress on the surface water within Timiș and Bega catchments is induced by the total number of inhabitants (almost 700000 people) and the urban inhabitants (428168 people, by National Institute of Statistics (INS) from Romania, 2011) from cities like Timișoara, Lugoj, Buziaș, Făget, Recaș, Ciacova, Caransebeș, and Oțelu Roșu, by the organic loading that they generate through the industrial activities, land use, and animal husbandry in animal farming complexes, and finally through the degree of improvement of the hydrographical network, as a result of human activity.

At the monitoring sections situated downstream of the wastewater discharge high values of nitrogen compounds have been identified, more exactly of the nitrate, nitrite, and ammonium ions, which influence the quality of the watercourses, especially Timiş and Bega that flow into the Tisa River and Danube River on the territory of Serbia.

The waters of Timiş River and Bega River at the exit from Romania country are much polluted because the rivers quality state suffers a slight depreciations downstream thanks to effects of the urban sewage, of the urban wastewaters, of the agricultural wastes, and of the natural causes such as erosion in the hydrographical basins of these main rivers from Banat [29].

Water pollution by nitrates reaches high levels due to the introduction of intensive farming methods, with increased use of chemical fertilizers and higher concentrations of animals in smaller areas, especially in animal farming complexes from the Timiş-Bega hydrographical basin. In this basin the values of these parameters vary from one monitoring station to another due to the hydrological regime of the surface water but also to the origin and the behavior of the physical, chemical, and biological parameters.

The anthropogenic factor has an important role in the formation and the influence of leakage water processes on the rivers of this hydrographical system. Starting from 1716 and up to the present, it has mostly influenced the water discharge, by achieving several types of hydraulic structures, among which the most important are the regulation of maximum discharges on the main rivers and the most important tributaries, the performance of flood mitigation works, and river bed regulation, damming works on the most important rivers and tributaries, within the proximity of the most important localities [30, 31].

More than that, in Timiş-Bega hydrographical basin several significant water intakes and two secondary intakes (Slatina and Borlova) have been identified. The most important units that require large amounts of water within the basin are S.C. Aquacaraş S.A. from Caransebeş and Oţelu Roşu, S.C. Meridian 22 S.A. from Lugoj, S.C. Aquatim S.A. from Timişoara, and the National Administration for Land Improvement from Romania (ANIF).

The company which represents the main economic actor in the water supply field from the Timiş-Bega hydrographical basin is S.C. Aquatim S.A. This company operates with public water and wastewater services for Timişoara Municipality and many other localities [32].

During the analyzed period (2004–2014) the evaluation of the ecological status of surface water courses (rivers), existing within Timiş-Bega hydrographical system has revealed the fact that the most rivers have been found in good ecological status.

Concerning the evaluation of the chemical status, one could notice that most rivers have been found in good chemical status and only some of them have been characterized by a bad chemical status (ANPM-Timiş Environment Protection Agency, Timişoara).

Regarding the surface water courses that are heavily modified (rivers), which exist within this basin, it has been found that most of the water courses have had a moderate ecological potential, the difference being represented by the water courses that have had a good ecological potential; and from a chemical status point of view, more than half had a good chemical status and less than half have had a bad ecological status.

Also in the same period, the evaluation of the ecological potential of the three surface reservoirs existing within the basin analyzed has revealed the fact that all these have had a moderate ecological potential and that all have been found in a good chemical status.

However, the evolution trend within the last few years of the pollutant concentrations recorded at the monitoring stations on the basin rivers has had a significant decrease, due to the measures introduced by the national and European legislation, referring especially to the treatment of the urban waste waters and to the reduction of the pollution with nitrogen and phosphorous from the agricultural practices.

In the analyzed period, the limited excess on the water quality according to the Law 311/2004 was due to the zootechnical complexes (Recaş, Peciu Nou, Pădureni, Parţa, Ciacova, Voiteni, etc.), some of them owned by COMTIM, currently S.C. Smithfield Ferme S.R.L.) within the Timiş-Bega hydrographical basin, as well as to the spray irrigation of the fields with phenolic waters from S.C. Solventul S.A. from Margina, which at the moment, although it has suspended its activity, continues to influence the quality of ground waters from this region.

Another source which influences quite a lot the surface and underground waters quality from this hydrographical area is Waste Deposit Parţa, which does not have environment factor protection equipment [29].

4. Conclusions

The results of this paper present the water polluting and quality assessment of two transboundary rivers (Timiş and Bega) from two different hydrographical basins and show that WQI values of the Timiş River ranging from 86 to 58 and WQI values of the Bega River ranging from 85 to 61 denote degradation of water quality downstream of the rivers.

Water quality in the upstream sections of the Timiş and Bega has been in a better condition than the downstream river sections. There have been significant deterioration in values of the most important water quality parameters (DO, pH, BOD5, temperature, total P, $N-NO_2^-$, and slurry) downstream of the rivers, which indicates that the local pollutants may be contributing incrementally to the degradation of river quality.

The given WQI values control sections of the studied area are distributed on quality classes as follows: 90% in the 2nd class of quality (good) and 10% in the 3rd class of quality (medium) taking into account the mean annual values; 20% in the 2nd class of quality (good) and 80% in the 3rd class of quality (medium) taking into account the maximum annual; 90% in the 2nd class of quality (good) and 10% in the 3rd class of quality (medium) taking into account the minimum annual values.

The trend in the water quality index is determined by the economic activities in the agriculture, industrial, and residential areas in the sampling stations vicinity in the Timiş-Bega hydrographical basin. For these reasons a constant monitoring is necessary in order to ensure water quality of Timiş and Bega rivers at the optimum level according to the Water Framework Directive (2000/60/EC), especially because these rivers flow further through the territory of Serbia where they are discharging into Tisa and Danube rivers.

This article focuses on water resources management and shows the need to enforce the existing international bilateral agreements and to implement this European directive in order to improve the water quantity and quality received by the downstream country of a shared watershed, like Timiş-Bega hydrographical basin, shared by two countries, Romania (EU country) and Serbia (non-EU country).

According to the Water Framework Directive requirements, knowledge of anthropogenic pressure formed on water resources is highly imperative, in order to identify the quality of water bodies and ultimately for adopting appropriate measures to protect and conserve the water in this region of Romania which have so many transboundary rivers.

In order to protect the environment in general and preserve a good water quality in particularly, especially of the transboundary rivers from Banat, shared by two countries and by so many communities, it is necessary to implement an adequate wastewater management through the construction of modern and efficient waste water treatment plants.

Conflicts of Interest

The author declares that he has no conflicts of interest.

References

[1] S. Venkatramanan, S. Y. Chung, S. Y. Lee, and N. Park, "Assessment of river water quality via environmentric multivariate statistical tools and water quality index: A case study of Nakdong River Basin, Korea," *Carpathian Journal of Earth and Environmental Sciences*, vol. 9, no. 2, pp. 125–132, 2014.

[2] P. Mandal, R. Upadhyay, and A. Hasan, "Seasonal and spatial variation of Yamuna River water quality in Delhi, India," *Environmental Modeling & Assessment*, vol. 170, no. 1–4, pp. 661–670, 2010.

[3] L. Ferencz and A. Balog, "A Pesticide survey in soil, water and foodstuffs from Central Romania," *Carpathian Journal of Earth and Environmental Sciences*, vol. 5, no. 1, pp. 111–118, 2010.

[4] L. Pârvulescu and C. Hamchevici, "The relation between water quality and the distribution of Gammarus balcanicus schäferna 1922 (Amphipoda: Gammaridae) in the Anina mountains," *Carpathian Journal of Earth and Environmental Sciences*, vol. 5, no. 2, pp. 161–168, 2010.

[5] M. Pantelić, D. Dragan, S. Savić, V. Stojanović, and I. Nađ, "Statistical analysis of water quality parameters of Veliki Bački canal (Vojvodina, Serbia) in the period 2000–2009," *Carpathian*

[6] A. Milanović, M. Urošev, and D. Milijašević, "Use of the RHS method in Golijska Moravica river basin," *Bulletin of the Serbian Geographical Society*, vol. 86, no. 2, pp. 53–61, 2006.

[7] A. Milanović, D. Milijašević, and J. Brankov, "Assessment of polluting effects and surface water quality using water pollution index: a case study of hydro-system Danube-Tisa-Danube, Serbia," *Carpathian Journal of Earth and Environmental Sciences*, vol. 6, no. 2, pp. 269–277, 2011.

[8] M. Paiu and I. G. Breabăn, "Water quality index—an instrument for water resources management," in *Air and Water—Components of the Environment Conference*, G. Pandi and F. Moldovan, Eds., pp. 391–398, Presa Universitară Clujeană Press, Cluj Napoca, Romania.

[9] R. Munteanu, *Hydrographical Basin of Timiş River—Hydrological Study*, Mirton Publishing, Timişoara, Romania, 1998 (Romanian).

[10] A. M. Arba, *Water Resources from the Timiş-Bega Hydrographical System: Genesis, Hydrological Regime and Hydrological Risks*, West University of Timişoara Publishing, Timişoara, Romania, 2016 (Romanian).

[11] O. Ionus, "Water quality index—assessment method of the Motru River water quality (Oltenia, Romania)," *Annals of the University of Craiova, Series Geography*, vol. 13, pp. 74–83, 2010.

[12] S.-M. Liou, S.-L. Lo, and S.-H. Wang, "A generalized water quality index for Taiwan," *Environmental Modeling & Assessment*, vol. 96, no. 1–3, pp. 35–52, 2004.

[13] S. Tyagi, B. Sharma, P. Singh, and R. Dobhal, "Water quality assessment in terms of water quality index," *American Journal of Water Resources*, vol. 1, no. 3, pp. 34–38, 2013.

[14] M. A. House and J. B. Ellis, "The development of water quality indices for operational management," *Water Science and Technology*, vol. 19, no. 9, pp. 145–154, 1987.

[15] M. A. House, "Water quality indices as indicators of ecosystem change," *Environmental Modeling & Assessment*, vol. 15, no. 3, pp. 255–263, 1990.

[16] A. A. Bordalo, R. W. Teixeira, and W. J. Wiebe, "A water quality index applied to an international shared river basin: the case of the Douro River," *Journal of Environmental Management*, vol. 38, no. 6, pp. 910–920, 2006.

[17] B. Oram, Water Quality Index Calculator, According to the book Field Manual for Water Quality Monitoring, 2010, http://www .water-research.net/index.php/water-treatment/water-monitoring/ monitoring-the-quality-of-surfacewaters.

[18] D. Hallock, *A Water Quality Index for Ecology's Stream Monitoring Program*, Washington State Department of Ecology, Olympia, Wash, USA, 2002.

[19] The Water Law 107 of 25 September 1996, Official Monitor of Romania, no. 244 of 08/10/1996.

[20] The Law 310 of 30 June 2004, amending and supplementing The Water Law no. 107/1996, Official Monitor of Romania, no. 584 of 30/06/2004.

[21] The Law 311 of 28 June 2004, amending and supplementing Law no. 458/2002 on drinking water quality, Official Monitor of Romania, 1st part, no. 582 of 30/06/2004.

[22] The Order no. 161 of 16 February 2006, approving the Norms concerning the classification of surface water quality to determine the ecological status of water bodies, the Ministry of Environment and Water, Official Monitor, no. 511 of 13/06/2006.

Journal of Earth and Environmental Sciences, vol. 7, no. 2, pp. 255–264, 2012.

[23] Water Framework Directive (2000/60/EC), *Official Journal of the European Community*, vol. L327, pp. 1–73, 2000.

[24] V. Trufaș and C. Trufaș, *Hydrochemistry*, University of Bucharest Press, Bucharest, Romania, 1975 (Romanian).

[25] L. Oprean, E. Lengyel, and R. Iancu, "Monitoring and evaluation of Timiș River (Banat, Romania) water quality based on physicochemical and microbiological analysis," *Transylvanian Review of Systematical and Ecological Research*, vol. 15, no. 3, pp. 33–42, 2013.

[26] L. Șmuleac, S. Oncia, A. Ienciu, and R. Bertici, "Quality indices of the water in the middle Timiș River basin," *Annals of the University of Oradea, Environmental Protection Fascicle*, vol. 21, pp. 757–764, 2013.

[27] A. Ienciu, S. Oncia, L. Șmuleac, P. Fazakas, and C. A. Nicolici, "The quality of the Timis River Waters," *Research Journal of Agricultural Science*, vol. 45, no. 2, pp. 146–151, 2013.

[28] A. Balint, G. Cîrciu, E. Alexa, and A. Cozma, "Monitoring of nitrogen compound long ways Timis River basin," *Journal of Horticulture, Forestry and Biotechnology*, vol. 18, no. 1, pp. 144–150, 2014.

[29] I. Both, I. Borza, L. Copăcean, and P. Mergheș, "The impact of the antropic activities on the water quality of the rivers Timiș and Bega in the inferior sector," *Research Journal of Agricultural Science*, vol. 46, no. 2, pp. 38–45, 2014.

[30] A. M. Dunca, "Chronological study of the water resources management within the Banat historical region (1716–2016)," *Review of Historical Geography and Toponomastics, West University of Timișoara Press, Timișoara*, vol. 12, pp. 83–94, 2017.

[31] A. M. Dunca, "The history of hydraulic structures realized in Banat hydrographical area (Romania)," *Transactions on Hydrotechnics, Scientific Bulletin of Politehnica University of Timișoara*, vol. 61, no. 75, 1, pp. 53–60, 2017.

[32] B. Mitrică and I. Mocanu, "Drinking water supply and consumption territorial disparities in the Timiș Plain," *Annals of the University of Oradea, Geography Series*, vol. 21, no. 2, pp. 239–247, 2011.

19

Selective Catalytic Reduction of NO$_x$ with NH$_3$ on Cu-, Fe-, and Mn-Zeolites Prepared by Impregnation: Comparison of Activity and Hydrothermal Stability

Siva Sankar Reddy Putluru,[1,2] Leonhard Schill ![orcid],[1,2] Anker Degn Jensen,[2] and Rasmus S. N. Fehrmann ![orcid][1]

[1]Centre for Catalysis and Sustainable Chemistry, Department of Chemistry, Technical University of Denmark, Building 207, DK-2800 Kgs. Lyngby, Denmark
[2]CHEC Research Center, Department of Chemical and Biochemical Engineering, Technical University of Denmark, Building 229, DK-2880 Kgs. Lyngby, Denmark

Correspondence should be addressed to Rasmus S. N. Fehrmann; rf@kemi.dtu.dk

Academic Editor: Davut Avci

Cu-, Fe-, and Mn-zeolite (SSZ-13, ZSM-5, and BEA) catalysts have been prepared by incipient wetness impregnation and characterized by N$_2$ physisorption, H$_2$-TPR, NH$_3$-TPD, and XPS methods. Both metal and zeolite support influence the deNO$_x$ activity and hydrothermal stability. Cu-zeolites and Mn-zeolites showed medium temperature activity, and Fe zeolites showed high temperature activity. Among all the catalysts, Cu-SSZ-13 and Fe-BEA are the most promising hydrothermally resistant catalysts. Fresh and hydrothermally treated catalysts were further examined to investigate the acidic and redox properties and the zeolite surface composition. Increased total acidity after metal impregnation and loss of acidity due to hydrothermal treatment were observed in all the catalysts. Hydrothermal treatment resulted in migration of metal or in strong metal support interations, whereby changes in reduction patterns are observed.

1. Introduction

The selective catalytic reduction (SCR) of nitrogen oxides (NO$_x$) with ammonia as the reducing agent is important for minimizing harmful emissions from combustion and high temperature processes [1–3]. The process is currently being used extensively to reduce the NO$_x$ from stationary sources (especially power plants), and SCR technology has also obtained increasing attention for reduction of NO$_x$ from diesel vehicle emissions [4].

The most widely used SCR catalyst for reduction of NO$_x$ is vanadia supported on TiO$_2$ (anatase) promoted with WO$_3$ or MoO$_3$. However, V$_2$O$_5$/WO$_3$(MoO$_3$)-TiO$_2$ has some shortcomings regarding high activity for oxidation of SO$_2$ to SO$_3$, toxicity, and lack of high-temperature stability which highly limits its applicability in the automotive sector [4]. In

addition, there is the problem of the transformation of the anatase to the rutile phase at high temperatures leading to catalyst deactivation. The first SCR systems were installed on power plants in the late 1970's and early 1980's. It was realized already at this time that metal-zeolite catalysts based upon a metal like copper possessed high activity for the decomposition of NO, even in the absence of a reducing agent [5]. More recently, especially the interest in automotive SCR has initiated a lot of research within the metal-zeolite systems [4, 6]. Initial studies have reported that lack of hydrothermal stability at high temperatures is one of the commercially limiting factors. However, recent development and demonstration of small pore Cu-SSZ-13 and Cu-SAPO-34 catalysts have received substantial attention due to their outstanding activity and hydrothermal stability [6–14]. Recent reports have shown that Cu-SAPO catalysts suffer from poor hydrothermal stability at very low temperatures (70°C) caused by the breakage of the Si-O-Al bonds [15]. The hydrothermal stability at low temperatures can be somewhat increased by doping with Ce [16]. Furthermore, Cu-CHA has the advantage of generating N_2O only at low levels [6].

At low temperatures, metal-zeolite reaction mechanisms are considered to be Langmuir–Hinshelwood type [14, 17, 18]. Metal-zeolites contain metal ions which are very well dispersed to interact with NO to form NO_2 and zeolite acid sites to interact with NH_3 to form NH_4^+ ions. NO_2 and adjacent NH_4^+ ions further react to produce N_2 and H_2O [17–20]. It has been proposed that NO_2 is the reactive species for the SCR of NO_x with NH_3 on zeolite catalysts and that the oxidation of NO to NO_2 is the rate-limiting step [18]. The suggested mechanistic explanations for reactions over the promising Cu-CHA system at high temperatures are completely different from the Langmuir-Hinshelwood type ones used for the low temperature regime [8, 11]. Janssens et al. [14] proposed a catalytic cycle proceeding on a single Cu ion. The oxidation of NO to NO_2 is the rate-determining step, and there is no need for Brønsted acid site participation. The NO_2 then engages in a "fast SCR"-like step. Falsig et al. [13] have shown using density functional theory calculations that proximity to another Cu ion can decrease the activation energy of NO oxidation and is in good agreement with experimental studies [21].

In the present work, we report on the influence of Cu, Fe, and Mn metals supported on zeolite supports (SSZ-13, BEA, and ZSM-5) on the SCR activity and hydrothermal stability. The catalysts were characterized by means of nitrogen physisorption, NH_3-TPD, H_2-TPR, and XPS techniques. The characteristics were further related to the catalytic performance for SCR of NO with NH_3.

2. Experimental

2.1. Catalyst Preparation and Characterization. Pure ZSM-5 and BEA supports were obtained from Zeolyst International. SSZ-13 support was supplied by Haldor Topsøe A/S, and the synthesis was reported recently [22]. 3 wt.% Cu, 3 wt.% Fe, and 3 wt.% Mn-zeolite catalysts were prepared by incipient wetness impregnation using appropriate amounts of copper nitrate trihydrate, iron nitrate nonahydrate, and manganese acetate tetrahydrate (Aldrich, 99.9%) as precursors, respectively, and SSZ-13 (Si/Al = 13), ZSM-5 (Si/Al = 15), or BEA (Si/Al = 12.5) as supports. The prepared catalysts were oven-dried at 120°C for 12 h and finally calcined at 500°C for 5 h before use.

A portion of the catalysts was also treated under mild hydrothermal conditions at a total flow rate of 1100 mL/min of a gas containing 11 vol% H_2O in air at 650°C for 3 h.

BET surface areas of the samples were determined from nitrogen physisorption measurements on about 100 mg sample at liquid nitrogen temperature with a Micromeritics ASAP 2010 instrument. The samples were kept at 200°C for 1 h before the measurement.

NH_3-TPD (temperature-programmed desorption) experiments were performed on a Micromeritics Autochem-II instrument. In an usual experiment, 100 mg of the dehumidified sample was inserted into a quartz reactor and treated in a flow of 50 $cm^3 \cdot min^{-1}$ of He at 100°C for 1 h. Afterwards, the sample was treated with anhydrous NH_3 gas (Air Liquide, 5% NH_3 in He) for 2 h at 100°C. After ammonia adsorption, the sample was flushed with He (50 mL/min) for 100 min at 100°C. The NH_3 desorption operation was carried out by heating the sample from 100 to 700°C (10°C/min) under a flow of He (25 mL/min).

H_2-TPR (temperature-programmed reduction) measurements were performed on a Micromeritics Autochem-II instrument. 100 mg of the dried sample was placed in one arm of a U-shaped quartz tube on quartz wool. TPR analysis was performed under a flow of 50 mL/min of 4% H_2 and balance Ar (Air Liquide) from 50°C to 850°C with a ramp rate of 10°C/min. The hydrogen concentration was monitored using a thermal conductivity detector (TCD).

XPS (X-ray photoelectron spectroscopy) measurements were conducted on a Thermo scientific system at room temperature using Al K_{alpha} radiation (1484.6 eV). Before acquisition of the data, the sample was outgassed for about 1 h under vacuum to minimize surface contamination. The XPS instrument was calibrated using Au as the standard. To avoid surface charges caused by irradiation, a short exposure time is used to acquire the spectrum [23].

2.2. Catalytic Activity Measurements. The SCR activity measurements were carried out at atmospheric pressure in a fixed-bed quartz reactor loaded with about 10 mg of the catalyst (180–300 μm) positioned between two layers of quartz wool. The gas composition was 1000 ppm NO, 1100 ppm NH_3, 3.5% O_2, 2.3% H_2O, and N_2 (remaining). The total flow rate was 300 $cm^3 \cdot min^{-1}$ (ambient conditions). During the experiments, the temperature was raised in steps of 25°C from 200 to 600°C, while the NO and NH_3 concentrations were monitored by a Thermo Electron's Model 17C chemiluminescence NH_3-NO_x gas analyzer. N_2O concentration was measured by gas chromatography (Shimadzu-14B GC, TDC detection, Poraplot column). The activity was measured after reaching the steady state. Fresh and hydrothermally treated catalysts were compared by change in relative activity (%) of the corresponding catalysts.

Since the SCR reaction is known to be first-order with respect to NO under stoichiometric NH_3 conditions [24], the activity can be represented by the first-order rate constant k ($cm^3 \cdot g^{-1} \cdot s^{-1}$), which can be obtained from the conversion of NO by the following equation:

$$k = -\frac{F}{w} * \ln(1 - x) \qquad (1)$$

where F denotes the flow rate (cm^3/s), w denotes the catalyst mass, and X denotes the fractional conversion of NO.

3. Results and Discussion

3.1. Catalyst Characterization

3.1.1. Support Properties. The results of surface area ($m^2 \cdot g^{-1}$), pore size (Å), Si/Al ratios, and acidity ($\mu mol \cdot g^{-1}$) of the zeolite supports are summarized in Table 1. The surface

TABLE 1: Physical characterization of support materials.

Support	Si/Al	Pore size (Å)[a]	Surface area ($m^2 \cdot g^{-1}$)	Total acidity (μmol $NH_3 \cdot g^{-1}$)
SSZ-13	13.1	3.8×3.8	655	1277
BEA	12.5	6.6×7.7	640	1008
ZSM-5	15	5.3×5.6	430	1062

[a]http://www.iza-structure.org/default.htm.

areas of the supports are decreasing in the order SSZ-13 (655) > BEA (640) > ZSM-5 (430). Based on the pore size of the supports, SSZ-13 (3.8×3.8 Å) is a small pore zeolite, ZSM-5 (5.3×5.6 Å) is a medium pore zeolite, and BEA (6.6×7.7 Å) is a large pore zeolite.

3.1.2. NH_3-TPD. Temperature-programmed desorption (TPD) of a basic molecule, especially ammonia, is a convenient method for quick determination of the acidic properties of a solid acid material [25, 26]. From NH_3-TPD, total acidity, strength, and distribution can be obtained from the peak area, position, and shape, respectively. Figure 1 shows NH_3-TPD profiles of the zeolite supports in the temperature range 100–700°C. Pure zeolite supports show two ammonia desorption regions: one due to weak acid strength and the other due to moderate acid strength. The weak acid sites were observed at lower temperatures, i.e., around 200°C, while the moderate acid sites were observed between 400 and 500°C [25]. The moderate acid strength of the zeolite supports is in the order SSZ-13 (473°C) > BEA (457°C) > ZSM-5 (412°C). The total acidity (μmol·g^{-1}) of the zeolites follow the order SSZ-13 (1277) > ZSM-5 (1062) > BEA (1008). The amounts of acid sites are close to the respective number of Al atoms, namely, SSZ-13 (1200), ZSM-5 (1050), and BEA (1247).

The results of surface area and acidity measurements of the Cu-, Fe-, and Mn-zeolite catalysts are summarized in Table 2. Surface area of the metal-zeolite catalysts are in the range and order of SSZ-13 (585–549) > BEA (528–518) > ZSM-5 (369–345). The change in the surface area from metal to metal on a given zeolite support is very small.

Figure 2(a) shows NH_3-TPD profiles of the fresh and hydrothermally treated Cu-zeolite catalysts in the temperature range 100–700°C. After impregnating Cu on zeolites, extra desorption peaks were observed between weak and moderate acid sites. These extra desorption peaks are resulting from the decomposition of a copper ammonia complex (NH_3 bonded to Cu cationic species) [27]. The total acidity of the Cu-zeolites is Cu-SSZ-13 (1792) > Cu-BEA (1759) > Cu-ZSM-5 (1671). The difference between the amounts of NH_3 desorbed from the Cu-zeolite and the corresponding zeolite is ascribed to the formation of complexes by interaction between ammonia molecules and copper cations $[Cu^I(NH_3)_x]^+$ ($x \geq 2$) [28]. The increase in acidity of the Cu-zeolite catalysts with the zeolite support is Cu-BEA (751) > Cu-ZSM-5 (609) > Cu-SSZ-13 (515), corresponding to molar NH_3/Cu ratios of Cu-BEA (1.59) > Cu-ZSM-5 (1.29) > Cu-SSZ-13 (1.09). The fact that these ratios are below 2 gives strong evidence for Cu not exclusively being present as isolated cationic

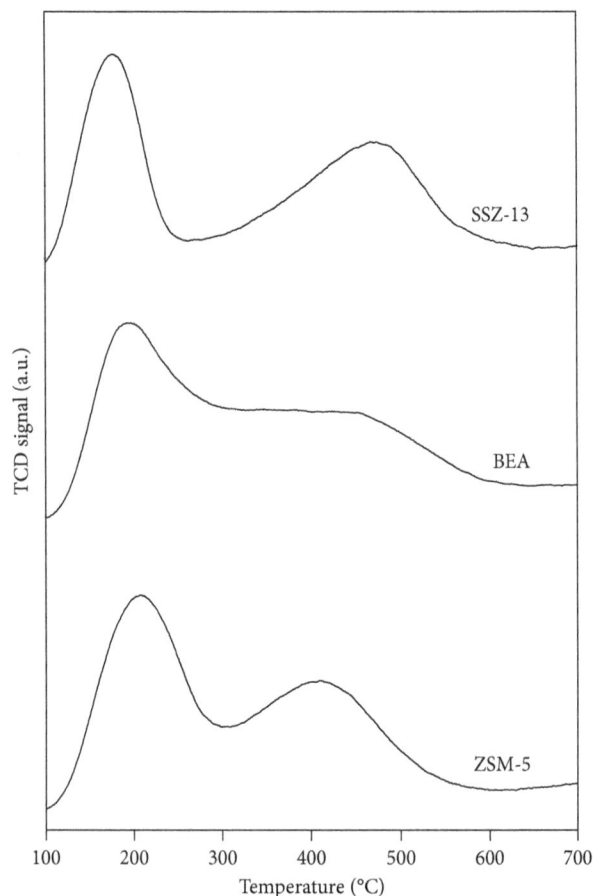

FIGURE 1: NH_3-TPD profiles of zeolite supports.

species. The order of the number of added acid sites suggests that Cu-BEA is the catalyst with most copper cations located at the exchange sites of the zeolite, however, since x in $[Cu^I(NH_3)_x]^+$ is unknown, the differences are minor, and no clear conclusion can be drawn.

Hydrothermally treated Cu-zeolites showed a decrease in total acidity along with the shift of T_{max} of the NH_3 desorption peak to the lower region. Such a decrease in total acidity is probably due to dealumination of the zeolites but could also be due to migration of active, cationic copper species from the framework to form metal oxide particles [4, 8]. The extent of dealumination can be well monitored through ^{27}Al NMR analysis [8, 29]. Migration of copper active species can be monitored with H_2-TPR analysis and XPS surface composition, see the following discussion. The total acidity of the Cu-zeolites after hydrothermal treatment is in the order of Cu-SSZ-13-HT(1580) > Cu-ZSM-5-HT(1350) > Cu-BEA-HT(1186). The relative loss of acidity of the hydrothermally treated catalysts

TABLE 2: Physical characterization of catalysts.

Catalyst	Surface area ($m^2 \cdot g^{-1}$)	Total acidity (μmol of $NH_3 \cdot g^{-1}$)		Metal/Si ratio[a] (based on XPS)		Change (%)
		Fresh	HT	Fresh	HT	
Cu-SSZ-13	549	1792	1580	0.0280	0.0292	4
Cu-BEA	523	1759	1186	0.0187	0.0300	38
Cu-ZSM-5	345	1671	1350	0.0189	0.0253	25
Fe-SSZ-13	585	1338	1098	0.0454	0.0491	8
Fe-BEA	518	1772	1195	0.0219	0.0231	5
Fe-ZSM-5	369	1519	1018	0.0247	0.0276	11
Mn-SSZ-13	577	1611	1358	0.0335	0.0488	31
Mn-BEA	528	1419	956	0.0242	0.0263	8
Mn-ZSM-5	355	1436	1065	0.0251	0.0298	16

[a]Cu/Si bulk composition is 0.032, Fe/Si bulk composition is 0.036, and Mn/Si bulk composition is 0.037.

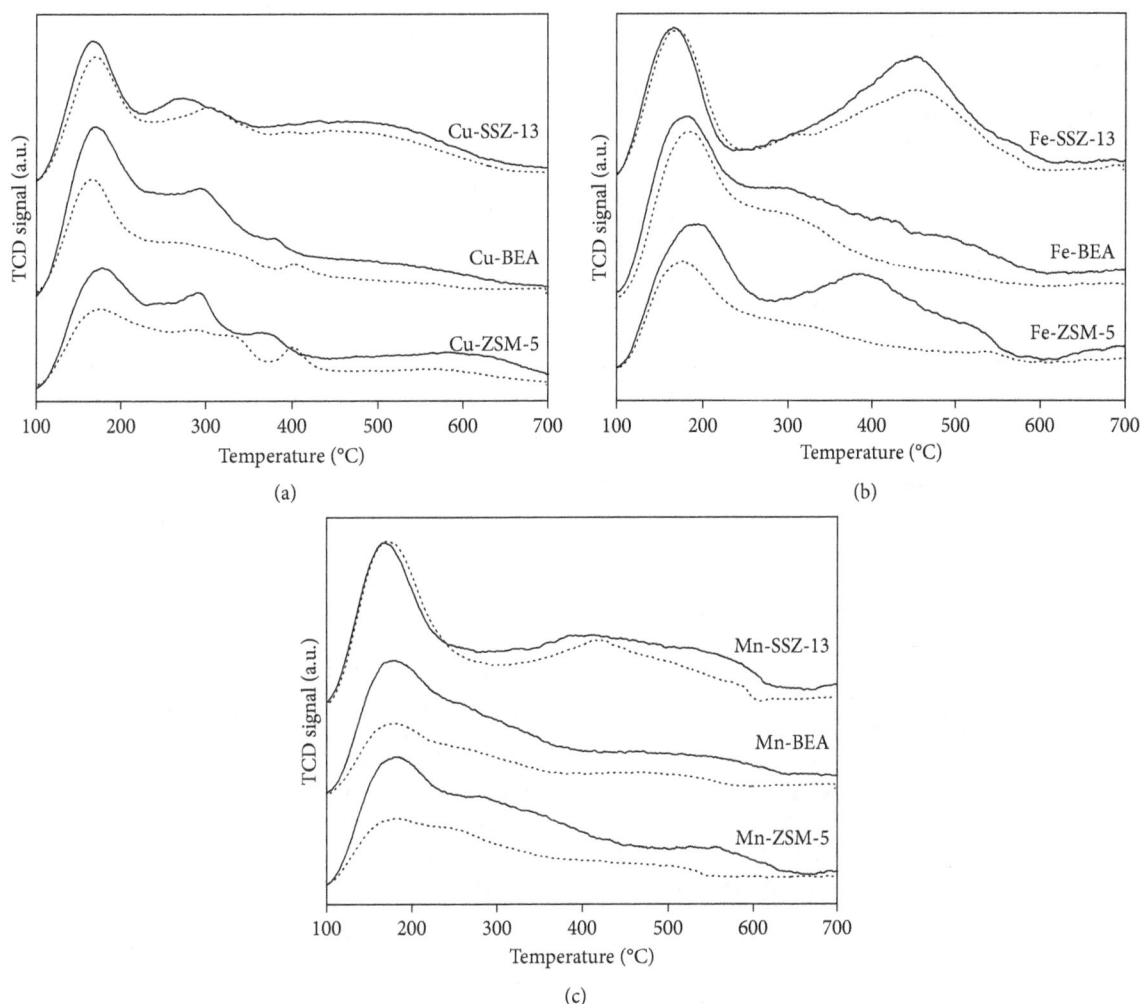

(a)

(b)

(c)

FIGURE 2: NH_3-TPD profiles of fresh and hydrothermally treated (dotted line) (a) Cu-, (b) Fe-, and (c) Mn-zeolite catalysts.

with respect to fresh Cu-zeolite catalysts is in the order Cu-BEA-HT (-573) > Cu-ZSM-5-HT (-321) > Cu-SSZ-13-HT (-212)>. The loss of acidity is also a function of pore size, as already known that small pore zeolites are more resistant to dealumination, whereby more retained acidity can be expected [29].

Figure 2(b) shows NH_3-TPD profiles of fresh and hydrothermally treated Fe-zeolite catalysts in the temperature range 100–700°C. All Fe-zeolites catalysts showed two ammonia desorption regions: one due to weak acid strength at the lower temperature and the other due to moderate acid strength at the high temperature. The total acidity of the Fe-zeolites is Fe-BEA(1772) > Fe-ZSM-5(1519) > Fe-SSZ-13(1338). The increase in acidity of the Fe-zeolite catalysts with the zeolite support is Fe-BEA (764) > Fe-ZSM-5 (457) > Fe-SSZ-13 (61). Hydrothermally treated Fe-zeolite catalysts showed a less intense high temperature desorption peak along with loss in total acidity. The total acidity of the Fe-zeolites after hydrothermal treatment is Fe-BEA-HT(1195) > Fe-SSZ-13-HT(1098) > Fe-ZSM-5-HT(1018). The loss of acidity of the hydrothermally treated catalysts with respect to fresh Fe-zeolite catalysts is in the order Fe-BEA-HT (-577) > Fe-ZSM-5-HT (-501) > Fe-SSZ-13-HT (-240)>. As seen in Cu catalysts, the relative loss of acidity is a function of pore size with greater stability for small pores.

Figure 2(c) shows NH_3-TPD profiles of the fresh and hydrothermally treated Mn-zeolite catalysts in the temperature range 100–700°C. All Mn-zeolites catalysts showed two ammonia desorption regions: one due to weak acid strength at lower temperature and the other due to moderate acid strength at high temperature. The total acidity of the Mn-zeolites is Mn-SSZ-13(1611) > Mn-ZSM-5(1436) > Mn-BEA (1419). The difference between the amounts of NH_3 desorbed from the Mn-zeolite and the corresponding pure zeolite is Mn-BEA (411) > Mn-ZSM-5 (374) > Mn-SSZ-13 (334). Baran et al. [30] measured the number of acid sites of 2 wt.% Mn/Si-BEA prepared by impregnation. The authors used pyridine as the probe molecule at 150°C with FTIR detection and determined the number of Brønsted and Lewis acid sites to be 26 and 136 μmol/g, respectively. The corresponding number for SiBEA is 8 and 3 μmol/g. The increase of the number of Lewis acid sites was ascribed to the existence of framework mononuclear Mn species. The creation of Brønsted acid sites was explained by the presence of bridging hydroxyl groups in Si-O(H)-Mn$^{(III)}$. The difference between the number of acid sites due to Mn reported in the current study and by Baran et al. [30] can probably be explained by the differences in Mn loading (3 vs. 2 wt.%), characterization technique, and temperature range. The total acidity of the Mn-zeolites after hydrothermal treatment is Mn-SSZ-13-HT(1358) > Mn-ZSM-5-HT(1065) > Mn-BEA-5-HT(956). The loss of acidity of the hydrothermally treated catalysts with respect to fresh Mn-zeolite catalysts is in the order Mn-BEA-HT (-463) > Mn-ZSM-5 (-371)-HT > Mn-SSZ-13-HT (-253). Overall, all the catalysts exhibited increase in acidity after metal impregnation (Cu, Fe, and Mn), and after hydrothermal treatment, most of the catalysts lost significant amount of acidic sites.

3.1.3. H_2-TPR. H_2-TPR is frequently used to study the redox property of the catalysts. Fresh and hydrothermally treated Cu-zeolite catalysts TPR patterns are shown in Figure 3(a). The fresh Cu-zeolite catalysts showed peaks between 160 and 230°C. According to the literature, bulk copper oxide exhibits only one TPR peak attributed to direct reduction of Cu^{2+} ions to Cu^0 [31]. A large number of literature report on Cu-zeolites showing that Cu may belong to cations or nanosized polynuclear oxocations ($n = 2, 3$) located at exchange sites and supported CuO aggregates. The former are

reduced into two steps: Cu^{2+} to Cu^+ at low temperature (<230°C) and then Cu^+ to Cu^0 at above 830°C. Cu-SSZ-13 and Cu-ZSM-5 catalysts showed a two-stage reduction with a shoulder peak that could be due to reduction of cationic Cu species at various exchange sites [29, 32]. Combined with the NH_3-TPD results, we conclude that Cu is probably both present in the form of cationic framework species as well as CuO_x particles. Figure 3(a) also shows the H_2-TPR patterns of hydrothermally treated Cu-zeolite catalysts (dotted line). Fresh and hydrothermally treated catalysts looked very similar except a slight shift of the reduction peaks to higher temperatures. Similar observations are also made on Cu-SAPO-34 and Cu-SSZ13 catalysts [8, 29]. Such a shift in reduction peaks to higher temperatures could be due to migration of copper from exchange sites to the surface or to strong interaction of Cu with the support [9, 29, 33]. This could indicate that the redox properties of the Cu-zeolite catalysts are altered by hydrothermal treatment.

The TPR profiles of Fe-zeolite catalysts are shown in Figure 3(b). The reduction of Fe species on the Fe-zeolites started at 200°C and continued till 850°C with profiles broadly divided into two regions: low temperature reduction region between 200 and 400°C and medium temperature reduction region between 400 and 700°C. According to the literature, the low temperature reduction peak can be attributed to the reduction of Fe^{3+} to Fe^{2+} species as well as the reduction of Fe_2O_3 to Fe_3O_4 [34, 35]. The medium temperature reduction peak corresponds to reduction of small nanoclusters of Fe_3O_4 to FeO and further reduction to Fe occurs at above 1000°C along with zeolite framework collapse [35, 36]. As evidenced in Figure 3(b), iron oxide reduction profiles are different for each catalyst. Hydrothermally treated catalysts showed a shift of the reduction patterns to higher temperatures. Such a change in the reduction temperature could be due to reduction of iron oxide clusters particles which have clustered during hydrothermal treatment [37].

The TPR profiles of Mn-zeolite catalysts are presented in Figure 3(c). The reduction of manganese oxide species is influenced by the MnO_x-support interaction. It can be seen that the reduction patterns are different for each support. The direct assignment of peaks is difficult. Three different reduction peaks are observed on all catalysts. These three different reduction peaks correspond to stepwise reduction of MnO_2 to MnO. The low temperature reduction peak below 250°C (MnO_2 to Mn_2O_3), medium temperature reduction peak around 300°C (Mn_2O_3 to Mn_3O_4), and a high temperature reduction peak between 350 and 450°C are observed (Mn_3O_4 to MnO) [38, 39]. Similar to Cu- and Fe-zeolites, hydrothermally treated Mn-zeolite catalysts also show a slight shift of reduction peaks to higher temperatures, which is regarded as an indication of strong interactions between oxide species and the support [39].

3.1.4. XPS. Surface composition (metal/Si ratio) of fresh and hydrothermally treated catalysts measured by XPS is summarized in Table 2. For all the fresh catalysts, except Fe-SSZ-

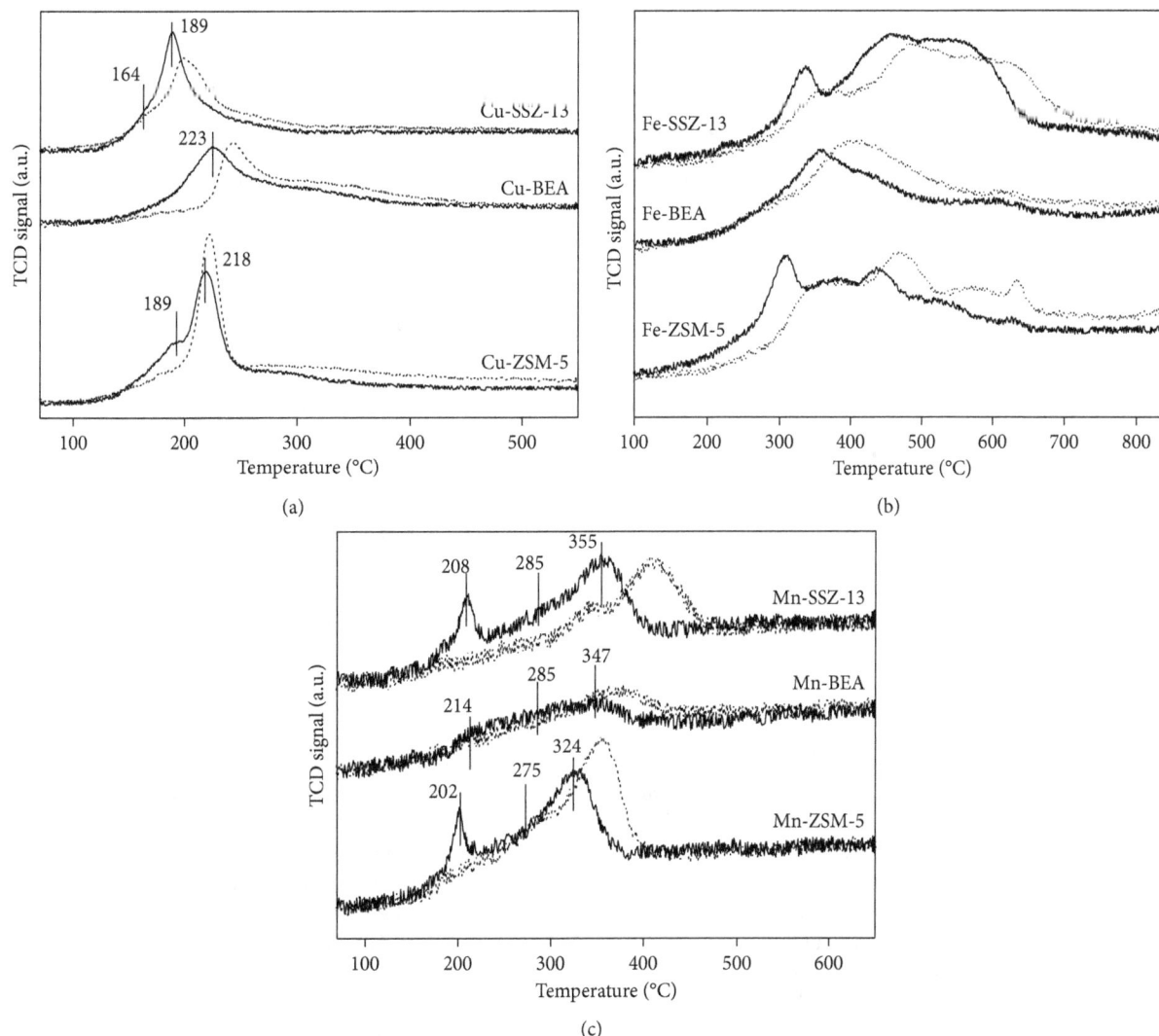

FIGURE 3: H$_2$-TPR profiles of fresh and hydrothermally treated (dotted line) (a) Cu-, (b) Fe-, and (c) Mn-zeolite catalysts.

13, the metal/Si ratios measured by XPS are lower than those corresponding to the bulk composition of the zeolites, indicating that the metal concentration in the surface of the zeolite is lower than those in the interior. The exceptional case of Fe-SSZ-13 could be due to Fe is not being incorporated in the zeolite due to the preparation method. Steric hindrance of highly solvated iron cations was reported to limit their diffusion into small pore zeolites [40, 41]. After hydrothermal treatment, the metal/Si ratio increased, suggesting that the metal migrated towards the external surface of the zeolite [42]. Overall, Fe-containing catalysts exhibit the smallest change with M/Si ratios increasing by only 5 to 11%. Manganese-containing catalysts show increases between 9 and 45%, while the copper-containing catalysts are the most affected ones with M/Si ratio increase between 4 and 60%. These values suggest that the iron species are less affected by hydrothermal treatment at 650°C and less prone to migration to the outer surface than the manganese and copper species are. Notably, there is a minor increase of the metal/Si ratio in the Cu-SSZ-13, Fe-BEA, and Mn-BEA catalysts.

3.2. NH$_3$-SCR Activity Results

3.2.1. Cu-Zeolites.
The catalytic activity of the fresh and hydrothermal treated Cu-zeolite catalysts was measured in the temperature range 250–550°C. In Figure 4, the catalytic activities obtained are shown as the first-order mass-based rate constant (cm$^3 \cdot$g$^{-1} \cdot$s^{-1}). All the catalysts showed increase in activity with increase in reaction temperature, reaching a maximum value and decreases thereafter due to the increased rate of ammonia oxidation. Cu-SSZ-13, Cu-BEA, and Cu-ZSM-5 catalysts showed maximum rate constants of 1443, 1688 ,and 1753 cm$^3 \cdot$g$^{-1} \cdot$s^{-1} at their T_{max} of 450, 550, and 425°C, respectively. The difference in T_{max} exhibited during the SCR of NO could be due to a change in ease of reduction of copper oxide and ammonia adsorption strength [32]. Irrespective of the T_{max}, all the Cu-zeolites catalysts showed high SCR activity in the temperature window of 350–550°C. The Cu-SSZ-13 catalyst exhibits a different activity profile with two maximum rate constants of 832 and 1443 cm$^3 \cdot$g$^{-1} \cdot$s^{-1} at low (325°C) and high reaction temperatures (450°C), respectively. The fact that the activity profile of Cu-SSZ-13 posses one global (at 450°C)

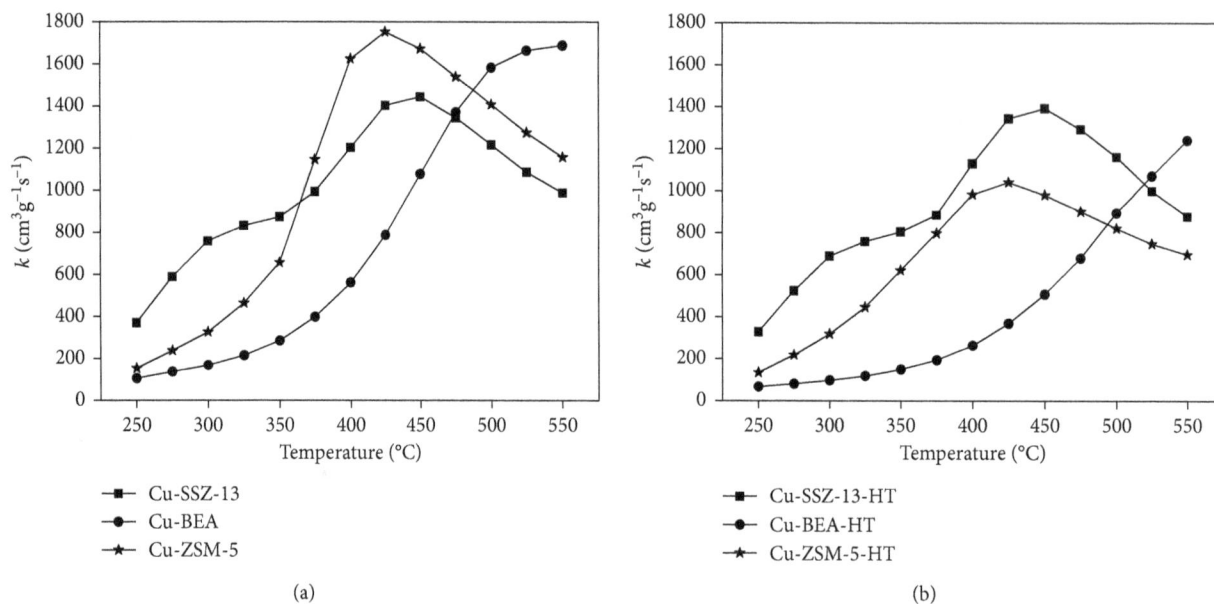

FIGURE 4: SCR activity of (a) fresh and (b) hydrothermally treated Cu-zeolites.

and one local maximum (at 325°C) could be due to location of Cu active sites in two different sites of SSZ-13 as also observed by Kwak et al. [29, 43].

Hydrothermally treated Cu-zeolite SCR activity profiles are shown in Figure 4. All the catalysts showed similar activity profile patterns as the fresh catalysts. Cu-SSZ-13-HT, Cu-BEA-HT, and Cu-ZSM-5-HT catalysts exhibit maximum rate constants of 1393, 1241, and 1041 $cm^3 \cdot g^{-1} \cdot s^{-1}$ at their T_{max} of 450, 550, and 425°C, respectively. Except for the Cu-SSZ-13, the catalysts lost significant SCR activity. The superior hydrothermal resistance of the small-pore Cu-SSZ-13 catalyst is better than medium and large pore zeolites [6, 7]. In the present investigation, along with the small-pore size effect, the superior hydrothermal stability of Cu-SSZ-13 may also be related to the retained total acidity and the similar redox pattern.

3.2.2. Fe-Zeolites. The catalytic activity of the fresh and the hydrothermally treated Fe-zeolite catalysts is shown in Figure 5. All the catalysts showed increase in activity with an increase in reaction temperature. Fe-zeolite catalysts are known to be performing well at high temperatures, e.g., in a temperature window of 400–600°C [37, 44]. Fe-SSZ-13, Fe-BEA, and Fe-ZSM-5 catalysts exhibit maximum rate constants of 102, 593, and 544 $cm^3 \cdot g^{-1} \cdot s^{-1}$, respectively, at their T_{max} of 525°C. Fe-CHA (chabazite zeolite) catalysts activity was reported to be lower than Fe-MOR and Fe-BEA catalysts [45]. Compared to the Cu-zeolites, the SCR activity of the Fe-zeolites is lower and T_{max} temperatures are higher. Especially, Fe-SSZ-13 exhibits much lower activity. Very poor activity of Fe-SSZ-13 shows that impregnation might not be the best method to incorporate Fe. Instead other catalyst preparation techniques like conventional ion exchange and chemical vapour ion exchange are recommended [4]. Hydrothermally treated Fe-zeolite SCR activity profiles are shown in Figure 5. Fe-SSZ-13-HT, Fe-

BEA-HT, and Fe-ZSM-5-HT catalysts exhibit maximum rate constants of 57, 610, and 474 $cm^3 \cdot g^{-1} \cdot s^{-1}$ at their T_{max} of 550, 550, and 525°C, respectively. Except for Fe-BEA, Fe-zeolite catalysts lost significant SCR activity. The hydrothermal resistance of Fe-BEA catalyst has earlier been reported [4, 46–48].

According to Balle et al. [48], the superior hydrothermal resistance of the Fe-BEA catalyst is due to the retained structure of the Fe sites, the slight decrease in surface area, and total acidity. The Fe-zeolite catalysts do not exhibit the small-pore size effect or the retained total acidity pattern to correlate the activity. Moderate hydrothermal resistance of Fe-BEA catalyst could be due to retained redox patterns under this milder hydrothermal condition. According to Shwan et al. [49], hydrothermal treatment is classified as milder and severe aging processes. Milder aging transforms isolated iron species to oligomeric iron clusters in the zeolite pores and further aging results in migration and formation of large iron oxide particles on the external surface of the zeolite. Oligomeric iron cluster formation inside the pores should not alter the surface Fe/Si ratio as measured by XPS. Migration of iron from the pores to the surface followed by the formation of large iron oxide particles does not necessarily increase the Fe/Si ratio as measured by XPS. This is because the FeO_x particle might become bigger than the penetration depth of the X-rays of a few nanometers. The fact that the surface Fe/Si ratios did not change might either be a sign of no iron migrating to the surface as is expected under mild hydrothermal treatment or that a possible migration to the surface is accompanied by formation of clusters too big to be fully recorded by XPS.

3.2.3. Mn-Zeolites. The catalytic activity of the fresh and hydrothermally treated Mn-zeolite catalysts is shown in Figure 6. The Mn-SSZ-13, Mn-BEA, and Mn-ZSM-5 catalysts exhibit maximum rate constants of 386, 153, and

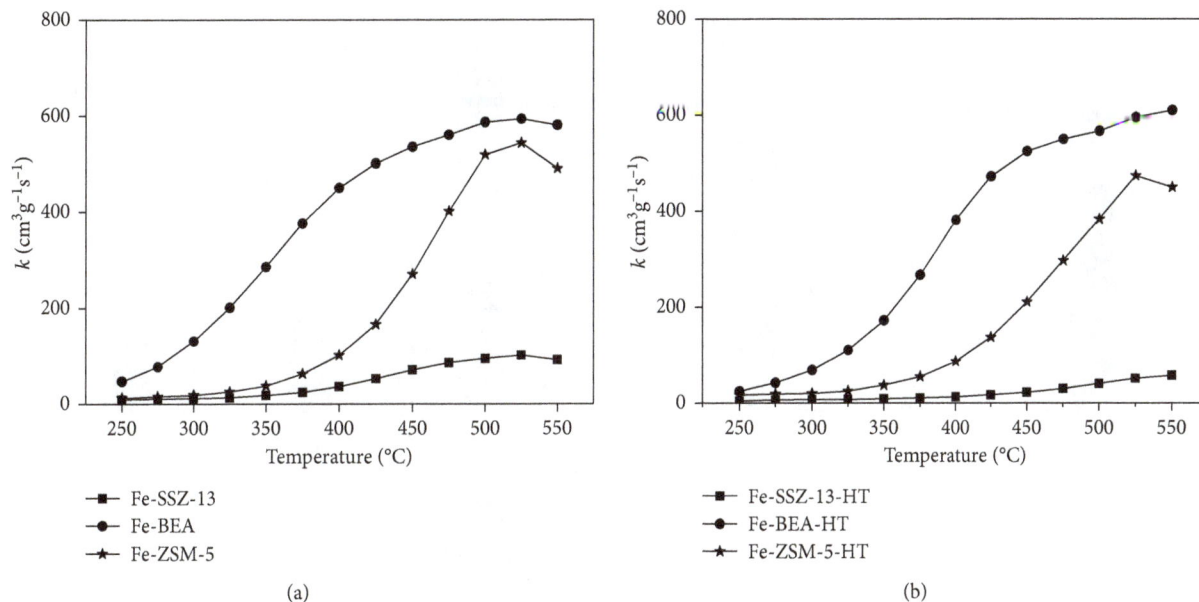

FIGURE 5: SCR activity of (a) fresh and (b) hydrothermally treated Fe-zeolites.

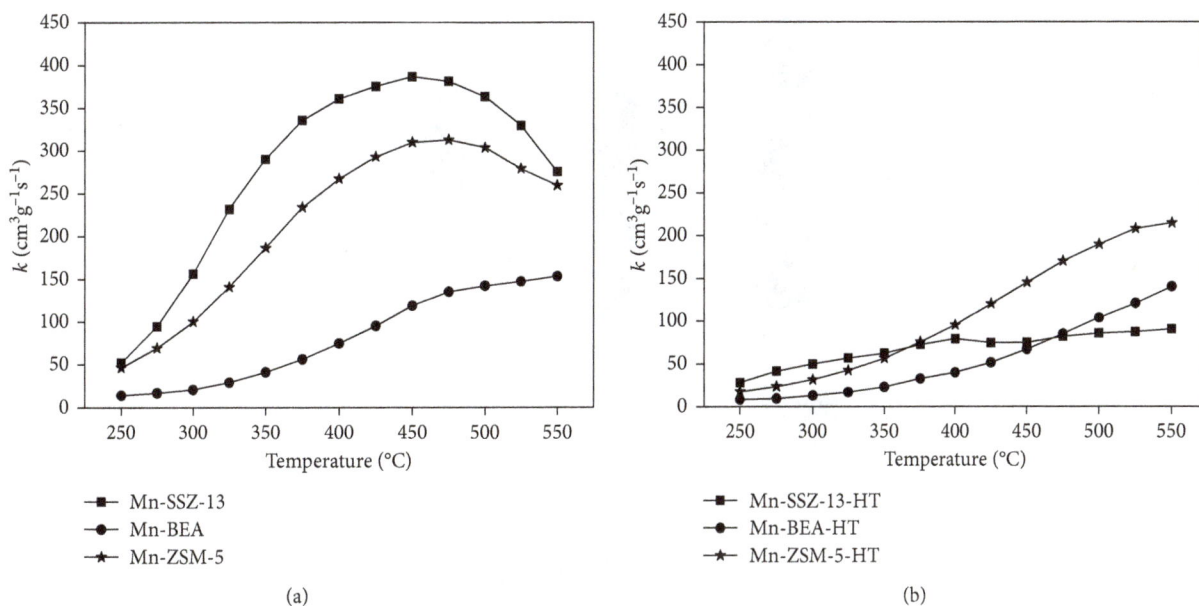

FIGURE 6: SCR activity of (a) fresh and (b) hydrothermally treated Mn-zeolites.

$313 \, cm^3 \cdot g^{-1} \cdot s^{-1}$ at their T_{max} of 450, 550, and 475°C, respectively. Compared to Cu-zeolites and Fe-zeolites, the SCR activity of Mn-zeolites is lower. Mn-SSZ-13-HT, Mn-BEA-HT, and Mn-ZSM-5-HT catalysts exhibit maximum rate constants of 90, 140, and $215 \, cm^3 \cdot g^{-1} \cdot s^{-1}$ at their T_{max} of 550°C, respectively. All Mn-zeolite catalysts lost significant SCR activity by the hydrothermal treatment. SCR activities of the present catalysts (Cu, Fe, and Mn) are not compared with the open literature because each metal and support might have its own optimum regarding loading and choice of the preparation method.

3.3. Resistance to Hydrothermal Treatment. The relative retained catalytic activity at 300, 400, and 500°C of Cu-, Fe-, and Mn-zeolites after hydrothermal treatment is shown in Figure 7. Cu-zeolite catalysts exhibit a relative activity in the order Cu-SSZ-13 > Cu-ZSM-5 > Cu-BEA at all three temperatures. The relative activity of the Cu-zeolite catalysts can be correlated with small-pore size, retained total acidity, and less migration of Cu species. At 500°C, Fe-zeolite catalysts follow the order of Fe-BEA > Fe-ZSM-5 > Fe-SSZ-13. At 400°C, the order is very similar, being Fe-BEA = Fe-ZSM-5 > Fe-SSZ-13, and at 300°C, only Fe-BEA exhibits and retains a

(a)

(b)

(c)

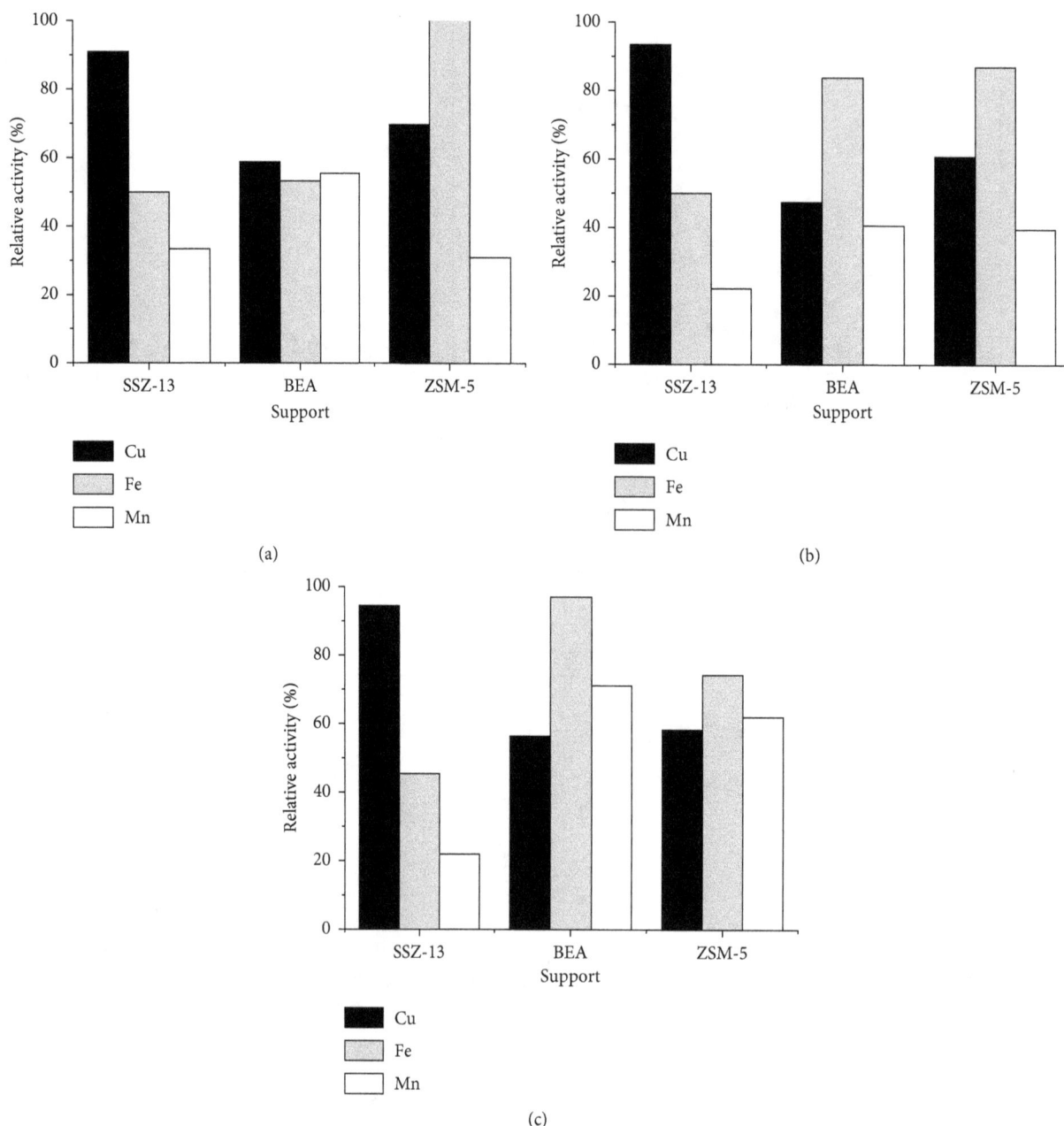

FIGURE 7: Relative catalytic activity (%) of Cu-zeolite, Fe-zeolite, and Mn-zeolite at (a) 300, (b) 400, and (c) 500°C.

relatively high degree of activity. Out of the manganese-containing catalysts, Mn-BEA is the most resistant one at all three temperatures, however, starting from a very low fresh activity. Its high resistance to hydrothermal treatment coincides with its H_2-TPR profile being the least affected one but is in contradiction to it showing the highest loss in acid sites. At 400 and 500°C, Mn-SSZ-13 which is the most active catalyst in the fresh state retains only about 20 % of its activity, while Mn-ZSM-5 retains about 40 and 62%. Fe and Mn-zeolite catalysts do not follow the small-pore size effect and the trend of retained total acidity. Among all investigated catalysts, the Cu-SSZ-13 and Fe-BEA exhibit the most promising stabilities and activities after hydrothermal treatment.

3.4. N_2O Formation. The catalytic selectivity in terms of N_2O formation was observed at 450°C. Cu-SSZ-13, Cu-ZSM-5, and Cu-BEA catalysts displayed N_2O concentration of 8 ppm, 15 ppm, and 25 ppm, respectively. Fe-SSZ-13, Fe-ZSM-5, and Fe-BEA catalysts did not display N_2O formation. Mn-SSZ-13, Mn-ZSM-5, and Mn-BEA catalysts displayed N_2O concentration of 12 ppm, 26 ppm, and 38 ppm, respectively. The highly active and hydrothermally resistant Cu-SSZ-13 catalyst performs also well regarding the level of N_2O formation.

4. Conclusions

Cu-, Fe-, and Mn-zeolite catalysts are active for the SCR of NO with NH_3. Increased total acidity after metal impregnation and loss of acidity due to hydrothermal treatment are observed. Cu-zeolites and Mn-zeolites showed medium temperature catalytic activity, and Fe-zeolites showed high temperature activity. The resistance against hydrothermal treatment is dependent on the type of metal and support. Among all the catalysts, Cu-SSZ-13 and Fe-BEA catalysts are the most promising hydrothermal-resistant catalysts. Hydrothermal resistance of Cu-SSZ-13 catalyst is a function of small pore size and retained acidic and redox properties, which in term can be seen with respect to minor increase in migration of Cu species to the surface of the zeolite. Fe- and Mn-zeolite catalysts do not follow the small-pore size or retained acidic redox properties effect.

Conflicts of Interest

The authors declare that they have no conflicts of interest.

Acknowledgments

This work has been financially supported by Energinet.dk through the PSO projects 2009-1-10521 and 2013-1-12096.

References

[1] E. T. C. Vogt, A. J. V. Dillen, J. W. Geus, and F. J. J. G. Janssen, "Selective catalytic reduction of NO_x with NH_3 over a V_2O_5/TiO_2 on silica catalyst," *Catalysis Today*, vol. 2, no. 5, pp. 569–579, 1988.

[2] M. Shigeru, Y. Hideto, T. Kazumasa, and K. Satoru, "Improvement of V_2O_5-TiO_2 catalyst for No_x reduction with NH_3 in flue gases," *Chemistry Letters*, vol. 10, no. 2, pp. 251–254, 1981.

[3] G. Busca, L. Lietti, G. Ramis, and F. Berti, "Chemical and mechanistic aspects of the selective catalytic reduction of NO_x by ammonia over oxide catalysts : a review," *Applied Catalysis B: Environmental*, vol. 18, no. 1-2, pp. 1–36, 1998.

[4] S. Brandenberger, O. Kröcher, A. Tissler, and R. Althoff, "The state of the art in selective catalytic reduction of NO_x by ammonia using metal-exchanged zeolite catalysts," *Catalysis Reviews*, vol. 50, no. 4, pp. 492–531, 2008, ISBN 0161-4940r1520-5703.

[5] T. Seiyama, "Catalytic activity of transition metal ion exchanged Y zeolites in the reduction of nitric oxide with ammonia," *Journal of Catalysis*, vol. 48, no. 1–3, pp. 1–7, 1977.

[6] J. H. Kwak, R. G. Tonkyn, D. H. Kim, J. Szanyi, and C. H. F. Peden, "Excellent activity and selectivity of Cu-SSZ-13 in the selective catalytic reduction of NO_x with NH_3," *Journal of Catalysis*, vol. 275, no. 2, pp. 187–190, 2010.

[7] D. W. Fickel, E. D'Addio, J. A. Lauterbach, and R. F. Lobo, "The ammonia selective catalytic reduction activity of copper-exchanged small-pore zeolites," *Applied Catalysis B: Environmental*, vol. 102, no. 3-4, pp. 441–448, 2011.

[8] L. Ma, Y. Cheng, G. Cavataio, R. W. McCabe, L. Fu, and J. Li, "Characterization of commercial Cu-SSZ-13 and Cu-SAPO-34 catalysts with hydrothermal treatment for NH_3-SCR of NO_x in diesel exhaust," *Chemical Engineering Journal*, vol. 225, pp. 323–330, 2013.

[9] J. S. McEwen, T. Anggara, W. F. Schneider et al., "Integrated operando X ray absorption and DFT characterization of Cu-SSZ-13 exchange sites during the selective catalytic reduction of NO_x with NH_3," *Catalysis Today*, vol. 184, no. 1, pp. 129–144, 2012.

[10] C. Paolucci, I. Khurana, A. A. Parekh et al., "SI-Dynamic multinuclear sites formed by mobilized copper ions in NO_x selective catalytic reduction," *Science*, vol. 357, no. 6354, pp. 898–903, 2017.

[11] Y. Xin, Q. Li, and Z. Zhang, "Zeolitic materials for $DeNO_x$ selective catalytic reduction," *ChemCatChem*, vol. 10, no. 1, pp. 29–41, 2017.

[12] K. A. Lomachenko, E. Borfecchia, C. Negri et al., "The Cu-CHA $deNO_x$ catalyst in action: temperature-dependent NH_3-assisted selective catalytic reduction monitored by operando XAS and XES," *Journal of the American Chemical Society*, vol. 138, no. 37, pp. 12025–12028, 2016.

[13] H. Falsig, P. N. R. Vennestrøm, P. G. Moses, and T. V. W. Janssens, "Activation of oxygen and NO in NH_3-SCR over Cu-CHA catalysts evaluated by density functional theory," *Topics in Catalysis*, vol. 59, no. 10–12, pp. 861–865, 2016.

[14] T. V. W. Janssens, H. Falsig, L. F. Lundegaard et al., "A consistent reaction scheme for the selective catalytic reduction of nitrogen oxides with ammonia," *ACS Catalysis*, vol. 5, no. 5, pp. 2832–2845, 2015.

[15] J. Wang, D. Fan, T. Yu et al., "Improvement of low-temperature hydrothermal stability of Cu/SAPO-34 catalysts by Cu^{2+} species," *Journal of Catalysis*, vol. 322, pp. 84–90, 2015.

[16] C. Niu, X. Shi, K. Liu, Y. You, S. Wang, and H. He, "A novel one-pot synthesized CuCe-SAPO-34 catalyst with high NH_3-SCR activity and H_2O resistance," *Catalysis Communications*, vol. 81, pp. 20–23, 2016.

[17] E. Y. Choi, I. S. Nam, and Y. G. Kim, "TPD study of mordenite-type zeolites for selective catalytic reduction of NO by NH_3," *Journal of Catalysis*, vol. 161, no. 2, pp. 597–604, 1996.

[18] M. Mizumoto, N. Yamazoe, and T. Seiyama, "Catalytic reduction of NO with ammonia over Cu(II) NaY," *Journal of Catalysis*, vol. 55, no. 2, pp. 119–128, 1978.

[19] C. Paolucci, A. A. Verma, S. A. Bates et al., "Isolation of the copper redox steps in the standard selective catalytic reduction on Cu-SSZ-13," *Angewandte Chemie–International Edition*, vol. 53, no. 44, pp. 11828–11833, 2014.

[20] M. Richter, A. Trunschke, U. Bentrup et al., "Selective catalytic reduction of nitric oxide by ammonia over egg-shell MnO_x/NaY composite catalysts," *Journal of Catalysis*, vol. 206, no. 1, pp. 98–113, 2002.

[21] F. Gao, E. D. Walter, M. Kollar, Y. Wang, J. Szanyi, and C. H. F. Peden, "Understanding ammonia selective catalytic reduction kinetics over Cu/SSZ-13 from motion of the Cu ions," *Journal of Catalysis*, vol. 319, pp. 1–14, 2014.

[22] F. Giordanino, P. N. R. Vennestrøm, L. F. Lundegaard et al., "Characterization of Cu-exchanged SSZ-13: a comparative FTIR, UV-Vis, and EPR study with Cu-ZSM-5 and Cu-β with similar Si/Al and Cu/Al ratios," *Dalton Transactions*, vol. 42, no. 35, pp. 12741–12761, 2013.

[23] N. Wilken, R. Nedyalkova, K. Kamasamudram et al., "Investigation of the effect of accelerated hydrothermal aging on the Cu sites in a Cu-BEA catalyst for NH_3-SCR applications," *Topics in Catalysis*, vol. 56, no. 1–8, pp. 317–322, 2013.

[24] R. Q. Long and R. T. Yang, "Catalytic performance and characterization of VO2+-exchanged titania-pillared clays for selective catalytic reduction of nitric oxide with ammonia," *Journal of Catalysis*, vol. 196, no. 1, pp. 73–85, 2000.

[25] S. S. R. Putluru, A. D. Jensen, A. Riisager, and R. Fehrmann, "Heteropoly acid promoted V_2O_5/TiO_2 catalysts for NO abatement with ammonia in alkali containing flue gases," *Catalysis Science and Technology*, vol. 1, no. 4, p. 631, 2011.

[26] S. A. Bates, W. N. Delgass, F. H. Ribeiro, J. T. Miller, and R. Gounder, "Methods for NH_3 titration of Brønsted acid sites in Cu-zeolites that catalyze the selective catalytic reduction of NO_x with NH_3," *Journal of Catalysis*, vol. 312, pp. 26–36, 2014.

[27] A. V. Salker and W. Weisweiler, "Catalytic behaviour of metal based ZSM-5 catalysts for NO_x reduction with NH_3 in dry and humid conditions," *Applied Catalysis A: General*, vol. 203, no. 2, pp. 221–229, 2000.

[28] S. Shwan, M. Skoglundh, L. F. Lundegaard et al., "Solid-state ion-exchange of copper into zeolites facilitated by ammonia at low temperature," *ACS Catalysis*, vol. 5, no. 1, pp. 16–19, 2015.

[29] J. H. Kwak, D. Tran, S. D. Burton, J. Szanyi, J. H. Lee, and C. H. F. Peden, "Effects of hydrothermal aging on NH_3-SCR reaction over Cu/zeolites," *Journal of Catalysis*, vol. 287, pp. 203–209, 2012.

[30] R. Baran, L. Valentin, J. M. Krafft, T. Grzybek, P. Glatzel, and S. Dzwigaj, "Influence of the nature and environment of manganese in Mn-BEA zeolites on NO conversion in selective catalytic reduction with ammonia," *Physical Chemistry Chemical Physics*, vol. 19, no. 21, pp. 13553–13561, 2017.

[31] Á. Szegedi, Z. Kónya, D. Méhn et al., "Spherical mesoporous MCM-41 materials containing transition metals: synthesis and characterization," *Applied Catalysis A: General*, vol. 272, no. 1-2, pp. 257–266, 2004.

[32] S. S. R. Putluru, A. Riisager, and R. Fehrmann, "Alkali resistant Cu/zeolite deNOx catalysts for flue gas cleaning in biomass fired applications," *Applied Catalysis B: Environmental*, vol. 101, no. 3-4, pp. 183–188, 2011.

[33] O. Bortnovsky, P. Sazama, and B. Wichterlova, "Cracking of pentenes to C2–C4 light olefins over zeolites and zeotypes: role of topology and acid site strength and concentration," *Applied Catalysis A: General*, vol. 287, no. 2, pp. 203–213, 2005.

[34] R. Q. Long and R. T. Yang, "Selective catalytic reduction of NO with ammonia over FE3+-exchanged mordenite (FE-MOR): catalytic performance, characterization, and mechanistic study," *Journal of Catalysis*, vol. 207, no. 2, pp. 274–285, 2002.

[35] H. Lee and H. Rhee, "Stability of Fe/ZSM-5 de-NO_x catalyst : effects of iron loading and remaining Brønsted acid sites," *Catalysis Letters*, vol. 61, no. 1-2, pp. 71–76, 1999.

[36] A. Guzman-vargas, G. Delahay, and B. Coq, "Catalytic decomposition of N_2O and catalytic reduction of N_2O and N_2O + NO by NH_3 in the presence of O_2 over Fe-zeolite," *Applied Catalysis B: Environmental*, vol. 42, pp. 369–379, 2003.

[37] S. Brandenberger, O. Kröcher, M. Casapu, A. Tissler, and R. Althoff, "Hydrothermal deactivation of Fe-ZSM-5 catalysts for the selective catalytic reduction of NO with NH_3," *Applied Catalysis B: Environmental*, vol. 101, no. 3-4, pp. 649–659, 2011.

[38] A. Sultana, M. Sasaki, and H. Hamada, "Influence of support on the activity of Mn supported catalysts for SCR of NO with ammonia," *Catalysis Today*, vol. 185, no. 1, pp. 284–289, 2012.

[39] A. Derylo-Marczewska, W. Gac, N. Popivnyak, G. Zukocinski, and S. Pasieczna, "The influence of preparation method on the structure and redox properties of mesoporous Mn-MCM-41

materials," *Catalysis Today*, vol. 114, no. 2-3, pp. 293–306, 2006.

[40] I. Melián-Cabrera, S. Espinosa, J. C. Groen, B. V. D. Linden, F. Kapteijn, and J. A. Moulijn, "Utilizing full-exchange capacity of zeolites by alkaline leaching: preparation of Fe-ZSM5 and application in N_2O decomposition," *Journal of Catalysis*, vol. 238, no. 2, pp. 250–259, 2006.

[41] A. Shishkin, H. Kannisto, P. A. Carlsson, H. Härelind, and M. Skoglundh, "Synthesis and functionalization of SSZ-13 as an NH_3-SCR catalyst," *Catalysis Science and Technology*, vol. 4, no. 11, pp. 3917–3926, 2014.

[42] P. Budi, E. Curry-hyde, and R. F. Howe, "Stabilization of CuZSM-5 NO_x reduction catalysts with lanthanum," *Catalysis Letters*, vol. 41, no. 1-2, pp. 47–53, 1996.

[43] J. Hun Kwak, H. Zhu, J. H. Lee, C. H. F. Peden, and J. Szanyi, "Two different cationic positions in Cu-SSZ-13?," *Chemical Communications*, vol. 48, no. 39, pp. 4758–4760, 2012.

[44] S. S. R. Putluru, A. D. Jensen, A. Riisager, and R. Fehrmann, "Alkali resistant Fe-zeolite catalysts for SCR of NO with NH_3 in flue gases," *Topics in Catalysis*, vol. 54, no. 16–18, pp. 1286–1292, 2011.

[45] R. Q. Long and R. T. Yang, "Reaction mechanism of selective catalytic reduction of NO with NH_3 over Fe-ZSM-5 catalyst," *Journal of Catalysis*, vol. 207, no. 2, pp. 224–231, 2002.

[46] C. He, Y. Wang, Y. Cheng, C. K. Lambert, and R. T. Yang, "Activity, stability and hydrocarbon deactivation of Fe/Beta catalyst for SCR of NO with ammonia," *Applied Catalysis A: General*, vol. 368, no. 1-2, pp. 121–126, 2009.

[47] J. A. Z. Pieterse, G. D. Pirngruber, J. A. van Bokhoven, and S. Booneveld, "Hydrothermal stability of Fe-ZSM-5 and Fe-BEA prepared by wet ion-exchange for N_2O decomposition," *Studies in Surface Science and Catalysis*, vol. 170, pp. 1386–1391, 2007.

[48] P. Balle, B. Geiger, D. Klukowski et al., "Study of the selective catalytic reduction of NO_x on an efficient Fe/HBEA zeolite catalyst for heavy duty diesel engines," *Applied Catalysis B: Environmental*, vol. 91, no. 3-4, pp. 587–595, 2009.

[49] S. Shwan, E. C. Adams, J. Jansson, and M. Skoglundh, "Effect of thermal ageing on the nature of iron species in Fe-BEA," *Catalysis Letters*, vol. 143, no. 1, pp. 43–48, 2013.

20

Highly Active Low Cobalt Content-Based Bulk MoS$_2$ Hydrodesulfurization Catalysts with a Unique Impact of H$_2$S

Hamdy Farag ,[1,2] **Abd-Alrahman Embaby,**[3] **Masahiro Kishida,**[2] **Abdel-Nasser A. El-Hendawy,**[4] **and Mohamed Mahmoud Nasef**[5]

[1]*Chemistry Department, Faculty of Science, Mansoura University, Mansoura 35516, Egypt*
[2]*Department of Material Process Engineering, Graduate School of Engineering, Kyushu University, Motooka 744, Fukuoka 819-0395, Japan*
[3]*Geology Department, Faculty of Science, Damietta University, Damietta 34517, Egypt*
[4]*Physical Chemistry Department, National Research Center, 12622 Dokki, Cairo, Egypt*
[5]*Chemical Engineering Department, Universiti Teknologi Petronas, 32610 Seri Iskandar, Perak, Malaysia*

Correspondence should be addressed to Hamdy Farag; hamdy.farag@gmail.com

Academic Editor: Mohammad A. Al-Ghouti

A series of unsupported MoS$_2$, Co$_9$S$_8$, and Co-promoted MoS$_2$ catalysts have been synthesized by tuned impregnation and successive thermal annealing methods using a continuous flow of a mixture of H$_2$ and H$_2$S gases. The resulting catalysts were evaluated in terms of their activity and selectivity for the hydrodesulfurization of dibenzothiophene (DBT) both in the absence and the presence of H$_2$S. The inclusion of Co onto MoS$_2$ affected both the hydrogenation and direct desulfurization reactions, with the latter (production of biphenyl) being magnified to a much greater degree than the former. Interestingly, low cobalt/molybdenum ratio of ca. 0.05 of the catalyst exhibited outstanding promotion efficiency in the hydrodesulfurization reaction. However, as cobalt is added, the synergy effect drastically decreased. H$_2$S in the reaction mixture led to a remarkable step up in the product from the direct desulfurization reaction route with the most notable increases occurring for the product from the hydrogenation reaction pathway. The HDS activity of such catalysts was much higher than that of the commercial CoMo/Al$_2$O$_3$. The promotion by H$_2$S was discussed.

1. Introduction

Significant research efforts have been directed towards the development of catalysts for better reducing the sulfur contents of petroleum fuel fractions [1–4]. Dwindling oil supplies, especially those extracted under vacuum distillation, and recent stringent environmental regulations aimed at limiting the sulfur contents of transportation fuels have resulted in a strong demand for improved hydrothermal treatment techniques. Clean fuels, such as sulfur- and nitrogen-free fuels, not only have the advantage of being environmentally benign transportation fuels but can also be used in several other emerging energy-related fields, including fuel cells [5]. The main catalysts used in hydrothermal treatment processes are Ni- and/or Co-containing Mo-based Al$_2$O$_3$ catalysts. Such catalysts are proved to be highly active for thiophenes and benzothiophenes sulfur elimination. However, they are not sufficiently active for desulfurizing compounds such as dibenzothiophene (DBT), and it is analogous. This is particularly problematic because these compounds are the major sulfur-containing species remaining in the middle distillate fractions and in the atmospheric or heavy residues. The MoS$_2$ phases are the main catalysts in hydrothermal treatment reactions such as hydrodesulfurization (HDS). The addition of Co to supported MoS$_2$ catalysts has been studied intensively, and its role as a promoter in HDS reactions is well established. MoS$_2$-based catalysts with no promoter have very low activities towards HDS reactions. The significant enhancement in the HDS performance of these catalysts following the incorporation of Co has been attributed to the synergy between the Co and Mo phases [6–9]. The details of this synergy have been investigated extensively in previous occasions [10–12]. However, the influence of the incorporation of Co on the selectivity of such MoS_2-based catalysts towards the HDS reactions is far from being fully understood [13–17]. The nature of the interactions between the Mo support and the substrate can have a significant impact on the efficiency of the synergy caused by Co and/or Ni promoters. On the contrary, the function of the promoter in the bulk MoS_2 in the HDS reaction remains beyond a complete coverage. Catalysts of this type have high activities in HDS reactions [18]. Although a large number of reports are available for CoMo-based catalysts, relatively very little information has been published pertaining to the unsupported CoMo catalysts. A better understanding of how Co or Mo modifies the activity of bulk MoS_2 or Co_9S_8 catalysts could provide valuable information for the development of new Mo-based catalysts for hydrothermal treatment processes. In this study, the DBT HDS reactions using bulk MoS_2, Mo-Co_9S_8, and Co-MoS_2 catalysts containing various amounts of Co, especially in low cobalt concentration range, were investigated. The potential of Co as a promoter of the bulk MoS_2 catalyst is dealt with. The effects of the hydrogen sulfide matrix on the catalytic performance of the HDS reaction of DBT were also evaluated.

2. Experimental

2.1. Materials and Methods. Cobalt acetate tetrahydrate $[Co(C_2H_3O_2)_2 \cdot 4H_2O]$, ammonium heptamolybdate tetrahydrate $[(NH_4)_6Mo_7O_{24} \cdot 4H_2O]$ (AH), ammonium tetrathiomolybdate $[NH_4]_2[MoS_4]$ (AT), molybdenum acetylacetonate $[C_{10}H_{14}MoO_6]$, DBT, and decane were supplied from Wako Chemical Industries, Ltd. All chemicals were exploited as received. The bulk MoS_2 catalysts, AHS and ATS, were prepared by heat annealing the corresponding Mo precursors (AH and AT) in the presence of a concurrent flow of a 1 : 9 (v/v) mixture of H_2S/H_2 gas at 830°C and 400°C, respectively. Co_9S_8 (CS) was also synthesized from the heat annealing of Co-acetate tetrahydrate at 400°C following the same procedure as before. The details of such processes have been described elsewhere [19]. A sample of the AHS material (obtained by sulfiding the ammonium heptamolybdate tetrahydrate precursor) was comminuted using a mill with an inner volume of 100 mL equipped with a media of zirconia beads, which was purged with He prior to being used. The resulting material was denoted as AHS-G. The BET surface areas of AHS, AHS-G, and ATS, which were measured using automatic Micromeritics ASAP 2010 instrument by N_2 adsorption-desorption technique at −196°C, were approximately >10, 115, and 65 m^2/g, respectively. The AHS-G and ATS sulfide samples were used as preliminary supports for the addition of the Co precursor. The Mo-sulfide phase was impregnated with a 1 : 1 (v/v) mixture of water/alcohol containing a specific amount of cobalt acetate tetrahydrate at an ambient temperature. The solution was then subjected to sonication for 3 h, followed by drying in air and thereafter

heating in a vacuum oven at 120°C overnight. All catalysts were thereafter subjected to sulfidation with a stream of a 1 : 9 (v/v) mixture of H_2S/H_2 gases (5 ml/min) at 400°C. The resulting samples were denoted as C-AHSG- and C-ATS-, I to III based on their Co loading. Another sample was synthesized in which the prepared Co_9S_8 phase was impregnated with a certain amount of an alcoholic solution of molybdenum acetylacetonate following the typical procedure mentioned before until obtaining the sulfide form of the catalyst. This sample (denoted as MPC) represents the Mo-promoted cobalt sulfide catalyst of ca. 0.4 of the Co/(Co + Mo) atomic ratio. The commercial CoMo/Al_2O_3 catalyst was also studied for comparison. All Mo-sulfide samples exhibited a hexagonal molybdenite-2H structure (JCPDS# 65-0160) according to the XRD patterns of AHS, AHS-G, and ATS catalysts shown in Figure 1, which is in agreement with literature [20].

No phases other than MoS_2 were detected. On the contrary, the diffraction patterns of the synthesized Co-sulfide phase matched well with the JCPDS# 73-1442 of Co_9S_8. The AHS-G and ATS catalysts had MoS_2 crystallite sizes of approximately ≈5 nm. Transmission electron microscopy images (obtained from TEM; JEOL-2000EX) for C-ATS-I were depicted in Figure 2. Five to 10 layers of MoS_2 is obviously noted. Interlayer spacing of ca. 0.62 nm was determined from the (002) XRD peak at 2θ of 14.2.

2.2. Activity Measurements. The catalysts were investigated for the DBT HDS reaction. DBT was selected as a model compound for these reactions because it is a representative of some of the sulfur-containing refractory compounds found in middle distillates and heavy residue. All tests were implemented under a 3 MPa of H_2 pressure and at 340°C. The experiments were conducted using a stainless steel batch microautoclave reactor (100 mL) equipped with a magnetic stirrer and stainless steel filter, which allowed convenient withdrawal of small samples from the reaction mixtures at regular time intervals. A decane solution of DBT (1 wt.%) was used as the reaction feedstock. Some reaction runs were conducted in the presence of Cu powder (ca. 0.7 g), which was used as a scrubber for the H_2S produced as a by-product during the HDS reaction. A blank run was also conducted with Cu in the absence of a catalyst, which confirmed that Cu did not exhibit any activity towards the DBT HDS reaction. The effect of the reaction matrix was evaluated by investigating the reaction over the present catalysts in both with and without the existence of H_2S in the feedstocks. Immediately before the reaction test, the catalyst was once more subjected to a sulfidation step with a mixture of H_2S/H_2 gases. The catalyst and the reaction mixture (a typical of 15 mL of 1 wt.% DBT in decane) were loaded in situ into the reactor, which was subsequently pressurized with H_2 and heated to 340°C under continuous stirring at 1000 rpm. Small samples of the reaction mixture (0.1– 0.2 mL) were withdrawn from the reactor periodically for analysis to determine the rate of conversion. Gas chromatography (Agilent HP 6890) and GC-mass spectrometry (GC-MS) equipped with an Agilent HP 5970 MS were used

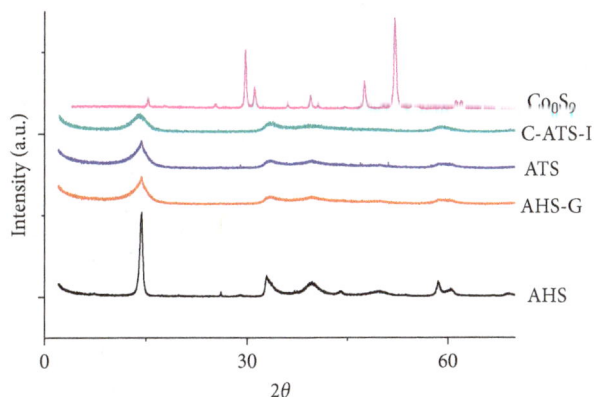

FIGURE 1: XRD patterns of the AHS, AHS-G, ATS, C-ATS-I, and Co_9S_8 catalysts.

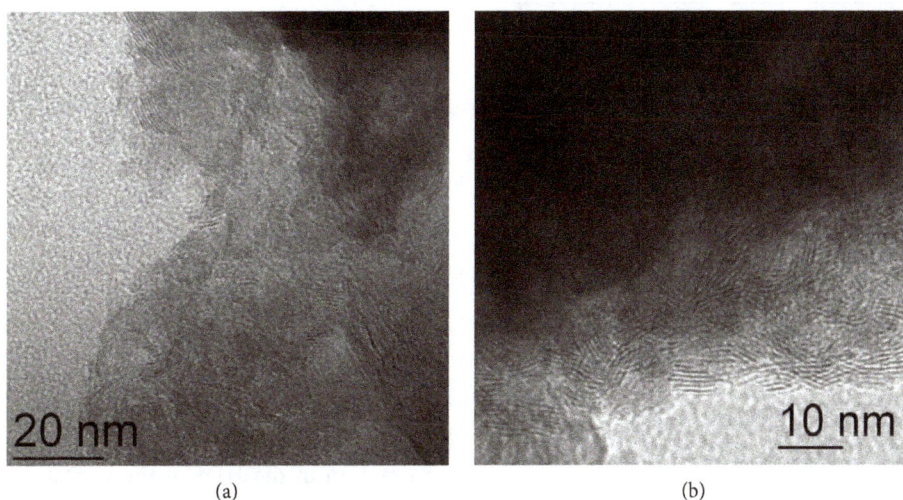

(a)   (b)

FIGURE 2: TEM images of the C-ATS-I catalyst.

3. Results and Discussion

3.1. Catalyst Performances. The main HDS reaction products obtained under the operating terms described above over the present catalysts were cyclohexylbenzene (CB), biphenyl (BP), 1,2,3,4-tetrahydrodibenzothiophene (H4-DBT), and H_2S. Trace quantities of the different isomers of partially hydrogenated DBT were also detected. Based on these products, we have proposed two reaction pathways for DBT HDS, which are depicted in Scheme 1.

The rate constants for these transformations were estimated by fitting the experimental data using a nonlinear least square analysis on the supposition that the HDS reaction behaves kinetically as pseudo-first-order [3, 4, 8, 9]. Figure 3 shows the fit curves for the DBT HDS over the C-ATS-I catalyst under different levels of H_2S.

The reaction system was classified into two reaction routes, including (1) hydrogenation (HYD), which would

to analyze the reactions. All GC analyses were conducted on a methylsiloxane capillary column ($0.32 \, mm \times 50 \, m$).

involve the initial formation of H4-DBT and its isomers by partial hydrogenation of DBT, followed by further reduction of these intermediates to give CB and (2) direct desulfurization (DDS), which would produce BP directly by the C-S bond scission. These data were treated kinetically and evaluated according to the model that was recently developed for consecutive-parallel reactions [21]. Based on this approach, we used the following differential equations to calculate the catalytic constants:

$$C_A = C_{A^0} \exp^{-k_0 t},$$

$$C_B = \frac{C_{A^0} k_1^0}{k_3^0 - k_0} \left[\exp^{-k_0 t} - \exp^{-k_3^0 t} \right],$$

$$C_C = \frac{C_{A^0} k_2^0}{k_4^0 - k_0} \left[\exp^{-k_0 t} - \exp^{-k_4^0 t} \right], \quad (1)$$

where $k_0 = k_1 K_1 + k_2 K_2$ and k_1 and k_2 denote the DDS and the HYD intrinsic kinetic rate constants, respectively. K_1 and K_2 point to the constants of DBT adsorption at equilibrium onto the DDS and HYD reaction sites, respectively.

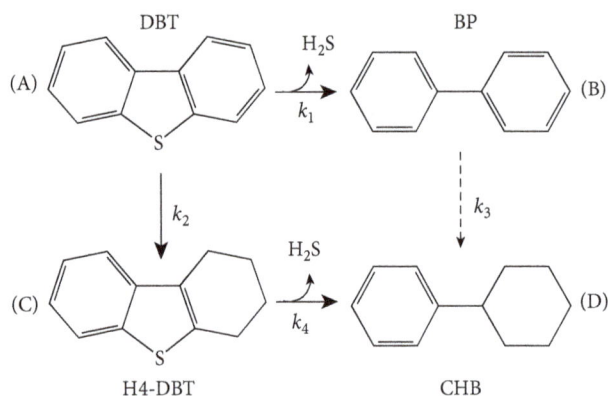

SCHEME 1: Reaction scheme for HDS of dibenzothiophene.

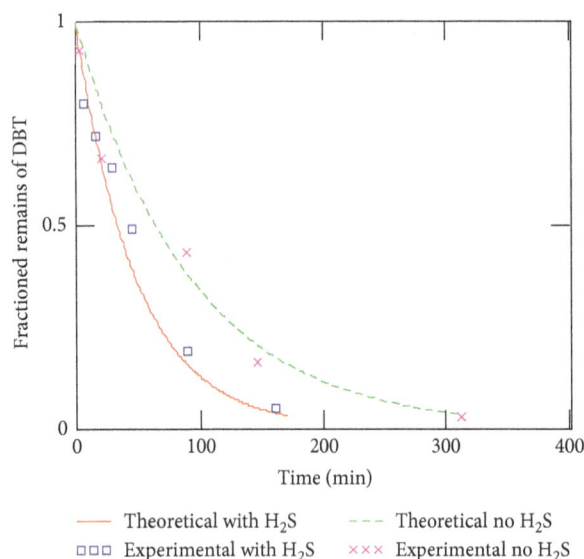

FIGURE 3: Pseudo-first-order plots for the HDS reactions of DBT over the C-ATS-I catalyst in the absence and presence of H_2S.

The kinetic parameters were adjusted for each compound according to the following formula: $k_n^0 = k_n K_n$, where K_n and k_n are the equilibrium adsorption and intrinsic kinetic rate constants of each compound, respectively. This treatment provided a better estimation of the individual contributions of the hydrogenation and direct desulfurization reaction pathways to the HDS reaction. Furthermore, this process allowed more reliable quantification of the individual reaction rates according to the contribution of each reaction pathway. The experimental data were solved and fitted to the model using the Mathcad program. The estimated activity data are listed in Table 1.

Figures 4 and 5 show the nonlinear curves that were fitted with the product yields from the HDS reaction over the C-ATS-1 catalyst in the presence and in the absence of H_2S, respectively, using the model described above. All catalysts tested in this study provided a reasonable fit to the data exposed from the HDS reaction. The C-ATS-I catalyst displayed much higher level of reaction activity than the ATS, AHS-G, C-AHS-G, MPC, and commercial

CoMo/Al_2O_3 catalysts. However, DBT HDS proceeded to a lower activity over the C-AHS-G catalysts than it did over the bare commuted MoS_2 catalyst. The difference in the activity of these two systems could be attributed to the delicate structure of the commuted MoS_2 as a primary support, which led to differences in the specific surface areas of these catalysts (the surface area (BET) of the AHS-G catalyst was $110\,m^2/g$, and therefore, it is much higher than that of the C-AHS-G-II catalyst, which was less than $5\,m^2/g$). Thus, commuted MoS_2 may not be a recommended preliminary support for Co promotion. Taken together, these results indicate that the selectivity profiles of the catalysts need to be analyzed in a greater detail to develop a better view of the changes in the activity following the impregnation of the catalysts with Co. In this study, the selectivity was estimated according to the determined ratio of the HYD to the DDS rate constant, that is, k_{HYD}/k_{DDS}. One should be careful when considering the catalytic selectivity according to the product distributions because the selectivity in this case underwent significant mobile changes depending on the level of conversion (Figure 4). When the bare MoS_2 catalysts were used, the HDS of DBT proceeded preferentially by the HYD route, as shown in Table 1. The relatively high selectivity observed in this case for the products resulting from the HYD route is contrary to the common view that DBT HDS proceeds mostly by a DDS route. These results therefore highlight the flexibility of the HDS reaction, in the sense that it can proceed by these routes with no path limitation to either of them.

The impregnation of Co onto the ATS catalyst (C-ATS series of catalysts) led to a significant development in the overall HDS activity. One may notice the obvious pioneer Co promotion in the HDS of these catalysts, especially for the C-ATS-I catalyst. This catalyst showed a remarkable tendency towards the DDS pathway (ca. 1~2-fold higher activity towards the DDS route over the HYD route (Table 1)) for the HDS of DBT, whilst the bare MoS_2 catalysts exhibited a preference for the HYD pathway. The change in the activity of the Co-containing catalyst was accompanied by an obvious shift in its selectivity towards the BP product. This tendency towards the DDS pathway has been reported previously for the conventional MoS_2-based catalysts as a result of the impregnation of a Co promoter [5]. These results therefore demonstrate that the HYD and DDS reaction pathways occur in parallel and can be independently manipulated. Notably, all Co catalysts prepared in this study underwent the HDS reaction preferentially by the DDS pathway, and a clear correlation between the selectivity and activity of the HDS reactions could be established (Figure 6(a)). It is noteworthy mentioning that the trend in the selectivity of these reactions was in accordance with similar reports from the literature for related systems [3]. Figure 6(b) displays the association between the reaction activity of the C-ATS catalysts and the Co/(Co + Mo) ratio. It is obvious that the C-ATS-I catalyst, which had 0.05 Co/(Co + Mo) ratio, exhibited the highest HDS activity of all of the catalysts belonging to this series. Interestingly, such ratio is far lower than those commonly reported for conventional Co-Mo-based catalysts, where the average is

TABLE 1: Activities of the catalysts towards the HDS of dibenzothiophene.

Catalyst	Co/(Co + Mo) atomic ratio	Level of H₂S[a]	HDS activity (sec⁻¹·g·cat.⁻¹·10⁻⁴)	Individual activity (sec⁻¹·g·cat.⁻¹·10⁻⁴)			Products at 50% conversion level		
				k_1^b	k_2^c	k_2/k_1	BP	CB	H4-DBT
AHS-G	—	L	24	10	14	1.4	23	25	2
		H	77	15	62	4.0	8	29	13
ATS	—	L	9	3.9	4.8	1.2	23	24	3
		H	38	6.7	31.4	4.7	8	28	14
CoMo/Al₂O₃[e]	0.27	L	211[d]	191	20	0.1	46	3	1
		H	128[d]	112	16	0.14	43	6	1
C-AHSG-I	0.02	H	38	21	17	0.8	29	16	5
C-AHSG-II	0.23	L	17	14	3	0.2	43	6	1
		H	33	20	13	0.7	28	17	5
C-ATS-I	0.05	L	184	127	57	0.4	37	12	1
		H	352	194	158	0.8	27	18	5
MPC	0.40	L	1.8[f]	1.6	0.2	0.12	43	6	1
	—	H	2.6[f]	1.8	0.8	0.4	37	12	1
CS	—	H	1.6[f]	0.8	0.8	1	24	22	4

[a]Level of H₂S: low (L) and high (H) denote pressures of ca. 3 and 20 kPa, respectively. [b]Apparent rate constant for the direct desulfurization route, 1st order. [c]Apparent rate constant for the hydrogenation route, 1st order. [d]Normalized to the metal sulfide content. [e]Commercial catalyst Co, 3.2%, and Mo, 13.7%, in terms of the weight percent. [f]Pseudo-zero-order rate constant, $\times 10^{17}$ molecule/(g·S). The uncertainty was within ±5–10%.

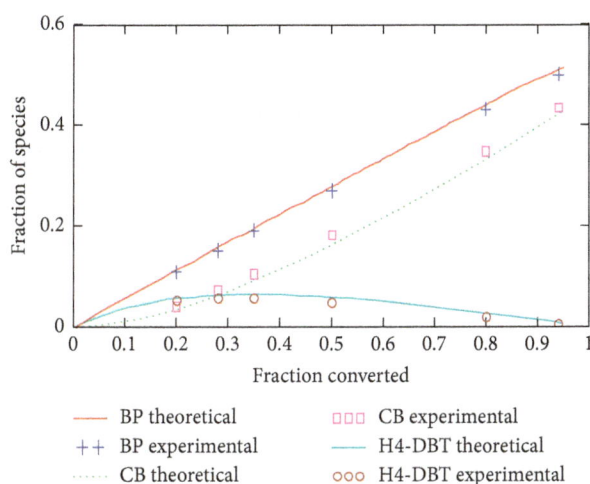

FIGURE 4: Transformation of dibenzothiophene over the C-ATS-I catalyst in the presence of H₂S.

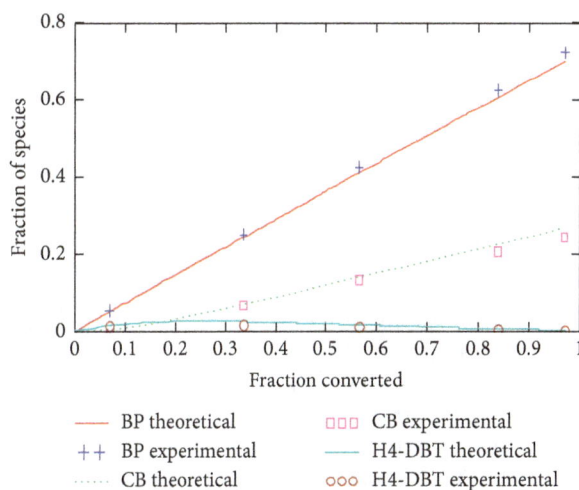

FIGURE 5: Transformation of dibenzothiophene over the C-ATS-I catalyst in the absence of H₂S.

generally around 0.3. However, the present results are generally in a trend with other studies where low Co concentration catalysts were reported to be active for hydrogenation and hydrogenolysis reactions [22, 23]. The selectivity for the HYD reaction was clearly effected by the amount of Co incorporated into the existing catalyst. The results in Figure 6(b) show that the tendency towards the HYD reaction decreased with the increase in Co content. These results therefore may indicate that only a limited number of active Mo and S sites can be further activated by the inclusion of Co. These sites were assumed to be in the type II CoMoS phase.

The active sites of HDS catalysts are coordinately unsaturated sites. The incorporation of promoters such as Co and Ni can lead to improvements in the supply of spillover hydrogen. This may lead to an increase in the DDS and HYD activities [9, 12, 15, 16, 21]. It is therefore envisaged that

these catalysts must contain different types of active sites to allow them to catalyze multiple reactions in parallel. The Co atoms incorporated into the MoS₂ catalysts could occupy the active sites on their surface. The enlargement in the catalytic activity of the Co-containing MoS₂ catalysts (i.e., C-ATS catalysts) compared with the bare MoS₂ catalyst could be attributed to the occurrence of a large synergistic effect (Table 2) between Co and Mo atoms. A synergistic effect of this type could potentially result in the formation of two different Co-based active sites for the DDS and HYD reactions. The Co atoms could therefore occupy different positions on the surface (i.e., edges and corners) of the unsaturated MoS₂ sites, resulting in new sites. However, not all of these sites would be accessible for the HDS reaction of the substrate. The results therefore imply that the higher activity of these compounds could be attributed to an increase in the effectiveness of the active sites. The corner-edge

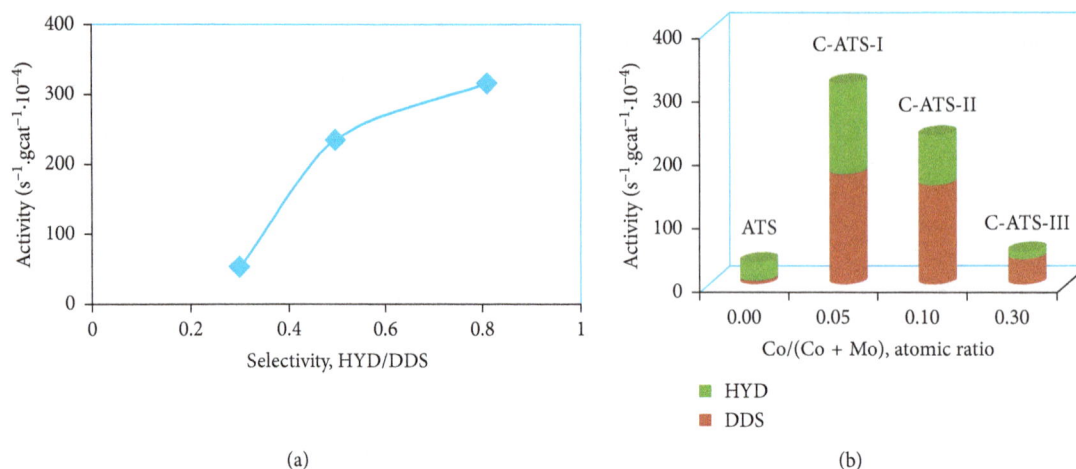

(a)

(b)

FIGURE 6: Effect of the Co content on the HDS activity (a) and the relationship between the activity and the selectivity (b).

TABLE 2: Changes in the selectivity of the HDS reaction for the MoS$_2$-based catalysts caused by Co and H$_2$S.

Reaction condition		Comparison of the average fold change (increase or decreases) in the activity of the ATS catalyst due to the inclusion of Co		
		Absence of H$_2$S		Presence of H$_2$S
	Catalyst	C-ATS-I		C-ATS-I
DBT	DDS	+33		+29
	HYD	+12		+5
		Comparison of the average fold change (increase or decrease) in the activity caused by H$_2$S		
	Catalyst	C-AHSG-II	AHS-G	C-ATS-I	ATS	CoMo/Al$_2$O$_3$
DBT	DDS	+1.4	+1.5	+1.5	+1.7	−1.7
	HYD	+4.3	+4.4	+2.8	+6.5	−1.3

+, increase; −, decrease.

model [24–26] can be used to explain the differences in the selectivity profiles of promoted and unpromoted MoS$_2$ catalysts for the DDS route. The Mo atoms at the corner sites of the stacked cluster would probably be unsaturated because of the steric configuration of the hexagonal crystals in the catalyst. There could also be a high proportion of coordinately unsaturated sites in this case. It can be concluded that the selectivity of the catalysts for the DDS reaction correlates highly with the number of MoS$_2$ layers. Nikulshin et al. [27, 28] in studies for the HDS over CoMo-based catalysts showed that the (Co/Mo) edge ratio is directly proportional with the catalyst HDS activity. They further revealed that, with increasing the cobalt content, the HDS selectivity turns slightly towards HYD but still the predominant route is the DDS. To date, our preliminary results have shown that the low content of the Co-promoted MoS$_2$ catalyst, especially the C-ATS catalysts, and led to an increase in the overall activity. On the contrary, MPC catalyst exhibited low HDS performance. Furthermore, the reaction followed pseudo-zero-order kinetics (Figures S1–S3).

3.2. Impact of H$_2$S. The H$_2$S matrix could also have a significant impact on the behaviors of the catalysts developed in this study for the HDS reaction. The activities and selectivities of the current catalysts towards DBT HDS varied considerably depending on the existence of H$_2$S in the reaction medium. The parameters derived from the kinetic treatments of these experimental results are presented in Table 1. Several clear and interesting trends can be observed. For example, the presence of H$_2$S resulted in a slight rise in the formation of BP (DDS route product). However, the inclusion of H$_2$S also drove to a remarkable positive change in the tendency of the reaction towards the HYD route and the production of CB. Table 2 shows that the presence of H$_2$S led to a rise in the activity of the present catalysts by increasing the rates of the HYD and DDS reactions by approximately 2.5- to 4-fold and 1.4- to 1.7-fold, respectively. The promotional trends of H$_2$S towards the DDS reaction were very similar for all of the prepared catalysts. For instance, the enhancement in the activity resulting from the inclusion of H$_2$S was much more pronounced for the HYD reaction route, especially in case of the bare MoS$_2$ catalyst. The HYD pathway contributed to more than 80% of DBT HDS over AHS-G and/or ATS catalysts. H$_2$S influence on the HDS reaction has been discussed extensively in various studies [29–35], which consensually suggested that H$_2$S suppresses the DDS route, whilst having very little impact on the HYD route. In contrast, several research groups, including our own, have reported that some transition metal sulfide catalysts can effectively promote the HYD reaction when they are carried out in the presence of H$_2$S [36–38]. Guernalec et al. [38] correlated this positive behavior to the increase of the -SH concentration on the catalyst surface. However, our obtained results represent a unique catalytic trend in which the use of H_2S may promote the HDS reaction over unsupported Co-promoted MoS_2 catalysts. The increase observed in the catalytic activity in presence of H_2S reported here is contrary to the well-known observation that H_2S severely inhibits the HDS reactions that are conducted over conventional catalysts. This is evident from the data highlighting the inhibition of the commercial $CoMo/Al_2O_3$ catalyst by H_2S presented in Tables 1 and 2. Interestingly, this reaction (impacted by H_2S) was found to be reversible indicating that H_2S may have caused no permanent changes to the structure of the catalyst. Taken together, these results demonstrate that our collective understanding of H_2S influence on the selectivity of the HDS reaction is rather incomplete. Crystallite sizes in the extent of 4-5 nm for the MoS_2 crystallites appeared to exert a specific catalytic performance, which definitely depends on how such catalyst is synthesized. The nature of substrate, catalyst, and reaction conditions therefore appear to be of considerable importance for quantitative evaluation of this phenomenon.

3.3. Synergy Effect.

The results of this study clearly show that the incorporation of Co had a decisive influence on the catalytic performance of the unsupported MoS_2 catalysts (series C-ATS). It is noteworthy mentioning that the results are consistent with the well-known promoting effects of conventional supported MoS_2 catalysts. Taken together, these results suggest that the HDS reaction resulted from divergent active sites. The synergy between the Co and Mo phases appeared to be on line with that claimed for the well-known Co-Mo-S type II materials, such as the supported Co-Mo catalysts [10], which are assumed to be responsible for catalyzing HDS reactions. All C-ATS catalysts prepared in the present study exhibited some degree of synergy between the Co and Mo atoms. This synergistic interaction enhanced the HDS reaction by promoting the DDS route to a higher extent than that of the HYD route. This result indicated that the Co-based sites were more effective for the C-S bond scission. Thus, it can be stated that the incorporation of Co not only improved the activity of the catalyst but also modified the selectivity along the two pathways. Figure 7 shows the relation between the synergy factor (SF, activity ratio of (Co-MoS_2/MoS_2)) and the Co-added ratio. Calculations based on the density functional theory (DFT) for such system have predicted that the sulfur and the metal edges of the catalyst may conduct the DDS reactions [34, 35]. These DFT studies have also shown that the Co atoms can be accommodated on the rims of the MoS_2 crystallites especially on the rims of the sulfur atoms.

It has been suggested that the HYD reactions most likely occur on the brim sites and the metal edges. In the present study, CB was isolated as the major product from the reactions catalyzed by the bare MoS_2 catalyst, which indicated that the HYD pathway was favored under these conditions. However, Table 1 shows that the share from the DDS route was higher than that of the HYD route when the HDS reaction was conducted over the C-ATS-I catalyst. The overall HDS activities of the ATS and C-ATS catalysts in the presence of H_2S were much higher than those in case of the

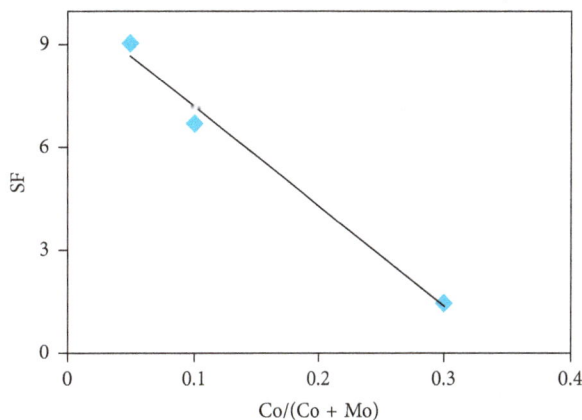

FIGURE 7: Cobalt content as a function of synergy promotion.

absence of H_2S. The major enhancement was in the HYD reactions of DBT. The MoS_2 and Co-containing MoS_2 catalysts contain acidic sites on their surfaces of two different types, including (i) Lewis acidic sites, that is, the sulfur vacancies on the Mo and Co atoms, and (ii) Brønsted acidic sites. The latter of these two sites would contain the -SH and -SH_2 groups. The -SH sites may act to eliminate sulfur from DBT and the HYD reactions (under certain circumstances), whilst the -SH_2 sites would favorably be involved in hydrogenation reactions. The involvement of these active sites in the HDS reaction would be dependent on their availability with the appropriate geometry to interact with the substrate [8, 35]. Given that the Co-S bond is relatively weaker than the Mo-S bond, the Co sites would be more acidic than the Mo sites [39]. This diversity in the acidity of the two sites could explain the higher DDS activity of the Co-containing MoS_2 catalysts compared with the unpromoted MoS_2 catalyst. The cleavage of the S-C bonds is most likely to occur with greater ease over the Co-promoted catalyst. This suggestion is also consistent with the results obtained for the commercial $CoMo/Al_2O_3$ catalyst (Table 1). It has therefore been suggested that the inclusion of H_2S may result in new active sites through interactions with vacant sulfur sites on the MoS_2 and Co-containing MoS_2 catalysts. The quality of these sites for promoting or suppressing the HYD and hydrogenolysis reactions would probably be dependent not only on their configuration and concentration but also on the structure of the substrate layers. The data shown in Table 2 provide a summary of the effects of Co and H_2S on the HDS reaction. The main points that could be drawn from the data are as follows:

(1) The incorporation of Co (low concentration range) into the unsupported MoS_2 catalyst led to a significant activity boost in the DBT HDS reaction, as well as resistance to the inhibition phenomena caused by H_2S.

(2) The Co atoms preferentially may occupy the edge sites of the MoS_2 catalyst [34, 35].

(3) H_2S inclusion in the HDS reaction feedstock led to an enhancement in the hydrogenation activities of all the studied catalysts.

(4) The incorporation of Co into the MoS_2 catalyst resulted in a drastic selectivity change towards the DDS reaction, which increased by an order of magnitude over the HYD reaction.

(5) The presumable $Co-SH_2$ active sites may be much more effective for the HDS reaction than the $Mo-SH_2$ active sites.

(6) Mo may, however slightly, promote the Co_9S_8 catalyst in the HDS reactions.

4. Conclusions

The catalytic results of the comminuted MoS_2, ATS, and Co-containing MoS_2 catalysts for the reaction of DBT showed that the nature of the preliminary MoS_2 support had a critical role on the activity of the Co-promoted one. The hydrogenation route was most predominant in the reaction over the bare MoS_2. H_2S inclusion enhanced the rates of both possible reaction routes but promoted the HYD pathway to a much greater extent. The activity of the MoS_2 catalysts increased significantly when the Co promoter was added. All catalysts synthesized in the current study exhibited a close trend towards the influence of H_2S in the HDS reaction. The C-ATS and C-AHSG catalysts showed a strong preference for the DDS pathways, with BP being produced as a major product. The boost in the activity observed with the Co-containing catalyst (C-ATS) was attributed to the obvious enhancement in the DDS route and to the significant increase in the HYD route reactions. The results from the kinetic analysis of the HDS reactions over the MoS_2 and MoS_2 promoted by Co catalysts suggested the existence of different discrete active sites, which were dependent on the identity of the promoter synergy and the reaction matrix. This study therefore provides important information for controlling the selectivity and activity characteristics of MoS_2-based catalysts by tailoring their properties through synthetic manipulation and promoter ratio and adapting proper reaction matrices.

Conflicts of Interest

The authors declare that they have no conflicts of interest.

Acknowledgments

This work was supported by Kyushu University, Japan, and Mansoura University, Egypt.

References

[1] I. Mochida and K.-H. Choi, "An overview of hydrodesulfurization and hydrodenitrogenation," *Journal of the Japan Petroleum Institute*, vol. 47, no. 3, pp. 145–163, 2004.

[2] D. D. Whitehurst, T. Isoda, and I. Mochida, "Present state of the art and future challenges in the hydrodesulfurization of polyaromatic sulfur compounds," *Advances in Catalysis*, vol. 42, pp. 345–471, 1998.

[3] Z. Wu, V. M. L. Whiffen, W. Zhu, D. Wang, and K. J. Smith, "Effect of annealing temperature on co–MoS_2 nanosheets for hydrodesulfurization of dibenzothiophene," *Catalysis Letters*, vol. 144, no. 2, pp. 261–267, 2014.

[4] J. Ancheyta-Juárez, E. Aguilar-Rodriguez, and D. Salazar-Sotelo, "Effect of hydrogen sulfide on the hydrotreating of middle distillates over Co–Mo/Al_2O_3 catalyst," *Applied Catalysis A: General*, vol. 183, no. 2, pp. 265–272, 1999.

[5] C. Song, "An overview of new approaches to deep desulfurization for ultra-clean gasoline, diesel fuel and jet fuel," *Catalysis Today*, vol. 86, no. 1-4, pp. 211–263, 2003.

[6] M. L. Vrinat and L. D. Mourgues, "On the role of cobalt in sulfided unsupported Co-Mo hydrodesulfurization catalysts: kinetic studies and scanning electron microscopic observations," *Applied Catalysis*, vol. 5, no. 5, pp. 43–57, 1983.

[7] G. Meijburg, *Production of Ultra Low-Sulfur Diesel in Hydrocracking with the Latest and Future Generation Catalysts Courier 46*, Akzo Nobel, Amsterdam, Netherlands, 2001.

[8] M. Egorova and R. Prins, "Competitive hydrodesulfurization of 4,6-dimethyldibenzothiophene, hydrodenitrogenation of 2-methylpyridine, and hydrogenation of naphthalene over sulfided $NiMo/\gamma$-Al_2O_3," *Journal of Catalysis*, vol. 224, no. 2, pp. 278–287, 2004.

[9] H. Topsoe, B. S. Clausen, and F. E. Massoth, HydrotreatingCatalysis (Catalysis-cience and Technology), Vol. 11, Springer, New York, NY, USA, 1996.

[10] Z. Vit, D. Gulková, L. Kaluza, and M. Zdrazil, "Synergetic effects of Pt and Ru added to Mo/AlO sulfide catalyst in simultaneous hydrodesulfurization of thiophene and hydrogenation of cyclohexene," *Journal of Catalysis*, vol. 232, no. 2, pp. 447–455, 2005.

[11] E. J. M. Hensen, H. J. A. Brans, G. M. H. J. Lardinois, V. H. J. De Beer, J. A. R. Van Veen, and R. A. van Santen, "Periodic trends in hydrotreating catalysis: thiophene hydrodesulfurization over carbon-supported 4d transition metal sulfides," *Journal of Catalysis*, vol. 192, no. 1, pp. 98–107, 2000.

[12] A. Ishihara, F. Dumeignil, J. Lee, K. Mitsuhashi, E. W. Qian, and T. Kabe, "Hydrodesulfurization of sulfur-containing polyaromatic compounds in light gas oil using noble metal catalysts," *Applied Catalysis A: General*, vol. 289, no. 2, pp. 163–173, 2005.

[13] M. Egorova and R. Prins, "Hydrodesulfurization of dibenzothiophene and 4,6-dimethyldibenzothiophene over sulfided $NiMo/\gamma$-Al_2O_3, $CoMo/\gamma$-Al_2O_3, and Mo/γ-Al_2O_3 catalysts," *Journal of Catalysis*, vol. 225, no. 2, pp. 417–427, 2004.

[14] M. Kouzu, K. Uchida, Y. Kurili, and F. Ikazaki, "Microcrystalline molybdenum sulfide prepared by mechanical milling as an unsupported model catalyst for the hydrodesulfurization of diesel fuel," *Applied Catalysis A: General*, vol. 276, no. 1-2, pp. 241–249, 2004.

[15] A. Niquille-Röthlisberger and R. Prins, "Hydrodesulfurization of 4,6-dimethyldibenzothiophene and dibenzothiophene over alumina-supported Pt, Pd, and Pt-Pd catalysts," *Journal of Catalysis*, vol. 242, no. 1, pp. 207–216, 2006.

[16] C. T. Tye and K. J. Smith, "Hydrodesulfurization of dibenzothiophene over exfoliated MoS_2 catalyst," *Catalysis Today*, vol. 116, no. 4, pp. 461–468, 2006.

[17] H. Nava, C. Ornelas, A. Aguilar, G. Berhault, S. Fuentes, and G. Alonso, "Cobalt-molybdenum sulfide catalysts prepared by

in situ activation of bimetallic (Co-Mo) alkylthiomolybdates," *Catalysis Letters*, vol. 86, no. 4, pp. 257–261, 2003.

[18] M. C. Kerby, T. F. Degnan, D. O. Marler, and J. S. Beck, "Advanced catalyst technology and applications for high quality fuels and lubricants," *Catalysis Today*, vol. 104, no. 1, pp. 55–63, 2005.

[19] H. Farag and H. Al-Megren, "Textural characterizations and catalytic properties of quasispherical nanosized molybdenum disulfide," *Journal of Colloid and Interface Science*, vol. 332, no. 2, pp. 425–431, 2009.

[20] International Centre for Diffraction Data, *JCPDS Powder Diffraction File*, International Centre for Diffraction Data, Swarthmore, PA, USA, 2000.

[21] H. Farag, "Hydrodesulfurization of dibenzothiophene and 4,6-dimethyldibenzothiophene over NiMo and CoMo sulfide catalysts: kinetic modeling approach for estimating selectivity," *Journal of Colloid and Interface Science*, vol. 348, no. 1, pp. 219–226, 2010.

[22] G. Delvaux, P. Grange, and B. Delmon, "X-ray photoelectron spectroscopic study of unsupported cobalt-molybdenum sulfide catalysts," *Journal of Catalysis*, vol. 56, no. 1, pp. 99–109, 1979.

[23] P. Grange, "Catalytic hydrodesulfurization," *Catalysis Reviews*, vol. 21, no. 1, pp. 135–181, 1980.

[24] M. Daage and R. R. Chianelli, "Structure-function relations in molybdenum sulfide catalysts: the "rim-edge" model," *Journal of Catalysis*, vol. 149, no. 49, pp. 414–427, 1994.

[25] F. E. Massoth, G. Muralidhar, and J. Shabtai, "Catalytic functionalities of supported sulfides II. Effect of support on Mo dispersion," *Journal of Catalysis*, vol. 85, no. 1, pp. 53–62, 1984.

[26] B. Scheffer, P. J. Mangnus, and J. A. Moulijn, "A temperature-programmed sulfiding study of NiO_3/Al_2O_3 catalysts," *Journal of Catalysis*, vol. 121, no. 1, pp. 18–30, 1990.

[27] P. A. Nikulshin, D. I. Ishutenko, A. A. Mozhaev, K. I. Maslakov, and A. A. Pimerzin, "Effects of composition and morphology of active phase of $CoMo/Al_2O_3$ catalysts prepared using Co_2Mo10–heteropolyacid and chelating agents on their catalytic properties in HDS and HYD reactions," *Journal of Catalysis*, vol. 312, pp. 152–169, 2014.

[28] A. Pimerzin, A. Mozhaev, A. Varakin, K. Maslakov, and P. Nikulshin, "Comparison of citric acid and glycol effects on the state of active phase species and catalytic properties of $CoPMo/Al_2O_3$ hydrotreating catalysts," *Applied Catalysis B: Environmental*, vol. 205, pp. 93–103, 2017.

[29] B. Vogelaar, N. Kagami, A. van Langeveld, S. Eijsbouts, and J. A. Moulijn, "Active sites and activity in HDS catalysis: the effect of H2 and H2S partial pressure," *American Chemical Society, Division of Fuel Chemistry*, vol. 48, no. 2, pp. 548-549, 2003.

[30] N. Hermann, M. Brorson, and H. Topsøe, "Activities of unsupported second transition series metal sulfides for hydrodesulfurization of sterically hindered 4,6-dimethyldibenzothiophene and of unsubstituted dibenzothiophene," *Catalysis Letters*, vol. 65, no. 4, pp. 169–174, 2000.

[31] T. Kabe, Y. Aoyama, D. Wang et al., "Effects of H2S on hydrodesulfurization of dibenzothiophene and 4,6-dimethyldibenzothiophene on alumina-supported NiMo and NiW catalysts," *Applied Catalysis A: General*, vol. 209, no. 1-2, pp. 237–247, 2001.

[32] H. M. Nakamura, K. Amemiya, and K. Ishida, "Inhibition effect of hydrogen sulfide and ammonia on $NiMo/Al_2O_3$, $CoMo/Al_2O_3$, $NiCoMo/Al_2O_3$ catalysts in hydrodesulfurization of dibenzothiophene and 4,6-dimethyldibenzothiophene," *Journal of the Japan Petroleum Institute*, vol. 48, no. 5, pp. 281–289, 2005.

[33] V. S. Rabarihoela-Rakotovao, G. Brunet, F. Perot, and F. Diehl, "Effect of H2S partial pressure on the HDS of dibenzothiophene and 4,6-dimethyldibenzothiophene over sulfided $NiMoP/Al_2O_3$ and $CoMoP/Al_2O_3$ catalysts," *Applied Catalysis A: General*, vol. 306, pp. 34–44, 2006.

[34] E. krebs, B. Silvi, and P. Raybaud, "Mixed sites and promoter segregation: a DFT study of the manifestation of Le Chatelier's principle for the Co(Ni)MoS active phase in reaction conditions," *Catalysis Today*, vol. 130, no. 1, pp. 160–169, 2008.

[35] Y. Sun and R. Prins, "Mechanistic studies and kinetics of the hydrodesulfurization of dibenzothiophene on $Co-MoS_2/\gamma-Al_2O_3$," *Journal of Catalysis*, vol. 267, no. 2, pp. 193–201, 2009.

[36] G. Nadege, C. Tivadar, R. Pascal, G. Christophe, and V. Michel, "Influence of H2S on the hydrogenation activity of relevant transition metal sulfides," *Catalysis Today*, vol. 98, no. 1-2, pp. 61–66, 2004.

[37] H. Farag, K. Sakanishi, M. Kouzu, A. Matsumura, Y. Sugimoto, and I. Salto, "Dual character of H2S as promoter and inhibitor for hydrodesulfurization of dibenzothiophene," *Catalysis Communications*, vol. 4, no. 7, pp. 321–326, 2003.

[38] N. Guernalec, C. Geantet, P. Raybaud, T. Cseri, M. Aouine, and M. Vrinat, "Dual effect of H2S on volcano curves in hydrotreating sulfide catalysis," *Oil and Gas Science and Technology- Revue de l'IFP*, vol. 61, no. 4, pp. 515–525, 2006.

[39] H. Toulhoat, P. Raybaud, S. Kasztelan, G. Kresse, and J. Hafner, "Transition metals to sulfur binding energies relationship to catalytic activities in HDS: back to sabatier with first principle calculations"," *Catalysis Today*, vol. 50, no. 3-4, pp. 629–636, 1999.

21

A Fusion Water Quality Soft-Sensing Method Based on WASP Model and Its Application in Water Eutrophication Evaluation

Xiaoyi Wang, Jie Jia, Tingli Su, Zhiyao Zhao ⓘ, Jiping Xu, and Li Wang ⓘ

School of Computer and Information Engineering, Beijing Technology and Business University, Beijing 100048, China

Correspondence should be addressed to Zhiyao Zhao; zhaozy@btbu.edu.cn

Academic Editor: Carlos A Martínez-Huitle

Water environment protection is of great significance for both economic development and improvement of people's livelihood, where modeling of water environment evolution is indispensable in water quality analysis. However, many water quality indexes related to water quality model cannot be measured online, and some model parameters always vary among different water areas. Thus, this paper proposes a water quality soft-sensing method based on the water quality mechanism model to simulate evolution of water quality indexes online, where unscented Kalman filter is utilized to estimate model parameters. Furthermore, a modified fuzzy comprehensive evaluation method is presented to evaluate the level of water eutrophication condition. Finally, the water quality data collected from Taihu Lake and Beihai Lake are used to validate the effectiveness and generality of the proposed method. The results show that the proposed soft-sensing method is able to describe the variation of related water quality indexes, with better accuracy compared to nonlinear least squares based method and traditional trial-and-error based method. On this basis, the water eutrophication condition can be also accurately evaluated.

1. Introduction

With the rapid development of modern society, production of industrial and sanitary sewage is daily increasing, and eutrophication phenomenon of lakes and reservoirs is becoming much more serious [1, 2]. Generally, the occurrence of eutrophication is related to excessive nitrogen, phosphorus, and other inorganic nutrients in water, where nitrogen and phosphorus are the main reasons accounting for the eutrophication of slow flow water, such as lakes, reservoirs, and bays [3–5]. Currently, eutrophication phenomenon exists in 54% of the lakes of Asia-Pacific region [6]. Therefore, how to economically and effectively handle the eutrophication problem has become an urgent priority.

In order to timely evaluate or predict the eutrophication condition, the variation of water quality indexes should be timely measured or learned [7, 8]. However, many water quality indexes, such as biochemical oxygen demand (BOD) and total nitrogen (TN), cannot be measured online [9]. Thus, the manner of soft-sensing is introduced to overcome this limitation in this paper. Soft-sensing is to establish a mathematical relation model between easily measured process variables and difficultly measured process variables based on mechanism analysis and sensor data mining [10]. The existing soft-sensing modeling approach can be classified into three types: mechanism modeling, identification modeling, and artificial intelligence-based modeling [11–13]. The mechanism modeling approach is to obtain a mathematical expression based on the analysis of the system's internal relations, which adopts the basic physical and chemical laws, such as material, energy, or momentum conservation relation [14]. The identification modeling approach is to establish a mathematical model based on the information of system input and output by certain parameter identification, filtering, or regression analysis methods, without understanding the mechanism of the dynamic process [15]. The artificial intelligence-based modeling approach is to get an underlying model of the real-world system or a portion of system based on artificial intelligent methods [16]. In addition, there also exist fusion methods derived from a combination of the above approaches; that is, the mechanism modeling approach is utilized to describe partial behavior of the studied system with known mechanism, and identification modeling approach or artificial intelligence-based approach is used to handle the remaining part.

Water Quality Analysis Simulation Program (WASP) mechanism model is a comprehensive water quality model that can be used to interpret the process of natural or artificial water quality deterioration [17]. It can simulate migration and transformation of conventional water quality indexes (including dissolved oxygen (DO), BOD, and nutrients) and toxic contaminants (including organic chemicals, metals, and sediment) in water [18]. Currently, WASP model has been widely applied to different water areas, such as Mobile Bay [19], Murderkill River [20], Lake Michigan [21], and Songhua River [22]. However, WASP also has a limitation. That is, some model parameters vary among different water areas, and their values are always determined by trial-and-error method [23]. This is insufficient for accurately modeling the water quality variation of a specific water area.

Therefore, this paper builds a fusion water quality soft-sensing method, where the WASP model is employed as a soft-sensing method and its unknown parameters are estimated by the unscented Kalman filter (UKF) [24]. Then, the variations of DO, BOD, nitrate nitrogen (NO_3-N, related to TN), ammonia nitrogen (NH_3-N), phytoplankton carbon (Phyt, related to chlorophyll_a (Chl_a)), and so forth can be simulated by the fusion water quality soft-sensing method. On this basis, a modified fuzzy comprehensive evaluation method is presented to evaluate the eutrophication condition of the rivers and lakes, combining both the simulated values of DO, BOD, TN, and Chl_a from the soft-sensing method and the online measured values of transparency (SD) and total phosphorus (TP) [25]. Finally, by taking Taihu Lake and Beihai Lake as examples, the effectiveness and generality of the proposed method are validated and the water eutrophication condition is evaluated. Comparative studies are also presented and discussed.

The remainder of this paper is organized as follows. Section 2 presents the methodology of the WASP based water quality soft-sensing method, where a simplified WASP model is presented and the procedure of unknown parameters estimation by UKF is listed. Section 3 introduces the modified fuzzy comprehensive evaluation method, where the modified methods of selecting water quality indexes and calculation of the corresponding weight are presented. Section 4 presents two case studies of Taihu Lake and Beihai Lake to validate the effectiveness and generality of the proposed fusion soft-sensing method and the water eutrophication evaluation method. Section 5 gives the conclusion and indicates future development.

2. Fusion Water Quality Soft-Sensing Method

2.1. Dynamic Model of Water Quality Based on WASP. Eutrophication module (shorten as EUTRO) is an essential part of the WASP model. This module describes the dynamic behavior of water quality indexes including DO, BOD, Phyt, NO_3-N, NH_3-N, organic nitrogen (ON), and organic phosphorus (OP). The interacting relations of them can be represented by four reaction systems, namely, phytoplankton kinetics, phosphorus cycle, nitrogen cycle, and DO balance.

In the EUTRO module, a simplified dynamic process of seven water quality indexes are presented as [18]

$$
\begin{aligned}
\dot{C}_{DO} &= k_1\left(C_{NO_3\text{-}N} - C_{DO}\right) - 1.047k_2\left(\frac{C_{DO}}{0.5 + C_{DO}}\right)C_{BOD} \\
&\quad - 4.94k_3 - 5.4k_4 - \left(14.53k_5 + 0.35\right)C_{Phyt} \\
\dot{C}_{BOD} &= 2.67k_6 C_{Phyt} - 1.047k_2\left(\frac{C_{DO}}{0.5 + C_{DO}}\right)C_{BOD} \\
&\quad - 2.99k_7\left(\frac{0.1}{0.1 + C_{DO}}\right)C_{NO_3\text{-}N} \\
\dot{C}_{Phyt} &= \left[k_5 - \left(0.125 + k_6\right) - 0.02\right]C_{Phyt} \\
\dot{C}_{NO_3\text{-}N} &= 1.08k_3\left(\frac{C_{DO}}{k_8 + C_{DO}}\right)C_{NH_3\text{-}N} + 0.645k_5 C_{Phyt} \\
&\quad - 1.045k_7\left(\frac{0.1}{0.1 + C_{NH_3\text{-}N}}\right)C_{BOD} \\
\dot{C}_{NH_3\text{-}N} &= 0.075\left(0.125 + k_6\right)C_{Phyt} + 1.08k_9\left(\frac{C_{Phyt}}{1 + C_{Phyt}}\right)C_{ON} \\
&\quad + 0.645k_5 C_{Phyt} - 1.08k_3\left(\frac{C_{DO}}{k_8 + C_{DO}}\right)C_{NH_3\text{-}N} \\
\dot{C}_{ON} &= 0.075\left(0.125 + k_6\right)C_{Phyt} \\
&\quad + 1.08k_9\left(\frac{C_{Phyt}}{1 + C_{Phyt}}\right)C_{OP} \\
\dot{C}_{OP} &= 0.011\left(0.125 + k_6\right)C_{Phyt} \\
&\quad - 1.08k_{10}\left(\frac{C_{Phyt}}{1 + C_{Phyt}}\right)C_{OP},
\end{aligned}
\tag{1}
$$

$$
\underbrace{\hspace{8cm}}_{\dot{x} = f(x)}
$$

where we let $x = (C_{DO}, C_{BOD}, C_{Phyt}, C_{NO_3\text{-}N}, C_{NH_3\text{-}N}, C_{ON}, C_{OP})^T$ represent concentrations of DO, BOD, Phyt, NO_3-N, NH_3-N, ON, and OP, respectively. In (1), the chemical meaning of unknown model parameters $\{k_1, k_2, \ldots, k_{10}\}$ is shown in Notations [18].

Let $\theta = (k_1, k_2, k_3, k_4, k_5, k_6, k_7, k_8, k_9, k_{10})^T$. Here, assume that the unknown parameters are piecewise-constant. Then, the process equation can be written as

$$
\begin{bmatrix} \dot{x} \\ \dot{\theta} \end{bmatrix} = \begin{bmatrix} f(x) \\ 0_{10 \times 1} \end{bmatrix}.
\tag{2}
$$

Further let $X = [x^T, \theta^T]^T$ and the process noise item be added. An augmented process equation is obtained as [26]

$$
\dot{X} = \begin{bmatrix} f(x) \\ 0_{10 \times 1} \end{bmatrix} + w,
\tag{3}
$$

where w is the process noise, satisfying that $w \sim N(0, Q)$, Q is the covariance matrix. Then, the observation equation is set as follows:

$$
y = \begin{bmatrix} h & 0 \end{bmatrix} \begin{bmatrix} x \\ \theta \end{bmatrix} + v,
\tag{4}
$$

where h is the observation matrix, describing the mapping from state indexes to observations; v is the measurement noise, satisfying that $v \sim N(0, R)$; R is the covariance matrix.

Up to now, we obtain a continuous-time dynamic model of water quality indexes as follows:

$$\dot{X} = \begin{bmatrix} f(x) \\ 0_{10 \times 1} \end{bmatrix} + w,$$

$$y = [h \quad 0] \begin{bmatrix} x \\ \theta \end{bmatrix} + v. \qquad (5)$$

To obtain the discrete-time model, the fourth-order Runge-Kutta is utilized, where the step length is l. The formula is as follows:

$$K_1 = f(X_k),$$

$$K_2 = f\left(X_k + \frac{l}{2}K_1\right),$$

$$K_3 = f\left(X_k + \frac{l}{2}K_2\right), \qquad (6)$$

$$K_4 = f(X_k + lK_3),$$

$$X_{k+1} = X_k + \frac{l}{6}(K_1 + 2K_2 + 2K_3 + K_4).$$

Then, (5) is discretized as

$$X(k) = f(X(k-1)) + w(k-1),$$

$$y(k) = [h \quad 0] X(k) + v(k). \qquad (7)$$

2.2. Unknown Parameter Estimation Based on UKF. In practice, UKF is an effective tool for state estimation [27]. Given the dynamic equation shown in (7) and a series of observations $Y(k) = [y(1), y(2), \ldots, y(k)]$, the state estimation procedure by the UKF is presented as follows [28].

Step 1 (UT transformation). In the original state distribution, some sampling points are selected according to certain rules, so that the mean and covariance of these sampling points are equal to the mean and covariance of the previous state distribution. These points are substituted into the nonlinear function, and the corresponding set of the nonlinear function values is obtained. Then, the mean and covariance of the nonlinear transformation are obtained from these sets of points.

First, compute the $2n + 1$ sigma points, namely, sampling points:

$$X_{k-1}^{(0)} = X_{k-1}^a,$$

$$X_{k-1}^{(i)} = X_{k-1}^a + \left(\sqrt{(n+\lambda)P_{k-1}}\right)_i, \qquad (8)$$

$$X_{k-1}^{(n+i)} = X_{k-1}^a - \left(\sqrt{(n+\lambda)P_{k-1}}\right)_i, \quad i = 1, \ldots, n,$$

where n is the number of the state dimension.

Second, calculate the associated weight w of the sampling points:

$$w_m^{(0)} = \frac{\lambda}{n+\lambda}$$

$$w_c^{(0)} = \frac{\lambda}{n+\lambda} + \left(1 - \alpha^2 + \beta\right)$$

$$w_m^{(i)} = w_c^{(i)} = \frac{1}{2(n+\lambda)}, \quad i = 1, \ldots, 2n \qquad (9)$$

$$\lambda = \alpha^2(n+\kappa) - n,$$

where P is the estimated covariance, satisfying $(\sqrt{P})^T(\sqrt{P}) = P$; $(\sqrt{P})_i$ is the square root of the ith column in P; λ is scaling parameter used to reduce the total prediction error; α controls the spread of sampling points; κ is the selected parameter, and its value is not bounded generally, but it is usually necessary to ensure the semidefinite of $(n + \lambda)P$ matrix. Under normal circumstances, $\alpha = 10^{-3}$ and $\kappa = 0$. β is a nonnegative weighting index, and the optimal value is 2 for a Gaussian distribution of X.

Step 2. Compute predicted state X_k^f and predicted covariance P_k^f:

$$\widehat{X}_k^{(i)} = f\left(\widehat{X}_{k-1}^i\right),$$

$$X_k^f = \sum_{i=0}^{2n} w_m^{(i)} \widehat{X}_k^{(i)}, \qquad (10)$$

$$P_k^f = \sum_{i=0}^{2n} w_c^{(i)} \left(\widehat{X}_k^{(i)} - X_k^f\right)\left(\widehat{X}_k^{(i)} - X_k^f\right)^T.$$

Step 3. Compute predicted $\widehat{y}_k$, measurement covariance $P_{y_k y_k}$, and cross-covariance of the state and measurement $P_{x_k y_k}$:

$$\widehat{Y}_k^{(i)} = \widehat{X}_k^i,$$

$$\widehat{y}_k = \sum_{i=0}^{2n} w_m^{(i)} \widehat{Y}_k^{(i)},$$

$$P_{y_k y_k} = \sum_{i=0}^{2n} w_c^{(i)} \left(\widehat{Y}_k^{(i)} - \widehat{y}_k\right)\left(\widehat{Y}_k^{(i)} - \widehat{y}_k\right)^T, \qquad (11)$$

$$P_{x_k y_k} = \sum_{i=0}^{2n} w_c^{(i)} \left(\widehat{X}_k^{(i)} - X_k^f\right)\left(\widehat{Y}_k^{(i)} - \widehat{y}_k\right)^T.$$

Step 4. Compute gain K_k, updated state X_k^a, and covariance P_k:

$$K_k = P_{x_k y_k} P_{y_k y_k}^{-1},$$

$$X_k^a = X_k^f + K_k(y_k - \widehat{y}_k), \qquad (12)$$

$$P_k = P_k^f - K_k P_{y_k y_k} K_k^T.$$

From the estimate of the augmented variable X, the estimated values of unknown parameters can be obtained as $\hat{\theta} - [\hat{k}_1, \hat{k}_2, \ldots, \hat{k}_{10}]$

3. Eutrophication Condition Evaluation Based on Modified Fuzzy Comprehensive Evaluation Method

On the basis of the output from the proposed fusion water quality soft-sensing method and online measurements of water quality indexes, a modified fuzzy comprehensive evaluation method is used to evaluate the water eutrophication condition.

Fuzzy comprehensive evaluation method adopts fuzzy mathematical theory to obtain a quantitative evaluation result of an object in view of the complexity of object and fuzziness of the evaluation index. The procedure of the proposed fuzzy comprehensive evaluation method is presented as follows.

Step 1 (water evaluation index selection). In the traditional fuzzy comprehensive evaluation method, the key evaluation indexes, which have great influence on the water environment, are obtained by empirical measures. Although these methods are relatively convenient, they lack objectivity and theoretical foundation. In order to compensate this limitation, a cumulative frequency method is introduced by calculating the cumulative frequency of excessive multiple of each water quality index [29].

The evaluation set is a collection of criteria for evaluating the object [30]. Suppose the water eutrophication condition can be classified to n levels, written as

$$V = \{V_1, V_2, \ldots, V_n\}. \tag{13}$$

Further suppose there exist M evaluation indexes, written as $\{C_1, C_2, \ldots, C_M\}$. On this basis, the cumulative frequency can be calculated as follows:

$$\beta_i = \frac{C_i}{(1/n) \sum_{j=1}^{n} \sigma_{ij}},$$

$$\tilde{\beta}_i = \{\beta_i \mid \beta_1 \geq \beta_i \geq \beta_M\}$$

$$i = 1, 2, \ldots, M; \quad j = 1, 2, \ldots, n, \tag{14}$$

$$K_i = \frac{\sum_{j=1}^{i} \tilde{\beta}_j}{\sum_{j=1}^{M} \tilde{\beta}_j} \times 100\%,$$

where i is the label of the evaluation index; j is the label of eutrophication level; C_i is the concentration value of the ith index; σ_{ij} is the standard value of the ith index in level j; β_i is the excessive multiple value of the ith index; K_i is the cumulative frequency of the first i indexes. According to the statistical analysis requirements in the selection of evaluation indexes, generally take [31]

$$K_i \geq 85\%. \tag{15}$$

Following (14) and (15), we have the selected key water evaluation indexes as $C = \{C_1, C_2, \ldots, C_m\}$, $m \leq M$.

Step 2 (establishment of the fuzzy relation matrix $R : C \rightarrow V$). By adopting fuzzy mathematical theory to evaluation study, the most critical issue is to establish the membership functions for the evaluation indexes. Triangular linear membership functions are commonly used in practice, which are also selected for determining the fuzzy relation matrix R in this paper [32]. The configuration of the membership function is as follows:

For $\forall V_j \in V$ $(j = 1, 2, \ldots, n)$,

(1) when $j = 1$, the membership function is

$$r_{ij} = \begin{cases} 1 & 0 \leq C_i \leq \sigma_{ij} \\ \dfrac{\sigma_{ij+1} - C_i}{\sigma_{ij+1} - \sigma_{ij}} & \sigma_{ij} < C_i \leq \sigma_{ij+1} \\ 0 & C_i > \sigma_{ij+1}, \end{cases} \tag{16}$$

(2) when $1 < j < n - 1$, the membership function is

$$r_{ij} = \begin{cases} 1 & C_i = \sigma_{ij} \\ \dfrac{C_i - \sigma_{ij-1}}{\sigma_{ij} - \sigma_{ij-1}} & \sigma_{ij-1} \leq C_i < \sigma_{ij} \\ \dfrac{\sigma_{ij+1} - C_i}{\sigma_{ij+1} - \sigma_{ij}} & \sigma_{ij} < C_i \leq \sigma_{ij+1} \\ 0 & C_i < \sigma_{ij-1}, \ C_i > \sigma_{ij+1}, \end{cases} \tag{17}$$

(3) when $j = n$, the membership function is

$$r_{ij} = \begin{cases} 1 & C_i \geq \sigma_{ij} \\ \dfrac{C_i - \sigma_{ij-1}}{\sigma_{ij} - \sigma_{ij-1}} & \sigma_{ij-1} \leq C_i < \sigma_{ij} \\ 0 & C_i < \sigma_{ij-1}, \end{cases} \tag{18}$$

where r_{ij} presents the membership degree of index C_i in level V_j; $\sigma_{ij-1}, \sigma_{ij}, \sigma_{ij+1}$ are the standard values of the ith index in level V_{j-1}, V_j, V_{j+1}, respectively. When C_i is given, the above membership functions can be applied to determine the membership degree of the evaluation index C_i for each level of water eutrophication.

Then, the fuzzy relation matrix $R : C \rightarrow V$ is constructed as

$$R = (r_{ij})_{m \times n} = \begin{pmatrix} r_{11} & r_{12} & \cdots & r_{1n} \\ r_{21} & r_{22} & \cdots & r_{2n} \\ \cdots & \cdots & \cdots & \cdots \\ r_{m1} & r_{m2} & \cdots & r_{mn} \end{pmatrix}, \tag{19}$$

where $\sum_{j=1}^{n} r_{ij} = 1$, $i = 1, 2, \ldots, m$.

Step 3 (weight determination of evaluation index). The determination of the evaluation index weight is one of the most important factors that directly affect the final evaluation results. In this paper, the clustering weight method is used to determine the weight of each evaluation index, combining

TABLE 1: Average values of unknown model parameters of each stage.

Stage	Symbol									
	k_1	k_2	k_3	k_4	k_5	k_6	k_7	k_8	k_9	k_{10}
Recovery	1.396	0.359	0.024	1.794	0.016	0.166	0.086	1.002	0.047	0.220
Biomass increase and accumulation	1.200	0.719	0.441	1.341	0.003	0.152	0.047	1.012	0.003	0.223
Dormancy	1.162	0.229	0.302	1.176	0.011	0.175	0.022	1.009	0.007	0.225

the index concentration value with the standard values which more objectively reflect the relative importance of each evaluation index in all the indexes. The method of calculating the index weight is as follows:

$$w_{ij} = \frac{C_i/\sigma_{ij}}{\sum_{i=1}^{m} C_i/\sigma_{ij}} \quad i = 1, 2, \ldots, m; \ j = 1, 2, \ldots, n, \quad (20)$$

where w_{ij} is the weight of the ith water evaluation index in the jth eutrophication level. Therefore, the index weight matrix W determined by the clustering method is

$$W = (W_1, W_2, \ldots, W_j, \ldots, W_n), \quad (21)$$

where W_j is the index weight matrix of the jth eutrophication level.

Step 4 (fuzzy synthesis operation). Combining the index weight W with the fuzzy relation matrix R, the multiplication and addition method of the weighted average is chosen to obtain fuzzy comprehensive evaluation result B based on all indexes. The advantage of this method is that it can balance all indexes according to the weight values to reflect the comprehensive condition of water quality [33]. The specific formula is as follows:

$$B = (b_j)_{1 \times n} = W \circ R,$$
$$b_j = \sum_{i=1}^{m} w_{ji} \cdot r_{ij}, \quad (22)$$

where the element b_j is membership of the water object with regard to jth water eutrophication level. The water eutrophication level can be obtained by the principle of maximum membership, where the specific formula is as follows:

$$j = \arg \max_j b_j. \quad (23)$$

4. Case Study

In this part, water quality data of Taihu Lake (case 1) and Beihai Lake (case 2) are utilized to verify the effectiveness and generality of the proposed method. First the UKF is used to estimate the unknown parameters in the soft-sensing method, where the estimated result is compared to the results obtained by the nonlinear least squares method and trial-and-error method. Then, simulated values of the water quality indexes are deduced, which are compared to the real measured water quality data. On this basis, the water

eutrophication evaluation is carried out by the modified fuzzy comprehensive evaluation method, depending on both simulated values of DO, BOD, TN, Chl_a and real measured values of SD, TP.

4.1. Case 1: Taihu Lake

4.1.1. Model Parameter Estimation Result and Analysis. The UKF is used to estimate the unknown model parameters of Taihu Lake shown in Notations. Besides, due to the limitation of measured data, the TN data are used instead of NO_3-N. Since the measured data include DO, BOD, Phyt, TN, and NH_3-N, the observation matrix in (7) is set as

$$h$$

$$= \begin{bmatrix} 1 & 0 & 0 & 0 & 0 & 0 & 0 & 0 & 0 & 0 & 0 & 0 & 0 & 0 & 0 & 0 & 0 \\ 0 & 1 & 0 & 0 & 0 & 0 & 0 & 0 & 0 & 0 & 0 & 0 & 0 & 0 & 0 & 0 & 0 \\ 0 & 0 & 1 & 0 & 0 & 0 & 0 & 0 & 0 & 0 & 0 & 0 & 0 & 0 & 0 & 0 & 0 \\ 0 & 0 & 0 & 1 & 0 & 0 & 0 & 0 & 0 & 0 & 0 & 0 & 0 & 0 & 0 & 0 & 0 \\ 0 & 0 & 0 & 0 & 1 & 0 & 0 & 0 & 0 & 0 & 0 & 0 & 0 & 0 & 0 & 0 & 0 \end{bmatrix}_{5 \times 17} . \quad (24)$$

Then the real-time estimated values of the ten unknown model parameters can be obtained, which are depicted in Figure 1.

The appearance of algal blooms is a feature of water eutrophication, whose formation process can be divided into three stages, namely, recovery, biomass increase and accumulation, dormancy [34]. In this paper, January to March is the recovery period of algae bloom, April to mid-October is the second stage, and the remaining part is the dormancy stage. Following the multistage principle, the average values of unknown model parameters of each stage are shown in Table 1.

Then, the estimated values of the unknown parameters in Table 1 are substituted into the water quality soft-sensing method, and the simulation process is carried out. For comparison, the simulated values of water quality indexes are compared to the real measured values and the simulated values obtained by the models with estimated parameters based on the trial-and-error method and nonlinear least squares method, respectively. Figure 2 depicts the results of DO concentration, BOD concentration, NH_3-N concentration, TN concentration, and Chl_a concentration. It should be noted that the amount of Chl_a is indirectly expressed as the concentration of Phyt; that is to say, through the ratio between Phyt and Chl_a, the concentration of Chl_a is obtained.

It can be drawn from Figure 2 that the simulated values are in good agreement with the measured values of water

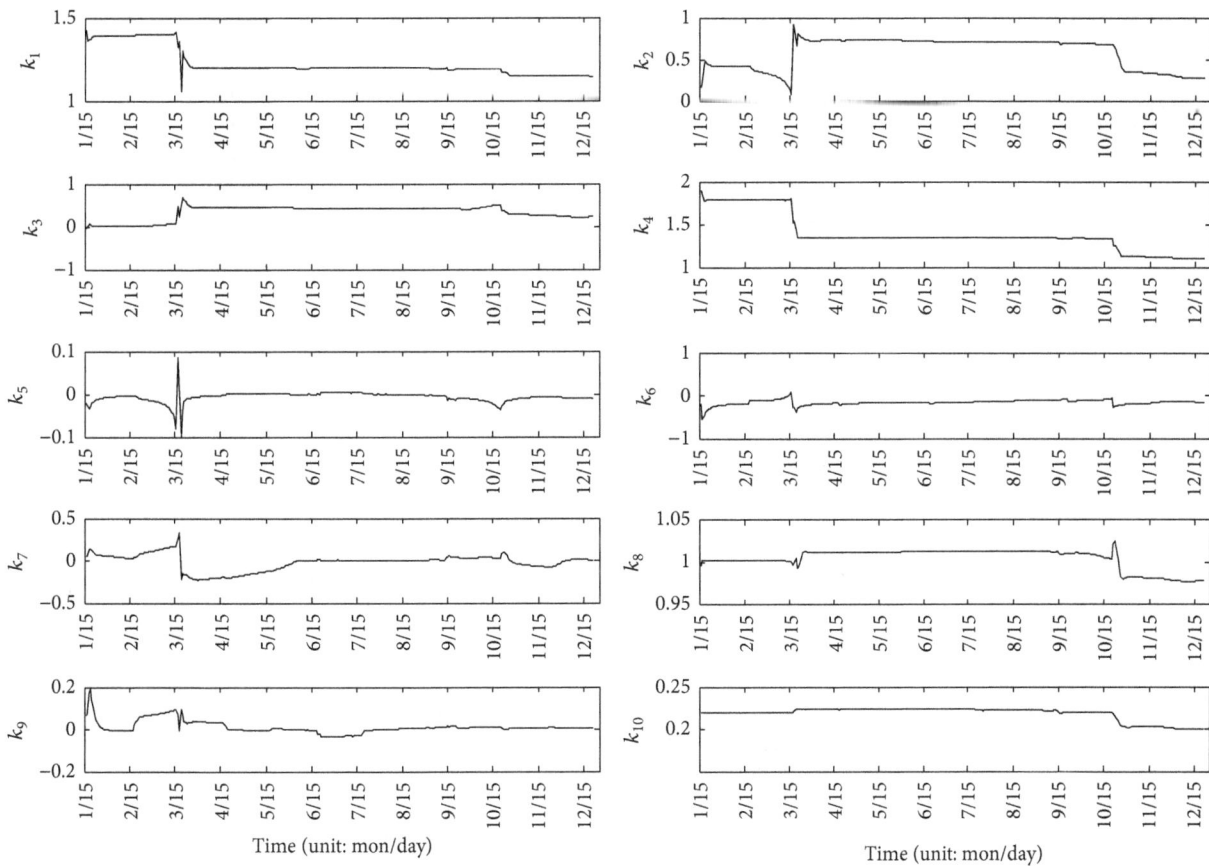

FIGURE 1: Real-time estimated values of ten unknown model parameters.

quality indexes, besides TN concentration. There are two main reasons for this. One is that the data of TN are used instead of NO_3-N during the experiment, and the other is the inaccuracies caused by external influences. However, the overall experimental results can verify the effectiveness of the fusion water quality soft-sensing method. Then the model accuracy based on the UKF is better than those based on the nonlinear least squares method and trial-and-error method. In order to quantitatively evaluate the error, Root Mean Square Error (RMSE) is utilized to indicate the deviation between the simulated values and the measured values of each water quality index. The specific formula is as follows:

$$\text{RMSE} = \sqrt{\frac{\sum_{i=1}^{n} \left(X_i - \widehat{X}_i \right)^2}{n}}, \tag{25}$$

where X_i is the measured value, $\widehat{X}_i$ is the simulated value, and n is the number of measurement. The result is shown in Table 2.

4.1.2. Water Eutrophication Evaluation Result and Analysis. The modified fuzzy comprehensive evaluation method is used to evaluate the water eutrophication status of Taihu Lake, by taking both the simulated values and measured values of the water quality indexes into consideration. For comparison,

the eutrophication evaluation result is compared to the result based on merely measured data.

Step 1 (water evaluation index selection). As the eutrophication mechanism is complex, scholars often choose different evaluation criteria to assess water quality status. By referring to [35, 36], five levels are classified to describe the eutrophication condition: I (none), II (mild), III (medium), IV (heavy), and V (extremely heavy); namely, $V = \{I, II, III, IV, V\}$.

Then, according to (14) and (15) to select the water quality evaluation indexes, SD, BOD, TN, TP, DO, and Chl_a are selected as evaluation indexes. By referring to the technological regulations for surface water resources quality assessment published by Ministry of Water Resources, People's Republic of China, and related references [37, 38], the standard values of water quality indexes for lakes and reservoirs in each eutrophication level can be determined, which are shown in Table 3. On this basis, triangular linear membership functions of the water quality indexes are constructed as shown in Figure 3.

Step 2 (determine the fuzzy relation matrix R and the weight matrix W). Given the data collected from a monitoring station in Taihu Lake, according to (16)–(18) and (20), the time-related fuzzy relation matrix $R(t)$ and weight matrix $W(t)$ are obtained, respectively.

TABLE 2: RMSE values of each water quality index obtained by different methods.

Method	DO (mg/L)	BOD (mg/L)	NH$_3$-N (mg/L)	TN (mg/L)	Chl_a (mg/L)
Trial-and-error method	1.47	0.429	0.14	0.52	0.007
Nonlinear least squares	1.09	0.297	0.12	0.45	0.004
UKF	0.72	0.215	0.11	0.39	0.003

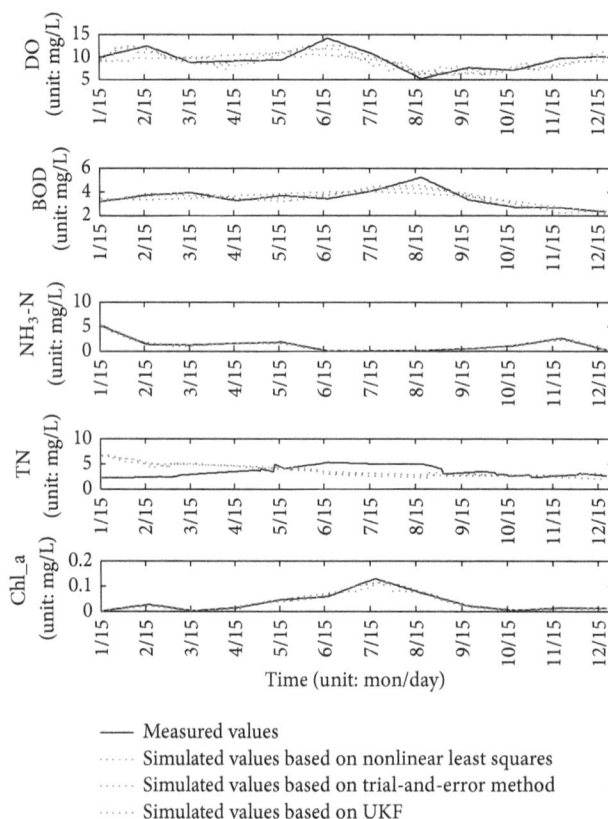

FIGURE 2: Results of DO concentration, BOD concentration, NH$_3$-N concentration, TN concentration, and Chl_a concentration.

Step 3 (fuzzy synthesis operation). The fuzzy comprehensive evaluation result $B(t)$ of the monitoring station in Taihu Lake is obtained by using (22). Figure 4 depicts the obtained memberships of level I, level II, level III, level IV, and level V based on the measured values and simulated values of UKF, respectively.

According to (23), the water eutrophication level based on measured values, simulated values of trial-and-error method, simulated values of nonlinear least squares, and simulated values of UKF is evaluated, and results are shown in Figure 5.

It can be seen from Figure 5 that, during the recovery and dormancy stage of algal blooms, the water eutrophication status is basically in level I, while in the biomass increase and accumulation stage, the water eutrophication status is mainly in level IV. Then, with the increase of temperature and rainfall in June and July, algae blooms gradually accumulate so that the water eutrophication status reaches level V.

However, we can see that the deviations appear in eutrophication levels based on measured and simulated values from Table 3 and Figure 5. The reason lies in two aspects: (1) Model approximation: the used WASP model is an approximation of real water quality evolution. This approximation will bring uncertainty and inaccuracy during simulating the evolution process, such as abrupt variation process of eutrophication degree. (2) Model parameter selection: in the simulation, the model parameters are selected as constant values, which can be viewed as a simplification, since the model parameters are time-variant in practice. This simplification will introduce errors when simulating the water quality evolution process.

Here, it should be noted that the modified fuzzy comprehensive evaluation method can be applied to more monitoring stations in Taihu Lake, which will lead to a more comprehensive evaluation result of its eutrophication level. In order to quantitatively represent the accuracy of evaluation results based on simulated values of different methods, the

TABLE 3: Standard values of water quality indexes for lakes and reservoirs.

Level	SD (m)	BOD (mg/L)	TN (mg/L)	TP (mg/L)	DO (mg/L)	Chl_a (mg/L)
I	10.00	0.18	0.02	0.001	15	0.001
II	5.00	0.24	0.31	0.004	8.5	0.002
III	1.50	1.20	1.20	0.023	5.0	0.004
IV	0.55	6.00	3.60	0.110	2.0	0.01
V	0.17	15.00	4.70	0.660	1.0	0.065

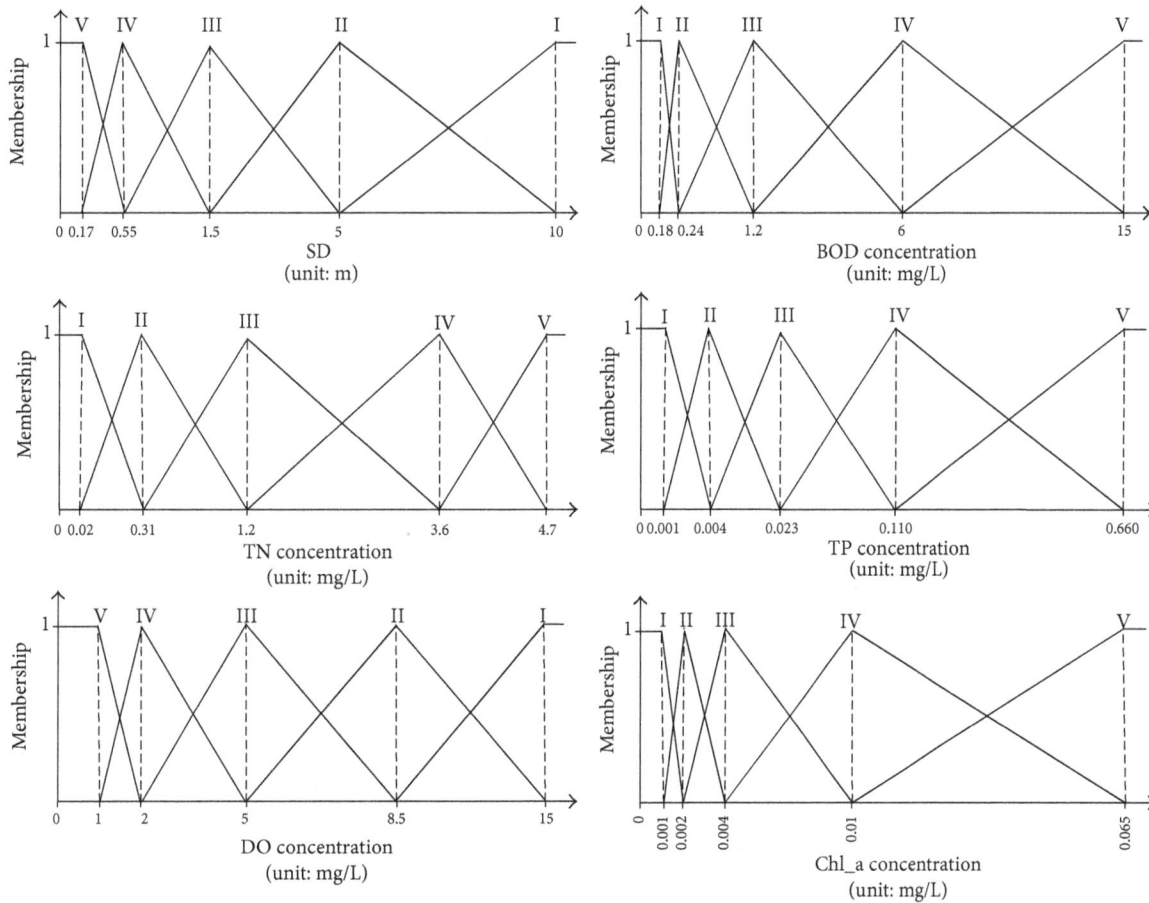

FIGURE 3: Triangular linear membership functions of the water quality indexes.

consistency percentage of evaluation results of each method is given in Table 4.

It can be drawn from Table 4 that the accuracy of the evaluation result based on simulated values of UKF is higher than those based on simulated values of trial-and-error method and simulated values of nonlinear least squares.

4.2. Case 2: Beihai Lake

4.2.1. Model Parameter Estimation Result and Analysis. The UKF is also used to estimate the unknown model parameters of Beihai Lake shown in Notations. And the observation matrix is the same as that of Taihu Lake. Then the real-time estimated values of the ten unknown model parameters can be obtained, which are depicted in Figure 6.

The algal blooms formation process of Taihu Lake also applies to Beihai Lake. Following the multistage principle, the average values of unknown model parameters of each stage are shown in Table 5.

Similarly, the estimated values of the unknown parameters in Table 5 are substituted into the water quality soft-sensing method, and the simulation process is carried out. Then Figure 7 depicts the results of DO concentration, BOD concentration, NH_3-N concentration, TN concentration, and Chl_a concentration based on measured values and simulated values of different methods.

From Figure 7, it can be seen that the simulated values are in good agreement with the measured values of each water quality index, which can verify the effectiveness and generality of the fusion water quality soft-sensing method.

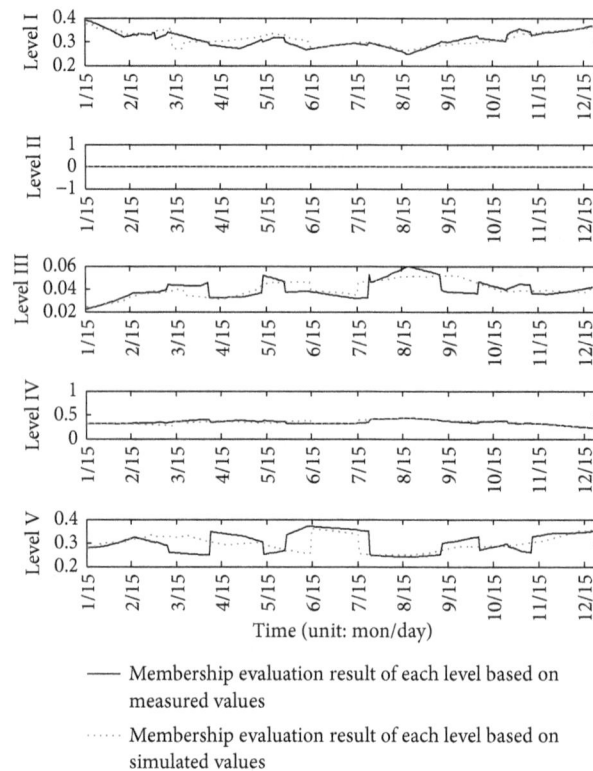

FIGURE 4: Membership evaluation result of each level based on the measured values and simulated values of UKF.

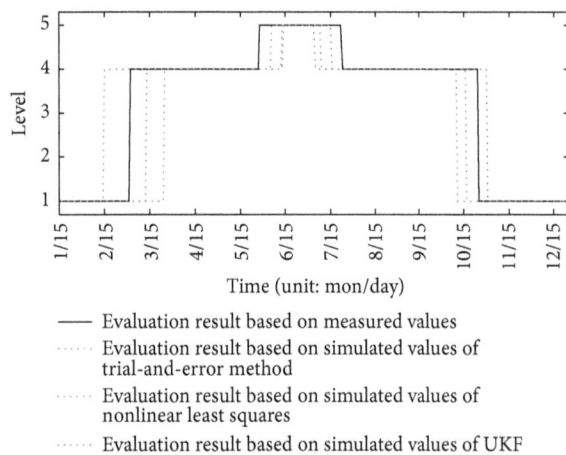

FIGURE 5: Evaluation result of the monitoring station.

TABLE 4: Consistency percentage of results based on simulated values of different methods.

Method	Trial-and-error method	Nonlinear least squares	UKF
Consistency percentage	0.81%	0.84%	0.88%

Then the model accuracy based on the UKF is better than those based on the nonlinear least squares method and trial-and-error method. Similarly, RMSE is utilized to indicate the deviation between the simulated values and the measured values of each water quality index. By using (25), the specific result is shown in Table 6.

4.2.2. Water Eutrophication Evaluation Result and Analysis. Following the proposed algorithm, the eutrophication evaluation result of Beihai Lake is shown in Figure 8.

According to (23), the water eutrophication level based on measured values, simulated values of trial-and-error method, simulated values of nonlinear least squares, and simulated values of UKF is evaluated, and results are shown in Figure 9.

It can be seen from Figure 9 that, during the recovery and dormancy stage of algal blooms, the water eutrophication status is basically in level II, while in the biomass increase and accumulation stage, the water eutrophication status is mainly in level IV. Analogously, the modified fuzzy comprehensive evaluation method can be applied to more monitoring stations in Beihai Lake.

In order to quantitatively represent the accuracy of evaluation results based on simulated values of different methods, the consistency percentage is also introduced and utilized for calculation, and the results are shown in Table 7.

It can be drawn from Table 7 that accuracy of evaluation result based on simulated values of UKF is higher than those based on simulated values of trial-and-error method and simulated values of nonlinear least squares, too.

TABLE 5: Average values of unknown model parameters of each stage.

Stage	Symbol									
	k_1	k_2	k_3	k_4	k_5	k_6	k_7	k_8	k_9	k_{10}
Recovery	2.681	1.044	0.688	1.285	0.094	−0.047	2.737	1.374	0.001	0.434
Biomass increase and accumulation	2.695	1.358	2.593	1.066	0.143	0.006	4.652	2.151	0.015	0.673
Dormancy	2.719	1.568	2.565	0.641	0.167	0.013	4.866	2.548	0.020	0.663

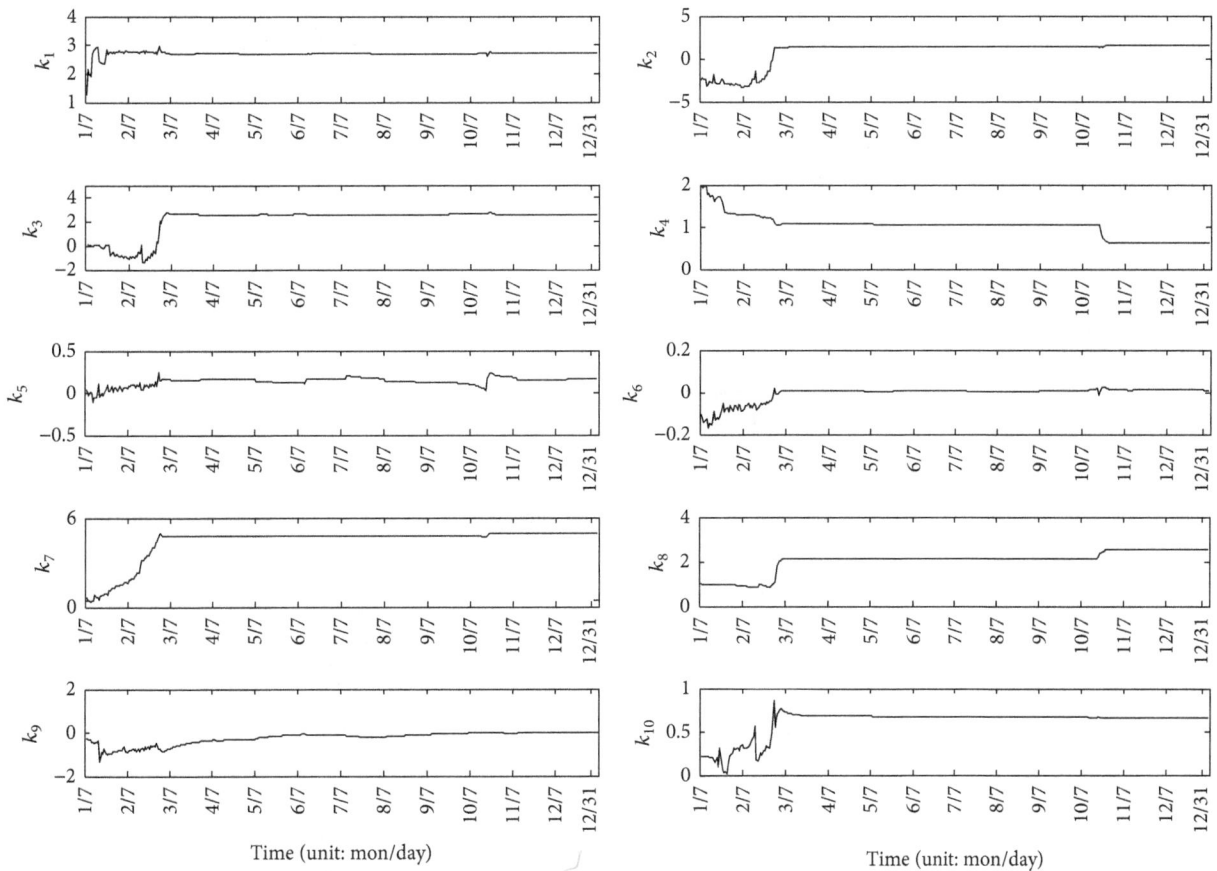

FIGURE 6: Real-time estimated values of ten unknown model parameters.

5. Conclusions

The fusion water quality soft-sensing method is constructed with a combination between the WASP mechanism model and UKF, and the modified fuzzy comprehensive evaluation method is presented to evaluate the level of water eutrophication. Then, taking Taihu Lake and Beihai Lake as examples, the results show that the simulated values of water quality indexes are in good agreement with the measured values, which can verify the effectiveness and generality of the fusion water quality soft-sensing method, and unknown parameter estimation based on UKF can further improve the accuracy of the model more than nonlinear least squares method and trial-and-error method. Besides, the modified fuzzy comprehensive evaluation method is used to assess the water eutrophication status, and the evaluation results of eutrophication level are consistent in most cases based on simulated values and measured values. Moreover, the modified fuzzy comprehensive evaluation method can be applied to more monitoring stations in Taihu Lake and Beihai Lake, which will lead to a more comprehensive evaluation result of their eutrophication level and provide a scientific reference for water environment management. In future research, more water quality indexes should be considered in the procedure of water eutrophication evaluation. Furthermore, some quantitative indicators, such as health degree [39, 40], should be introduced to evaluate water eutrophication.

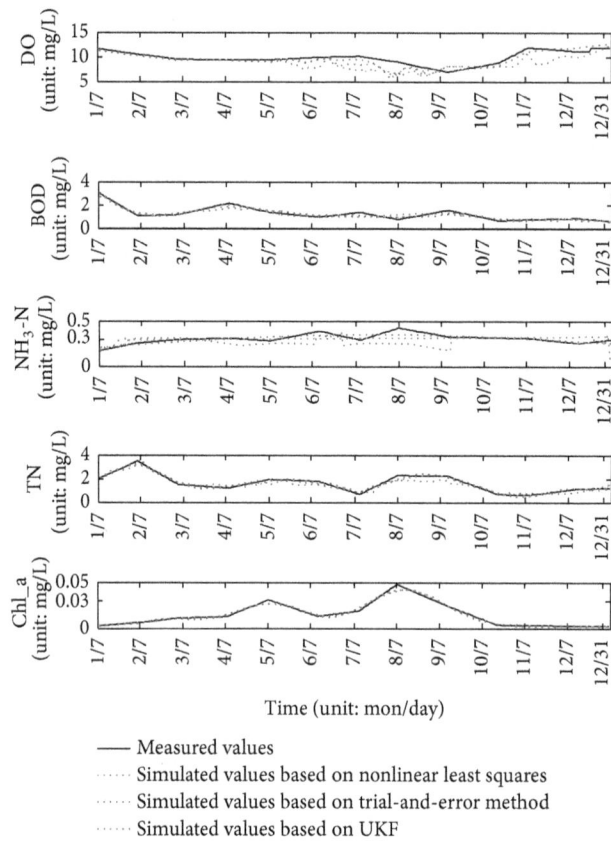

FIGURE 7: Results of DO concentration, BOD concentration, NH_3-N concentration, TN concentration, and Chl_a concentration.

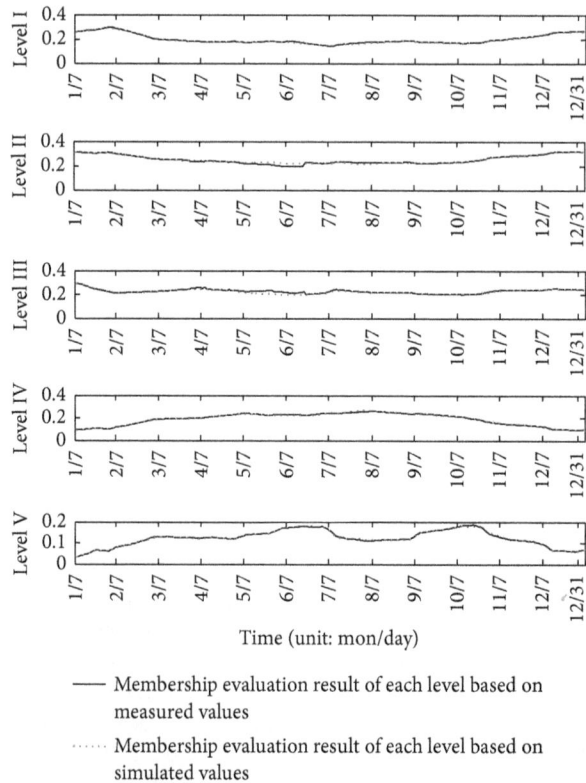

FIGURE 8: Membership evaluation result of each level based on the measured values and simulated values of UKF.

TABLE 6: RMSE values of each water quality index obtained by different methods.

Method	DO (mg/L)	BOD (mg/L)	NH$_3$-N (mg/L)	TN (mg/L)	Chl_a (mg/L)
Trial-and-error method	1.31	0.167	0.07	0.22	0.002
Nonlinear least squares	0.88	0.151	0.04	0.13	0.001
UKF	0.68	0.080	0.02	0.09	0.001

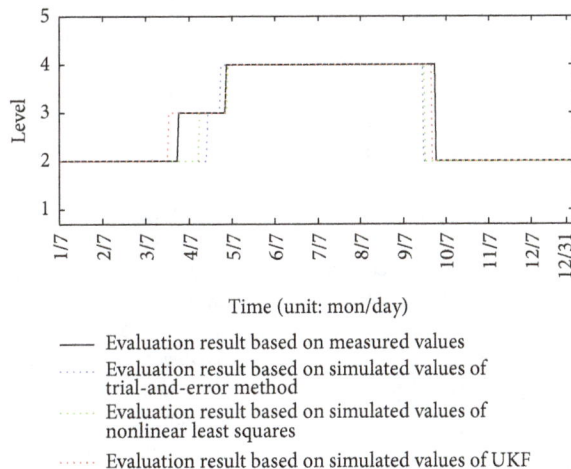

— Evaluation result based on measured values
····· Evaluation result based on simulated values of trial-and-error method
····· Evaluation result based on simulated values of nonlinear least squares
····· Evaluation result based on simulated values of UKF

FIGURE 9: Evaluation result of the monitoring station.

TABLE 7: Consistency percentage of results based on simulated values of different methods.

Method	Trial-and-error method	Nonlinear least squares	UKF
Consistency percentage	0.91%	0.94%	0.97%

Notations

k_1: Reaeration rate coefficient at 20°C
k_2: Deoxygenation rate coefficient at 20°C
k_3: Nitrification rate coefficient at 20°C
k_4: Sediment oxygen demand coefficient
k_5: The maximum growth rate of phytoplankton at 20°C
k_6: Phytoplankton mortality (nonzooplankton)
k_7: Denitrification rate coefficient at 20°C
k_8: Half saturation constant for oxygen limitation of nitrification
k_9: Mineralization rate coefficient of dissolved organic nitrogen at 20°C
k_{10}: Mineralization rate coefficient of dissolved organic phosphorus at 20°C.

Conflicts of Interest

The authors declare no conflicts of interest.

Authors' Contributions

Xiaoyi Wang and Zhiyao Zhao conceived and designed the framework of the paper; Jie Jia and Tingli Su performed the experiment; Jiping Xu and Li Wang collected references and processed the data; Xiaoyi Wang and Jie Jia wrote the paper; Zhiyao Zhao revised the paper.

Acknowledgments

This work was financially supported by the National Key R&D Program of China (2017YFC1600605), the National Natural Science Foundation of China (61703008), and the Research Foundation for Youth Scholars of Beijing Technology and Business University (QNJJ2018-18).

References

[1] M. Lürling, E. Mackay, K. Reitzel, and B. M. Spears, "Editorial – A critical perspective on geo-engineering for eutrophication management in lakes," *Water Research*, vol. 97, pp. 1–10, 2016.

[2] R. Xia, Y. Zhang, A. Critto et al., "The potential impacts of climate change factors on freshwater eutrophication: Implications for research and countermeasures of water management in China," *Sustainability*, vol. 8, no. 3, article no. 229, 2016.

[3] X. Yao, Y. Zhang, L. Zhang, and Y. Zhou, "A bibliometric review of nitrogen research in eutrophic lakes and reservoirs," *Journal of Environmental Sciences*, 2016.

[4] W. Xiaoyi, Y. Junyang, S. Yan, S. Tingli, W. Li, and X. Jiping, "Research on hybrid mechanism modeling of algal bloom formation in urban lakes and reservoirs," *Ecological Modelling*, vol. 332, pp. 67–73, 2016.

[5] T. Kim and C. Chae, "Environmental impact analysis of acidification and eutrophication due to emissions from the production of concrete," *Sustainability*, vol. 8, no. 12, p. 578, 2016.

[6] M. A. Darwish, H. K. Abdulrahim, A. S. Hassan, and B. Shomar, "Reverse osmosis desalination system and algal blooms Part I: harmful algal blooms (HABs) species and toxicity," *Desalination and Water Treatment*, vol. 57, no. 54, pp. 25859–25880, 2016.

[7] W. Duan, B. He, D. Nover et al., "Water quality assessment and pollution source identification of the eastern poyang lake basin using multivariate statistical methods," *Sustainability*, vol. 8, no. 2, article no. 133, 2016.

[8] Y. Chen, R. Zou, H. Su et al., "Development of an integrated water quality and macroalgae simulation model for tidal marsh eutrophication control decision support," *Water (Switzerland)*, vol. 9, no. 4, article no. 277, 2017.

[9] C. H. Hansen, S. J. Burian, P. E. Dennison, and G. P. Williams, "Spatiotemporal variability of lake water quality in the context of remote sensing models," *Remote Sensing*, vol. 9, no. 5, article no. 19, 2017.

[10] G. A. Susto, A. Schirru, S. Pampuri, and S. McLoone, "Supervised aggregative feature extraction for big data time series regression," *IEEE Transactions on Industrial Informatics*, vol. 12, no. 3, pp. 1243–1252, 2016.

[11] B. Bidar, J. Sadeghi, F. Shahraki, and M. M. Khalilipour, "Data-driven soft sensor approach for online quality prediction using state dependent parameter models," *Chemometrics Intelligent Laboratory Systems*, vol. 162, pp. 130–141, 2017.

[12] J.-J. Zhu and P. R. Anderson, "Assessment of a soft sensor approach for determining influent conditions at the MWRDGC Calumet WRP," *Journal of Environmental Engineering (United States)*, vol. 142, no. 6, Article ID 04016023, 2016.

[13] H. Gharehbaghi and J. Sadeghi, "A novel approach for prediction of industrial catalyst deactivation using soft sensor modeling," *Journal of Catalysis*, vol. 6, no. 7, article no. 93, 2016.

[14] J. L. Zhang, Y. P. Li, G. H. Huang, B. W. Baetz, and J. Liu, "Uncertainty analysis for effluent trading planning using a Bayesian estimation-based simulation-optimization modeling approach," *Water Research*, vol. 116, pp. 159–181, 2017.

[15] Y. Parvini, J. B. Siegel, A. G. Stefanopoulou, and A. Vahidi, "Supercapacitor electrical and thermal modeling, identification, and validation for a wide range of temperature and power applications," *IEEE Transactions on Industrial Electronics*, vol. 63, no. 3, pp. 1574–1585, 2016.

[16] D. T. Bui, Q. T. Bui, Q. P. Nguyen, B. Pradhan, H. Nampak, and P. T. Trinh, "A hybrid artificial intelligence approach using GIS-based neural-fuzzy inference system and particle swarm optimization for forest fire susceptibility modeling at a tropical area," *Agricultural Forest Meteorology*, vol. 233, pp. 32–44, 2017.

[17] J. C. Quijano, Z. Zhu, V. Morales, B. J. Landry, and M. H. Garcia, "Three-dimensional model to capture the fate and transport of combined sewer overflow discharges: A case study in the Chicago Area Waterway System," *Science of the Total Environment*, vol. 576, pp. 362–373, 2017.

[18] T. A. Wool, R. B. Ambrose, J. L. Martin, and E. A. Comer, *Water Quality Analysis Simulation Program (WASP) Version 6.0 DRAFT: User's Manual*, US Environmental Protection Agency, Athens, Georgia, 2006.

[19] T. A. Wool, S. R. Davie, Y. M. Plis, and J. Hamrick, "The development of a hydrodynamic and water quality model to support tmdl determinations and water quality management of a stratified shallow estuary: mobile bay, alabama," *Proceedings of the Water Environment Federation*, vol. 15, no. 4, pp. 378–392, 2003.

[20] United States Environmental Protection Agency Region 3, "Final regulation total maximum daily loads for the Murderkill River watershed," 2015, http://www.dnrec.state.de.us/water2000/Sections/Watershed/TMDL/SecOrder.pdf.

[21] Water resources and TMDL center, "Lake Michigan shoreline TMDL for E. coil bacteria modeling framework report," 2003, http://www.ogdendunes.net/images/tmdl_lakemich_report.pdf.

[22] S. Yu, L. He, and H. Lu, "An environmental fairness based optimisation model for the decision-support of joint control over the water quantity and quality of a river basin," *Journal of Hydrology*, vol. 535, pp. 366–376, 2016.

[23] T. Kim and Y. P. Sheng, "Estimation of water quality model parameters," *KSCE Journal of Civil Engineering*, vol. 14, no. 3, pp. 421–437, 2010.

[24] H. Zhou, H. Huang, H. Zhao, X. Zhao, and X. Yin, "Adaptive unscented Kalman filter for target tracking in the presence of nonlinear systems involving model mismatches," *Remote Sensing*, vol. 9, no. 7, article no. 657, 2017.

[25] J.-F. Chen, H.-N. Hsieh, and Q. H. Do, "Evaluating teaching performance based on fuzzy AHP and comprehensive evaluation approach," *Applied Soft Computing*, vol. 28, pp. 100–108, 2015.

[26] Z. Y. Zhao, Q. Quan, and K. Y. Cai, "A modified profust-performance-reliability algorithm and its application to dynamic systems," *Journal of Intelligent & Fuzzy Systems*, vol. 32, pp. 643–660, 2017.

[27] S. K. Biswas, L. Qiao, and A. G. Dempster, "A novel a priori state computation strategy for the unscented kalman filter to improve computational efficiency," *IEEE Transactions on Automatic Control*, vol. 62, no. 4, pp. 1852–1864, 2017.

[28] G. Xu-sheng, G. Wen-ming, D. Zhe, and L. Wei-dong, "Research on WNN soft fault diagnosis for analog circuit based on adaptive UKF algorithm," *Applied Soft Computing*, vol. 50, pp. 252–259, 2017.

[29] S. B. Morwal, S. G. Narkhedkar, B. Padmakumari et al., "Cloud characteristics over the rain-shadow region of North Central peninsular India during monsoon withdrawal and post-withdrawal periods," *Climate Dynamics*, vol. 46, no. 1-2, pp. 495–514, 2016.

[30] E. M. Ali and H. M. Khairy, "Environmental assessment of drainage water impacts on water quality and eutrophication level of Lake Idku, Egypt," *Environmental Pollution*, vol. 216, pp. 437–449, 2016.

[31] S. J. Zhang, L. Zhao, X. T. Li, and B. Cheng, "A sequential and partial ambiguity resolution strategy for improving the initialization performance of medium-baseline relative positioning," *Earth Planets & Space*, vol. 68, pp. 29–38, 2016.

[32] S. K. Singh and S. P. Yadav, "Modeling and optimization of multi objective non-linear programming problem in intuitionistic fuzzy environment," *Applied Mathematical Modelling*, vol. 39, no. 16, pp. 4617–4629, 2015.

[33] X. Peng, B. Zhao, R. Yan, H. J. Tang, and Y. Zhang, "Bag of Events: an efficient probability-based feature extraction method for AER image sensors," *IEEE Transactions on Neural Networks & Learning Systems*, vol. 28, pp. 791–803, 2016.

[34] R. Vijay, S. M. Pinto, V. K. Kushwaha, S. Pal, and T. Nandy, *A Multi-Temporal Analysis for Change Assessment And Estimation of Algal Bloom in Sambhar Lake*, vol. 188, Environmental Monitoring & Assessment, Rajasthan, India, 2016.

[35] H. Yan, Y. Huang, G. Wang et al., "Water eutrophication evaluation based on rough set and petri nets: A case study in Xiangxi-River, Three Gorges Reservoir," *Ecological Indicators*, vol. 69, pp. 463–472, 2016.

[36] J. H. Andersen, J. Aroviita, J. Carstensen et al., "Approaches for integrated assessment of ecological and eutrophication status of surface waters in Nordic Countries," *AMBIO*, vol. 45, no. 6, pp. 681–691, 2016.

[37] People's Republic of China, *Technological regulations for surface water resources quality assessment (SL 395-2007)*, vol. 43, Ministry of Water Resources, 2007.

[38] S. Tahsin and N. B. Chang, "Fast eutrophication assessment for stormwater wet detention ponds via fuzzy probit regression analysis under uncertainty," *Environmental Monitoring & Assessment*, vol. 188, pp. 1–18, 2016.

[39] Z. Y. Zhao, Q. Quan, and K. Y. Cai, "A health performance prediction method of large-scale stochastic linear hybrid systems with small failure probability," *Reliability Engineering & System Safety*, vol. 165, pp. 74–88, 2017.

[40] Z. Zhao, Q. Quan, and K.-Y. Cai, "A profust reliability based approach to prognostics and health management," *IEEE Transactions on Reliability*, vol. 63, no. 1, pp. 26–41, 2014.

22

Modelling and Interpretation of Adsorption Isotherms

Nimibofa Ayawei, Augustus Newton Ebelegi, and Donbebe Wankasi

Department of Chemical Sciences, Niger Delta University, Wilberforce Island, Bayelsa State, Nigeria

Correspondence should be addressed to Nimibofa Ayawei; ayawei4acad@gmail.com

Academic Editor: Wenshan Guo

The need to design low-cost adsorbents for the detoxification of industrial effluents has been a growing concern for most environmental researchers. So modelling of experimental data from adsorption processes is a very important means of predicting the mechanisms of various adsorption systems. Therefore, this paper presents an overall review of the applications of adsorption isotherms, the use of linear regression analysis, nonlinear regression analysis, and error functions for optimum adsorption data analysis.

1. Introduction

The migration of pollutant(s) in aqueous media and subsequent development of containment measures have resulted in the use of adsorption among other techniques [1, 2]. Adsorption equilibrium information is the most important piece of information needed for a proper understanding of an adsorption process.

A proper understanding and interpretation of adsorption isotherms is critical for the overall improvement of adsorption mechanism pathways and effective design of adsorption system [3].

In recent times, linear regression analysis has been one of the most applied tools for defining the best fitting adsorption models because it quantifies the distribution of adsorbates, analyzes the adsorption system, and verifies the consistency of theoretical assumptions of adsorption isotherm model [4].

Because of the inherent bias created by linearization, several error functions have been used to address this shortfall. Concomitant with the evolution of computer technology, the use of nonlinear isotherm modelling has been extensively used.

2. One-Parameter Isotherm

2.1. Henry's Isotherms. This is the simplest adsorption isotherm in which the amount of surface adsorbate is proportional to the partial pressure of the adsorptive gas [4].

This isotherm model describes an appropriate fit to the adsorption of adsorbate at relatively low concentrations such that all adsorbate molecules are secluded from their nearest neighbours [5].

Thus, the equilibrium adsorbate concentrations in the liquid and adsorbed phases are related to the linear expression:

$$q_e = K_{HE}C_e, \tag{1}$$

where q_e is amount of the adsorbate at equilibrium (mg/g), K_{HE} is Henry's adsorption constant, and C_e is equilibrium concentration of the adsorbate on the adsorbent.

3. Two-Parameter Isotherm

3.1. Hill-Deboer Model. The Hill-Deboer isotherm model describes a case where there is mobile adsorption as well as lateral interaction among adsorbed molecules [6, 7].

The linearized form of this isotherm equation is as follows [8]:

$$\ln\left[\frac{C_e(1-\theta)}{\theta}\right] - \frac{\theta}{1-\theta} = -\ln K_1 - \frac{K_2\theta}{RT}, \tag{2}$$

where K_1 is Hill-Deboer constant (Lmg^{-1}) and K_2 is the energetic constant of the interaction between adsorbed molecules (KJmol^{-1}). Equilibrium data from adsorption experiments can be analyzed by plotting $\ln[C_e(1-\theta)/\theta] - \theta/(1-\theta)$ versus θ [8–10].

3.2. Fowler-Guggenheim Model. Fowler-Guggenheim proposed this isotherm equation which takes into consideration the lateral interaction of the adsorbed molecules [11]. The linear form of this isotherm model is as follows [8]:

$$\ln\left[\frac{C_e(1-\theta)}{\theta}\right] = -\ln K_{\text{FG}} + \frac{2w\theta}{RT}, \qquad (3)$$

where K_{FG} is Fowler-Guggenheim equilibrium constant (Lmg^{-1}), θ is fractional coverage, R is universal gas constant ($\text{KJmol}^{-1}\,\text{K}^{-1}$), T is temperature (k), and w is interaction energy between adsorbed molecules (KJmol^{-1}).

This isotherm model is predicated on the fact that the heat of adsorption varies linearly with loading. Therefore, if the interaction between adsorbed molecules is attractive, then the heat of adsorption will increase with loading because of increased interaction between adsorbed molecules as loading increases (i.e., w = positive). However, if the interaction among adsorbed molecules is repulsive, then the heat of adsorption decreases with loading (i.e., w = negative). But when w = 0 then there is no interaction between adsorbed molecules, and the Fowler-Guggenheim isotherm reduces to the Langmuir equation.

A plot of $\ln[C_e(1-\theta)/\theta]$ versus θ is used to obtain the values for K_{FG} and w.

It is important to note that this model is only applicable when surface coverage is less than 0.6 ($\theta < 0.6$).

Kumara et al. analyzed the adsorption data for the phenolic compounds onto granular activated carbon with the Fowler-Guggenheim isotherm and reported that the interaction energy (w) was positive which indicates that there is attraction between the adsorbed molecules [8].

3.3. Langmuir Isotherm. Langmuir adsorption which was primarily designed to describe gas-solid phase adsorption is also used to quantify and contrast the adsorptive capacity of various adsorbents [12]. Langmuir isotherm accounts for the surface coverage by balancing the relative rates of adsorption and desorption (dynamic equilibrium). Adsorption is proportional to the fraction of the surface of the adsorbent that is open while desorption is proportional to the fraction of the adsorbent surface that is covered [13].

The Langmuir equation can be written in the following linear form [14]:

$$\frac{C_e}{q_e} = \frac{1}{q_m K_e} + \frac{C_e}{q_m}, \qquad (4)$$

where C_e is concentration of adsorbate at equilibrium ($\text{mg}\,\text{g}^{-1}$).

K_L is Langmuir constant related to adsorption capacity ($\text{mg}\,\text{g}^{-1}$), which can be correlated with the variation of the suitable area and porosity of the adsorbent which implies that large surface area and pore volume will result in higher adsorption capacity.

The essential characteristics of the Langmuir isotherm can be expressed by a dimensionless constant called the separation factor R_L [15].

$$R_L = \frac{1}{1 + K_L C_o}, \qquad (5)$$

where K_L is Langmuir constant ($\text{mg}\,\text{g}^{-1}$) and C_o is initial concentration of adsorbate ($\text{mg}\,\text{g}^{-1}$).

R_L values indicate the adsorption to be unfavourable when $R_L > 1$, linear when $R_L = 1$, favourable when $0 < R_L < 1$, and irreversible when $R_L = 0$.

Dąbrowski studied the adsorption of direct dye onto a Novel Green Adsorbate developed from Uncaria Gambir extract; their equilibrium data were well described by the Langmuir isotherm model [14].

3.4. Freundlich Isotherm. Freundlich isotherm is applicable to adsorption processes that occur on heterogonous surfaces [15]. This isotherm gives an expression which defines the surface heterogeneity and the exponential distribution of active sites and their energies [16].

The linear form of the Freundlich isotherm is as follows [17]:

$$\log q_e = \log K_F + \frac{1}{n}\log C_e, \qquad (6)$$

where K_F is adsorption capacity (L/mg) and $1/n$ is adsorption intensity; it also indicates the relative distribution of the energy and the heterogeneity of the adsorbate sites.

Boparai et al. investigate the adsorption of lead (II) ions [17] from aqueous solutions using coir dust and its modified extract resins. Although several isotherm models were applied, the equilibrium data was best represented by Freundlich and Flory-Huggins isotherms due to high correlation coefficients [18].

3.5. Dubinin-Radushkevich Isotherm. Dubinin-Radushkevich isotherm model [19] is an empirical adsorption model that is generally applied to express adsorption mechanism with Gaussian energy distribution onto heterogeneous surfaces [20].

This isotherm is only suitable for intermediate range of adsorbate concentrations because it exhibits unrealistic asymptotic behavior and does not predict Henry's laws at low pressure [21].

The model is a semiempirical equation in which adsorption follows a pore filling mechanism [22]. It presumes a multilayer character involving Van Der Waal's forces, applicable for physical adsorption processes, and is a fundamental equation that qualitatively describes the adsorption of gases and vapours on microporous sorbents [23].

It is usually applied to differentiate between physical and chemical adsorption of metal ions [22]. A distinguishing feature of the Dubinin-Radushkevich isotherm is the fact that it is temperature dependent; hence when adsorption data at different temperatures are plotted as a function of logarithm of amount adsorbed versus the square of potential energy, all suitable data can be obtained [13].

Dubinin-Radushkevich isotherm is expressed as follows [16]:

$$\ln q_e = \ln q_m - \beta E^2$$

$$\epsilon = RT \ln\left(1 + \frac{1}{C_e}\right)$$

$$E = \frac{1}{\sqrt{2B}},$$

$$(7)$$

where ϵ is Polanyi potential, β is Dubinin-Radushkevich constant, R is gas constant ($8.31\,\mathrm{Jmol^{-1}\,k^{-1}}$), T is absolute temperature, and E is mean adsorption energy.

Ayawei et al. and Vijayaraghavan et al. applied the Dubinin-Radushkevich isotherm in their investigation of Congo red adsorption behavior on Ni/Al-CO$_3$ and sorption behavior of cadmium on nanozero-valent iron particles, respectively [16, 22].

3.6. Temkin Isotherm.

Temkin isotherm model takes into account the effects of indirect adsorbate/adsorbate interactions on the adsorption process; it is also assumed that the heat of adsorption (ΔH_{ads}) of all molecules in the layer decreases linearly as a result of increase surface coverage [24]. The Temkin isotherm is valid only for an intermediate range of ion concentrations [25]. The linear form of Temkin isotherm model is given by the following [22]:

$$q_e = \frac{Rt}{b} \ln K_T + \frac{RT}{b} \ln C_e,$$

$$(8)$$

where b is Temkin constant which is related to the heat of sorption ($\mathrm{Jmol^{-1}}$) and K_T is Temkin isotherm constant ($\mathrm{Lg^{-1}}$) [26].

Hutson and Yang applied Temkin isotherm model to confirm that the adsorption of cadmium ion onto nanozero-valent iron particles follows a chemisorption process. Similarly, Elmorsi et al. used the Temkin isotherm model in their investigation of the adsorption of methylene blue onto miswak leaves [18].

3.7. Flory-Huggins Isotherm.

Flory-Huggins isotherm describes the degree of surface coverage characteristics of the adsorbate on the adsorbent [27].

The linear form of the Flory-Huggins equation is expressed as

$$\ln\left(\frac{\theta}{C_o}\right) = \ln K_{FH} + n \ln (1 - \theta),$$

$$(9)$$

where θ is degree of surface coverage, n is number of adsorbates occupying adsorption sites, and K_{FH} is Flory-Huggins equilibrium constant ($\mathrm{Lmol^{-1}}$).

This isotherm model can express the feasibility and spontaneity of an adsorption process.

The equilibrium constant K_{FH} is used to calculate spontaneity Gibbs free energy as shown in the following expression [28]:

$$\Delta G^o = RT \ln (K_{FH}),$$

$$(10)$$

where ΔG^o is standard free energy change, R is universal gas constant $8.314\,\mathrm{Jmol^{-1}\,K^{-1}}$, and T is absolute temperature.

Hamdaoui and Naffrechoux used the Flory-Huggins isotherm model in their study of the biosorption of Zinc from aqueous solution using coconut coir dust [29].

3.8. Hill Isotherm.

The Hill isotherm equation describes the binding of different species onto homogeneous substrates. This model assumes that adsorption is a cooperative phenomenon with adsorbates at one site of the adsorbent influencing different binding sites on the same adsorbent [30].

The linear form of this isotherm is expressed as follows [29]:

$$\log \frac{q_e}{q_H - q_e} = n_H \log (C_e) - \log (K_D),$$

$$(11)$$

where K_D, n_H, and q_H are constants.

Hamdaoui and Naffrechoux investigated the equilibrium adsorption of aniline, benzaldehyde, and benzoic acid on granular activated carbon (GAC) using the Hill isotherm model; according to their report, the Hill model was very good in comparison with previous models with $R^2 = 0.99$ for all adsorbates [29].

3.9. Halsey Isotherm.

The Halsey isotherm is used to evaluate multilayer adsorption at a relatively large distance from the surface [16]. The adsorption isotherm can be given as follows [31]:

$$q_e = \frac{1}{n_H} I_n K_H - \frac{1}{n_H} \ln C_{qe},$$

$$(12)$$

where K_H and n are Halsey isotherm constant and they can be obtained from the slope and intercept of the plot of $\ln q_e$ versus $\ln C_e$.

Fowler and Guggenheim reported the use of Halsey isotherm in their equilibrium studies of methyl orange sorption by pinecone derived activated carbon. The fitting of their experimental data to the Halsey isotherm model attests to the heteroporous nature of the adsorbent [31]. Similarly, Song et al. applied the Halsey isotherm for the study of coconut shell carbon prepared by KOH activation for the removal of pb^{2+} ions from aqueous solutions. The Halsey isotherm fits the experimental data well due to high correlation coefficient (R^2), which may be attributed to the heterogeneous distribution of activate sites and multilayer adsorption on coconut shell carbons [16].

3.10. Harkin-Jura Isotherm.

Harkin-Jura isotherm model assumes the possibility of multilayer adsorption on the surface of absorbents having heterogeneous pore distribution [32]. This model is expressed as follows:

$$\frac{1}{q_e^2} = \frac{B}{A} - \left(\frac{1}{A}\right) \log C_e,$$

$$(13)$$

where B and A are Harkin-Jura constants that can be obtained from plotting $1/q_e^2$ versus $\log C_e$.

Foo and Hameed reported that the Harkin-Jura isotherm model showed a better fit to the adsorption data than Freundlich, Halsey, and Temkin isotherm models in their investigation of the adsorptive removal of reactive black 5 from wastewater using Bentonite clay [32].

3.11. Jovanovic Isotherm.

The Jovanovic model is predicated on the assumptions contained in the Langmuir model, but in addition the possibility of some mechanical contacts between the adsorbate and adsorbent [33].

The linear form of the Jovanovic isotherm is expressed as follows [34]:

$$\ln q_e = \ln q_{max} - K_J C_e, \qquad (14)$$

where q_e is amount of adsorbate in the adsorbent at equilibrium ($mg\,g^{-1}$), q_{max} is maximum uptake of adsorbate obtained from the plot of $\ln q_e$ versus C_e, and K_J is Jovanovic constant.

Kiseler reported the use of Jovanovic isotherm model while determining adsorption isotherms for L-Lysine imprinted polymer. Their report showed that the best prediction of retention capacity was obtained by applying the Jovanovic isotherm model [33].

3.12. Elovich Isotherm. The equation that defines this model is based on a kinetic principle which assumes that adsorption sites increase exponentially with adsorption; this implies a multilayer adsorption [35]. The equation was first developed to describe the kinetics of chemisorption of gas onto solids [36].

The linear forms of the Elovich model are expressed as follows [37]:

$$\frac{q_e}{q_m} = K_E C_e e^{\frac{q_e}{q_m}} \qquad (15)$$

but the linear form is expressed as follows [8]:

$$\ln \frac{q_e}{C_e} = \ln K_e q_m - \frac{q_e}{q_m}. \qquad (16)$$

Elovich maximum adsorption capacity and Elovich constant can be calculated from the slope and intercept of the plot of $\ln(q_e/C_e)$ versus q_e.

Rania et al. reported the use of Elovich isotherm model in their work titled "Equilibrium and Kinetic Studies of Adsorption of Copper (II) Ions on Natural Sorbent." Their investigation showed that the value of the regression coefficient (R^2) for the Elovich model was 0.808 which is higher than that of Langmuir; therefore the adsorption of copper (II) onto Chitin was best described by the Elovich isotherm.

3.13. Kiselev Isotherm. The Kiselev adsorption isotherm equation also known as localized monomolecular layer model [38] is only valid for surface coverage $\theta > 0.68$ and its linearized expression is as follows:

$$\frac{1}{C_e(1-\theta)} = \frac{K_1}{\theta} + K_i K_n, \qquad (17)$$

where K_i is Kiselev equilibrium constant (Lmg^{-1}) and K_n is equilibrium constant of the formation of complex between adsorbed molecules.

Equilibrium data from adsorption processes can be modelled by plotting $1/C_e(1-\theta)$ versus $1/\theta$ [8, 38–40].

4. Three-Parameter Isotherms

4.1. Redlich-Peterson Isotherm. The Redlich-Peterson isotherm is a mix of the Langmuir and Freundlich isotherms. The numerator is from the Langmuir isotherm and has the benefit of approaching the Henry region at infinite dilution [41].

This isotherm model is an empirical isotherm incorporating three parameters. It combines elements from both Langmuir and Freundlich equations; therefore the mechanism of adsorption is a mix and does not follow ideal monolayer adsorption [42].

This model is defined by the following expression:

$$q_e = \frac{AC_e}{1 + BC_e^{\beta}}, \qquad (18)$$

where A is Redlich-Peterson isotherm constant (Lg–1), B is constant (Lmg^{-1}), β is exponent that lies between 0 and 1, C_e is equilibrium liquid-phase concentration of the adsorbent (mgl^{-1}), and q_e is equilibrium adsorbate loading on the adsorbent ($mg\,g^{-1}$).

At high liquid-phase concentrations of the adsorbate, (16) reduces to the Freundlich equation:

$$q_e = \frac{A}{B} C_e^{1-\beta}, \qquad (19)$$

where $A/B = K_F$ and $(1-\beta) = 1/n$ of the Freundlich isotherm model.

When $\beta = 1$, (18) reduces to Langmuir equation with $b = B$ (Langmuir adsorption constant (Lmg^{-1} which is related to the energy of adsorption.

$A = bq_{ml}$ where q_{ml} is Langmuir maximum adsorption capacity of the adsorbent ($mg\,g^{-1}$); when $\beta = 0$, (18) reduces to Henry's equation with $1/(1 + b)$ representing Henry's constant.

The linear form of the Redlich-Peterson isotherm can be expressed as follows [34]:

$$\ln \frac{C_e}{q_e} = \beta \ln C_e - \ln A \qquad (20)$$

A plot of $\ln(C_e/q_e)$ versus $\ln C_e$ enables the determination of Redlich-Peterson constants, where β is slope and A is intercept [30, 42–45].

This isotherm model has a linear dependence on concentration in the numerator and an exponential function in the denomination which altogether represent adsorption equilibrium over a wide range of concentration of adsorbate which is applicable in either homogenous or heterogeneous systems because of its versatility [46, 47].

4.2. Sips Isotherm. Sips isotherm is a combination of the Langmuir and Freundlich isotherms and it is given the following general expression [48]:

$$q_e = \frac{K_s C_e^{\beta s}}{1 - a_s C_e^{\beta s}}, \qquad (21)$$

where K_s is Sips isotherm model constant (Lg^{-1}), β_s is Sips isotherm exponent, and a_s is Sips isotherm model constant (Lg^{-1}). The linearized form is given as follows [12]:

$$\beta_s \ln C_e = -\ln\left(\frac{K_s}{q_e}\right) + \ln(a_s). \qquad (22)$$

This model is suitable for predicting adsorption on heterogeneous surfaces, thereby avoiding the limitation of increased adsorbate concentration normally associated with the Freundlich model [19]. Therefore at low adsorbate concentration this model reduces to the Freundlich model, but at high concentration of adsorbate, it predicts the Langmuir model (monolayer adsorption). The parameters of the Sips isotherm model are p^H, temperature, and concentration dependent [12, 49] and isotherm constants differ by linearization and nonlinear regression [50].

4.3. *Toth Isotherm.* The Toth isotherm is another empirical modification of the Langmuir equation with the aim of reducing the error between experimental data and predicted value of equilibrium data [51]. This model is most useful in describing heterogeneous adsorption systems which satisfy both low and high end boundary of adsorbate concentration [52]. The Toth isotherm model is expressed as follows [52]:

$$\frac{q_e}{q_m} = \theta = \frac{K_e C_e}{\left[1 + (K_L C_e)^n\right]^{1/n}}, \qquad (23)$$

where K_L is Toth isotherm constant ($mg\,g^{-1}$) and n is Toth isotherm constant ($mg\,g^{-1}$).

It is clear that when $n = 1$, this equation reduces to Langmuir isotherm equation. Therefore the parameter n characterizes the heterogeneity of the adsorption system [51] and if it deviates further away from unity (1), then the system is said to be heterogeneous. The Toth isotherm may be rearranged to give a linear form as follows:

$$\ln \frac{q_e^n}{q_m^n - q_e^n} = n \ln K_L + n \ln C_e. \qquad (24)$$

The values of parameters of the Toth model can be evaluated by nonlinear curve fitting method using sigma plot software [53].

This isotherm model has been applied for the modelling of several multilayer and heterogeneous adsorption systems [53, 54].

4.4. *Koble-Carrigan Isotherm.* Koble-Carrigan isotherm model is a three-parameter equation which incorporates both Langmuir and Freundlich isotherms for representing equilibrium adsorption data [55]. The linear form of this module is represented by the following equation [56]:

$$\frac{1}{q_e} = \left(\frac{1}{A_k C_e^p}\right) + \frac{B_k}{A_k}, \qquad (25)$$

where A_k is Koble-Carrigan's isotherm constant, B_k is Koble-Carrigan's isotherm constant, and p is Koble-Carrigan's isotherm constant.

All three Koble-Carrigan isotherm constants can be evaluated with the use of a solver add-in function of the Microsoft Excel [56]. At high adsorbate concentrations, this model reduces to Freundlich isotherm. It is only valid when the constant "p" is greater than or equal to 1. When "p" is less than unity (1), it signifies that the model is incapable of defining the experimental data despite high concentration coefficient or low error value [54].

4.5. *Kahn Isotherm.* The Kahn isotherm model is a general model for adsorption of biadsorbate from pure dilute equations solutions [57].

This isotherm model is expressed as follows [58]:

$$Q_e = \frac{Q_{max} b_k C_e}{(1 + b_k C_e) a_k}, \qquad (26)$$

where a_k is Kahn isotherm model exponent, b_k is Khan isotherm model constant, and Q_{max} is Khan isotherm maximum adsorption capacity ($mg\,g^{-1}$).

Nonlinear methods have been applied by several researchers to obtain the Khan isotherm model parameters [59, 60].

4.6. *Radke-Prausniiz Isotherm.* The Radke-Prausnitz isotherm model has several important properties which makes it more preferred in most adsorption systems at low adsorbate concentration [61].

The isotherm is given by the following expression:

$$q_e = \frac{q_{MRP} K_{RP} C_e}{(1 + K_{RP} C_e)^{MRP}}, \qquad (27)$$

where q_{MRP} is Radke-Prausnitz maximum adsorption capacity ($mg\,g^{-1}$), K_{RP} is Radke-Prausnitz equilibrium constant, and MRP is Radke-Prausnitz model exponent.

At low adsorbate concentration, this isotherm model reduces to a linear isotherm, while at high adsorbate concentration it becomes the Freundlich isotherm and when $M_{RP} = 0$, it becomes the Langmuir isotherm. Another important characteristic of this isotherm is that it gives a good fit over a wide range of adsorbate concentration. The Radke-Prausnitz model parameters are obtained by nonlinear statistical fit of experimental data [61, 62].

4.7. *Langmuir-Freundlich Isotherm.* Langmuir-Freundlich isotherm includes the knowledge of adsorption heterogeneous surfaces. It describes the distribution of adsorption energy onto heterogeneous surface of the adsorbent [54]. At low adsorbate concentration this model becomes the Freundlich isotherm model, while at high adsorbate concentration it becomes the Langmuir isotherm. Langmuir-Freundlich isotherm can be expressed as follows:

$$q_e = \frac{q_{MLF} \left(K_{LF} C_e\right)^{MLF}}{1 + \left(K_{LF} C_e\right)^{MLF}}, \qquad (28)$$

where q_{MLF} is Langmuir-Freundlich maximum adsorption capacity ($mg\,g^{-1}$), K_{LF} is equilibrium constant for heterogeneous solid, and M_{LF} is heterogeneous parameter and it lies between 0 and 1. These parameters can be obtained by using the nonlinear regression techniques [63].

4.8. Jossens Isotherm.

The Jossens isotherm model predicts a simple equation based on the energy distribution of adsorbate-adsorbent interactions at adsorption sites [64]. This model assumes that the adsorbent has heterogeneous surface with respect to the interactions it has with the adsorbate. The Jossen isotherm can be represented as follows:

$$C_e = \frac{q_e}{H} \exp\left(F q_e^p\right), \tag{29}$$

where H is Jossens isotherm constant (it corresponds to Henry's constant), p is Jossens isotherm constant and it is characteristic of the adsorbent irrespective of temperature and the nature of adsorbents, and F is Jossens isotherm constant.

The equation reduces to Henry's law at low capacities. However, upon rearranging (29) [65],

$$\ln\left(\frac{C_e}{q_e}\right) = -\ln(H) + F q_e^p. \tag{30}$$

The values of H and F can be obtained from either a plot of $\ln(C_e/Q_e)$ versus q_e or using a least square fitting procedure.

A good representation of equilibrium data using this equation was reported for phenolic compounds on activated carbon [66] and on amberlite XAD-4 and XAD-7 macroreticular resins [67].

5. Four-Parameter Isotherms

5.1. Fritz-Schlunder Isotherm.

Fritz and Schlunder derived an empirical equation which can fit a wide range of experimental results because of the large number of coefficients in the isotherm [68].

This isotherm model has the following equation:

$$q_e = \frac{q_{\mathrm{mFS5}} K_{\mathrm{FS}} C_e}{1 + q_m C_e^{\mathrm{MFS}}}, \tag{31}$$

where q_{mFS} is Fritz-Schlunder maximum adsorption capacity (mg g^{-1}), K_{FS} is Fritz-Schlunder equilibrium constant (mg g^{-1}), and MFS is Fritz-Schlunder model exponent.

If $M_{\mathrm{FS}} = 1$, then the Fritz-Schlunder model becomes the Langmuir model, but, for high adsorbate concentrations, the model reduces to Freundlich model.

Fritz-Schlunder isotherm parameters can be determined by nonlinear regression analysis [69, 70].

5.2. Baudu Isotherm.

Bauder observed that the estimation of the Langmuir coefficients, b and q_{ml}, by measurement of tangents at different equilibrium concentrations shows that they are not constants in a broad range [71]; therefore the Langmuir isotherm has been reduced to the Bauder isotherm [62]:

$$q_e = \frac{q_m b o C_e^{1+x+y}}{1 + b o C_e^{1+x}}, \tag{32}$$

where q_m id Bauder maximum adsorption capacity (mg g^{-1}), bo is equilibrium constant, x is Baudu parameter, and Y is Baudu parameter.

For lower surface coverage the Bauder isotherm model reduces to Freundlich model.

Due to the inherent bias resulting from linearization this isotherm parameters are determined by nonlinear regression analysis [72].

5.3. Weber-Van Vliet Isotherm.

Weber and Van Vliet postulated an empirical relation with four parameters that provided excellent description of data patterns for a wide range of adsorption systems [73].

The isotherm developed by weber and Van Vliet has the following form:

$$C_e = p_1 q_e^{\left(p_2 q_e^{p_3} + p_4\right)}, \tag{33}$$

where C_e is equilibrium concentration of the adsorbate (mg g^{-1}), q_e is adsorption capacity mg g^{-1}, p_1, p_2, p_3, and p_4 are Weber-Van Vliet isotherm parameters

The isotherm parameters (p_1, p_2, p_3, and p_4) can be defined by multiple nonlinear curve fitting techniques which is predicated on the minimization of sum of square of residual [72–74].

5.4. Marczewski-Jaroniec Isotherm.

The Marczewski-Jaroniec isotherm is also known as the four-parameter general Langmuir equation [75]. It is recommended on the basis of the supposition of local Langmuir isotherm and adsorption energies distribution in the active sites on adsorbent [76].

The isotherm equation is expressed as follows:

$$q_e = q_{MMJ} \left(\frac{\left(K_{MJ} C_e\right)^{nMJ}}{1 + \left(K_{MJ} C_E\right)^{nMJ}}\right)^{M_{MJ}/n_{MJ}}, \tag{34}$$

where n_{MJ} and M_{MJ} are parameters that characterize the heterogeneity of the adsorbent surface, M_{MJ} describes the spreading of distribution in the path of higher adsorption energy, and n_{MJ} describes the spreading in the path of lesser adsorption energies.

The isotherm reduces to Langmuir isotherm when n_{MJ} and $M_{MJ} = 1$, when $n_{MJ} = M_{MJ}$; it reduces to Langmuir-Freundlich model.

6. Five-Parameter Isotherms

Fritz and Schlunder developed a five-parameter empirical model that is capable of simulating the model variations more precisely for application over a wide range of equilibrium data [74].

The isotherm equation is

$$q_e = \frac{1_m \mathrm{FS}_s K_1 C_e^{\alpha_{\mathrm{FS}}}}{1 + K_2 C_e^{\beta_{\mathrm{FS}}}}, \tag{35}$$

where q_{mFS5} is Fritz-Schlunder maximum adsorption capacity (mg g^{-1}) and K_1, K_2, α_{FS}, and β_{FS} are Fritz-Schlunder parameters.

This isotherm is valid only in the range of L_{FS} value less than or equal to 1.

This model approaches Langmuir model while the value of both exponents α_{FS} and β_{FS} equals 1 and for higher adsorbate concentrations it reduces to Freundlich model.

7. Error Analysis

In recent times linear regression analysis has been among the most pronounced and viable tools frequently applied for analysis of experimental data obtained from adsorption process. It has been used to define the best fitting relationship that quantify the distribution of adsorbates and also in the verification of the consistency of adsorption models and the theoretical assumptions of adsorption models [77, 78].

Studies have shown that the error structure of experimental data is usually changed during the transformation of adsorption isotherms into their linearized forms [79]. It is against this backdrop that nonlinearized regression analysis became inevitable, since it provides a mathematically rigorous method for determining adsorption parameters using original form of isotherm equations [80, 81].

Unlike linear regression, nonlinear regression usually involved the minimization of error distribution between the experimental data and the predicted isotherm based on its convergence criteria [82]. This operation is no longer computationally difficult because of availability of computer algorithms [21].

7.1. The Sum Square of Errors (ERRSQ). The sum of square of errors (ERRSQ) is said to be the most widely used error function [83]. This method can be represented by the following expression [45]:

$$\sum_{i=1}^{n} \left(q_{e,1,\mathrm{calc}} - q_{e,i,\mathrm{meas}} \right)^2, \tag{36}$$

where $q_{e,i,\mathrm{calc}}$ is the theoretical concentration of adsorbate on the adsorbent, which have been calculated from one of the isotherm models.

$q_{e,i,\mathrm{meas}}$ is the experimentally measured adsorbed solid phase concentration of the adsorbate adsorbed on the adsorbent.

One major disadvantage of this error function is that at higher end of liquid-phase adsorbate concentration ranges the isotherm parameters derived using this error function will provide a better fit as the magnitude of the errors and therefore the square of errors tend to increase illustrating a better fit for experimental data obtained at the high end of concentration range [45, 84].

7.2. Hybrid Fractional Error Function (HYBRID). The hybrid fractional error function (HYBRID) was developed by Kapoor and Yang, to improve the fit of the sum square of errors (ERRSQ) [85] at low concentrations by dividing it by the measured value. This function includes the number of data points (n), minus the number of parameters (p) or isotherm equation as a divisor [85].

The equation for this error function is

$$\mathrm{HYBRID} = \frac{100}{n-p} \sum_{i=1}^{n} \left[\frac{\left(q_{e,i,\mathrm{meas}} - q_{e,i,\mathrm{calc}} \right)^2}{q_{e,i,\mathrm{meas}}} \right]. \tag{37}$$

7.3. Average Relative Error (ARE). The average relative error was developed by Marquardt [86] with the aim of minimizing the fractional error distribution across the entire concentration range. It is given by the following expression:

$$\mathrm{ARE} = \frac{100}{n} \sum_{i=1}^{n} \left[\frac{q_{e,i,\mathrm{calc}} - q_{e,i,\mathrm{meas}}}{q_{e,i,\mathrm{meas}}} \right]. \tag{38}$$

7.4. Marquardt's Percent Standard Deviation (MPSD). The Marquardt's percent standard deviation error function is similar to a geometric mean error distribution modified according to the degree of freedom of the system [87]. It is given by the following expression:

$$\mathrm{MPSD} = \sqrt{\frac{1}{n-p} \sum_{i=1}^{n} \left(\frac{\left(q_{e,i,\mathrm{exp}} - q_{e,\mathrm{calc}} \right)}{q_{e,\mathrm{exp}}} \right)^2}. \tag{39}$$

7.5. Sum of Absolute Errors (EABS). This model is similar to the sum square error (ERRSQ) function. In this case isotherm parameters determined using this error function would provide a better fit as the forward high concentration data [88]. It is represented by the following equation:

$$\mathrm{EABS} = \sum_{I=1}^{p} \left[q_{e,\mathrm{meas}} - q_{e,\mathrm{calc}} \right]. \tag{40}$$

7.6. Sum of Normalized Errors (SNE). Since each of the error criteria is likely to produce a different set of parameters of the isotherm, a standard procedure known as sum of the normalized errors is adopted to normalize and to combine the error in order to make a move meaningful comparison between the parameters sets. It has been used by several researchers to determine the best fitting isotherm model [88–91].

Calculation procedure is as follows:

(i) Selection of an isotherm model and error function and determination of the adjustable parameters which minimized the error function

(ii) Determination of all other error functions by referring to the parameters set

(iii) Computation of other parameter sets associated with their error function values

(iv) Normalization and selection of maximum parameters sets with respect to the largest error measurement

(v) Summation of all these normalized errors for each parameter set

7.7. Coefficient of Determination (R^2) Spearman's Correlation Coefficient (R_s) and Standard Deviation of Relative Errors (S_{RE}). The coefficient of determination represents the variance about the mean; it is used to analyze the fitting degrees of isotherms and kinetic models with experimental data [12, 92]. The coefficient of determination (R^2) is defined by the following equation [93]:

$$R^2 = \frac{\sum \left(q_{ecal} - q_{mexp}\right)^2}{\sum \left(q_{ecal} - q_{mexp}\right)^2 + \left(q_{cal} - q_{mexp}\right)^2}, \quad (41)$$

where q_{exp} is amount of adsorbate adsorbed by adsorbent during the experiment (mg g^{-1}), q_{cal} is amount of adsorbate obtained by kinetic isotherm models (mg g^{-1}), and q_{mexp} is average of q_{exp} (mg g^{-1}).

7.8. Nonlinear Chi-Square Test (X^2). This function is very important in the determination of the best fit of an adsorption system. It can be obtained by judging the sum square difference between experimental and calculated data, with each square difference divided by its corresponding values [90]. The value of this function can be obtained from the following equation:

$$\sum_{i=1}^{n} \frac{\left(q_{ecal} - q_{emeas}\right)^2}{q_{emeas}}. \quad (42)$$

7.9. Coefficient of Nondetermination. This function is very valuable tool for describing the extent of relationship between the transformed experimental data and the predicted isotherms and minimization of error distribution [93].

$$\text{Coefficient of nondetermination} = 1.00 - R^2, \quad (43)$$

where R^2 is coefficient of determination.

8. Conclusion

The level of accuracy obtained from adsorption processes is greatly dependent on the successful modelling and interpretation of adsorption isotherms.

While linear regression analysis has been frequently used in accessing the quality of fits and adsorption performance because of its wide applicability in a variety of adsorption data, nonlinear regression analysis has also been widely used by a number of researchers in a bid to close the gap between predicted and experimental data. Therefore, there is the need to identify and clarify the usefulness of both linear and nonlinear regression analysis in various adsorption systems.

Conflicts of Interest

The authors (Ayawei Nimibofa, Ebelegi Newton Augustus, and Wankasi Donbebe) declare that there are no conflicts of interest regarding the publication of this paper.

References

[1] N. Ayawei, M. Horsfall Jnr, and I. Spiff, "Rhizophora mangle waste as adsorbent for metal ions removal from aqueous solution," *European Journal of Scientic Research*, vol. 9, no. 1, p. 21, 2005.

[2] N. D. Shooto, N. Ayawei, D. Wankasi, L. Sikhwivhilu, and E. D. Dikio, "Study on cobalt metal organic framework material as adsorbent for lead ions removal in aqueous solution," *Asian Journal of Chemistry*, vol. 28, no. 2, pp. 277–281, 2016.

[3] M. I. El-Khaiary, "Least-squares regression of adsorption equilibrium data: comparing the options," *Journal of Hazardous Materials*, vol. 158, no. 1, pp. 73–87, 2008.

[4] S. D. Fost and M. O. Aly, *Adsorption Processes for Water Treatment*, Betterworth Publications, Stoneharm, Massachusetts, Mass, USA, 1981.

[5] D. M. Ruthven, *Principle of Adsorption and Adsorption Processes*, John Willey and Sons, New Jersey, NJ, USA, 1984.

[6] T. L. Hill, "Statistical mechanics of multimolecular adsorption II. Localized and mobile adsorption and absorption," *The Journal of Chemical Physics*, vol. 14, no. 7, pp. 441–453, 1946.

[7] J. H. De Boer, *The Dynamical Character of Adsorption*, Oxford University Press, Oxford, England, 1953.

[8] P. S. Kumara, S. Ramalingamb, S. D. Kiruphac, A. Murugesanc, and S. Vidhyarevicsivanesam, "Adsorption behaviour of Nickel (II) onto cashew nut shell," in *Equilibrium, Thermodynamics, Kinetics, Mechanism and Process design. Chemical Engineering Journal*, vol. 1169, pp. 122–131, Adsorption behaviour of Nickel (II) onto cashew nut shell, Equilibrium, 2010.

[9] M. A. Hubbe, J. Park, and S. Park, "Cellulosic substrates for removal of pollutants from aqueous systems: A review. Part 4. Dissolved petrochemical compounds," *BioResources*, vol. 9, no. 4, pp. 7782–7925, 2014.

[10] O. Redlich and D. L. Peterson, "A useful adsorption isotherm," *The Journal of Physical Chemistry*, vol. 63, no. 6, p. 1024, 1959.

[11] P. Sampranpiboon, P. Charnkeitkong, and X. Feng, "Equilibrium isotherm models for adsorption of zinc (II) ion from aqueous solution on pulp waste," *WSEAS Transactions on Environment and Development*, vol. 10, pp. 35–47, 2014.

[12] T. M. Elmorsi, "Equilibrium isotherms and kinetic studies of removal of methylene blue dye by adsorption onto miswak leaves as a natural adsorbent," *Journal of Environmental Protection*, vol. 2, no. 6, pp. 817–827, 2011.

[13] A. Günay, E. Arslankaya, and I. Tosun, "Lead removal from aqueous solution by natural and pretreated clinoptilolite: adsorption equilibrium and kinetics," *Journal of Hazardous Materials*, vol. 146, no. 1-2, pp. 362–371, 2007.

[14] A. Dąbrowski, "Adsorption—from theory to practice," *Advances in Colloid and Interface Science*, vol. 93, no. 1–3, pp. 135–224, 2001.

[15] N. Ayawei, S. S. Angaye, D. Wankasi, and E. D. Dikio, "Synthesis, characterization and application of Mg/Al layered double hydroxide for the degradation of congo red in aqueous solution," *Open Journal of Physical Chemistry*, vol. 5, no. 03, pp. 56–70, 2015.

[16] N. Ayawei, A. T. Ekubo, D. Wankasi, and E. D. Dikio, "Adsorption of congo red by Ni/Al-CO$_3$: equilibrium, thermodynamic and kinetic studies," *Oriental Journal of Chemistry*, vol. 31, no. 30, pp. 1307–1318, 2015.

[17] H. K. Boparai, M. Joseph, and D. M. O'Carroll, "Kinetics and thermodynamics of cadmium ion removal by adsorption onto nano zerovalent iron particles," *Journal of Hazardous Materials*, vol. 186, no. 1, pp. 458–465, 2011.

[18] N. D. Hutson and R. T. Yang, "Theoretical basis for the Dubinin-Radushkevitch (D-R) adsorption isotherm equation," *Adsorption*, vol. 3, no. 3, pp. 189–195, 1997.

[19] C. C. Travis and E. L. Etnier, "A survey of sorption relationships for reactive solutes in soil," *Journal of Environmental Quality*, vol. 10, no. 1, pp. 8–17, 1981.

[20] O. Çelebi, Ç. Üzüm, T. Shahwan, and H. N. Erten, "A radiotracer study of the adsorption behavior of aqueous Ba^{2+} ions on nanoparticles of zero-valent iron," *Journal of Hazardous Materials*, vol. 148, no. 3, pp. 761–767, 2007.

[21] C. Theivarasu and S. Mylsamy, "Removal of malachite green from aqueous solution by activated carbon developed from cocoa (Theobroma Cacao) shell—A kinetic and equilibrium studies," *E-Journal of Chemistry*, vol. 8, no. 1, pp. S363–S371, 2011.

[22] K. Vijayaraghavan, T. V. N. Padmesh, K. Palanivelu, and M. Velan, "Biosorption of nickel(II) ions onto Sargassum wightii: application of two-parameter and three-parameter isotherm models," *Journal of Hazardous Materials*, vol. 133, no. 1–3, pp. 304–308, 2006.

[23] U. Israel and U. M. Eduok, "Biosorption of zinc from aqueous solution using coconut (Cocos nuciferaL) coirdust," *Archives of Applied Science Research*, vol. 4, pp. 809–819, 2012.

[24] D. Ringot, B. Lerzy, K. Chaplain, J.-P. Bonhoure, E. Auclair, and Y. Larondelle, "In vitro biosorption of ochratoxin A on the yeast industry by-products: comparison of isotherm models," *Bioresource Technology*, vol. 98, no. 9, pp. 1812–1821, 2007.

[25] H. Shahbeig, N. Bagheri, S. A. Ghorbanian, A. Hallajisani, and S. Poorkarimi, "A new adsorption isotherm model of aqueous solutions on granular activated carbon," *World Journal of Modelling and Simulation*, vol. 9, no. 4, pp. 243–254, 2013.

[26] M. R. Samarghandi, M. Hadi, S. Moayedi, and F. B. Askari, "Two-parameter isotherms of methyl orange sorption by pinecone derived activated carbon," *Iranian Journal of Environmental Health Science and Engineering*, vol. 6, no. 4, pp. 285–294, 2009.

[27] M. T. Amin, A. A. Alazba, and M. Shafiq, "Adsorptive removal of reactive black 5 from wastewater using bentonite clay: isotherms, kinetics and thermodynamics," *Sustainability*, vol. 7, no. 11, pp. 15302–15318, 2015.

[28] N. A. Ebelegi, S. S. Angaye, N. Ayawei, and D. Wankasi, "Removal of congo red from aqueous solutions using fly ash modified with hydrochloric acid," *British Journal of Applied Science and Technology*, vol. 20, no. 4, pp. 1–7, 2017.

[29] O. Hamdaoui and E. Naffrechoux, "Modeling of adsorption isotherms of phenol and chlorophenols onto granular activated carbon. Part I. Two-parameter models and equations allowing determination of thermodynamic parameters," *Journal of Hazardous Materials*, vol. 147, no. 1-2, pp. 381–394, 2007.

[30] F. Rania and N. S. Yousef, "Equilibrium and Kinetics studies of adsorption of copper (II) on natural Biosorbent," *International Journal of Chemical/Engineering and Applications*, vol. 6, no. 5, 2015.

[31] R. H. Fowler and E. A. Guggenheim, *Statistical Thermodynamics*, Cambridge University Press, London, England, 1939.

[32] K. Y. Foo and B. H. Hameed, "Insights into the modeling of adsorption isotherm systems," *Chemical Engineering Journal*, vol. 156, no. 1, pp. 2–10, 2010.

[33] S. K. Knaebel, "Adsorbent selection," *International Journal of Trend in Research and Development*, Adsorption Research, Incorporated Dublin, Ohio. 43016, 2004.

[34] A. V. C. Kiseler, "vapour adsorption in the formation of adsorbate Mollecule Complexes on the surface," *Kolloid Zhur*, vol. 20, pp. 338–348, 1958.

[35] M. Gubernak, W. Zapała, and K. Kaczmarski, "Analysis of amylbenzene adsorption equilibria on an RP-18e chromatographic column," *Acta Chromatographica*, no. 13, pp. 38–59, 2003.

[36] O. Hamdaoui and E. Naffrechoux, "Modeling of adsorption isotherms of phenol and chlorophenols onto granular activated carbon. Part II. Models with more than two parameters," *Journal of Hazardous Materials*, vol. 147, no. 1-2, pp. 401–411, 2007.

[37] A. Achmad, J. Kassim, T. K. Suan, R. C. Amat, and T. L. Seey, "Equilibrium, kinetic and thermodynamic studies on the adsorption of direct dye onto a novel green adsorbent developed from Uncaria gambir extract," *Journal of Physical Science*, vol. 23, no. 1, pp. 1–13, 2012.

[38] T. W. Weber and R. K. Chakravorti, "Pore and solid diffusion models for fixed-bed adsorbers," *Journal of American Institute of Chemical Engineers*, vol. 20, no. 2, pp. 228–238, 1974.

[39] C. Song, S. Wu, M. Cheng, P. Tao, M. Shao, and G. Gao, "Adsorption studies of coconut shell carbons prepared by KOH activation for removal of lead(ii) from aqueous solutions," *Sustainability (Switzerland)*, vol. 6, no. 1, pp. 86–98, 2014.

[40] A. A. Israel, O. Okon, S. Umoren, and U. Eduok, "Kinetic and equilibrium studies of adsorption of lead (II) Ions from aqueous solution using coir dust (Cocos nucifera L) and it's modified extract resins," *The Holistic Approach to Environment*, 2013.

[41] M. Davoundinejad and S. A. Gharbanian, "Modelling of adsorption isotherm of benzoic compounds onto GAC and introducing three neww isotherm models using new concept of adsorption effective surface (AEC)," *Academic Journals*, vol. 18, no. 46, pp. 2263–2275, 2013.

[42] F. Brouers and T. J. Al-Musawi, "On the optimal use of isotherm models for the characterization of biosorption of lead onto algae," *Journal of Molecular Liquids*, vol. 212, pp. 46–51, 2015.

[43] F.-C. Wu, B.-L. Liu, K.-T. Wu, and R.-L. Tseng, "A new linear form analysis of Redlich-Peterson isotherm equation for the adsorptions of dyes," *Chemical Engineering Journal*, vol. 162, no. 1, pp. 21–27, 2010.

[44] L. S. Chan, W. H. Cheung, S. J. Allen, and G. McKay, "Error analysis of adsorption isotherm models for acid dyes onto bamboo derived activated carbon," *Chinese Journal of Chemical Engineering*, vol. 20, no. 3, pp. 535–542, 2012.

[45] J. C. Y. Ng, W. H. Cheung, and G. McKay, "Equilibrium studies of the sorption of Cu(II) ions onto chitosan," *Journal of Colloid and Interface Science*, vol. 255, no. 1, pp. 64–74, 2002.

[46] F. Gimbert, N. Morin-Crini, F. Renault, P.-M. Badot, and G. Crini, "Adsorption isotherm models for dye removal by cationized starch-based material in a single component system: error analysis," *Journal of Hazardous Materials*, vol. 157, no. 1, pp. 34–46, 2008.

[47] R. Sips, "On the structure of a catalyst surface," *The Journal of Chemical Physics*, vol. 16, no. 5, pp. 490–495, 1948.

[48] G. P. Jeppu and T. P. Clement, "A modified Langmuir-Freundlich isotherm model for simulating pH-dependent adsorption effects," *Journal of Contaminant Hydrology*, vol. 129-130, pp. 46–53, 2012.

[49] C. Chen, "Evaluation of equilibrium sorption isotherm equations," *Open Chemical Engineering Journal*, vol. 7, no. 1, pp. 24–44, 2012.

[50] J. Toth, "State equation of the solid gas interface layer," *Acta Chimica (Academiae Scientiarum) Hungaricae*, vol. 69, pp. 311–317, 1971.

[51] T. Jafari Behbahani and Z. Jafari Behbahani, "A new study on asphaltene adsorption in porous media," *Petroleum and Coal*, vol. 56, no. 5, pp. 459–466, 2014.

[52] M. S. Padder and C. B. C. Majunder, *Studies on Removal of As(II) and S(V) onto GAC/MnFe, 804 Composite: Isotherm Studies and Error Analysis*, 2012.

[53] T. Benzaoui, A. Selatnia, and D. Djabali, "Adsorption of copper (II) ions from aqueous solution using bottom ash of expired drugs incineration," *Adsorption Science and Technology*, pp. 1–16, 2017.

[54] R. A. Koble and T. E. Corrigan, "Adsorption isotherms for pure hydrocarbons," *Industrial and Engineering Chemistry*, vol. 44, no. 2, pp. 383–387, 1952.

[55] S. Alahmadi, S. Mohamad, and M. J. Maah, "Comparative study of tributyltin adsorption onto mesoporous silica functionalized with calix(4) arene, p-tert-butylcalix(4) arene and p-sulfonatocalix(4) arene," *Molecules*, vol. 19, no. 4, pp. 4524–4547, 2014.

[56] A. R. Khan, R. Ataullah, and A. Al-Haddad, "Equilibrium adsorption studies of some aromatic pollutants from dilute aqueous solutions on activated carbon at different temperatures," *Journal of Colloid and Interface Science*, vol. 194, no. 1, pp. 154–165, 1997.

[57] O. Amrhar and M. S. NassaliElyoubi, "Two and three-parameter isothermal modeling for adsorption of Crystal Violet dye onto Natural Illitic Clay: nonlinear regression analysis," *Journal of Chemical and Pharmaceutical Research*, vol. 7, no. 9, pp. 892–903, 2015.

[58] O. Amrhar, H. Nassai, and M. S. Elyoubi, "Application of non-linear regression analysis to select the optimum absorption isotherm for Methylene Blue adsorption onto Natural Illitic Clay," *Bulletia de la societe Royale des Science de Kiege*, vol. 84, pp. 116–130, 2015.

[59] G. Varank, A. Demir, K. YetiImezsoy, S. Top, E. Sekman, and M. S. Bilgili, "Removal of 4-nitrophenol from aqueous solution by natural low cost adsorbents," *Indian Journal of Chemical Technology*, vol. 19, pp. 7–25, 2011.

[60] F. B. Aarden, *Adsorption onto Heterogeneous porous materials, Equilibria and Kinetics Eindorea, Technische Universiteit Eindoven*, 2001.

[61] B. Subramanyam and D. Ashutosh, "Adsorption isotherm modeling of phenol onto natural soils—applicability of various isotherm models," *International Journal of Environmental Research*, vol. 6, no. 1, pp. 265–276, 2012.

[62] S. A. Al-Jlil and M. S. Latif, "Evaluation of Equilibrium isotherms models for the adsorption of Cu and Ni from wastewater on Benronite clay," *Material and Technology*, vol. 47, no. 4, pp. 481–486, 2013.

[63] L. Jossens, J. M. Prausnitz, W. Fritz, E. U. Schlünder, and A. L. Myers, "Thermodynamics of multi-solute adsorption from dilute aqueous solutions," *Chemical Engineering Science*, vol. 33, no. 8, pp. 1097–1106, 1978.

[64] M. F. Dilekoglu, "Use of generic algorithm optimzation Techniques in the adsorption of phenol on Banana and grapetruit peels," *Journal of the Chemical Society of Pakistan*, vol. 38, no. 6, 2016.

[65] R. Juang, F. Wu, and R. Tseng, "Adsorption isotherms of phenolic compounds from aqueous solutions onto activated carbon fibers," *Journal of Chemical and Engineering Data*, vol. 41, no. 3, pp. 487–492, 1996.

[66] A. Etaya, N. Koto, and J. Yamanato, "Liquid phase adsorption equilibrium of rhenol and it's derivatives on macroreticular adsorbents," *J. Chem. Eng. JPN*, vol. 17, p. 389, 1984.

[67] W. Fritz and E.-U. Schluender, "Simultaneous adsorption equilibria of organic solutes in dilute aqueous solutions on activated carbon," *Chemical Engineering Science*, vol. 29, no. 5, pp. 1279–1282, 1974.

[68] Z. L. Yaneva, B. K. Koumanova, and N. V. Georgieva, "Linear regression and nonlinear regression methods for equilibrium modelling of p-nitrophenol biosorption by Rhyzopus oryzen: comparison of error analysic criteria," *Journal of Chemistry*, vol. 2013, Article ID 517631, 10 pages, 2013.

[69] N. Singh and C. Balomajumder, "Removal of cyanide from aqueous media by adsorption using Al-activated carbon: parametric experiments equilibrium, kinetics and thermodynamic analysis," in *Proceedings of the 2nd International Conference on Science, Technology and Management*, University of Delhi, New Delhi, India, 2015.

[70] M. S. Podder and C. B. Majumder, *Simultaneous Biosorption and Bioaccumulation: A Novel Technique for the Efficient Removal of Arsenic*, Department of Chemical Engineering, Institute of Technology, Roorkee, India, 2017.

[71] G. McKay, A. Mesdaghinia, S. Nasseri, M. Hadi, and M. Solaimany Aminabad, "Optimum isotherms of dyes sorption by activated carbon: fractional theoretical capacity and error analysis," *Chemical Engineering Journal*, vol. 251, pp. 236–247, 2014.

[72] B. M. van Vliet, W. J. Weber Jr., and H. Hozumi, "Modeling and prediction of specific compound adsorption by activated carbon and synthetic adsorbents," *Water Research*, vol. 14, no. 12, pp. 1719–1728, 1980.

[73] K. Vijayaraghavan, "Biosorption of Lan thamide (preseodymium) using ulva lactuca: mechanistic study and application of two, three, four and five parameter isotherm models," *Journal of Environment and Biotechnology Research*, vol. 1, no. 1, pp. 1–8, 2015.

[74] G. R. Parker Jr., "Optimum isotherm equation and thermodynamic interpretation for aqueous 1,1,2-trichloroethene adsorption isotherms on three adsorbents," *Adsorption*, vol. 1, no. 2, pp. 113–132, 1995.

[75] N. Sivarajasekar and R. Baskar, "Adsoprtion of basic red onto activated carbon derived from immature cotton seeds: Isotherm studies and ether analysis," *Desalination and Water Treatment*, vol. 52, pp. 1–23, 2014.

[76] C. Chen, "Evaluation of equilibrium sorption isotherm equations," *Open Chemical Engineering Journal*, vol. 7, pp. 24–44, 2003.

[77] T. F. Edgar and D. M. Himmelblau, *Optimization of Chemical Processes*, 1989.

[78] O. T. Hanna and O. C. Sandall, *Computerization Methods in Chemical Engineering*, Printice-Hall International, New Jessey, NH, USA, 1995.

[79] K. V. Kumar, "Comparative analysis of linear and non-linear method of estimating the sorption isotherm parameters for malachite green onto activated carbon," *Journal of Hazardous Materials*, vol. 136, no. 2, pp. 197–202, 2006.

[80] D. H. Lataye, I. M. Mishra, and I. D. Mall, "Adsorption of 2-picoline onto bagasse fly ash from aqueous solution," *Chemical Engineering Journal*, vol. 138, no. 1–3, pp. 35–46, 2008.

[81] K. V. Kumar, K. Porkodi, and F. Rocha, "Isotherms and thermodynamics by linear and non-linear regression analysis for the sorption of methylene blue onto activated carbon: Comparison of various error functions," *Journal of Hazardous Materials*, vol. 151, no. 2-3, pp. 794–804, 2008.

[82] K. V. Kumar and S. Sivanesan, "Pseudo second order kinetics and pseudo isotherms for malachite green onto activated carbon: comparison of linear and non-linear regression methods,"

Journal of Hazardous Materials, vol. 136, no. 3, pp. 721–726, 2006.

[83] V. S. Mane, I. Deo Mall, and V. Chandra Srivastava, "Kinetic and equilibrium isotherm studies for the adsorptive removal of Brilliant Green dye from aqueous solution by rice husk ash," *Journal of Environmental Management*, vol. 84, no. 4, pp. 390–400, 2007.

[84] J. F. Porter, G. McKay, and K. H. Choy, "The prediction of sorption from a binary mixture of acidic dyes using single- and mixed-isotherm variants of the ideal adsorbed solute theory," *Chemical Engineering Science*, vol. 54, no. 24, pp. 5863–5885, 1999.

[85] A. Kapoor and R. T. Yang, "Correlation of equilibrium adsorption data of condensible vapours on porous adsorbents," *Gas Separation and Purification*, vol. 3, no. 4, pp. 187–192, 1989.

[86] D. Marquardt, "An algorithm for least-squares estimation of nonlinear parameters," *SIAM Journal on Applied Mathematics*, vol. 11, no. 2, pp. 431–441, 1963.

[87] J. Ng, W. H. Cheung, and G. McKay, "Equilibrium studies for the sorption of lead from effluents using chitosan," *Chemosphere*, vol. 52, no. 6, pp. 1021–1030, 2003.

[88] S. Kundu and A. K. Gupta, "Arsenic adsorption onto iron oxide-coated cement (IOCC): regression analysis of equilibrium data with several isotherm models and their optimization," *Chemical Engineering Journal*, vol. 122, no. 1-2, pp. 93–106, 2006.

[89] B. Boulinguiez, P. Le Cloirec, and D. Wolbert, "Revisiting the determination of langmuir parameters-application to tetrahydrothiophene adsorption onto activated carbon," *Langmuir*, vol. 24, no. 13, pp. 6420–6424, 2008.

[90] F. J. Rivas, F. J. Beltrán, O. Gimeno, J. Frades, and F. Carvalho, "Adsorption of landfill leachates onto activated carbon. Equilibrium and kinetics," *Journal of Hazardous Materials*, vol. 131, no. 1-3, pp. 170–178, 2006.

[91] D. Karadag, Y. Koc, M. Turan, and M. Ozturk, "A comparative study of linear and non-linear regression analysis for ammonium exchange by clinoptilolite zeolite," *Journal of Hazardous Materials*, vol. 144, no. 1-2, pp. 432–437, 2007.

[92] Y. S. Ho, "Second-order kinetic model for the sorption of cadmium onto tree fern: a comparison of linear and non-linear methods," *Water Research*, vol. 40, no. 1, pp. 119–125, 2006.

[93] K. V. Kumar, K. Porkodi, and F. Rocha, "Comparison of various error functions in predicting the optimum isotherm by linear and non-linear regression analysis for the sorption of basic red 9 by activated carbon," *Journal of Hazardous Materials*, vol. 150, no. 1, pp. 158–165, 2008.

Permissions

List of Contributors

Rotcharin Sawisai, Ratchaneekorn Wanchanthuek, Widchaya Radchatawedchakoon and Uthai Sakee
Creative Chemistry and Innovation Research Unit, Center of Excellence for Innovation in Chemistry (PERCH-CIC), Department of Chemistry, Faculty of Science, Mahasarakham University, Mahasarakham 44150,Thailand

Juan Frau
Departament de Química, Universitat de les Illes Balears, Palma de Mallorca 07122, Spain

Daniel Glossman-Mitnik
Departament de Química, Universitat de les Illes Balears, Palma de Mallorca 07122, Spain
Laboratorio Virtual NANOCOSMOS, Departamento de Medio Ambiente y Energía, Centro de Investigación en Materiales Avanzados, Miguel de Cervantes 120 Complejo Industrial Chihuahua, 31136 Chihuahua, CHIH, Mexico

Hongkui Ge and Zhihui Yang
China University of Petroleum, Beijing, China

Yinghao Shen
China University of Petroleum, Beijing, China
Texas Tech University, Lubbock, TX, USA

Zhaopeng Zhu
PetroChina Jilin Oilfield Company, Songyuan, China

Peng Shi
Research Institute of Shaanxi Yanchang Petroleum "Group" Co., Ltd., Xian, Shaanxi, China

Penka Vasileva and Teodora Alexandrova
Department of General and Inorganic Chemistry, Faculty of Chemistry and Pharmacy, Laboratory of Nanoparticle Science and Technology, University of Sofia "St. Kliment Ohridski", 1 J. Bourchier Blvd., 1164 Sofia, Bulgaria

Irina Karadjova
Department of Analytical Chemistry, Faculty of Chemistry and Pharmacy, University of Sofia "St. Kliment Ohridski", 1 J. Bourchier Blvd., 1164 Sofia, Bulgaria

Chengli Zhang and Peng Wang
College of Petroleum Engineering, Northeast Petroleum University, Daqing, Heilongjiang 163318, China

Guoliang Song
College of Mathematics and Statistics, Northeast Petroleum University, Daqing, Heilongjiang 163318, China

Minh Cam Le, Khu Le Van, Thu Ha T. Nguyen and Ngoc Ha Nguyen
Theoretical and Physical Chemistry Division, Faculty of Chemistry, Hanoi National University of Education, Hanoi 1000, Vietnam

Amira Satirawaty Mohamed Pauzan and Normala Ahad
Faculty of Resource Science and Technology, Universiti Malaysia Sarawak, 94300 Kota Samarahan, Sarawak, Malaysia

Xi Chen
School of Geography Science, Nanjing Normal University, Nanjing 210023, China

Yan-hua Wang, Zu-cong Cai and Hao Yang
School of Geography Science, Nanjing Normal University, Nanjing 210023, China
Jiangsu Center for Collaborative Innovation in Geographical Information Resource Development and Application, Nanjing 210023, China

Chun Ye
Chinese Research Academy of Environmental Sciences, Beijing 100012, China

Wei Zhou
Institute of Soil Science, Chinese Academy of Sciences, Nanjing 210008, China

Xiao Han
State Key Laboratory of Atmospheric Boundary Layer Physics and Atmospheric Chemistry, Institute of Atmospheric Physics, Chinese Academy of Sciences, Beijing 100029, China
College of Earth Science, University of Chinese Academy of Sciences, Beijing 100049, China

Nichaonn Chumuang
Department of Chemistry, Faculty of Science, Kasetsart University, Bangkok 10900,Thailand

Vittaya Punsuvon
Department of Chemistry, Faculty of Science, Kasetsart University, Bangkok 10900,Thailand
Center of Excellence-Oil Palm, Kasetsart University, Bangkok 10900,Thailand
Center for Advance Studies in Tropical Natural Resource, National Research University, Kasetsart University, Bangkok 10900,Thailand

Yong Zhang and JuanjuanMa
College ofWater Resource Science and Engineering, Taiyuan University of Technology, Taiyuan 030024, China

Feng Yu and Wenping Cheng
College of Chemistry and Chemical Engineering, Taiyuan University of Technology, Taiyuan 030024, China

Jiancheng Wang
State Key Laboratory Breeding Base of Coal Science and Technology Co-Founded by Shanxi Province and the Ministry of Science and Technology, Taiyuan University of Technology, Taiyuan 030024, China

Jun He, Laizhou Song, Hongxia Yang, Xiaohui Ren and Lifei Xing
School of Environmental and Chemical Engineering, Yanshan University, Qinhuangdao 066004, China

Samuel Tetteh and Michael Akrofi Anang
Chemistry Department, School of Physical Sciences, College of Agriculture and Natural Sciences, University of Cape Coast, Cape Coast, Ghana

Andrews Quashie
Institute of Industrial Research, Council for Scientific and Industrial Research, Accra, Ghana

Aldo J. Kitalika, Revocatus L. Machunda, Hans C. Komakech and Karoli N. Njau
Department ofWater and Environmental Science and Engineering, Nelson Mandela African Institute of Science and Technology, Tengeru, Arusha, Tanzania

Boniswa P. Goso and Omobola O. Okoh
Department of Chemistry, University of Fort Hare, Private Bag X1314, Alice 5700, South Africa

Anthony I. Anukam
Department of Chemistry, University of Fort Hare, Private Bag X1314, Alice 5700, South Africa
Fort Hare Institute of Technology, University of Fort Hare, Private Bag X1314, Alice 5700, South Africa

Sampson N. Mamphweli
Fort Hare Institute of Technology, University of Fort Hare, Private Bag X1314, Alice 5700, South Africa

Jianfeng Su
College of Materials Science and Engineering, Taiyuan University of Science and Technology, Taiyuan 030024, China
Shanxi Provincial Key Laboratory of Metallurgical Device Design Theory and Technology, Taiyuan University of Science and Technology, Taiyuan 030024, China

Yaqin Tian and Yugui Li
College of Materials Science and Engineering, Taiyuan University of Science and Technology, Taiyuan 030024, China
Shanxi Provincial Key Laboratory of Metallurgical Device Design Theory and Technology, Taiyuan University of Science and Technology, Taiyuan 030024, China
Collaborative Innovation Center of Taiyuan Heavy Machinery Equipment, Taiyuan University of Science and Technology, Taiyuan 030024, China

Hongping An and Jie Ren
College of Materials Science and Engineering, Taiyuan University of Science and Technology, Taiyuan 030024, China
Collaborative Innovation Center of Taiyuan Heavy Machinery Equipment, Taiyuan University of Science and Technology, Taiyuan 030024, China

Wei Guan
Chongqing Key Laboratory of Environmental Materials & Remediation Technologies, Chongqing University of Arts and Sciences, Chongqing 402160, China

Shichao Tian
Shenzhen Environmental Science and New Energy Technology Engineering Laboratory, Tsinghua-Berkeley Shenzhen Institute, Shenzhen 518055, China

Shuguang Cao and Bo Lin
School of Mechanical Engineering and Automation, Fuzhou University, Fuzhou, Fujian 350108, China

Guoqiang Han
School of Mechanical Engineering and Automation, Fuzhou University, Fuzhou, Fujian 350108, China
Fujian Institute of Research on the Structure of Matter, Chinese Academy of Science, Fuzhou, Fujian 350002, China

Andreea-Mihaela Dunca
Department of Geography, Faculty of Chemistry, Biology, Geography, West University of Timişoara, Blvd. V. Pârvan No. 4, Timis,oara, 300223 Timiş, Romania

Rasmus S. N. Fehrmann
Centre for Catalysis and Sustainable Chemistry, Department of Chemistry, Technical University of Denmark, Building 207, DK-2800 Kgs. Lyngby, Denmark

Siva Sankar Reddy Putluru and Leonhard Schill
Centre for Catalysis and Sustainable Chemistry, Department of Chemistry, Technical University of Denmark, Building 207, DK-2800 Kgs. Lyngby, Denmark
CHEC Research Center, Department of Chemical and Biochemical Engineering, Technical University of Denmark, Building 229, DK-2880 Kgs. Lyngby, Denmark

Anker Degn Jensen
CHEC Research Center, Department of Chemical and Biochemical Engineering, Technical University of Denmark, Building 229, DK-2880 Kgs. Lyngby, Denmark

Hamdy Farag
Chemistry Department, Faculty of Science, Mansoura University, Mansoura 35516, Egypt

Masahiro Kishida
Department of Material Process Engineering, Graduate School of Engineering, Kyushu University, Motooka 744, Fukuoka 819-0395, Japan

Abd-Alrahman Embaby
Geology Department, Faculty of Science, Damietta University, Damietta 34517, Egypt

Abdel-Nasser A. El-Hendawy
Physical Chemistry Department, National Research Center, 12622 Dokki, Cairo, Egypt

Mohamed Mahmoud Nasef
Chemical Engineering Department, Universiti Teknologi Petronas, 32610 Seri Iskandar, Perak, Malaysia

Xiaoyi Wang, Jie Jia, Tingli Su, Zhiyao Zhao, Jiping Xu and Li Wang
School of Computer and Information Engineering, Beijing Technology and Business University, Beijing 100048, China

Nimibofa Ayawei, Augustus Newton Ebelegi and Donbebe Wankasi
Department of Chemical Sciences, Niger Delta University, Wilberforce Island, Bayelsa State, Nigeria

Index

9 781632 388391